GAUGE THEORIES IN PARTICLE PHYSICS

Graduate Student Series in Physics

Other books in the series

Introduction to Gauge Field Theory
D BAILIN and A LOVE

Supersymmetric Gauge Field Theory and String Theory
D BAILIN and A LOVE

Mechanics of Deformable Media
A B BHATIA and R N SINGH

Symmetries in Quantum Mechanics
M CHAICHIAN and R HAGEDORN

Hadron Interactions
P D B COLLINS and A D MARTIN

The Physics of Structurally Disordered Matter: An Introduction
N E CUSACK

Basic Ideas and Concepts in Nuclear Physics
K HEYDE

Collective Effects in Solids and Liquids
N H MARCH and M PARRINELLO

Geometry, Topology and Physics
M NAKAHARA

Supersymmetry, Superfields and Supergravity: An Introduction
P P SRIVASTAVA

Superfluidity and Superconductivity
D R TILLEY and J TILLEY

GRADUATE STUDENT SERIES IN PHYSICS

Series Editor:
Professor Douglas F Brewer, MA, DPhil
Emeritus Professor of Experimental Physics, University of Sussex

GAUGE THEORIES IN PARTICLE PHYSICS

A PRACTICAL INTRODUCTION
THIRD EDITION

Volume 1
From Relativistic Quantum Mechanics to QED

IAN J R AITCHISON
Department of Physics
University of Oxford

ANTHONY J G HEY
Department of Electronics and Computer Science
University of Southampton

Taylor & Francis
Taylor & Francis Group
New York London

Taylor & Francis is an imprint of the
Taylor & Francis Group, an **informa** business

Published in 2003 by
Taylor & Francis Group
270 Madison Avenue
New York, NY 10016

Published in Great Britain by
Taylor & Francis Group
2 Park Square
Milton Park, Abingdon
Oxon OX14 4RN

International Standard Book Number-10: 0-7503-0864-8 (Softcover)
International Standard Book Number-13: 978-0-7503-0864-9 (Softcover)

Library of Congress Cataloging-in-Publication Data

Catalog record is available from the Library of Congress

informa
Taylor & Francis Group
is the Academic Division of Informa plc.

Visit the Taylor & Francis Web site at
http://www.taylorandfrancis.com

To Jessie
and to
Jean, Katherine and Elizabeth

CONTENTS

PREFACE TO THE THIRD EDITION

We had two main reasons for undertaking a second major rewriting of our book, which led to this new two-volume third edition. First, we felt there was now a need for a more extended treatment of quantum field theory, and of its applications to the 'Standard Model' of particle physics, at a pedagogical level consistent with that of the two earlier editions; which is to say, at a *practical* and *accessible* level, but one stopping well short of the full theorists' Monty. Second, while recognizing that the book belongs in a series intended for graduate students, we wanted to ease the path into the subject by including more background material, and at the same time make the book more approachable for students in the later stages of undergraduate courses. Indeed, we hope that much of this first volume will prove suitable for advanced undergraduate courses, for example those now offered in the fourth year at UK universities.

Taking the second point first, the more gradual introduction comprises Part 1, together with new supporting appendices F (contour integration), G (Green functions) and H (non-relativistic scattering theory). The 'from atoms to quarks' story in chapter 1 remains much as before, with some updating of the lepton and quark flavour sections. Then follows a completely new chapter 2, an overview of particle interactions in the Standard Model, which we believe should—taken together with chapter 1—be useful as early as the third year of undergraduate studies. The essential element here is a presentation of Yukawa's fundamental theory of force as virtual quantum exchange, which is fuller than most qualitative introductions at this level. We want to give students an appropriate understanding of some simple formulae for exchange amplitudes, the possession of which should help them to appreciate the dynamical unity of the subject, even at this early stage.

The road to QED then begins in chapter 3 (previously 2), in which Maxwell's theory and its incorporation into quantum mechanics are presented in gauge theory terms. Here we emphasize both *in*variance and *co*variance, under both Lorentz and gauge transformations; this too is supported by an extended appendix D on special relativity. Together these could serve as the basis for a short course on these more theoretical aspects of electromagnetism, from the gauge theory viewpoint. Part 1 ends with a chapter (4) on relativistic quantum mechanics (RQM). Additions to the corresponding material in the earlier editions include the Dirac calculation of $g = 2$, and a section on Lorentz transformations of Dirac spinors. We think this chapter now offers a self-contained introduction to RQM, such as might form part of advanced undergraduate theoretical courses, or introductory graduate courses.

Quantum field theory proper is introduced in Part 2. Readers familiar with the two earlier editions will sense a certain evolution. In our first edition (1982), we decided—in the interest of accessibility—to 'dispense with much of the machinery of Lagrangian-based relativistic quantum field theory', adopting instead a 'wavefunction' approach. Though successful at the time (which was quite soon after the initial establishment of the Standard Model), this approach could not do justice to several important conceptual issues—for example, how forces are due to virtual emission and absorption processes, and what symmetry is related to electric charge conservation. In the second edition (1989), we 'reversed ourselves on the crucial issue of whether to include the methods of quantum field theory or not'. We were 'convinced that the basic physics of quantum fields—the heart of the twentieth century answer to all those ancient questions concerning the nature of matter and force—should be far more readily accessible than it is'. And so we tried to provide, in one chapter, a self-contained elementary introduction to quantum field theory. That chapter indeed survives, essentially intact, as chapter 5 and part of chapter 7 in the present volume. But to these are now added three more chapters on quantum field theory: 6, in Part 2, and 10 and 11 in Part 4.

Why do we devote still more space to quantum field theory? The answer is a *practical* one. Precision experiments at LEP and other laboratories, notably in the years subsequent to 1989, tested the Standard Model beyond the 'tree' level; they were sensitive to quantum fluctuations predicted by the dynamics of relativistic quantum field theory (i.e. to 'loops'). As a striking example of this, we mention the remarkably precise prediction of the top quark (t) mass via the sensitivity of data to loop corrections involving *virtual* t-production, with an accuracy comparable to that achieved in experiments which produced the t-quark as a *real* particle. Thus we have an experimental arena in which, as with QED and atomic physics in an earlier generation, some appreciation of the subtle intricacies of quantum field theory is vital to a full understanding of the data. This is why we think that a book of this sort should now go further than tree-level calculations, and include at least the basics of one-loop physics—which is, of course, that of *renormalization*.

We have tried to provide an account of this topic which remains faithful to the pedagogical philosophy of the earlier editions, and thus appropriate to the needs of graduate students in experimental particle physics (and perhaps their supervisors too). We decided, for instance, to stick with the 'canonical' operator formalism, rather than embark at this stage on the entirely separate formalism of path integral quantization. And we regulate divergences with a physical cut-off, rather than by dimensional regularization.

This approach is satisfactory for scalar field theory, and (just about) for QED. A crucial new element, in fact, appears in chapter 6, where we treat interacting scalar fields, specifically the hypothetical 'ABC' model. Though a deviation from the strict 'gauge theories' remit, this allows us to present the logical derivation of the Feynman rules from the Dyson expansion (at least in outline), without the considerable complications of spin, and of gauge field quantization. In this same chapter, the reader is taken through prototype calculations of a decay rate and of a cross section, within the ABC theory. This chapter, therefore, represents the new technical 'spring' of the book. Part 2 ends, in chapter 7, with a discussion of the special difficulties encountered when quantizing gauge fields—even the familiar electromagnetic one. This leads to the recognition that the photon propagator is gauge-dependent (a fact we previously glossed over), and of what gauge invariance then entails.

Part 3 consists of elementary (tree-level) applications, much as before: first, in chapter 8, to scalar and spinor electrodynamics; then in chapter 9 to deep inelastic scattering and related processes. These two chapters are essentially numbers 5, 6 and 7 of the second edition.

The final two chapters, in Part 4, address one-loop amplitudes and renormalization—beginning, once more, with the technically simpler ABC theory, and leading on to QED. We have two aims here. First, to leave the reader with a good qualitative understanding of renormalization (including the notion of an effective field theory), at least insofar as we understand it ourselves. And second, we want to show the reader how one relatively simple one-loop calculation is done (the photon self-energy), and what a rich store of physics is contained in the result.

After the introductory survey, then, this first volume proceeds from relativistic quantum mechanics to one-loop QED. As the reader learns, QED is an Abelian gauge field theory. Non-Abelian theories will be the subject of the second volume, in particular those theories which describe the strong interactions (QCD) and the weak interactions (the Glashow–Salam–Weinberg theory). As in the second edition, some emphasis will be placed on exhibiting the confrontation between theory and experiment, both in the early discoveries and in the later precision tests. On the theoretical side, the discussion of the central topic of spontaneous symmetry breaking will be amplified, and chiral symmetry breaking will be included; renormalization group ideas will also be introduced. As before, we shall essentially 'draw a line' at the Standard Model.

Acknowledgments

Many people helped us with both the first and second editions, and their efforts still influence this one. These include John C Taylor, Chris Llewellyn Smith, Graham Ross, and Jack Paton at Oxford. IJRA records with great sadness the loss of Caroline Fraser's enthusiastic encouragement and acute criticism; she died in 1996. AJGH wishes to thank Garry McEwen for his meticulous vigilance in reading the original text and for his suggestions for improving its accessibility. We wish him well in his retirement. Others who read much, or in some cases, all of the previous editions include Jan Ambjorn, David Bailin, Richard Hughes, and especially the late Euan Squires.

In the second edition, the field theory material had its roots in lectures given first by AJGH, and then by IJRA, at successive Rutherford Appleton Laboratory Summer Schools for Young High Energy Physicists, first under the direction of Roger Cashmore, and then of John Dainton. This was developed in lectures given by IJRA to graduate students in both theoretical and experimental particle physics at Oxford. Subsequently, at the invitation of David Saxon and the SUSSP Committee, IJRA gave eight lectures on 'Standard Model Foundations' at the Forty Second Scottish Universities Summer School in Physics, in 1993; in preparing those lectures, he benefitted greatly from many conversations with Richard Ball and Robert Thorne on the subject of renormalization, the influence of which is felt in chapters 10 and 11. More recently, IJRA presented a condensed version of the approach followed here, at the 2001 European School of High Energy Physics. He is grateful to John March-Russell and Nick Ellis for the invitation to give these lectures, to Egil Lillestol and Claire Earnshaw for organizing such a successful school, and to the student participants for their lively involvement. For IJRA, the whole project was made possible by a year's leave of absence (2001–2) from the University of Oxford, and from Worcester College, Oxford, to both of which institutions he records his gratitude.

Once more we owe a very special debt of thanks to George Emmons, formerly Principal of the Merchant Navy College, Greenhithe, Kent, and now enjoying the status to which IJRA aspires in a year's time. He has been actively involved in this new edition right from the start. He made many very useful comments on early drafts; he checked the text as it approached final form, read carefully through the LaTeX-ed manuscript, and joined us in correcting the proofs, as he did for the second edition. He has made a major contribution to this one. We are also very grateful to Keith Hamilton for carefully reading chapters 2, 6, 10 and 11 at the proof stage, for spotting a number of misprints we had missed, and for helpful comments.

We, of course, bear full responsibility for the final version, and are only too well aware that errors will persist. We should be most grateful to have them brought to our attention. The publishers are able to support the book at the web site http://bookmarkphysics.iop.org/bookpge.htm?&book=1130p. Corrections will be listed here, and other relevant material in due course. The most convenient form of communication is by e-mail to 0750308648@iop.org.

Finally, we thank Nikki Fathers at Oxford, and David Rayner, John Tighe and especially Antonio Gatti at Southampton, for their help in converting the manuscript into electronic form.

Ian J R Aitchison and Anthony J G Hey
August 2002

PART 1

INTRODUCTORY SURVEY, ELECTROMAGNETISM AS A GAUGE THEORY, AND RELATIVISTIC QUANTUM MECHANICS

1

QUARKS AND LEPTONS

1.1 The Standard Model

The traditional goal of particle physics has been to identify what appear to be structureless units of matter and to understand the nature of the forces acting between them. Stated thus, the enterprise has a two-fold aspect: matter on the one hand, forces on the other. The expectation is that the smallest units of matter should interact in the simplest way; or that there is a deep connection between the basic units of matter and the basic forces. The joint matter/force nature of the enquiry is perfectly illustrated by Thomson's discovery of the electron and Maxwell's theory of the electromagnetic field, which together mark the birth of modern particle physics. The electron was recognized both as the 'particle of electricity'—or as we might now say, as an elementary source of the electromagnetic field, with its motion constituting an electromagnetic current—and also as an important constituent of matter. In retrospect, the story of particle physics over the subsequent one hundred years or so has consisted in the discovery and study of two new (non-electromagnetic) forces—the *weak* and the *strong* forces—and in the search for 'electron-figures' to serve both as constituents of the new layers of matter which were uncovered (first nuclei, and then hadrons) and also as sources of the new force fields. This effort has recently culminated in decisive progress: the identification of a collection of matter units which are indeed analogous to the electron; and the highly convincing experimental verification of theories of the associated strong and weak force fields, which incorporate and generalize in a beautiful way the original electron/electromagnetic field relationship. These theories are collectively (and rather modestly, if uninformatively) called 'the Standard Model', to which this book is intended as an elementary introduction.

In brief, the picture is as follows. There are two types of matter units: *leptons* and *quarks*. Both have spin-$\frac{1}{2}$ (in units of $\hbar$) and are structureless at the smallest distances currently probed by the highest-energy accelerators. The leptons are direct generalizations of the electron, the term denoting particles which, if charged, interact both electromagnetically and weakly, and, if neutral, only weakly. The involvement of the electron in weak, as well as electromagnetic, interactions became clear quite early on, with the study of nuclear β-decay—though it took many years before it was conclusively established that the electron emitted in this process was invariably accompanied by a neutral leptonic partner, the antineutrino $\bar{\nu}_e$. By contrast, the quarks—which are the fermionic constituents of hadrons, and thence of nuclei—are more like 'strongly interacting analogues' of the electron, since they interact via all three interactions, strong, electromagnetic and weak. This, indeed, is the basis for distinguishing between these two types of matter unit, a distinction which will presumably disappear if it eventually proves possible to unify all three types of force. At all events, the weak and electromagnetic interactions of both quarks and leptons are described in a (partially) unified way by the *electroweak theory* (see volume 2), which is a generalization of *quantum electrodynamics* or QED (chapters 8, 9 and 11); while the strong interactions of quarks are described by *quantum chromodynamics* or QCD (see volume 2), which is also analogous to QED. Both quarks and leptons act as sources of the electroweak field, while quarks alone act as the sources for the QCD field: we may speak of weak and strong 'charges' and 'currents' by analogy with similar concepts in QED.

According to this picture, the observed 'levels of structure'—atomic, nuclear and hadronic—are built up in a series of stages. Quarks interacting via QCD bind to form hadrons, such as neutrons, protons and pions. It has to be admitted at once that the details of this process are not yet fully understood, owing to the calculational difficulties of dealing with a strongly interacting quantum field theory. Residual interquark forces are then thought to give rise to the rather complicated observed forces between neutrons and protons, which are responsible for binding these *nucleons* to form nuclei. This stage is also not well understood, but is regarded as somewhat analogous to the way in which complicated interatomic forces are ultimately due to residual electromagnetic interactions between the constituents—for of course it is these latter interactions which bind electrons and nuclei into atoms. The reasons why, despite these gaps in the argument, we do nevertheless believe in QCD will be discussed in volume 2. As regards the weak force, it appears that, as its name already indicates, it is not strong enough to produce any binding effects among the systems studied so far.

The reader will certainly have noticed that the most venerable force of all—gravity—is conspicuously absent from our story so far. Apart from a brief appearance in section 2.7, it will continue to be so. In practical terms this is quite reasonable, since its effect is very many orders of magnitude smaller than even the weak force, at least until the interparticle separation reaches distances far smaller than those we shall be discussing. Conceptually, also, gravity still seems to be somewhat distinct from the other forces which, as we have already indicated, are encouragingly similar. There are no particular fermionic sources carrying 'gravity charges': it seems that *all* matter gravitates. This of course was a motivation for Einstein's geometrical approach to gravity. Despite remarkable recent developments in the theory of *strings* (Green *et al* 1987, Polchinski 1998), it is fair to say that the vision of the unification of all the forces, which possessed Einstein, is still some way from realization. To include any further discussion of these ideas would take us far beyond our stated aim.

This book is not intended as a completely self-contained textbook on particle physics, which would survey the broad range of observed phenomena and outline the main steps by which the picture described here has come to be accepted. For this we must refer the reader to other sources (e.g. Perkins 2000, Martin and Shaw 1992). However, in the remainder of this chapter we shall offer a brief summary of the salient experimental facts which have established this current view of matter and forces, and in the next chapter an introduction to the conceptual framework needed to understand it.

1.2 Levels of structure: from atoms to quarks

What kind of experiments are done to probe the structure of microscopically small systems? We may distinguish three basic types: *scattering*, *spectroscopy* and *break-up* experiments, the exploitation of which constitutes a common theme in the chain of discovery: atom $\to$ nucleus $\to$ hadron $\to$ quark.

1.2.1 Atoms $\to$ nucleus

Around the turn of the 19th century, the experiments of Thomson and others established the charge and mass of electrons. In particular, the charge was exactly equal in magnitude to that of a singly ionized atom. Thus part, at least, of the atom had been 'broken off'. Then, in 1909, Geiger and Marsden, of the University of Manchester in England, reported an unusual experimental finding. Working on a programme initiated by Rutherford, they were measuring how α particles are reflected by thin metal foils. In perhaps the understatement of the century, they remarked:

> It seems surprising that some of the α particles, as the experiment shows, can be turned within a layer of 6×10^{-5} cm of gold through an angle of 90°, and even more. (Geiger and Marsden 1909)

Many years later Rutherford, in a famous phrase, said that when he became aware of this result he was as amazed as if he had seen a shell bounce back from hitting a sheet of paper. It took him two years to find the explanation. The atom was known to be electrically neutral, and to contain negatively charged electrons of mass very much less than that of the atom. The question was: how is the heavy, positively charged matter distributed? It seems to have been generally thought of as being uniformly spread over the atomic volume, and a specific model of this type was proposed by J J Thomson in 1910. We shall call this the 'soft' model. In such a model, Rutherford (1911) noted, α particles are deflected very little in any single scattering process, so that large deflections can occur only as a result, as he put it, 'of a multitude of small scatterings by the atoms of matter traversed'. However, 'a simple calculation based on the theory of probability shows that the chance of an α particle being deflected through 90° is vanishingly small'.

Rutherford went on:

> It seems reasonable to suppose that the deflection through a large angle is due to a single atomic encounter, for the chance of a second encounter of a kind to produce a large deflection must in most cases be exceedingly small. A simple calculation shows that the atom must be a seat of an intense electric field in order to produce such a large deflection at a single encounter.

Thus was the nuclear atom born, in which the atom contained a 'hard' constituent—the small, massive, positively charged nucleus. In more precise terms, according to the soft model the distribution of scattered particles falls off exponentially with the angle of deflection; the departure of the observed distribution from this exponential form is the signal for hard scattering. Geiger and Marsden found that about one in 20 000 α particles was turned through 90° or more in passing through a thin gold foil; by contrast a 'soft' model would predict a fraction of something like 1 in 10^{3500}! (Eisberg 1961, chapter 4).

Rutherford went much further than this qualitative explanation of course. He calculated the angular distribution expected from his nuclear model, obtaining the famous $\sin^{-4}\phi/2$ law, where ϕ is the scattering angle. Two years later, Geiger and Marsden (1913) published a second paper which was a long and detailed verification of this prediction and of other features of the scattering. They recorded observed scintillations versus angle in the form of a table, and commented: 'In columns IV and VI the ratios of the numbers of scintillations to $1/\sin^4\phi/2$ are entered. It will be seen that in both sets the values are approximately constant'. Column IV was scattering from silver, column VI was gold. It is slightly curious that they chose not to plot a graph of their angular distribution, but we have taken the liberty of doing it for them: figure 1.1 records some points from their data for silver.

At almost the same time, Bohr (1913)—then also in Rutherford's group at Manchester—proposed his famous model for the *dynamics* of the nuclear atom, based on an inspired blend of classical mechanics (which Rutherford had also used in deriving his angular distribution) and early quantum theory. This succeeded in giving an excellent account of the *spectroscopy* of hydrogen, as it was then known. Figure 1.2(a) shows the gross structure of the hydrogen energy levels; the ground state is marked at zero energy and the threshold for ionizing the electron is about 13.6 eV. Also shown in figure 1.2(b) are some of the low-lying levels of the two-electron atom, helium.

Evidence for such quantized energy levels had originally come from the study of emission spectra. Beginning with the experiments of Franck and Hertz (1914), they were further confirmed in a new type of scattering experiment, in which electrons were *inelastically* scattered from atoms. Figure 1.3 shows the result of a typical experiment, the relative number of scattered atoms being plotted versus the energy of the scattering electron. The peak at 200 eV corresponds to elastic scattering, since this was the beam energy. The peak just below 180 eV corresponds to an inelastic process in which the beam electrons have delivered

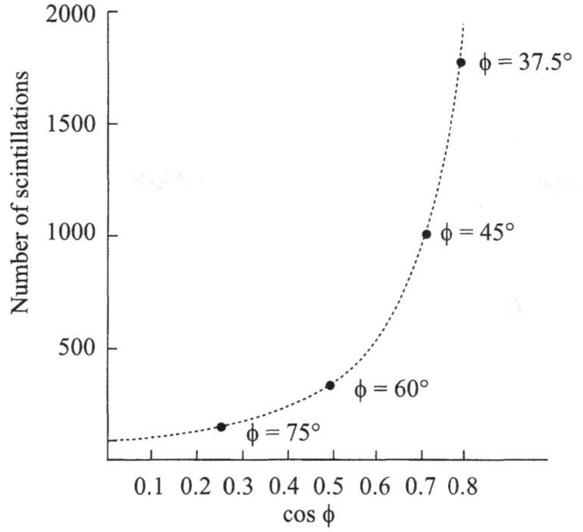

Figure 1.1. Four of the data points reported by Geiger and Marsden (1913), replotted here as the number of scintillations (for scattering of α particles by a silver foil) versus the cosine of the scattering angle ϕ. The curve is the Rutherford prediction $\sin^{-4}\phi/2$, normalized to the data point at $\phi = 60°$.

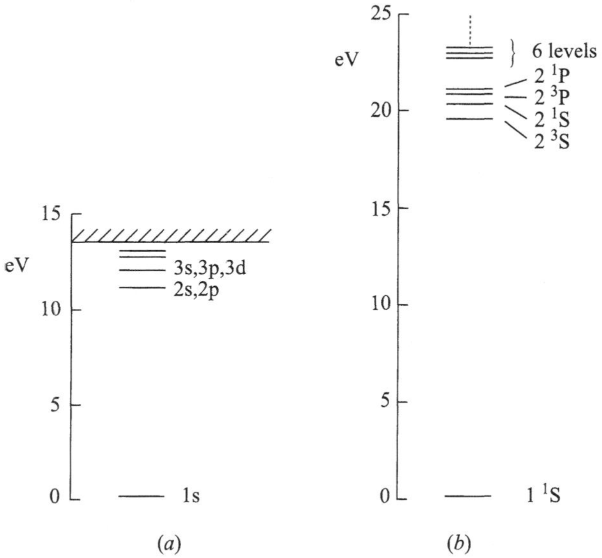

Figure 1.2. (*a*) Gross structure of hydrogen energy levels. (*b*) Gross structure of low-lying levels of helium.

some 21 eV of energy to the helium atoms, thereby exciting them to the first group of excited levels (unresolved in figure 1.3). Similarly, the peak at about

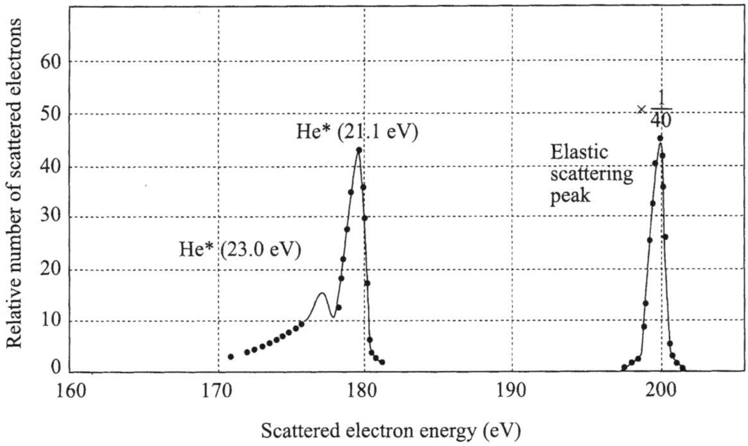

Figure 1.3. Elastic and inelastic electron scattering from helium atoms, taken from figure 1.13 of French and Taylor (1978).

177 eV corresponds to the excitation of the states near 23 eV.

With the development of quantum mechanics in the 1920s, such spectra began to be quantitatively understood in terms of the internal excitations (vibrations and rotations) of the light constituents, the electrons. The spatial distribution of the electrons was also mapped out by x-ray scattering, and extended typically over sizes of the order of a few ångstroms. The uncertainty principle gives an estimate for the typical electron momenta when they are confined to such a linear domain:

$$p \sim \hbar/10^{-10}\ \text{m} \sim 2\ \text{keV}/c$$

so that (with $mc^2 \approx 0.5$ MeV for the electron) the ratio $v/c \sim 4\times10^{-3}$. The gross features of atomic structure were thus described by the non-relativistic quantum mechanics of point-like electrons interacting with each other, and with a point-like nucleus, via Coulomb forces.

1.2.2 Nuclei → nucleons

But is the nucleus really point-like? As accelerator technology developed, it became possible to get beams of much higher energy or momentum. With the de Broglie relation $\lambda = h/p$ the resolving power of the beam became much finer, and *deviations* from the simple Rutherford formula for charged particle scattering (which assumed that both the α particle and the nucleus were point-like) began to be observed. The most precise series of experiments was done by Hofstadter and his collaborators at SLAC in the 1950s (Hofstadter 1963). They used an electron beam to avoid the complication of the poorly understood strong

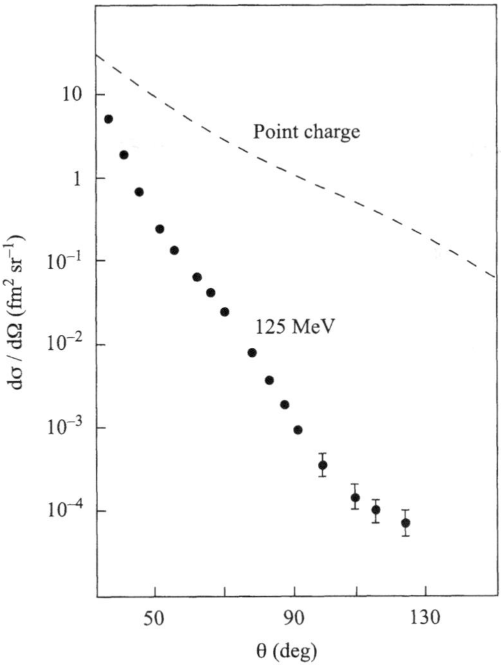

Figure 1.4. Angular distribution of 125 MeV electrons elastically scattered from gold, taken from Hofstadter (1957): data points are full circles; the broken curve is the 'point-nucleus' prediction.

interactions between α particles and nuclei. Figure 1.4 shows how the angular distribution of electrons scattered *elastically* from gold at an energy of 125 MeV falls *below* the 'point-nucleus' prediction. Qualitatively this is due to wave mechanical diffraction effects over the finite volume of the nucleus. The observed distribution is a product of two factors: the first is the distribution corresponding to scattering from a single point-like target (generalizing Rutherford's formula to include corrections for relativistic kinematics, quantum mechanics, electron spin, target recoil and so on), while the second is called the *form factor*, which is characteristic of the spatial extension of the target's charge density (Perkins 1987, section 6.2, see also sections 8.4 and 8.8 in this volume). Figure 1.5 shows the charge density of $^{12}_{6}C$ extracted (Hofstadter 1957) from such elastic electron scattering data. It is clear that this nucleus has a charge radius of about 1–2 fm (1 fm $= 10^{-15}$ m); in heavier nuclei of mass number A the radius goes roughly as $A^{1/3}$ fm.

The finite spatial extension of the nucleus is already an indication that it is not stuctureless. As with atoms, *inelastic* electron scattering from nuclei reveals that the nucleus can be excited into a sequence of quantized energy states. Figure 1.6

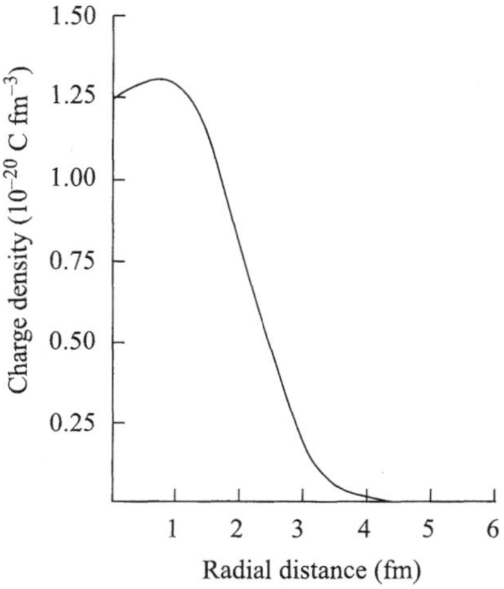

Figure 1.5. Charge density of ${}^{12}_{6}$C, extracted from electron scattering data (Hofstadter 1957).

shows the energy spectrum of electrons with incident energy 150 MeV scattered at 90° from carbon. In addition to the pronounced elastic peak, three other peaks can be seen, corresponding (after recoil corrections) to electron energy losses of about 4.4, 7.7 and 9.6 MeV. These energies correspond to the first three excited states of carbon as found in independent spectroscopic studies and summarized in figure 1.7.

It is clear that nuclei must contain constituents distributed over a size of order a few fermis, whose internal quantized motions lead to the observed nuclear spectra involving energy differences of order a few MeV. Chadwick's discovery of the neutron in 1932 established that, to a good approximation, the constituents of nuclei are neutrons and protons—generically called *nucleons*. Since the neutron is electrically neutral, it was clear that a novel force had to be operating over the range of nuclear dimensions in order to bind nucleons into nuclei. It must also be considerably stronger than the electromagnetic force, since it has to counterbalance the typical 'uncertainty' energy, which is now of order

$$\frac{1}{2M}\left(\frac{\hbar}{R}\right)^2 \sim 20 \text{ MeV}$$

for M = nucleon mass and $R \sim 1$ fm; the (repulsive) potential energy between two protons at this separation is about ten times smaller. Nevertheless, the nucleons move non-relativistically since 20 MeV is well below the nucleon rest energy of about 940 MeV.

A most important qualitative point now emerges: the typical scales of *size* and *energy* are quite different for atoms and nuclei. The excitation energies in atoms are, in general, insufficiently great to excite the nucleus; hence the

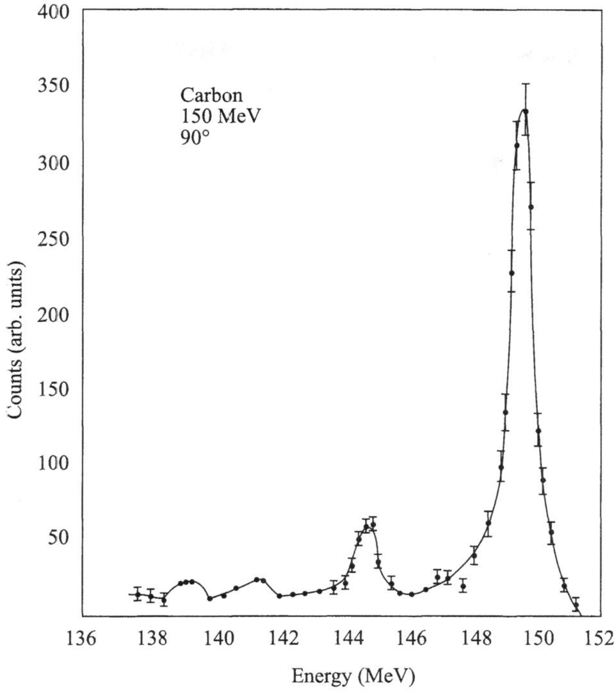

Figure 1.6. Elastic and inelastic scattering from carbon nuclei (Fregeau and Hofstadter 1955).

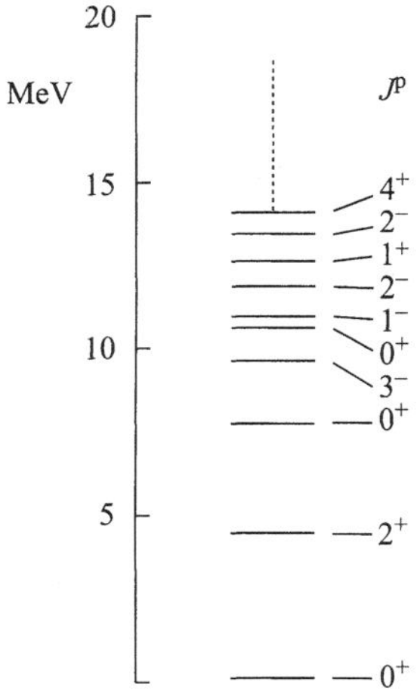

Figure 1.7. Energy levels of ${}^{12}_{6}C$.

latter appears as a small inert core inside the atom, and only when excited by appropriately higher-energy beams does it reveal its internal structure. The nuclear degrees of freedom are 'frozen' on the scale of atomic physics.

One further detail about nuclear spectra is of concern to us. Figures 1.8(*a*)–(*c*) show how, after allowing for Coulomb energy differences, the spectra of nuclei which differ from each other by the replacement of a proton by a neutron (or vice versa) contain striking similarities. In the case of figure 1.8(*a*) we have examples of 'mirror nuclei', in which a nucleus with Z protons and N neutrons is compared with another consisting of N protons and Z neutrons. In figure 1.8(*b*) the lowest (ground) state of ${}^{18}_{9}\mathrm{F}$ corresponds to a state which is excluded, by the Pauli principle, for the neighbouring nuclei in which a spare n–p is replaced by a p–p or n–n pair; otherwise the levels of these three 'analogue' nuclei match very well. Figure 1.8(*c*) shows near coincidences between the energy levels of four nuclei, with mass number 21. ${}^{21}_{10}\mathrm{Ne}$ and ${}^{21}_{11}\mathrm{Na}$ are a mirror pair, but beyond a certain excitation energy their levels also correspond to states in neighbouring nuclei, with $Z = 9$ and $Z = 12$. A pair of nuclei with corresponding levels, as in figure 1.8(*a*), is said to form a *doublet*; three such nuclei, as in figure 1.8(*b*), to form a *triplet*; and four, as in figure 1.8(*c*), a *quartet*. Even without the benefit of Chadwick's discovery, these figures would surely indicate that nuclei are made of two basic objects, one charged and one neutral, with *the forces between being very nearly independent of their charge*. We will make use of just this sort of argument in the following section.

1.2.3 Nucleons → quarks

Are nucleons point-like? The answer, in broad outline, is a remarkable repetition of the pictures just sketched for atoms and nuclei. First, the elastic scattering experiments of electrons from nucleons by Hofstadter and co-workers (Hofstadter 1963) revealed that the proton has a well-defined form factor, indicating an approximately exponential distribution of charge with a root mean square radius of about 0.8 fm. The magnetic moments of nucleons also have a spatial distribution; when scaled by their total (integrated) values μ_{p} and μ_{n}, the proton and neutron magnetic moment distributions fall on very nearly the same curve as the proton charge distribution (Perkins 1987, section 6.5). These distributions have been well measured down to distances of order 0.04 fm, corresponding to a momentum transfer to the nucleon of order 5 GeV/c.

As with both the atom and the nucleus, the clearly defined spatial extension measured by elastic electron scattering already indicates that the nucleon is not structureless. Further information bearing on nucleon structure can be obtained, as in the atomic and nuclear cases, by studying inelastic electron scattering to see if there are signs of a *nucleon spectroscopy*, which could be interpreted in terms of internal motions of its constituents. Figure 1.9 shows the energy spectrum of electrons of incident energy 4.879 GeV scattered at 10° from protons. Once again we see the usual large elastic peak (shifted from

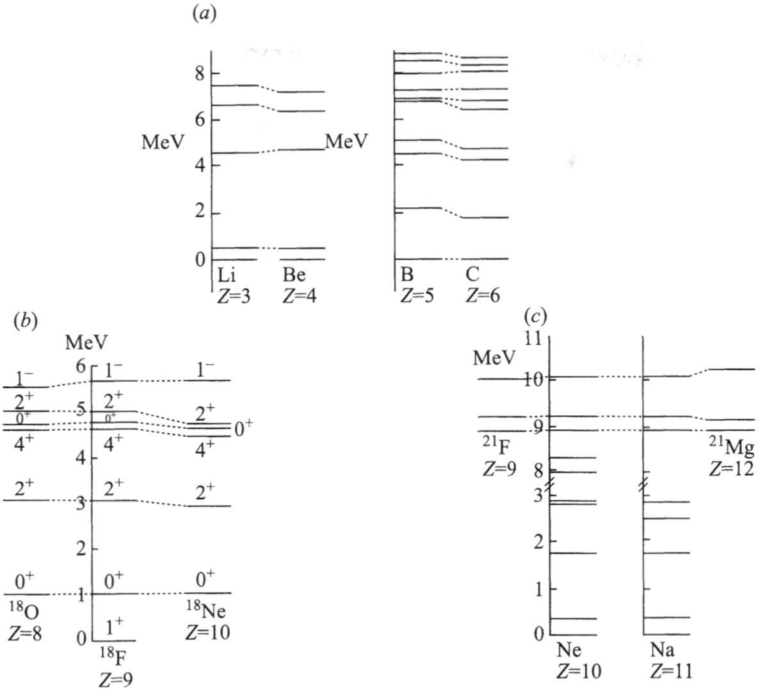

Figure 1.8. Energy levels (adjusted for Coulomb energy and neutron–proton mass differences) of nuclei of the same mass number but different charge, showing (*a*) 'mirror' doublets, (*b*) triplets and (*c*) doublets and quartets.

4.879 GeV by recoil effects), but there is also clear evidence for other peaks which presumably correspond to excitations of the recoiling system. However, the detailed interpretation of figure 1.9 is a complicated story: not only are there, in some cases, several excited recoil states contributing to a 'single' peak, but even the apparently featureless regions of the curve conceal considerable structure. Only the first of the peaks beyond the elastic peak, that at $E' \approx 4.2$ GeV (which corresponds to an energy of about 1.23 GeV for the recoiling system), has a simple interpretation. It corresponds exactly to a long-established resonant state observed in π–N scattering and denoted by the symbol Δ (Perkins 1987, section 4.9). This particular state comes in four charge combinations, corresponding to the accessible πN channels π^+p, π^+n (π^0p), π^-p (π^0n) and π^-n.

The detailed deciphering of the spectra such as the one in figure 1.9 involved much careful experimentation in the field of *baryon spectroscopy* in the 1960s and 1970s. The result of all the analysis revealed an elaborate level sequence, paralleling in a remarkable way the spectra established for atoms and, even more so, nuclei. The result is partially summarized in figures 1.10(*a*) and (*b*). One

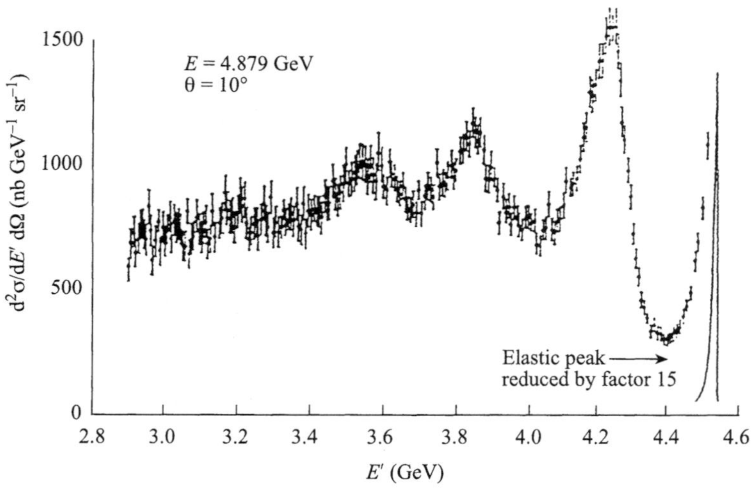

Figure 1.9. Elastic and inelastic electron scattering from protons (Bartel *et al* 1968). E and E' are the energies of the incident and scattered electrons.

series of levels, figure 1.10(*a*), comes in *two* charge combinations (charged and neutral) and is built on the proton and neutron as ground states. The two nearby states around 1.5 GeV correspond to the peak at $E' \approx 3.8$ GeV in figure 1.9 (but the latter figure shows no trace of the level at about 1.4 GeV in figure 1.10(*a*)!). This sequence is strikingly similar to the *doublet* sequences we saw in nuclear physics, figure 1.8(*a*). Another series, figure 1.10(*b*), comes in *four* charge combinations and is built on the Δ's as lowest levels. This one recalls the *quartets* of figure 1.8(*c*). It is noteworthy that these are all the observed nucleon-like excitations—for example, no triply charged states are seen.

There are also mesons. Originally predicted by Yukawa (1935) as the quantum of the short-range nuclear force field (see section 2.2), the pion was eventually discovered in 1947. In the 1960s it too turned out to be merely the ground state in a sequence of further excited states, forming charge triplets as shown in figure 1.11 (cf figure 1.8(*b*)).

The conclusion that we are looking, once again, at the constituents of composite systems seems irresistible. A specific proposal along these lines was made independently in 1964 by Gell-Mann (1964) and Zweig (1964). Though based on somewhat different (and much more fragmentary) evidence, their suggestion has turned out to be essentially correct. They proposed that the nucleon-like states are made of three spin-$\frac{1}{2}$ constituents called (by Gell-Mann) quarks, while the mesons are quark–anti-quark bound states. Such states are collectively called *hadrons*. As in the nuclear case, the simplest interpretation of the hadronic charge multiplets displayed in figures 1.10 and 1.11 is that the states are built out of two types of constituent, which differ by one unit of

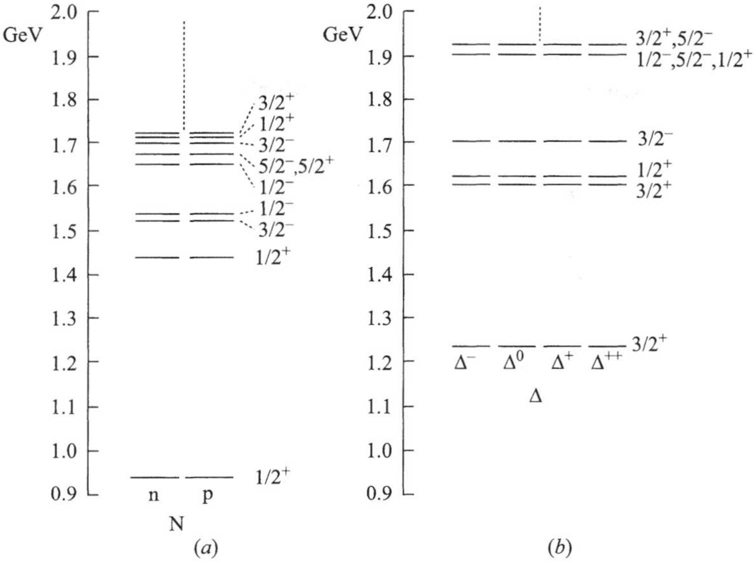

Figure 1.10. Baryon energy levels: (*a*) doublets (N); (*b*) quartets (Δ).

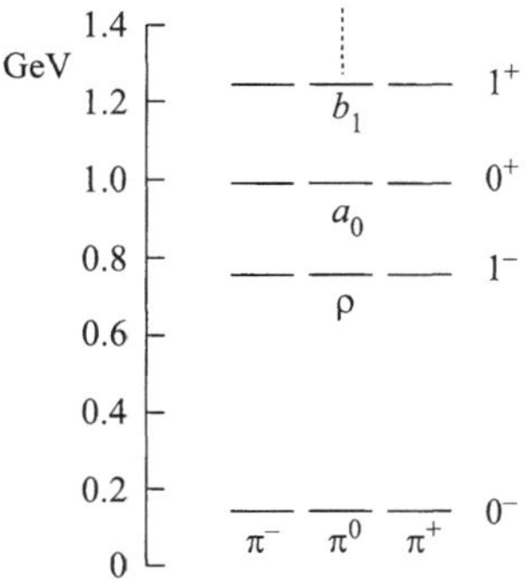

Figure 1.11. Meson triplets.

charge. However, the proton itself, for example, has unit charge—hence it would appear that its constituents must be fractionally charged. The most economical assignment turns out to be that the two constituents have charges $\frac{2}{3}e$ and $-\frac{1}{3}e$, and these are called the 'up' quark (u) and the 'down' quark (d) respectively. The series of proton levels is then the states of the uud system, the neutron levels are udd, the Δ^{++} levels are uuu, the π^+ levels are u$\bar{\text{d}}$ and so on. (The Δ^+ series differs from the p series in the detailed form of the wavefunctions for the uud constituents.) The forces between u's and d's must be charge-independent in order to lead to the multiplet structure of figures 1.10 and 1.11.

We may at this point recall that typical energy level differences in nuclei are measured in MeV. Thus in the vast majority of nuclear phenomena, the neutrons and protons remain in their unexcited ground states, hadronic excitations being typically of order 100 MeV: the hadronic degrees of freedom are largely 'frozen' in nuclear physics, just as the nuclear ones are frozen in atomic physics. The hadronic level is therefore effectively separated from the nuclear one, though in neither size nor energy scales is the separation as marked as that between nuclei and atoms. Nuclear physics thus becomes a semi-autonomous discipline.

The nucleon spectroscopy data are indirect evidence for constituents. Is there any way of 'seeing' the quarks directly? Here we naturally think of the original Rutherford-type experiment, in which the existence of the nuclear constituent of the atom was inferred from the anomalously large number of particles scattered at large angles. To probe the structure of nucleons in this way, it is not convenient to use α particles, since the interpretation of the scattering data is then complicated by the fact that the α's interact strongly with the nucleon. Instead, electrons are used, which interact via the well-understood electromagnetic force. Figure 1.12 shows the result of extending the inelastic electron scattering measurements (shown in figure 1.9 for $\theta = 10°$) to larger scattering angles. We know that the height of the elastic peak will fall rapidly due to the exponential fall-off of the nucleon form factor (recall figures 1.4 and 1.5 for the nuclear case). The same is true for the other distinct peaks, indicating that the well-defined excited nucleon states have definite spatial extension. But in the region of largest energy transfer, something dramatically new happens: the curve does *not* fall as the angle increases! This is illustrated in figure 1.12. In other words, for large enough energy transfer, the electrons bounce backwards quite happily, just as Geiger and Marsden's α particles did, suggesting that the nucleon contains 'hard' constituents. In chapter 9 we shall analyse the *deep inelastic scattering* experiments with the proper technical detail, and show how they do imply the existence of spin-$\frac{1}{2}$ hadronic constituents, which are point-like down to currently probed scales and which possess the other attributes of the quarks hypothesized from the hadronic spectroscopy.

Despite mounting evidence, of the type described, from both scattering and spectroscopy during the 1960s and early 1970s, many physicists at that time continued to regard quarks more as useful devices for systematizing a mass of complicated data than as genuine items of physical reality. One reason for their scepticism we must now confront, for it constitutes a major new twist in the story of the layers of matter. Gell-Mann ended his 1964 paper with the remark: 'A search for stable quarks of charge $-\frac{1}{3}$ or $+\frac{2}{3}$ and/or stable di-quarks of charge $-\frac{2}{3}$ or $+\frac{1}{3}$ or $+\frac{4}{3}$ at the highest energy accelerators would help to reassure us of the non-existence of real quarks'.

Indeed, with one possible exception (La Rue *et al* 1977, 1981), this 'reassurance' has been handsomely provided! *Unlike* the constituents of atoms and nuclei, quarks have *not* been observed as free isolated particles. When hadrons of the highest energies currently available are smashed into each other,

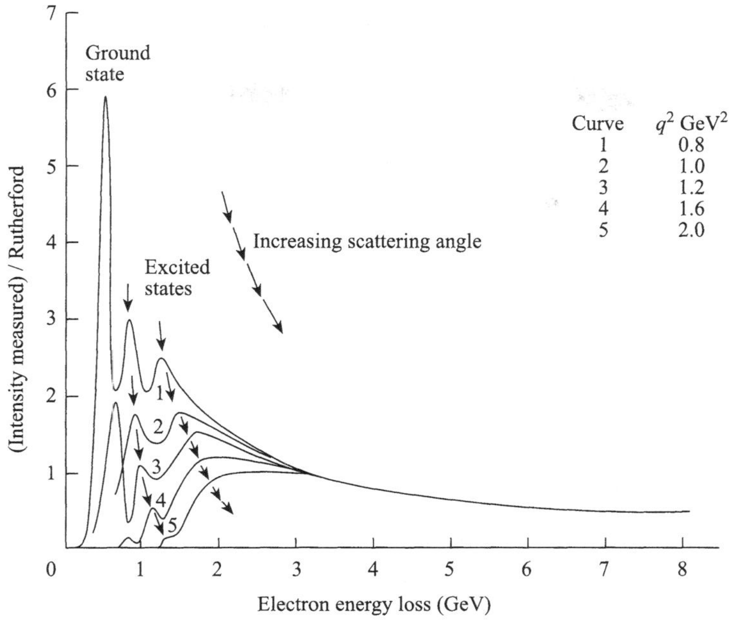

Figure 1.12. Inelastic electron–proton scattering cross section as a function of energy transfer ν for different values of q^2, the invariant square of the 4-momentum transfer: q^2 increases as the scattering angle increases. (From Perkins D 1972 *Introduction to High Energy Physics* 1st edn, courtesy Addison Wesley Publishing Company.)

what is observed downstream is only lots more hadrons, not fractionally charged quarks. The explanation for this novel behaviour of quarks—that they appear perfectly real when clustered into hadrons, but are not to be seen in isolation—is now believed to lie in the nature of the interquark force (QCD). We shall briefly touch on this force again in section 2.6 and take it up in more detail in volume 2. The consensus at present is that QCD does imply the '*confinement*' of quarks—that is, they do not exist as isolated single particles, only as groups confined to hadronic volumes.

We may summarize the story so far by saying that the basic matter units participating in electromagnetic, strong and weak interactions of nucleonic and pionic matter are just four: $\{(\nu_e, e^-), (u, d)\}$, together with their anti-particles. These are what are now called the 'first generation' of quarks and leptons. Unfortunately, this simple world is not the one in which we live. Instead, two further 'generations' of lepton and quark pairs are necessary to describe the real world, as we now briefly explain.

1.3 The generations and flavours of quarks and leptons

1.3.1 Lepton flavour

The first member of another generation (as it turned out) to be found was the muon (Street and Stevenson 1937, Anderson and Neddermeyer 1937). Following the convention for the electron, the μ^- is the particle and the μ^+ the anti-particle. At first, the muon was identified with the particle postulated by Yukawa only two years earlier (1935) as the field quantum of the 'strong nuclear force field', the exchange of which between two nucleons would account for their interaction (see section 2.2). In particular, its mass (105.7 MeV) was nicely within the range predicted by Yukawa. However, experiments by Conversi *et al* (1947) established that the muon could not be Yukawa's quantum since it did not interact strongly; it was therefore a lepton (not a meson!). The μ^- seems to behave in exactly the same way as the electron, interacting only electronically and weakly, with interaction strengths identical to those of an electron.

In 1975 Perl *et al* (1975) discovered yet another 'replicant' electron, the τ^- with a mass of 1.78 GeV. Once again, the weak and electromagnetic interactions of the τ^- (τ^+) are identical to those of the e^- (e^+).

At this stage one might well wonder whether we are faced with a 'lepton spectroscopy', of which the e^-, μ^- and τ^- are but the first three states. Yet this seems not to be the correct interpretation. First, no other such states have (yet) been seen. Second, all these leptons have the same spin ($\frac{1}{2}$), which is certainly quite unlike any conventional excitation spectrum. And third, no γ-transitions are observed to occur between the states, though this would normally be expected. For example, the branching fraction for the process

$$\mu^- \to e^- + \gamma \qquad (\textit{not}\text{ observed}) \tag{1.1}$$

is currently quoted as less than 1.2×10^{-11} at the 90% confidence level (Groom *et al* 2000). Similarly there are (much less stringent) limits on $\tau^- \to \mu^- + \gamma$ and $\tau^- \to e^- + \gamma$.

If the e^- and μ^- states in (1.1) were, in fact, the ground and first excited states of some composite system, the decay process (1.1) would be expected to occur as an electromagnetic transition, with a relatively high probability because of the large energy release. Yet the experimental upper limit on the rate is very tiny. In the absence of any mechanism to explain this, one *systematizes* the situation, empirically, by postulating the existence of a selection rule forbidding the decay (1.1). In taking this step, it is important to realize that 'absolute forbidden-ness' can never be established experimentally: all that can be done is to place a (very small) upper limit on the branching fraction to the 'forbidden' channel, as here. The possibility will always remain open that future, more sensitive, experiments will reveal that some processes, assumed to be forbidden, are in fact simply extremely rare.

Of course, such a proposed selection rule would have no physical content if it only applied to the one process (1.1); but in fact it does turn out to be generally true, applying not only to the electromagnetic interaction of the massive leptons, but to their weak interactions also. The upshot is that we can consistently account for observations (and non-observations) involving e's, μ's and τ's by assigning to each a new additive quantum number (called 'lepton flavour') which is assumed to be conserved. Thus we have electron flavour L_e such that $L_e(e^-) = 1$ and $L_e(e^+) = -1$; muon flavour L_μ such that $L_\mu(\mu^-) = 1$ and $L_\mu(\mu^+) = -1$; and tau flavour L_τ such that $L_\tau(\tau^-) = 1$ and $L_\tau(\tau^+) = -1$. Each is postulated to be conserved in all leptonic processes. So (1.1) is then forbidden, the left-hand side having $L_e = 0$ and $L_\mu = 1$, while the right-hand side has $L_e = 1$ and $L_\mu = 0$.

Just as, in the first generation of quarks and leptons, the charged electron is partnered by a neutral (uncharged) neutrino, so in the second the muon has a neutral partner too, and likewise the tau in the third generation. Neutrinos interact via the weak interactions only, and are assigned lepton flavour quantum numbers in such a way as to conserve each separate leptonic flavour, this being the way in which the data can most conveniently be systematized. Thus, starting with ordinary β-decay

$$n \to p + e^- + \bar{\nu}_e, \tag{1.2}$$

we assign $L_e = -1$ to the neutrino emitted in (1.2) since $L_e = 0$ in the initial state and $L_e(e^-) = +1$; so the neutrino in (1.2) is an 'anti-neutrino of electron type' (or 'of electron flavour'). The neutrino partnering the μ^- appears in the decay of the π^-:

$$\pi^- \to \mu^- + \bar{\nu}_\mu \tag{1.3}$$

where the $\bar{\nu}_\mu$ is an anti-neutrino of muon type ($L_\mu(\bar{\nu}_\mu) = -1$, $L_e(\bar{\nu}_\mu) = 0 = L_\tau(\bar{\nu}_\mu)$).

How do we know that $\bar{\nu}_\mu$ and $\bar{\nu}_e$ are not the same? An important experiment by Danby *et al* (1962) provided evidence that they are not. They found that the neutrinos accompanying muons from π-decay always produced muons on interacting with matter, never electrons. Thus, for example, the lepton flavour conserving reaction

$$\bar{\nu}_\mu + p \to \mu^+ + n \tag{1.4}$$

was observed, but the lepton flavour violating reaction

$$\bar{\nu}_\mu + p \to e^+ + n \qquad (\textit{not}\ \text{observed}) \tag{1.5}$$

was not. As with (1.1), 'non-observation' of course means, in practice, an upper limit on the cross section. Both types of neutrino occur in the β-decay of the muon itself:

$$\mu^- \to \nu_\mu + e^- + \bar{\nu}_e, \tag{1.6}$$

in which $L_\mu = 1$ is initially carried by the μ^- and finally by the ν_μ, and the L_e's of the e^- and $\bar{\nu}_e$ cancel each other out.

In the same way, the ν_τ is associated with the τ^-, and we have arrived at two further generations of charged and neutral leptons:

$$(\nu_\mu, \mu^-) \qquad \text{and} \qquad (\nu_\tau, \tau^-) \tag{1.7}$$

together with their anti-particles.

We should at this point note that another type of weak interaction is known, in which—for example—the $\bar{\nu}_\mu$ in (1.4) scatters elastically from the proton, instead of changing into a μ^+:

$$\bar{\nu}_\mu + \mathrm{p} \rightarrow \bar{\nu}_\mu + \mathrm{p}. \tag{1.8}$$

This is an example of what is called a 'neutral current' process, (1.4) being a 'charged current' one. In terms of the Yukawa exchange mechanism for particle interactions, to be described in the next chapter, (1.4) proceeds via the exchange of charged quanta ($\mathrm{W}^\pm$), while in (1.8) a neutral quantum (Z^0) is exchanged.

As well as their flavour, one other property of neutrinos is of great interest, namely their mass. As originally postulated by Pauli, the neutrino emitted in β-decay had to have very small mass, because the maximum energy carried off by the e^- in (1.2) was closely equal to the difference in rest energies of the neutron and proton. It was subsequently widely assumed (perhaps largely for simplicity) that all neutrinos were strictly massless, and it is fair to say that the conventional Standard Model makes this assumption. Yet there is, in fact, no convincing reason for this (as there is for the masslessness of the photon—see chapter 7), and there is gathering evidence that neutrinos may indeed have small, but non-zero, masses. It turns out that the question of neutrino masslessness is directly connected to another one: whether lepton flavour is, in fact, conserved. If neutrinos are massless, as in the conventional Standard Model, the neutrinos of different flavour cannot 'mix', in the sense of quantum-mechanical states; but mixing can occur if neutrinos have mass. By 2001, experiments had confirmed the phenomenon of 'neutrino oscillations', implying that neutrino mixing does occur (see, for example, the review by K Nakamura on 'Solar Neutrinos' in Groom *et al* 2000).

As an example of the kind of physics this would entail, consider the μ^--decay of (1.6). From a (hypothetical, at present) intense beam of μ^-'s, the ν_μ's will produce μ^-'s in a detector via the reaction $\nu_\mu + \mathrm{n} \rightarrow \mu^- + \mathrm{p}$, analogous to (1.4), but no μ^+'s will be produced, since the $\bar{\nu}_\mathrm{e}$'s cannot produce them. However, if the $\bar{\nu}_\mathrm{e}$ state can mix with the $\bar{\nu}_\mu$ state, the $\bar{\nu}_\mu$ component will be capable of producing μ^+'s via (1.4). Thus both μ^+ and μ^- will be produced from the μ^- source, a very clear signal of neutrino flavour mixing.

Assuming that all three neutrinos are involved, their mixing will be described by a unitary 3×3 matrix, as in all such cases in quantum mechanics. Establishing the values of the entries in this matrix promises to be an area of intense experimental activity in the future. The exciting possibility exists that such data may provide clues to physics 'beyond the Standard Model'. Further discussion must be postponed until we treat weak interactions in more detail in the second volume.

We now turn to the other generations of quarks, and their flavour quantum numbers.

1.3.2 Quark flavour

From the first generation of basic fermions $\{(\nu_e, e^-), (u, d)\}$, there is already a hint of some kind of 'parallelism' between the lepton and quark doublets, in that the two members of each doublet have charges differing by one unit. Indeed, accompanying the second-generation lepton doublet is an analogous quark doublet (c, s), where 'c' refers to the charm quark and 's' to the strange quark, with charges $\frac{2}{3}e$ and $-\frac{1}{3}e$ respectively, exactly the same as the charges of the u and d quarks. These quark attributes 'charm' and 'strangeness' are called quark flavours, as are 'up-ness' and 'down-ness' in the first generation. Quark flavour plays a role quite similar to that of lepton flavour—with some important differences, as we shall see.

The first evidence for new states of hadronic matter (which we would now call strange particles) was obtained by Rochester and Butler in 1947, the same year in which the pion was discovered. Using a cloud chamber to observe penetrating cosmic rays, Rochester and Butler (1947) obtained two pictures showing forked tracks. One was of a neutral particle decaying into two charged particles (which we now identify as $K^0 \rightarrow \pi^+ + \pi^-$), while the other was of a charged particle decaying into another charged particle and at least one neutral (now identified as $K^+ \rightarrow \mu^+ + \nu_\mu$). The decaying particles had lifetimes on the scale 10^{-9}–10^{-10} s, and masses of order 500 MeV (with a large error).

Other examples were slow to appear, but by the early 1950s the experimental situation was becoming clearer. Rochester and Butler's observations were confirmed, as was the existence of additional unstable particles heavier than nucleons and decaying to them, which we now call strange baryons. For example, events of the type

$$\pi^- + p \rightarrow x^0 + X^0 \tag{1.9}$$

were well established. Figure 1.13 shows an example of such a process. The two broken lines on the reconstruction (figure 1.13(*b*)) represent the neutral particles x^0 and X^0 which leave no ionizing track in figure 1.13(*a*). They are detected via their decay modes $x^0 \rightarrow \pi^+ + \pi^-$, $X^0 \rightarrow \pi^- + p$. By measuring the momenta of these decay products, the masses of the x^0 and X^0 particles were found to be some 500 and 1100 MeV respectively; the x^0 is therefore the neutral Rochester–Butler particle. The data showed two curious ('strange') features. First, the new particles were produced with large cross sections, characteristic of a strong interaction, yet their lifetimes were at least ten orders of magnitude too large to be due to strong interactions, and were instead typical of weak interactions. Second, it seemed that the new states were invariably produced in association with each other, never

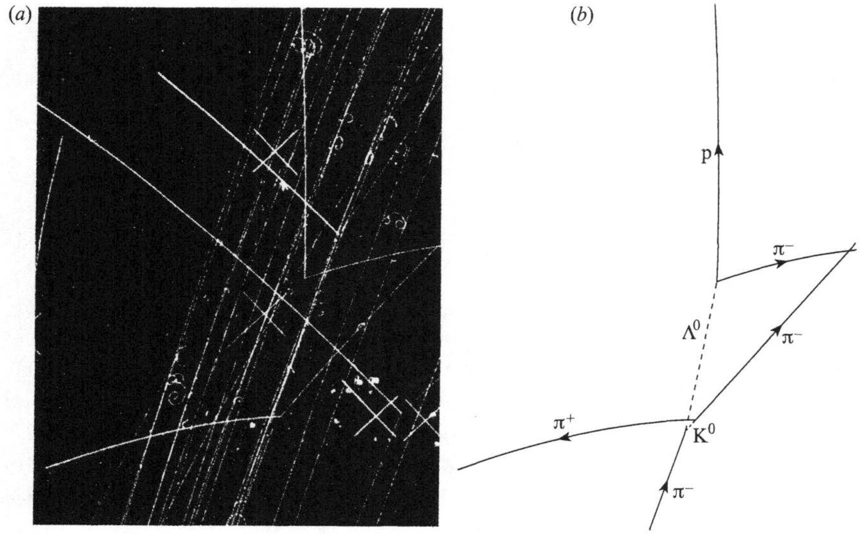

Figure 1.13. Associated production of strange particles.

singly: for example, the apparently quite possible (indeed, on energy grounds, more favourable) processes

$$\pi^- + \mathrm{p} \rightarrow \mathrm{x}^0 + \mathrm{n} \qquad (\textit{not}\ \text{observed}) \tag{1.10}$$

$$\pi^- + \mathrm{p} \rightarrow \pi^0 + \mathrm{X}^0 \qquad (\textit{not}\ \text{observed}) \tag{1.11}$$

were not observed. The non-observation of (1.10) appeared all the more peculiar (in fact, 'strange') given the fact that (as indicated in (1.9)) both the x^0 and X^0 *did* decay quite happily, by themselves, into familiar particles!

The first step in the resolution of these puzzles was taken by Pais (1952), who suggested a new selection rule which was valid in strong and electromagnetic interactions, but not in weak ones. This rule had the consequence that the new particles could only be produced (via a strong interaction) in association with each other, from an initial state of known particles such as pions and nucleons. They could decay singly, however, via a weak interaction, consistent with the long lifetime. The present scheme of quantum numbers was proposed independently by Gell-Mann (1956) and Nishijima (1955). As applied to the process (1.9), their proposal was that that the x^0 and X^0 each carried equal and opposite values of a new quantum number, called 'strangeness' (S) by Gell-Mann, which was additively conserved in strong and electromagnetic interactions, but which was violated in weak interactions. Existing known particles, such as the pions and nucleons, did not carry this quantum number. In today's nomenclature, the X^0 is the Λ^0 baryon, assigned $S = -1$ in the Gell-Mann–Nishijima scheme, and the x^0 is the K^0, assigned $S = +1$. So strangeness is conserved in the strong production process $\pi^-\mathrm{p} \to \mathrm{K}^0\Lambda^0$ but not in the weak decay processes $\mathrm{K}^0 \to \pi^+\pi^-$ and $\Lambda^0 \to \mathrm{p}\pi^-$.

Hadronic flavour selection rules become, in fact, much simpler when expressed in terms of the quark constituents. Pions and nucleons are made of u's and d's which have $S = 0$. The Λ^0 with quark composition uds has $S = -1$, and we assign $S = -1$ to the s-quark. The K^0 has $S = +1$ and quark composition $\bar{\mathrm{s}}\mathrm{d}$, consistent with the anti-quark $\bar{\mathrm{s}}$ having $S = +1$, as expected. The general rules governing quark flavour changes are then as follows. In strong and electromagnetic interactions individual quarks and anti-quarks cannot change their flavour, and they can only be created or destroyed in matching particle–anti-particle pairs. For example, the quark description of the production process $\pi^- + \mathrm{p} \to \mathrm{K}^0 + \Lambda^0$ of (1.9) is

$$(\bar{\mathrm{u}}\mathrm{d}) + (\mathrm{uud}) \to (\bar{\mathrm{s}}\mathrm{d}) + (\mathrm{uds}). \tag{1.12}$$

Note that the 'u' and 'd' flavours are also conserved here: on the left-hand side of (1.12), $N_\mathrm{u} = -1 + 2 = 1$, $N_\mathrm{d} = 2$, $N_\mathrm{s} = 0$, while on the right-hand side we have $N_\mathrm{u} = 1$, $N_\mathrm{d} = 2$ and $N_\mathrm{s} = -1 + 1 = 0$. On the other hand, the flavour of an individual quark or anti-quark can change in a weak interaction process such as neutron β-decay $\mathrm{n} \to \mathrm{p} + \mathrm{e}^- + \bar{\nu}_\mathrm{e}$, in which the initial quark state is (udd) and the final one is (uud), so that a d-quark has become a u-quark. At the quark level, then, neutron β-decay is a manifestation of the process $\mathrm{d} \to \mathrm{u} + \mathrm{e}^- + \bar{\nu}_\mathrm{e}$, which is quite analogous to the purely leptonic process (1.6).

Given the existence of the s quark, the notion of lepton–quark parallelism would lead one to expect a fourth quark, 'c', the partner of the s, so as to complete a second generation $\{(\nu_\mu, \mu^-), (\mathrm{c}, \mathrm{s})\}$, parallel to the first $\{(\nu_\mathrm{e}, \mathrm{e}^-), (\mathrm{u}, \mathrm{d})\}$. We shall come to the c quark shortly, but we first pause to appreciate an apparent difference between the weak interactions of the s quark in the second generation, and those of its leptonic 'partners'. At the quark level, a strangeness-changing weak interaction such as $\Lambda^0(\mathrm{uds}) \to \mathrm{p}(\mathrm{uud}) + \pi^-(\mathrm{d}\bar{\mathrm{u}})$ must involve a transition betwen an s and a u quark, which are in *different* generations. It is simpler to consider the strangeness-changing 'semi-leptonic' mode of the Λ^0,

$$\Lambda^0(\mathrm{uds}) \to \mathrm{p}(\mathrm{uud}) + \mathrm{e}^- + \bar{\nu}_\mathrm{e} \tag{1.13}$$

which is analogous to neutron decay (1.2). However, whereas (1.2) involves only first-generation quarks and leptons ($\mathrm{d} \to \mathrm{u} + \mathrm{e}^- + \bar{\nu}_\mathrm{e}$), (1.13) at the quark level is due to the 'inter-generational' transition $\mathrm{s} \to \mathrm{u} + \mathrm{e}^- + \bar{\nu}_\mathrm{e}$. True, this inter-generational transition is found experimentally to be somewhat weaker in intrinsic strength than the 'intra-generational' transition $\mathrm{d} \to \mathrm{u} + \mathrm{e}^- + \bar{\nu}_\mathrm{e}$. Nevertheless, both are in fact very similar processes. However, the analogous inter-generational leptonic transition $\mu^- \to \nu_\mathrm{e} + \mathrm{e}^- + \bar{\nu}_\mathrm{e}$ is not (as yet) observed - precisely as is 'encoded' in the leptonic flavour selection rules of the previous section (it would violate L_μ and L_e).

We must note, however, that if the ν_μ state can mix with the ν_e state, then $\mu^- \to \nu_\mathrm{e} + \mathrm{e}^- + \bar{\nu}_\mathrm{e}$ will occur! The reader may at this point be wondering why, if neutrino mixing is already established, such processes (violating the simple lepton number conservation laws of the previous section) have not been observed. The answer is that neutrino mixing amplitudes must be proportional to neutrino masses (since the phenomenon disappears at zero mass), so that a dimensionless parameter characterizing the small admixture of such an amplitude would be expected to be of order (neutrino mass)/(energy released in the decay). Present indications are that neutrino masses are considerably less than 1 eV, so that this ratio will be very small, and $\mu^- \to \nu_e + e^- + \bar{\nu}_e$ will be highly suppressed. This is another example of the interest attached to the measurement of such rare (possibly not forbidden) decay modes.

We return now to the c quark. The existence of a (then) 'fourth quark', carrying a new strongly-conserved quantum number C ('charm') was predicted on theoretical grounds well before its experimental discovery. Bjorken and Glashow (1964) had discussed the possible existence of a fourth quark on the basis of quark–lepton symmetry, but a strong theoretical argument for the existence of the c-quark, within the framework of gauge theories of electroweak interactions, was given in a landmark paper by Glashow, Iliopoulos and Maiani (1970). We shall discuss the Glashow, Iliopoulos and Maiani ('GIM') argument in volume 2. Glashow, Iliopoulos and Maiani themselves estimated that the c-quark mass should lie in the range 3–4 GeV. Subsequently, Gaillard and Lee (1974) performed a full (one-loop) calculation in the then newly-developed renormalizable electroweak theory, and predicted $m_c \approx 1.5$ GeV. The prediction was spectacularly confirmed in November of the same year with the discovery (Aubert *et al* 1974, Augustin *et al* 1974) of the J/ψ system, which was soon identified as a $c\bar{c}$ composite, with a mass in the vicinity of 3 GeV. Subsequently, mesons such as $D^0(c\bar{u})$ and $D^+(c\bar{d})$ carrying $C = 1$ were identified (Goldhaber *et al* 1976, Peruzzi *et al* 1976), consolidating this identification.

Figure 1.14(a) shows the low-lying energy levels of charmonium, as $c\bar{c}$ quickly became known, with a similar plot for positronium in figure 1.14(b). Beyond reasonable doubt, charmonium must be a fermion–anti-fermion composite (indeed, this realization was for many physicists the decisive moment when they finally believed in the reality of quarks inside hadrons!). The J^{PC} quantum numbers of charmonium match those of positronium quite remarkably (here P and C are the parity and charge-conjugation quantum numbers) . But the analogy seems to go further: the striking similarity in the level ordering suggest that the *strong* force between $c\bar{c}$ must behave dynamically rather like the *electromagnetic* force in e^+e^-. Of course, the scales differ by a dramatic factor of 10^8, indicating the quite different masses and excitation energies in the two problems. Nevertheless we shall indicate in section 2.6 how QCD explains this and other 'strong/electromagnetic' analogies.

This all relates to the strong interactions of the c quark; what about its weak interactions? The answer is rather simple: the (c, s) pair behave remarkably like the (u, d) pair. The dominant transition is the intra-generational one $c \to s + e^+ + \nu_e$, analogous to $d \to u + e^- + \bar{\nu}_e$, but there is also the somewhat weaker

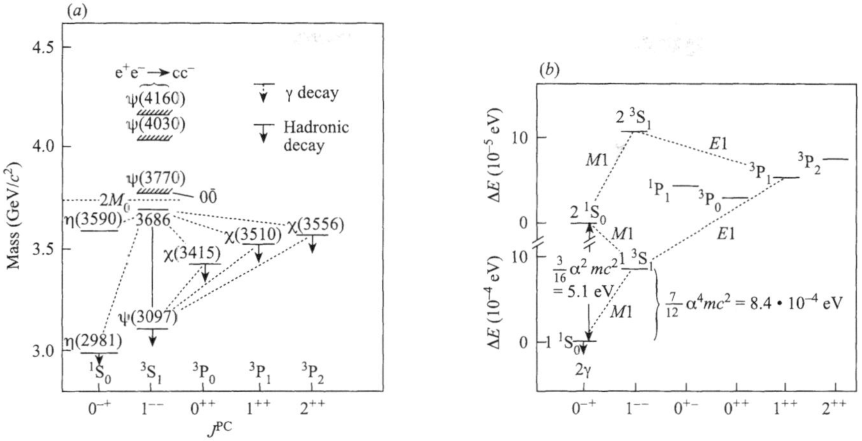

Figure 1.14. Energy levels of (*a*) charmonium ($c\bar{c}$) and (*b*) positronium (e^+e^-) from Perkins D 1987 *Introduction to High Energy Physics* 3rd edn, courtesy Addison-Wesley Publishing Company.)

inter-generational mode $c \rightarrow d + e^+ + \nu_e$, which has indeed very much the same strength as the inter-generational transition involving the s, $s \rightarrow u + e^- + \bar{\nu}_e$ as already discussed. The subtle intertwining of these (c, s) and (u, d) transitions in the GIM scheme will be discussed in volume 2.

The second generation was completed in 1974, but already in 1975 there was evidence for a third one, with the discovery of the τ^- mentioned earlier. In view of all the foregoing, one expects a complete third generation $\{(\nu_\tau, \tau^-), (\mathrm{t}, \mathrm{b})\}$ where 't' stands for the top quark with a new flavour $T = +1$, and 'b' for the bottom quark with a new flavour $\tilde{B} = -1$ (we use $\tilde{B}$ to distinguish this flavour quantum number from baryon number B). As before, T and $\tilde{B}$ are conserved in strong and electromagnetic interactions, but not in weak ones. The discovery of the b quark (charge $-\frac{1}{3}e$) resulted from the observation of massive mesonic states generally known as Υ ('upsilon') (Herb *et al* 1977, Innes *et al* 1977), which were identified as $\mathrm{b}\bar{\mathrm{b}}$ composites. Subsequently, $\tilde{B}$-carrying mesons were found. Finally, firm evidence for the expected t quark (charge $+\frac{2}{3}e$, with flavour $T = +1$) was obtained by the CDF and DO collaborations at Fermilab in 1995 (Abe *et al* 1995, Abachi *et al* 1995); see Perkins (2000, section 4.13) for details about the discovery of the top quark.

The description of inter-generational weak quark transitions must obviously now be extended to mixing across all three families. The three states u, d and s—defined by their strong interaction identities—must for weak interactions be replaced by a quantum-mechanical superposition, described by another 3×3 unitary matrix, part of which of course relates to the simpler two-family situation already outlined. Formally, therefore, this is the same situation that we have discussed for the neutrinos, though in the (u, d, s) case the numerical values of most of the entries are by now quite accurately known (and of course not the same as those for the neutrino matrix). Further discussion is postponed until volume 2.

Having discussed at some length the various quark quantum numbers, we should not end this section without some mention of another important physical attribute of quarks, namely their masses. In fact, since quarks do not, apparently, exist as free isolated particles, it is not straightforward to decide how to define their mass. Nevertheless, good descriptions of at least the heavier mesonic states (such as $c\bar{c}$ and $b\bar{b}$) are obtained by supposing that the major contribution to the mesonic mass is given by an effective quark mass contributing additively. This gives $m_c \sim 1.5$ GeV and $m_b \sim 4.5$ GeV, in rough terms. The top quark is very much more massive that any of the others: $m_t \sim 175$ GeV. This is inferred from the observed decay pattern. With such a large mass the top quark can decay to a bottom quark plus a W boson, with a lifetime of order 5×10^{-25} s. Though a weak decay, the very large energy release makes the lifetime much shorter than the typical timescale of hadronic interactions, so that no $t\bar{t}$ (toponium) states have time to form.

As far as the 'light' quarks (u, d and s) are concerned, comparison with the corresponding hadronic systems would suggest masses in the ranges $m_{u,d} \sim 0.3$ GeV, $m_s \sim 0.5$ GeV. However, there are good reasons to think that nearly all of this mass is 'effective', in the sense of being generated dynamically by the strong QCD interactions. Lattice gauge theory offers the prospect of being able to calculate the dynamical component reliably, and hence of extracting the 'intrinsic' (non-dynamically-generated) quark masses. As with the leptons, there is as yet no theory capable of explaining the 'spectrum' of quark masses.

We end this chapter by gathering together the three generations of quarks and leptons in table 1. The mass values are taken from Groom *et al* (2000). Note that recent (2001) data imply non-zero masses for neutrinos; see also the notes in Groom *et al* (2000) concerning both quark and lepton masses.

Table 1. Properties of quarks and leptons.

Leptons						Quarks						
Particle	Mass (MeV/c^2)	Q/e	L_e	L_μ	L_τ	Particle	Mass (GeV/c^2)	Q/e	S	C	$\tilde{B}$	T
ν_e	$< 3 \times 10^{-6}$	0	1	0	0	u	$1–5 \times 10^{-3}$	2/3	0	0	0	0
e^-	0.511	−1	1	0	0	d	$3–9 \times 10^{-3}$	−1/3	0	0	0	0
ν_μ	< 0.19	0	0	1	0	c	1.15–1.35	2/3	0	1	0	0
μ^-	105.66	−1	0	1	0	s	$75–170 \times 10^{-3}$	−1/3	−1	0	0	0
ν_τ	< 18.2	0	0	0	1	t	174.3 ± 5.1	2/3	0	0	0	1
τ^-	1777.0	−1	0	0	1	b	4–4.4	−1/3	0	0	−1	0

2

PARTICLE INTERACTIONS IN THE STANDARD MODEL

2.1 Introduction

In the world of the classical physicist, matter and force were clearly separated. The nature of matter was intuitive, based on everyday macroscopic experience; force, however, was more problematical. Contact forces between bodies were easy to understand, but forces which seemed capable of acting at a distance caused difficulties.

> That gravity should be innate, inherent and essential to matter, so that one body can act upon another at a distance, through a vacuum, without the mediation of anything else, by and through which action and force may be conveyed from one to the other, is to me so great an absurdity, that I believe no man who has in philosophical matters a competent faculty of thinking can ever fall into it. (Letter from Newton to Bentley)

Newton could find no satisfactory mechanism or physical model, for the transmission of the gravitational force between two distant bodies; but his dynamical equations provided a powerful predictive framework, given the (unexplained) gravitational force law; and this eventually satisfied most people.

The 19th century saw the precise formulation of the more intricate force laws of electromagnetism. Here too the distaste for action-at-a-distance theories led to numerous mechanical or fluid mechanical models of the way electromagnetic forces (and light) are transmitted. Maxwell made brilliant use of such models as he struggled to give physical and mathematical substance to Faraday's empirical ideas about lines of force. Maxwell's equations were indeed widely regarded as describing the mechanical motion of the ether—an amazing medium, composed of vortices, gear wheels, idler wheels and so on. But in his 1864 paper, the third and final one of the series on lines of force and the electromagnetic field, Maxwell himself appeared ready to throw away the mechanical scaffolding and let the finished structure of the *field equations* stand on its own. Later these field equations were derived from a Lagrangian (see chapter 7), and many physicists came to agree with Poincaré that this 'generalized mechanics' was more satisfactory than a multitude of different ether models; after all, the same mathematical equations can describe, when suitably interpreted, systems of masses, springs and dampers, or of inductors, capacitors and resistors. With this step, the concepts of mechanics were enlarged to include a new fundamental entity, the electromagnetic field.

The action-at-a-distance dilemma was solved, since the electromagnetic field permeates all of space surrounding charged or magnetic bodies, responds locally to them, and itself acts on other distant bodies, propagating the action to them at the speed of light: for Maxwell's theory, besides unifying electricity and magnetism, also predicted the existence of electromagnetic waves which should travel with the speed of light, as was confirmed by Hertz in 1888. Indeed, light *was* a form of electromagnetic wave.

As we have seen, electrons were isolated, around 1900, as the 'particles of electricity'—the prototype of all 'charged or magnetic bodies'. It seems almost to be implied by the local field concept, and the desire to avoid action at a distance, that the fundamental carriers of electricity should themselves be point-like, so that the field does not, for example, have to interact with different parts of an electron simultaneously. Thus the point-like nature of elementary matter units seems intuitively to be tied to the local nature of the force field via which they interact. Nevertheless, the very property of 'locality' itself led to difficulties, such as the infinite self-energy of a point-like charge (see, for example, Feynman 1964 section 28-1).

Very soon after these successes of classical physics, however, another world began to make its appearance—the quantum one. First the photoelectric effect and then—much later—the Compton effect showed unmistakeably that these electromagnetic *waves* somehow also had a *particle*-like aspect, the photon. At about the same time, the intuitive understanding of the nature of matter began to fail as well: supposedly *particle*-like things, like electrons, displayed *wave*-like properties (interference and diffraction). Thus the conceptual distinction between matter and forces, or between particle and field, was no longer so clear. On the one hand, electromagnetic forces, treated in terms of fields, now had a particle aspect; and on the other hand, particles now had a wave-like or field aspect. 'Electrons', writes Feynman (1965) at the beginning of volume 3 of his *Lectures on Physics*, 'behave just like light'.

How can we build a theory of electrons and photons which does justice to all the 'point-like', 'local', 'wave/particle' ideas just discussed? Consider the apparently quite simple process of spontaneous decay of an excited atomic state in which a photon is emitted:

$$A^* \to A + \gamma. \tag{2.1}$$

Ordinary non-relativistic quantum mechanics cannot provide a first-principles account of this process, because the degrees of freedom it normally discusses are those of the '*matter*' units alone—that is, in this example, the electronic degrees of freedom. However, it is clear that something has changed radically in the *field* degrees of freedom. On the left-hand side, the matter is in an excited state and the electromagnetic field is somehow not manifest; on the right, the matter has made a transition to a lower-energy state and the energy difference has gone into creating a quantum of electromagnetic radiation. In chapter 5 we shall try to give an elementary and self-contained introduction to the quantum theory of fields. We shall see that a field—which has infinitely many degrees of freedom—can be thought of as somewhat analogous to a vibrating solid (which has merely a very large number). The field can exist in well-defined states of definite quantized energy—just like the energy levels of systems of quantum particles. In the process (2.1) the electromagnetic field was originally in its ground state, and was raised finally to an excited state by the transfer of energy from the matter degrees of freedom. The excited field state is described quantum mechanically by the presence of the photon: in the ground state there are no photons.

A fully quantum mechanical treatment of the spontaneous emission process (2.1) was first given by Dirac (1927), following the procedure to be described in chapter 5, as applied there to the simpler case of scalar fields. This will be applied to QED in chapters 7, 8, 9 and 11, and to the other interactions in volume 2. In the present chapter, we begin our journey at a more descriptive and pictorial level, making essential use of Yukawa's remarkable insight into the quantum nature of force.

2.2 The Yukawa theory of force as virtual quantum exchange

Yukawa's revolutionary paper (Yukawa 1935) proposed a theory of the strong interaction between a proton and a neutron, and also considered its possible extension to neutron β-decay. He built his theory by analogy with electromagnetism, postulating a new field of force with an associated new field quantum, analogous to the photon. In doing so, he showed with particular clarity how, in quantum field theory, *particles interact by exchanging virtual quanta*, which *mediate* the force.

Before proceeding, we should emphasize that we are not presenting Yukawa's ideas as a viable candidate theory of strong and weak interactions. Crucially, Yukawa assumed that the nucleons and his quantum (later identified with the pion) were point-like, but we have seen in section 1.2 that both nucleons and pions are quark composites with spatial extension. The true 'strong' interaction relates to the quarks, as we shall see in section 2.6. There are also other details of his theory which were (we now know) mistaken, as we shall discuss. Yet his approach was profound, and—as happens often in physics—even though the initial application was ultimately superseded, the ideas have broad and lasting validity.

Yukawa began by considering what kind of static potential might describe the n–p interaction. It was known that this interaction decreased rapidly for interparticle separation $r \geq 2$ fm. Hence, the potential could not be of coulombic type $\propto 1/r$. Instead, Yukawa postulated an n–p potential energy of the form

$$U(\boldsymbol{r}) = \frac{-g_{\mathrm{s}}^2}{4\pi}\frac{\mathrm{e}^{-r/a}}{r} \tag{2.2}$$

where 'g_s' is a constant analogous to the electric charge e, $r = |\boldsymbol{r}|$ and 'a' is a range parameter ($\sim$ 2 fm). This static potential satisfies the equation

$$\left(\nabla^2 - \frac{1}{a^2}\right) U(\boldsymbol{r}) = g_s^2 \delta(\boldsymbol{r}) \tag{2.3}$$

(see appendix G) showing that it may be interpreted as the mutual potential energy of one point-like test nucleon of 'strong charge' g_s due to the presence of another point-like nucleon of equal charge g_s at the origin, a distance r away. Equation (2.3) should be thought of as analogous to Poisson's equation in electrostatics (equation (G.3))

$$\nabla^2 V(r) = -\rho(r)/\epsilon_0 \tag{2.4}$$

the delta function in (2.3) (see appendix E) expressing the fact that the 'strong charge density' acting as the source of the field is all concentrated into a single point, at the origin.

Yukawa now sought to generalize (2.3) to the non-static case, so as to obtain a field equation for $U(\boldsymbol{r}, t)$. For $r \neq 0$, he proposed the free-space equation (we shall keep factors of c and $\hbar$ explicit for the moment)

$$\left(\nabla^2 - \frac{\partial^2}{c^2 \partial t^2} - \frac{1}{a^2}\right) U(\boldsymbol{r}, t) = 0 \tag{2.5}$$

which is certainly relativistically invariant (see appendix D). Thus far, U is still a classical field. Now Yukawa took the decisive step of treating U quantum mechanically, by looking for a (de Broglie-type) *propagating wave solution* of (2.5), namely

$$U \propto \exp(\mathrm{i}\boldsymbol{p} \cdot \boldsymbol{r}/\hbar - \mathrm{i}Et/\hbar). \tag{2.6}$$

Inserting (2.6) into (2.5) one finds

$$\frac{E^2}{c^2\hbar^2} = \frac{\boldsymbol{p}^2}{\hbar^2} + \frac{1}{a^2} \tag{2.7}$$

or, taking the positive square root,

$$E = \left[c^2 \boldsymbol{p}^2 + \frac{c^2\hbar^2}{a^2}\right]^{1/2}.$$

Comparing this with the standard E–$\boldsymbol{p}$ relation for a massive particle in special relativity (appendix D) , the fundamental conclusion is reached that the quantum of the finite-range force field U has a mass m_U given by

$$m_U^2 c^4 = \frac{c^2\hbar^2}{a^2} \qquad \text{or} \qquad m_U = \frac{\hbar}{ac}. \tag{2.8}$$

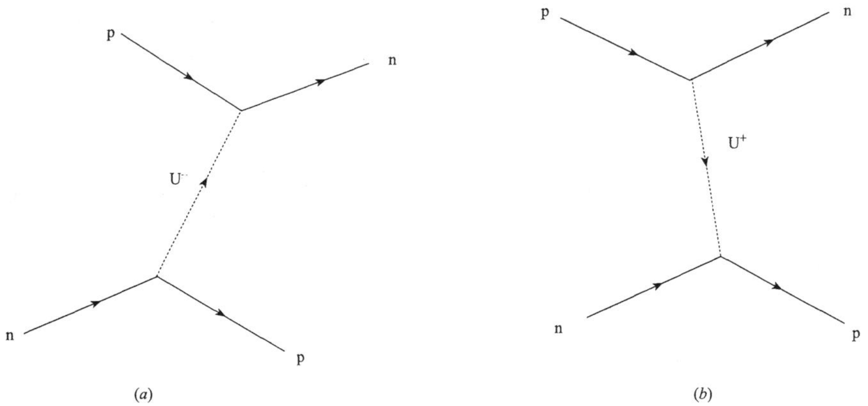

Figure 2.1. Yukawa's single-U exchange mechanism for the n–p interaction. (*a*) U^- exchange. (*b*) U^+ exchange.

This means that the *range parameter* in (2.2) is related to the *mass of the quantum* m_U by

$$a = \frac{\hbar}{m_U c}. \tag{2.9}$$

Inserting $a \sim 2$ fm gives $m_U \sim 100$ MeV, Yukawa's famous prediction for the mass of the nuclear force quantum.

Next, Yukawa envisaged that the U-quantum would be emitted in the transition $n \to p$, via a process analogous to (2.1):

$$n \to p + U^- \tag{2.10}$$

where charge conservation determines the U^- charge. Yet there is an obvious difference between (2.10) and (2.1): (2.10) violates energy conservation since $m_n < m_p + m_U$ if $m_U \sim 100$ MeV, so it cannot occur as a real emission process. However, Yukawa noted that if (2.10) were combined with the inverse process

$$p + U^- \to n \tag{2.11}$$

then an n–p interaction could take place by the mechanism shown in figure 2.1(*a*); namely, by the emission and subsequent absorption—that is, by the *exchange*—of a U^- quantum. He also included the corresponding U^+ exchange, where U^+ is the anti-particle of the U^-, as shown in figure 2.1(*b*).

An energy-violating transition such as (2.10) or (2.11) is known as a 'virtual' transition in quantum mechanics. Such transitions are routinely present in quantum-mechanical time-dependent perturbation theory and can be understood in terms of an 'energy–time uncertainty relation'

$$\Delta E \Delta t \geq \hbar/2. \tag{2.12}$$

It is important to emphasize that, although (2.12) is similar in appearance to the familiar position–momentum uncertainty relation, its origin and interpretation is quite different. In the case of x and p, both quantities are dynamical variables, whose incompatibility (non-commutativity) leads directly to the result $\Delta x \Delta p \geq \hbar/2$. However, although 'energy' is certainly a dynamical variable, 'time' is not: it is simply a parameter. There is no operator, conjugate to the Hamiltonian, whose eigenvalue is 'the time' (see, for example, Gottfried 1966, p 248).

The relation (2.12) may be interpreted as follows (we abridge the careful discussion in section 44 of Landau and Lifshitz (1977)). Imagine an 'energy-measuring device' set up to measure the energy of a quantum system. To do this, the device must interact with the quantum system for a certain length of time Δt. If the energy of a sequence of identically-prepared quantum systems is measured, only in the limit $\Delta t \to \infty$ will the same energy be obtained each time. For finite Δt, the measured energies will necessarily fluctuate by an amount ΔE as given by (2.12); in particular, the shorter the time over which the energy measurement takes place, the larger the fluctuations in the measured energy.

Wick (1938) applied (2.12) to Yukawa's theory, and thereby shed new light on the relation (2.9). Suppose a device is set up capable of checking to see whether energy is, in fact, conserved while the $\mathrm{U}^{\pm}$ crosses over in figure 2.1. The crossing time t must be at least r/c, where r is the distance apart of the nucleons. However, the device must be capable of operating on a time scale smaller than t (otherwise it will not be in a position to detect the $\mathrm{U}^{\pm}$), but it need not be very much less than this. Thus the energy uncertainty in the reading by the device will be

$$\Delta E \sim \frac{\hbar c}{r}. \tag{2.13}$$

As r decreases, the uncertainty ΔE in the measured energy increases. If we require $\Delta E = m_{\mathrm{U}}c^2$, then

$$r \sim \frac{\hbar}{m_{\mathrm{U}}c} \tag{2.14}$$

just as in (2.9). The 'r' in (2.14) is the extent of the separation allowed between the n and the p, such that—in the time available—the $\mathrm{U}^{\pm}$ can 'borrow' the necessary energy to come into existence and cross from one to the other. In this sense, r is the effective range of the associated force, as in (2.9).

Despite the similarity to virtual intermediate states in ordinary quantum mechanics, the Yukawa–Wick process is nevertheless truly revolutionary because it postulated an energy fluctuation ΔE great enough to create an as yet unseen new particle, a new state of matter.

Returning briefly to the historical development, we noted earlier that when the muon (mass 105.7 MeV) was discovered in 1937, just two years after Yukawa's paper, it was identified with his nuclear force quantum. But this interpretation failed when it was established in 1947 that the μ^- interacted too weakly to account for nuclear binding. In the same year, Lattes *et al* (1947) found evidence for a particle somewhat heavier than the muon and decaying into

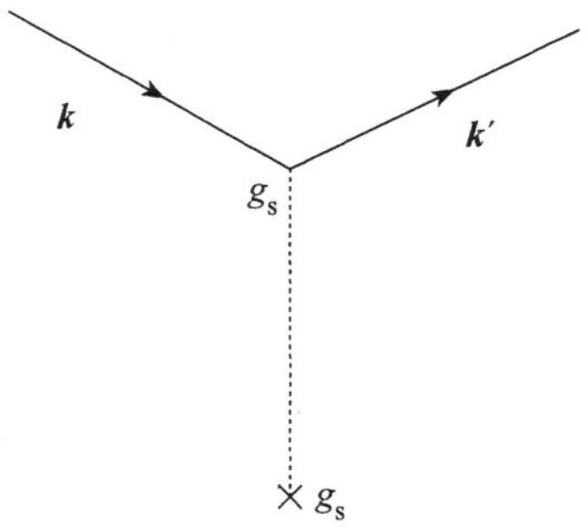

Figure 2.2. Scattering by a static point-like U-source.

it. These heavier particles—pions, of mass ∼140 MeV—were found to interact strongly (Perkins 1947, Occhialini and Powell 1947) and were accordingly identified with Yukawa's quantum. Today, strong interactions are regarded as occurring fundamentally at the quark level (see section 2.6). Nevertheless, the long-range part of the nucleon–nucleon interaction is still usefully described in terms of pion exchange.

We proceed to explore further aspects of Yukawa's force mechanism.

2.3 The one-quantum exchange amplitude

Consider a particle, carrying 'strong charge' g_s, being scattered by an infinitely massive (static) point-like U-source also of 'charge' g_s as pictured in figure 2.2. From the previous section, we know that the potential energy in the Schrödinger equation for the scattered particle is precisely the $U(\boldsymbol{r})$ from (2.2). Treating this to its lowest order in $U(\boldsymbol{r})$ ('Born Approximation'—see appendix H), the scattering amplitude is proportional to the Fourier transform of $U(\boldsymbol{r})$:

$$f(\boldsymbol{q}) = \int \mathrm{e}^{\mathrm{i}\boldsymbol{q}\cdot\boldsymbol{r}} U(\boldsymbol{r})\, \mathrm{d}^3\boldsymbol{r} \tag{2.15}$$

where $\boldsymbol{q}$ is the wavevector (or momentum, with $\hbar = 1$) transfer $\boldsymbol{q} = \boldsymbol{k} - \boldsymbol{k}'$. The transform is evaluated in appendix G, equation (G.24), with the result

$$f(\boldsymbol{q}) = -\frac{g_s^2}{\boldsymbol{q}^2 + m_{\mathrm{U}}^2}. \tag{2.16}$$

This implies that the amplitude (in this static case) for the one-U exchange amplitude is proportional to $-1/(\boldsymbol{q}^2 + m_{\mathrm{U}}^2)$, where $\boldsymbol{q}$ is the momentum carried by the U-quantum.

In this scattering by an infinitely massive source of potential, the energy of the scattered particle cannot change. In a real scattering process such as that in figure 2.1, both energy and momentum can be transferred by the U-quantum—that is, $\boldsymbol{q}$ is replaced by the four-momentum $q = (q_0, \boldsymbol{q})$, where $q_0 = k_0 - k_0'$. Then, as indicated in appendix G, the factor $-1/(\boldsymbol{q}^2 + m_U^2)$ is replaced by $1/(q^2 - m_U^2)$ and the amplitude for figure 2.1 is, in this model,

$$\frac{g_s^2}{q^2 - m_U^2}. \tag{2.17}$$

It will be the main burden of chapters 5, 6 and 7 to demonstrate just how this formula is arrived at, using the formalism of quantum field theory. In particular, we shall see in detail how the *propagator* $(q^2 - m_U^2)^{-1}$ arises. For the present, we can already note (from appendix G) that such propagators are, in fact, momentum–space Green functions.

In chapter 6 we shall also discuss other aspects of the physical meaning of the propagator, and we shall see how diagrams which we have begun to draw in a merely descriptive way become true 'Feynman diagrams', each diagram representing by a precise mathematical correspondence a specific expression for a quantum amplitude, as calculated in perturbation theory. The expansion parameter of this perturbation theory is the dimensionless number $g_s^2/4\pi$ appearing in the potential $U(\boldsymbol{r})$ (cf (2.2)). In terms of Feynman diagrams, we shall learn in chapter 6 that one power of g_s is to be associated with each 'vertex' at which a U-quantum is emitted or absorbed. Thus successive terms in the perturbation expansion correspond to exchanges of more and more quanta. Quantities such as g_s are called 'coupling strengths'.

It is not too early to emphasize one very important point to the reader: true Feynman diagrams are *representations of momentum–space amplitudes*. They are not representations of space–time processes: *all* space–time points are integrated over in arriving at the formula represented by a Feynman diagram. In particular, the two 'intuitive' diagrams of figure 2.1, which carry an implied 'time-ordering' (with time increasing to the right), are *both* included in a single Feynman diagram with propagator (2.17), as we shall see in detail (for an analogous case) in section 7.1.

We now indicate how these general ideas of Yukawa apply to the actual interactions of quarks and leptons.

2.4 Electromagnetic interactions

From the foregoing viewpoint, electromagnetic interactions are essentially a special case of Yukawa's picture, in which g_s^2 is replaced by the appropriate electromagnetic charges, and $m_U \to m_\gamma = 0$ so that $a \to \infty$ and the potential (2.2) returns to the Coulomb one, $-e^2/4\pi r$. A typical one-photon exchange scattering process is shown in figure 2.3, for which the generic amplitude (2.17) becomes

$$e^2/q^2. \tag{2.18}$$

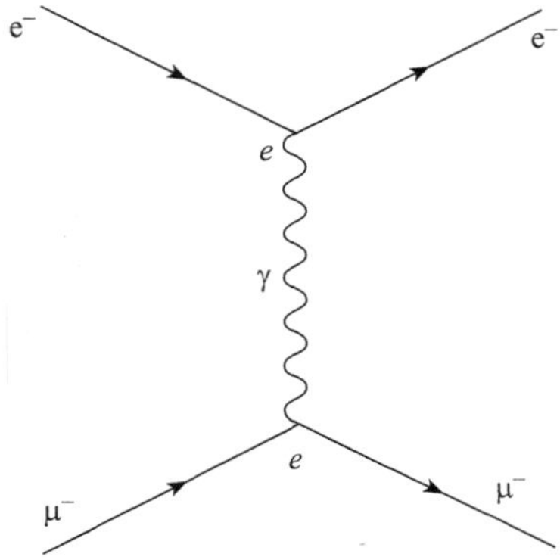

Figure 2.3. One photon exchange mechanism between charged leptons.

Note that we have drawn the photon line 'vertically', consistent with the fact that both time-orderings of the type shown in figure 2.1 are included in (2.18). In the case of electromagnetic interactions, the coupling strength is e and the expansion parameter of perturbation theory is $e^2/4\pi \equiv \alpha \sim 1/137$ (see appendix C).

We can immediately use (2.18) to understand the famous angular variation $\sim \sin^{-4}\theta/2$ of Rutherford scattering, as shown in figure 1.1 (we call the scattering angle θ now). Treating the nucleus as infinitely heavy, the α-particle scatters elastically so that $q_0 = 0$ and $q^2 = -(\boldsymbol{k}-\boldsymbol{k}')^2$ where $\boldsymbol{k}$ and $\boldsymbol{k}'$ are the incident and final α-particle momenta. So $q^2 = -2\boldsymbol{k}^2(1-\cos\theta) = -4\boldsymbol{k}^2\sin^2\theta/2$ where we have used the elastic scattering condition $\boldsymbol{k}^2 = \boldsymbol{k}'^2$. Inserting this into (2.18) and remembering that the cross section is proportional to the square of the amplitude (appendix H) we obtain the distribution $\sin^{-4}\theta/2$. Thus, such a distribution is a clear signature that *the scattering is proceeding via the exchange of a massless quantum*.

Unfortunately, the detailed implementation of these ideas to the electromagnetic interactions of quarks and leptons is complicated, because the electromagnetic potentials are the components of a 4-vector (see chapter 3), rather than a scalar as in (2.18), and the quarks and leptons all have spin-$\frac{1}{2}$, necessitating the use of the Dirac equation (chapter 4). Nevertheless, (2.18) remains the essential 'core' of electromagnetic amplitudes.

As far as the electromagnetic field is concerned, its 4-vector nature is actually a fundamental feature, having to do with a *symmetry* called *gauge invariance*, or (better) *local phase invariance*. As we shall see in chapters 3 and 7, the form of the electromagnetic interaction is very strongly constrained by this symmetry. In fact, turning the argument around, one can (almost) understand the *necessity* of electromagnetic interactions as being due to the requirement of gauge invariance. Most significantly, we shall see in section 7.3.1 how the *masslessness of the photon* is also related to gauge invariance.

In chapter 8 a number of elementary electromagnetic processes will be fully

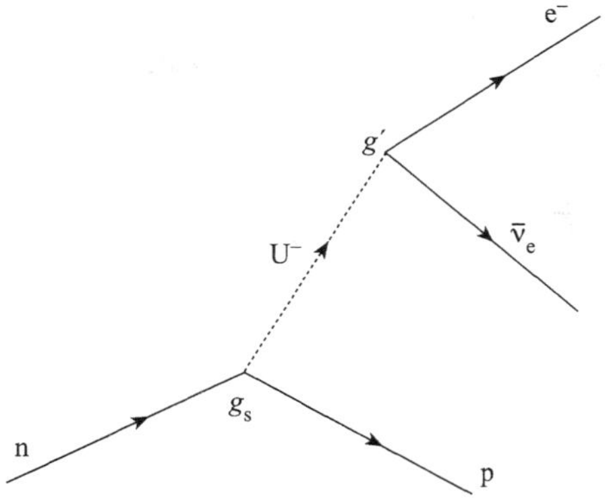

Figure 2.4. Yukawa's U-exchange mechanism for neutron β-decay.

analysed, and in chapter 11 we shall discuss higher-order corrections in QED.

2.5 Weak interactions

In a bold extension of his 'strong force' idea, Yukawa extended his theory to describe neutron β-decay as well, via the hypothesized process shown in figure 2.4 (here and in figure 2.5 we revert to the more intuitive 'time-ordered' picture—the reader may supply the diagrams corresponding to the other time-ordering). As indicated on the diagram, Yukawa assigned the strong charge g_s at the n–p end, and a different 'weak' charge g' at the lepton end. Thus the same quantum mediated both strong and weak transitions, and he had an embryonic 'unified theory' of strong and weak processes! Furthermore, interpreting U^- as the π^-, he predicted the existence of the weak decay $\pi^- \to e^- + \bar{\nu}_e$.

This decay does indeed occur (the main pion decay mode is (1.3), but this would be forbidden by overall energy conservation in neutron decay). But—apart from the now familiar problem with the compositeness of the nucleons and pions—this kind of unification is not chosen by Nature. Not unreasonably in 1935, Yukawa was assuming that the range $\sim m_{\mathrm{U}}^{-1}$ of the strong force in n–p scattering (figure 2.1) was the same as that of the weak force in neutron β-decay (figure 2.4); after all, the latter (and more especially positron emission) was viewed as a nuclear process. But this is now known not to be the case: in fact, the range of the weak force is much smaller than nuclear dimensions —or, equivalently (see (2.8)), the masses of the mediating quanta are much greater than that of the pion.

β-decay is now understood as occurring at the quark level via the W^--exchange process shown in figure 2.5(a). Similarly, positron emission proceeds via figure 2.5(b). Other 'charged current' processes mentioned in the previous chapter all involve $W^{\pm}$-exchange, generalized appropriately to include flavour

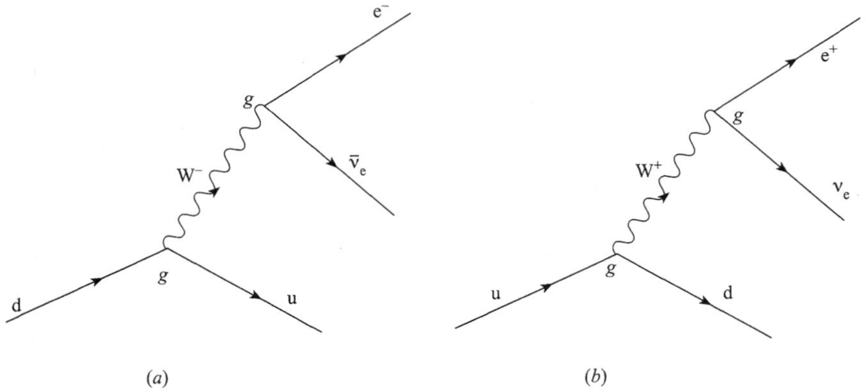

Figure 2.5. (*a*) β-decay and (*b*) e^+ emission at the quark level, mediated by $W^{\pm}$.

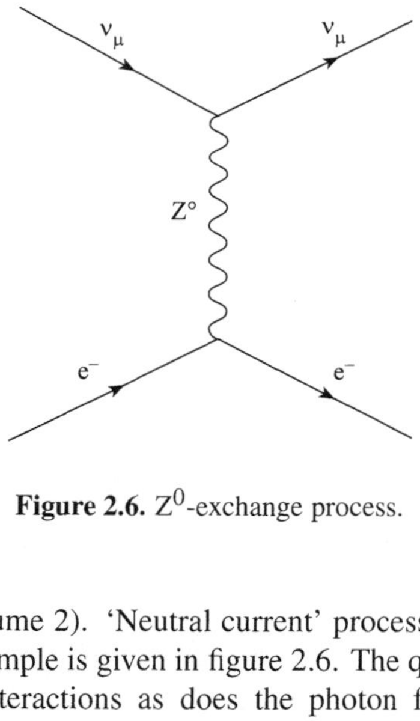

Figure 2.6. Z^0-exchange process.

mixing effects (see volume 2). 'Neutral current' processes involve exchange of the Z^0-quantum; an example is given in figure 2.6. The quanta $W^{\pm}$, Z^0 therefore mediate these weak interactions as does the photon for the electromagnetic one. Like the photon, the W and Z fields are 4-vectors but, unlike the photon, the masses of the W and Z are far from zero—in fact $M_W \approx 80$ GeV and $M_Z \approx 91$ GeV. So the range of the force is $\sim M_W^{-1} \sim 2.5 \times 10^{-18}$ m, much less than typical nuclear dimensions ($\sim$ few $\times 10^{-15}$ m). This, indeed, is one way of understanding why the weak interactions appear to be so weak: this range is so tiny that only a small part of the hadronic volume is affected.

Thus Nature has not chosen to unify the strong and weak forces via a common mediating quantum. Instead, it has turned out that the weak and strong forces (see section 2.6) are both *gauge theories*, generalizations of electromagnetism, as will be discussed in volume 2. This raises the possibility

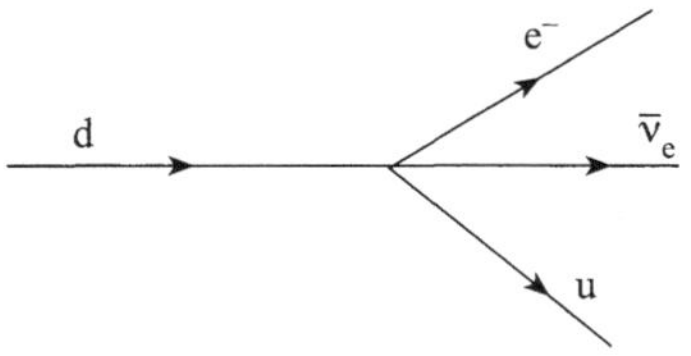

Figure 2.7. Point-like four-fermion interaction.

that it may be possible to 'unify' all three forces.

Some initial idea of how this works in the 'electroweak' case may be gained by considering the amplitude for figure 2.5(a) in the low $-q^2$ limit. In a simplified version analogous to (2.18) which ignores the spin of the W and of the leptons, this amplitude is

$$g^2/(q^2 - M_W^2) \tag{2.19}$$

where g is a 'weak charge' associated with W-emission and absorption. In actual β-decay, the square of the 4-momentum transfer q^2 is tiny compared to M_W^2, so that (2.19) becomes independent of q^2 and takes the constant value $-g^2/M_W^2$. This corresponds, in configuration space, to a point-like interaction (the Fourier transform of a delta function is a constant). Just such a point-like interaction, shown in figure 2.7, had been postulated by Fermi (1934a, b) in the first theory of β-decay: it is a 'four-fermion' interaction with strength G_F. The value of G_F can be determined from measured β-decay rates. The dimensions of G_F turn out to be energy $\times$ volume, so that $G_F/(\hbar c)^3$ has dimension (energy^{-2}). In our units $\hbar = c = 1$, the numerical value of G_F is

$$G_F \sim (300\,\text{GeV})^{-2}. \tag{2.20}$$

If we identify this constant with g^2/M_W^2 we obtain

$$g^2 \sim M_W^2/(300\,\text{GeV})^{-2} \sim 0.064 \tag{2.21}$$

a value quite similar to that of the electromagnetic charge e^2 as determined from $e^2 = 4\pi\alpha \sim 0.09$. Though this is qualitatively correct, we shall see in volume 2 that the actual 'unified' relation, in the electroweak theory, between the weak and electromagnetic coupling strengths is somewhat more complicated than the simple equality '$g = e$'. (Note that a corresponding connection with Fermi's theory was also made by Yukawa!)

We can now understand the 'weakness' of the weak interactions from another viewpoint. For $q^2 \ll M_W^2$, the ratio of the electromagnetic amplitude (2.18) to the weak amplitude (2.19) is of order q^2/M_W^2, given that $e \sim g$. Thus despite having an intrinsic strength similar to that of electromagnetism, weak interactions will appear very weak at low energies such that $q^2 \ll M_W^2$. At energies approaching M_W, however, weak interactions will grow in importance relative to electromagnetic ones and, when $q^2 \gg M_W^2$, weak and electromagnetic interactions will contribute roughly equally.

'Similar' coupling strengths are still not 'unified', however. True unification only occurs after a more subtle effect has been included, which goes beyond the one-quantum exchange mechanism. This is the *variation* or 'running' of the coupling strengths as a function of energy (or distance), caused by higher-order processes in perturbation theory. This will be discussed in chapter 11 for the case of QED, and in volume 2 for the other gauge couplings. It turns out that the possibility of unification depends crucially on an important difference between the weak interaction quanta $W^{\pm}$ (to take the present example) and the photons of QED, which has not been apparent in the simple β-decay processes considered so far. The W's are themselves 'weakly charged', acting as both carriers and sources of the weak force field, and they therefore interact directly amongst themselves even in the absence of other matter. By contrast, photons are electromagnetically neutral and have no direct self-interactions. In theories where the gauge quanta self-interact, the coupling strength decreases as the energy increases, while for QED it increases. It is this differing 'evolution' that tends to bring the strengths together, ultimately.

Even granted that the coupling strengths can be unified, however, and the fact that both are 4-vector fields, the idea of full electroweak unification appears to founder immediately on the markedly *different* ranges of the two forces or, equivalently, of the masses of the mediating quanta ($m_\gamma = 0$, $M_W \sim 80$ GeV!). This difficulty becomes even more pointed when we recall that, as previously mentioned, the masslessness of the photon is related to gauge invariance in electrodynamics: how then can there be any similar kind of gauge symmetry for weak interactions, given the distinctly non-zero masses of the mediating quanta? Nevertheless, in one of the great triumphs of 20th century theoretical physics, it *is* possible to see the two theories as essentially similar gauge theories, the gauge symmetry being 'spontaneously broken' in the case of weak interactions. An indication of how gauge quanta might acquire mass will be given in section 11.4 but a fuller explanation, with application to the electroweak theory, is reserved for volume 2.

2.6 Strong interactions

We turn to the contemporary version of Yukawa's theory of strong interactions, now viewed as occurring between quarks rather than nucleons. As already remarked in connection with figure 1.14, the strong interquark force appears to be in some way similar to QED. Further evidence for this comes when we look at the result of colliding two nucleons together. Regarding the nucleons as composites of point-like quarks, we would expect to see prominent events at large scattering angles corresponding to 'hard' q–q collisions (recall Rutherford's shell bouncing off the paper!). Now the result of such a hard collision would normally be to scatter the quarks to wide angles, 'breaking up' the nucleons in the process. However, as already noted in section 1.2, quarks are *not* observed as free particles. Instead, what appears to happen is that, as the two quarks separate from each other, their mutual potential energy increases—so much so that, at a certain stage in the evolution of the scattering process, the energy stored in the potential converts into a new $q\bar{q}$ pair. This process continues, with in general many pairs being produced as the original and subsequent pairs pull apart. By a mechanism which is still not quantitatively understood in detail, the produced quarks and antiquarks (and the original quarks in the nucleons) bind themselves into hadrons within an interaction volume of order 1 fm^3, so that no free quarks are finally observed, consistent with 'confinement'. Very strikingly, these hadrons emerge in quite well-collimated 'jets', suggesting rather vividly their ancestry in the original separating $q\bar{q}$ pair. Suppose, then, that we plot the angular distribution of such '*two jet events*': it should tell us about the dynamics of the original interaction at the quark level.

Figure 2.8 shows such an angular distribution, this time from proton–antiproton scattering, so that the fundamental interaction in this case is the elastic scattering process $\bar{q}q \to \bar{q}q$. Here θ is the scattering angle in the $\bar{q}q$ centre of mass system (CMS) and—amazingly—the θ-distribution follows almost exactly the 'Rutherford' form $\sin^{-4}\theta/2$!

We saw how, in the Coulomb case, this distribution could be understood as arising from the propagator factor $1/q^2$, which itself comes from the $1/r$ potential associated with the massless quantum involved, namely the photon. In the present case, the same $1/q^2$ factor is responsible: here, in the $\bar{q}q$ centre of mass system, $\boldsymbol{k}$ and $-\boldsymbol{k}$ are the momenta of the initial $\bar{q}$ and q, while $\boldsymbol{k}'$ and $-\boldsymbol{k}'$ are the corresponding final momenta (see figure 2.9). Once again, for elastic scattering there is no energy transfer, and $q^2 = -\boldsymbol{q}^2 = -(\boldsymbol{k} - \boldsymbol{k}')^2 = -4\boldsymbol{k}^2\sin^2\theta/2$ as before, leading to the $\sin^{-4}\theta/2$ form on squaring $1/q^2$.

Such observations as the charmonium level scheme and the angular distribution of two-jet events clearly imply that the interquark potential must have at least an important part which behaves as $1/r$ and comes (in Yukawa's terms) from the exchange of a massless quantum. This quantum is the *gluon*, the mediator of the strong interquark force.

Like the photon, then, gluons are massless and they also have spin-1. To this extent, gluons are more like photons than are the W's and Z's, but the precise form of the QCD interaction, though analogous to QED, is intrinsically more complicated. QCD will be discussed in volume 2 and, for the moment, we only want to mention the nature of the complicating factor, which relates to the concept of 'charge'. In the case of electromagnetism, the force between two particles is determined by their charges, which are reckoned in terms of simple multiples of one basic unit of charge (including zero, of course). In QCD the notion of 'charge' becomes considerably extended. The attribute analogous to charge is called 'colour', and it comes in more than one variety (as the name implies). As far as quarks are concerned, each one exists in three 'colour varieties', for

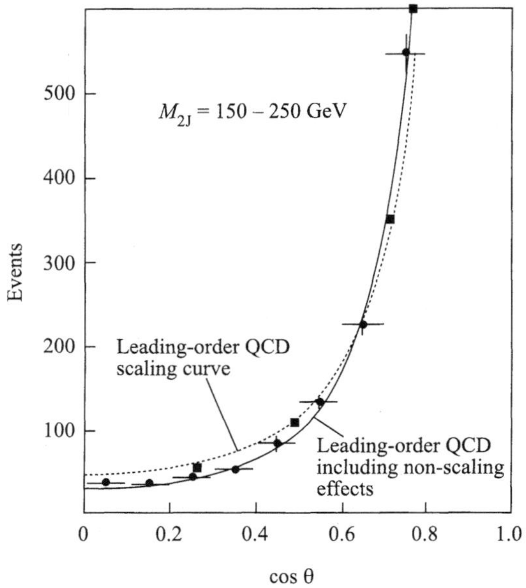

Figure 2.8. Angular distribution of two-jet events in p$\bar{\text{p}}$ collisions (Arnison *et al* 1985) as a function of $\cos\theta$, where θ is the CMS scattering angle. The points shown as squares are the Geiger–Marsden values from figure 1.1, scaled by an overall constant. The broken curve is the prediction of QCD, obtained in the lowest order of perturbation theory (one-gluon exchange); it is virtually indistinguishable from the Rutherford (one-photon exchange) shape $\sin^{-4}\theta/2$. The full curve includes higher order QCD corrections.

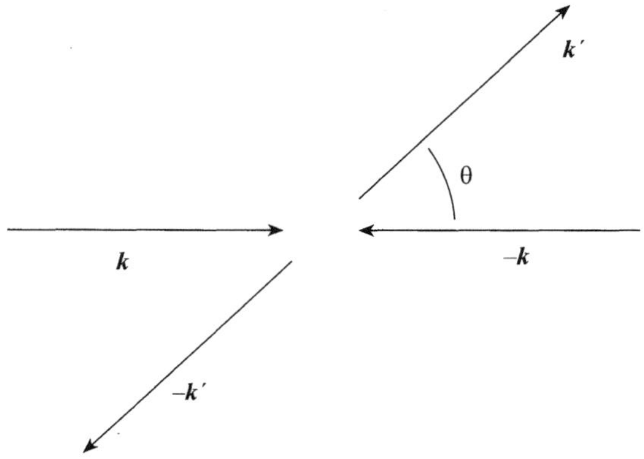

Figure 2.9. Kinematics for q–$\bar{\text{q}}$ scattering in the CMS.

example u_r (red up quark), u_b (blue up) and u_g (green up). The QCD force is sensitive to this *colour* label. Indeed, the gluons effectively 'carry colour' back

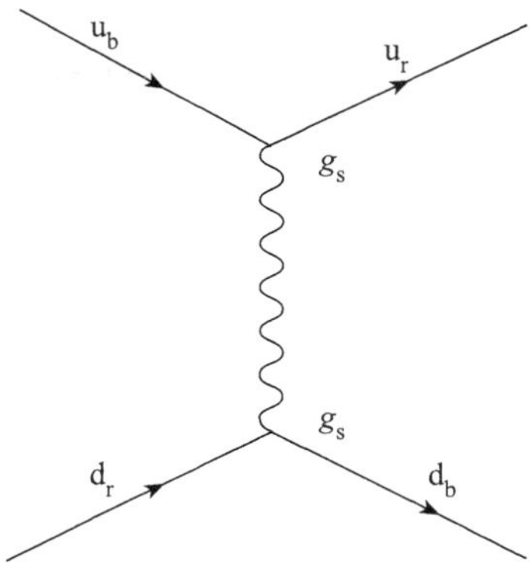

Figure 2.10. Strong scattering via gluon exchange.

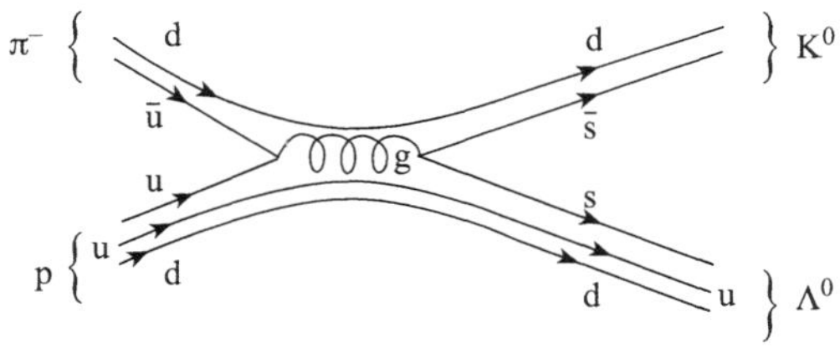

Figure 2.11. Simple quark mechanism for associated production.

and forth between the quarks, as shown in the basic gluon exchange process of figure 2.10. The name 'quantum chromodynamics' arises from this picture of the interaction as a 'coloured' version of QED. And, like QED and the W–Z theory, QCD is a gauge theory.

Because the gluons carry colour, they can interact with themselves, like the W's and Z's. It is believed that gluons too are confined by QCD, so that—like quarks—they are not seen as isolated free particles. Nevertheless, they also seem to 'hadronize' after being produced in a primitive short-distance collision process, as we saw happen in the case of q's and $\bar{q}$'s. Such 'gluon jets' provide indirect evidence for the existence and properties of gluons, as we shall see in volume 2.

One very important point to note is that neither the photon nor the gluons mediate any interactions which change quark flavour. This is the reason for the rule stated in section 1.3.2 concerning no quark flavour changes in strong and electromagnetic interactions. From the perspective of QCD, then, a simple mechanism explaining the strong associated production process $\pi^- p \to K^0 \Lambda^0$ is shown in figure 2.11. Note that the gluons responsible for binding the quarks into the initial and final hadrons have all been omitted from this diagram, which serves only to capture the 'flow', and creation/annihilation, of quark flavour.

This is an appropriate stage at which to emphasize what appears to be a crucial distinction between the three 'charges' (electromagnetic e, weak g and strong g_s) on the one hand, and the various flavour quantum numbers on the other. The former have a dynamical significance, whereas the latter do not. In the case of electric charge, for example, this means simply that a particle carrying this property responds in a definite way to the presence of an electromagnetic field and itself creates such a field. No such force fields are known for any of the flavour numbers, which are (at present) purely empirical classification devices, without dynamical significance.

The exchange of massless gluons leads to a $1/r$ potential for which, as we have seen, there is good evidence. However, this cannot be the whole story for the interquark potential, because in fact the level scheme of the heavy 'quarkonia' are not exactly Coulombic. More seriously, a simple $1/r$ potential is inconsistent with confinement. A *phenomenological* $q\bar{q}$ (or $\bar{q}q$) potential which is often used has the form

$$V = -\frac{a}{r} + br \tag{2.22}$$

where the first term, which dominates at small r, arises from a single-gluon exchange so that $a \sim g_s^2$, where the strong (QCD) charge is g_s. The second term models confinement at larger values of r. Such a potential provides quite a good understanding of the gross structure of the $c\bar{c}$ and $b\bar{b}$ systems (see problem 2.4). A typical value for b is 0.85 GeV fm^{-1} (which corresponds to a constant force of about 14 tonnes!). Thus at $r \sim 2$ fm, there is enough energy stored to produce a pair of the lighter quarks. This 'linear' part of the potential cannot be obtained by considering the exchange of one, or even a finite number of, gluons: in other words, not within an approach based on perturbation theory.

It is interesting to note that the linear part of the potential may be regarded as the solution of the *one-dimensional* form of $\nabla^2 V = 0$, namely $\mathrm{d}^2V/\mathrm{d}r^2 = 0$; this is in contrast to the Coulombic $1/r$ part, which is a solution (except at $r = 0$) to the full three-dimensional Laplace equation. This suggests that the colour field lines connecting two colour charges spread out into all of space when the charges are close to each other, but are somehow 'squeezed' into an elongated one-dimensional 'string' as the distance between the charges becomes greater than about 1 fm.

Numerical simulations of QCD, in which the space–time continuum is represented as a discrete lattice of points, indicate that such a linear potential does arise when QCD is treated non-perturbatively. It remains a challenge for theory to demonstrate that confinement follows from QCD.

2.7 Gravitational interactions

It is natural to expect that the 'Yukawa', or quantum field, framework should be applicable to gravity, as to the other forces; but in fact no consistent quantum field theory of gravity has been found, as we shall explain shortly. This is, perhaps, particularly disappointing since, as we shall mention in section 3.3,

Einsteinian gravity can be regarded as a gauge theory (Utiyama 1956, Kibble 1961), somewhat analogous to the gauge theories describing the electromagnetic, weak and strong interactions in the Standard Model. Instead, only the apparently more general framework of *string theory* offers hope of combining quantum mechanics and Einsteinian gravity satisfactorily. Quantum gravity is therefore well beyond the scope of this book. Nevertheless, we shall complete our qualitative survey of particle interactions by trying to give some impression of the issues, and the difficulties, that seem to arise when we consider applying quantum field theory to gravity.

At first sight, of course, gravity and electromagnetism seem encouragingly similar, at the classical level. The Newtonian potential energy between two point masses m_1 and m_2 a distance r apart is

$$V_{\mathrm{N}}(r) = \frac{-G_{\mathrm{N}} m_1 m_2}{r} \tag{2.23}$$

where G_{N} is Newton's gravitational constant; the similarity of (2.23) to the Coulomb potential $-e^2/4\pi r$ is obvious. And certainly, if we interpret (2.23) as the non-relativistic limit of a Yukawa-like exchange interaction, the exchanged quantum (the 'graviton') has to be massless, like the photon. But the appearance of the *particle masses* in (2.23) makes a crucial difference. In the electromagnetic case, we have seen that the expansion parameter of perturbation theory is the dimensionless number α, related to the coupling strength e by $\alpha = e^2/4\pi \sim 1/137$, in units $\hbar = c = 1$; in 'ordinary' units we recall that $\alpha = e^2/4\pi\hbar c$ (see appendix C). Because of the masses appearing in (2.23), the Newtonian strength parameter G_{N} cannot be dimensionless, even in units such that $\hbar = c = 1$. In fact, it is easy to see that G_{N} carries dimensions of energy $\times$ length $\times$ mass^{-2}. Since $\hbar c$ has dimensions of energy $\times$ length, we learn that $G_{\mathrm{N}}/\hbar c$ has dimensions of inverse mass squared. Reverting to our usual $\hbar = c = 1$ system, the numerical value is

$$G_{\mathrm{N}} \sim (1.2 \times 10^{19}\ \mathrm{GeV})^{-2}. \tag{2.24}$$

In 'elementary particle' terms, the energy appearing in (2.24) is enormous: we can immediately conclude that, even for particles with masses (or, equivalently, energies) of order 1 TeV, the gravitational interaction is completely insignificant by comparison with electromagnetism, for example. Calculating quantum corrections to this tiny classical effect is therefore hardly an urgent problem at the phenomenological level. (For macroscopic aggregates of matter, of course, every microscopic constituent contributes its full share to the total mass, giving a huge numerical multiplicative factor in the effective coupling between the heavy masses—but then quantum effects are negligible.)

Nevertheless, there is clearly a point of principle here: we should be able to calculate quantum corrections to classical gravity, or understand why we cannot. Besides, there are conceivable regimes in which such corrections will matter. For example, at energies E of order 10^{19} GeV, according to (2.24), the effective coupling—replacing $m_1 m_2$ by E^2—will be of order unity and gravity will be 'strong'.

What is the significance of this particular energy? Consider two equal masses M, a distance r apart. When r becomes comparable to the particles' Compton wavelength, we might expect quantum corrections to become important (assuming as a crude approximation that we can still use (2.23)). Also, relativistic effects will affect the classical energy at distances such that (2.23) is of order the particles' mass, M. So we might anticipate serious relativistic quantum field corrections to classical gravity when the gravitational energy of two masses M, placed a distance M^{-1} apart, is of order M itself—that is, when

$$\frac{G_{\mathrm{N}} M^2}{M^{-1}} \sim M. \tag{2.25}$$

Relation (2.25) gives

$$M \sim G_{\mathrm{N}}^{-1/2} \equiv M_{\mathrm{P}} \sim 1.2 \times 10^{19}\ \mathrm{GeV} \tag{2.26}$$

where M_{P} is the *Planck mass*, or energy, which appeared in (2.24). The corresponding distance $M_{\mathrm{P}}^{-1} = G_{\mathrm{N}}^{1/2}$ is the *Planck length*, approximately equal to 2×10^{-35} m. Impossibly remote as these scales may seem, they do imply, for instance, that without a theory of quantum gravity we cannot discuss the evolution of the universe at epochs of order the *Planck time* (the Planck length divided by the speed of light, approximately equal to 5×10^{-44} s), when the whole thing was of order the Planck size.

Thus far we have been discussing Newtonian gravity, but we know that this is only an approximation to Einstein's theory, in which gravity is interpreted in terms of the geometry of space–time. A famous consequence of Einstein's theory is the existence of black holes, which have associated with them a *horizon*: once an in-falling body passes this horizon, it cannot return, nor can any signal from it reach the region exterior to the horizon. For a black hole of mass M, the horizon is at a distance $2G_{\mathrm{N}}M$ (with $c = 1$). If we set $M = M_{\mathrm{P}}$, we see that for a black hole with the Planck mass, its horizon is roughly equal to its Compton wavelength M_{P}^{-1}. This is another way of thinking about the 'extreme physics' at such scales.

What happens, then, when we try to set up Feynman-type perturbation theory in terms of graviton exchanges? Certainly it would seem that the effective coupling '(mass or energy)$\times G_{\mathrm{N}}^{1/2}$' is extremely small at laboratory energy scales. But we hit a severe technical problem as soon as we go beyond the one-graviton exchange process and consider multi-graviton exchanges, as a matter of principle. Although we have not so far discussed anything other than one-quantum exchanges in the case of Standard Model forces, for them too we must clearly be able to include the effects of multi-quantum exchanges. We shall explain how to do this in chapter 10, for a simple scalar field theory. There we will learn that the amplitudes for multi-quantum exchange processes are given by integrals over the momenta of the 'virtual' exchanged particles. Now, while the momenta of the 'external' particles (i.e. the ones scattering off each other) are constrained to take laboratory values, no such constraint applies to the virtual particles being exchanged (if only a *single* particle is exchanged in a Feynman graph, its momentum is fixed by momentum conservation in terms of the momenta of the scattering particles). It turns out that the integrals nearly always *diverge* as the momenta of the exchanged particles tend to infinity. Nevertheless, as we shall explain in chapter 10, the theory can be reformulated, by a process called *renormalization*, in such a way that all such multi-quantum (or higher-order) processes become finite and calculable—a quite remarkable fact, and one that is of course absolutely crucial in the case of the Standard Model interactions, where the relevant data are precise enough to test the accuracy of the theory beyond lowest order, particularly in the case of QED (see chapter 11). But no one has been able to do a similar reformulation so as to control the higher-order corrections in quantum gravity: the theory is *non-renormalizable*, which means that no consistent treatment of multi-graviton exchanges exists within the quantum field framework.

A simple test for whether a theory is renormalizable or not will be discussed in section 11.8: if the coupling strength has dimensions of a mass to an inverse power, then the theory is non-renormalizable. As we have seen, $G_{\rm N}$ does have dimensions of an inverse (square) mass, and the theory is indeed non-renormalizable.

Interestingly, the coupling strength $G_{\rm F}$ of the original Fermi theory of weak interactions also has dimensions of an inverse square mass as we saw in (2.20), thus creating an energy scale $G_{\rm F}^{-1/2}$ below which the force is especially 'weak'; and this theory too is non-renormalizable, as will be discussed in section 11.8. But of course the Fermi theory is not the 'true' theory of the weak interactions, which appears as part of the Standard Model: the latter theory is actually renormalizable, as we shall see in volume 2. Perhaps the way in which the 'disease' of the Fermi theory is cured may provide a clue for dealing with quantum gravity.

Non-renormalizable divergences occurring as virtual momenta tend to infinity can alternatively be interpreted, in quantum theory, in terms of 'bad' behaviour of the theory at ultra-short distances. Consider the Fermi theory in this light. As we move from energies $\ll G_{\rm F}^{-1/2}$, or distances $\gg G_{\rm F}^{1/2}$, down to shorter distances of order $G_{\rm F}^{1/2}$, weak interactions are revealed as not being point-like after all (as in Fermi's theory), but Yukawa-like with a range $\sim M_{\rm W}^{-1} \sim 4 \times G_{\rm F}^{1/2}$. The extreme (and unphysical) point-like behaviour has been replaced by a finite ranged force, which ameliorates the short-distance behaviour. Now of course we cannot just apply the same logic to gravity, since if we were to interpret $G_{\rm N}^{1/2} = M_{\rm P}^{-1}$ as a range, gravity would be absurdly short-ranged! But it is very plausible that in gravity's case too, the quantity $G_{\rm N}^{1/2}$—the Planck length—does signal the scale at which the short-distance behaviour of the theory should be radically modified, so as to remove the uncontrollable divergences.

At this stage we have to remember that the quantum field theories we have been discussing in the context of the Standard Model are all formulated in terms of *local* interactions of *point-like* particles. They contain nothing in the nature of a natural 'cut-off' which would prevent virtual momenta tending to infinity or probing arbitrarily short distances. Indeed, it is not at all easy to introduce such a short-distance 'fuzziness'. Lorentz invariance certainly holds to a very good approximation, and if we introduce any non-locality (or spread) in space, we cannot avoid doing so in time as well, which will tend to conflict with causality. The situation is even worse if we include the more general invariances of Einstein's theory.

In fact, *string theory* (Polchinski 1998) is the only known way of spreading out the gravitational interaction, and improving the short-distance behaviour so as to yield a consistent theory. In this theory, all our 'elementary particles' are not point-like after all, but are one-dimensional objects of size $\sim M_{\mathrm{P}}^{-1}$. Their interactions are (ever so slightly!) non-local, indeed space–time itself has a non-local character: 'distance' loses meaning at scales shorter than M_{P}^{-1}.

Enormous effort has gone into the development of string theories, ever since their discovery in the early 1970s (in connection with the 'stringy' nature of the longer-ranged part of the QCD force). Examples of such theories exist which include the particles and interactions of the Standard Model, all now appearing as a low-energy limit (i.e. at energies $\ll M_{\mathrm{P}}$). These examples all make use of two further exotic (and as yet unobserved) ingredients: *supersymmetry* (which involves a symmetry between the bosons and the fermions of the theory); and *compactified extra space dimensions*. String theory has also begun to address some of the deeper questions of quantum gravity, such as the quantum mechanics of black holes.

The path to understanding all this is unfortunately long and arduous. Physics is a cumulative discipline, and it is hard if not impossible to jump to the contemporary end without mastering what has gone before. Fortunately the effort expended in the latter has worthwhile rewards: there is every reason to believe that, whatever complexities lie in store at the Planck scale, quantum field theory will provide a reliable calculational framework at energies well below it. What lies between currently accessible energies and the Planck scale remains to be seen, of course. Many theorists have been troubled by the enormous unexplained difference between the two scales with which we seem to be confronted, namely $M_{\mathrm{W}} \sim 100$ GeV and $M_{\mathrm{P}} \sim 10^{19}$ GeV. We shall discuss this further in volume 2; meanwhile, see problem 2.6 for a radical speculation!

2.8 Summary

Returning then to our proper subject, the Standard Model, we are left with a relatively simple picture of quarks and leptons and their non-gravitational interactions. The quarks are the basic source particles of the gluon fields in QCD, and they bind together to make hadrons. The weak interactions involve *pairs* of quarks and leptons—for instance the quark doublet (u, d) and the lepton doublet (ν_e, e^-) of the first generation. These are sources for the $W^\pm$ and Z^0 fields. Charged fermions (quarks and leptons) are sources for the photon field. All the mediating quanta have spin-1. The weak and strong force fields are generalizations of electromagnetism; all three are examples of gauge theories, but realized in subtly different ways.

In the following chapters our aim will be to lead the reader through the mathematical formalism involved in giving precise quantitative form to what we have so far described only qualitatively and to provide physical interpretation where appropriate. In the remainder of part 1 of the present volume, we first show how Schrödinger's quantum mechanics and Maxwell's electromagnetic theory may be combined as a gauge theory—in fact the simplest example of such a theory. We then introduce relativistic quantum mechanics for spin-0 and spin-$\frac{1}{2}$ particles, and include electromagnetism via the gauge principle. In part 2, we develop the formalism of quantum field theory, beginning with scalar fields and moving on to QED; this is then applied to many simple ('tree level') QED processes in part 3. In the final part 4, we present an introduction to renormalization at the one-loop level, including renormalization of QED. The non-Abelian gauge theories of QCD and the electroweak theory are reserved for volume 2.

Problems

2.1 Using the concept of strangeness conservation in strong interactions, explain why the threshold energy (for π^- incident on stationary protons) for

$$\pi^- + \mathrm{p} \to \mathrm{K}^0 + \text{anything}$$

is less than for

$$\pi^- + \mathrm{p} \to \bar{\mathrm{K}}^0 + \text{anything}$$

assuming both processes proceed through the strong interaction.

2.2 Note: the invariant square p^2 of a 4-momentum $p = (E/c, \boldsymbol{p})$ is defined as $p^2 = E^2/c^2 - \boldsymbol{p}^2$.

(i) An electron of 4-momentum k scatters from a stationary proton of mass M via a one-photon exchange process, producing a final hadronic state of 4-momentum p', the final electron 4-momentum being k'. Show that

$$p'^2 = q^2 + 2M(E - E') + M^2c^2$$

where $q^2 = (k - k')^2$, and E, E' are the initial and final electron energies in this frame (i.e. the one in which the target proton is at rest). Show that if the electrons are highly relativistic then $q^2 = -4EE' \sin\theta/2$, where θ is the scattering angle in this frame. Deduce that for elastic scattering E' and θ are related by

$$E' = E \Big/ \left(1 + \frac{2E}{Mc^2}\sin^2\theta/2\right).$$

(ii) Electrons of energy 4.879 GeV scatter elastically from protons, with $\theta = 10°$. What is the observed value of E'?

(iii) In the scattering of these electrons, at 10°, it is found that there is a peak of events at $E' = 4.2$ GeV; what is the invariant mass of the produced hadronic state (in MeV/c^2)?

(iv) Calculate the value of E' at which the 'quasi-elastic peak' will be observed, when electrons of energy 400 MeV scatter at an angle $\theta = 45°$ from a He nucleus, assuming that the struck nucleon is at rest inside the nucleus. Estimate the broadening of this final peak caused by the fact that the struck nucleon has, in fact, a momentum distribution by virtue of being localized within the nuclear size.

2.3

(i) In a simple non-relativistic model of a hydrogen-like atom, the energy levels are given by

$$E_n = \frac{-\alpha^2 Z^2 \mu c^2}{2n^2}$$

where Z is the nuclear charge and μ is the reduced mass of the electron and nucleus. Calculate the splitting in eV between the $n = 1$ and $n = 2$ states in positronium, which is an e^+e^- bound state, assuming this model holds.

(ii) In this model, the e^+e^- potential is the simple Coulomb one

$$-\frac{e^2}{4\pi\epsilon_0 r} = -\alpha\frac{\hbar c}{r}.$$

Suppose that the potential between a heavy quark Q and an anti-quark $\bar{Q}$ was

$$-\alpha_s\frac{\hbar c}{r}$$

where α_s is a 'strong fine structure constant'. Calculate values of α_s (different in (a) and (b)) corresponding to the information

(a) the splitting between the $n = 2$ and $n = 1$ states in charmonium ($c\bar{c}$) is 588 MeV, and $m_c = 1870$ MeV/c^2;

(b) the splitting between the $n = 2$ and $n = 1$ states in the upsilon series ($b\bar{b}$) is 563 MeV, and $m_b = 5280$ MeV/c^2.

(iii) In positronium, the $n = 1\,^3S_1$ and $n = 1\,^1S_0$ states are split by the hyperfine interaction, which has the form $\frac{7}{48}\alpha^4 m_e c^2 \boldsymbol{\sigma}_1 \cdot \boldsymbol{\sigma}_2$ where m_e is the electron mass and $\boldsymbol{\sigma}_1, \boldsymbol{\sigma}_2$ are the spin matrices for the e^- and e^+ respectively. Calculate the expectation value of $\boldsymbol{\sigma}_1 \cdot \boldsymbol{\sigma}_2$ in the $^3S_1$ and $^1S_0$ states, and hence evaluate the splitting between these levels (calculated in lowest order perturbation theory) in eV. [*Hint*: the total spin $\boldsymbol{S}$ is given by $\boldsymbol{S} = \frac{\hbar}{2}(\boldsymbol{\sigma}_1 + \boldsymbol{\sigma}_2)$. So $\boldsymbol{S}^2 = \frac{\hbar^2}{4}(\boldsymbol{\sigma}_1^2 + \boldsymbol{\sigma}_2^2 + 2\boldsymbol{\sigma}_1 \cdot \boldsymbol{\sigma}_2)$. Hence the eigenvalues of $\boldsymbol{\sigma}_1 \cdot \boldsymbol{\sigma}_2$ are directly related the those of $\boldsymbol{S}^2$.]

(iv) Suppose an analogous 'strong' hyperfine interaction existed in the $c\bar{c}$ system, and was responsible for the splitting between the $n = 1\,^3S_1$ and $n = 1\,^1S_0$ states, which is 116 MeV experimentally (i.e. replace α by α_s and m_e by $m_c = 1870\ \text{MeV}/c^2$). Calculate the corresponding value of α_s.

2.4 The potential between a heavy quark Q and an anti-quark $\bar{Q}$ is found empirically to be well represented by

$$V(r) = -\frac{\alpha_s}{r} + br$$

where $\alpha_s/\hbar c \approx 0.5$ and $b\hbar c \approx 0.18\ \text{GeV}^2$. Indicate the origin of the first term in $V(r)$, and the significance of the second.

An estimate of the ground-state energy of the bound $Q\bar{Q}$ system may be made as follows. For a given r, the total energy is

$$E(r) = 2mc^2 - \frac{\alpha_s}{r} + br + \frac{p^2}{m}$$

where m is the mass of the Q (or $\bar{Q}$) and p is its momentum (assumed non-relativistic). Explain why p may be roughly approximated by $\hbar/r$, and sketch the resulting $E(r)$ as a function of r. Hence show that, in this approximation, the radius of the ground state, r_0, is given by the solution of

$$\frac{2\hbar^2}{mr_0^3} = \frac{\alpha_s}{r_0^2} + b.$$

Taking $mc^2 = 1.5$ GeV as appropriate to the $c\bar{c}$ system, verify that for this system

$$(\hbar c/r_0) \approx 0.67\ \text{GeV}$$

and calculate the energy of the $c\bar{c}$ ground state in GeV, according to this model.

An excited $c\bar{c}$ state at 3.686 GeV has a total width of 278 keV, and one at 3.77 GeV has a total width of 24 MeV. Comment on the values of these widths.

2.5 The Hamiltonian for a two-state system using the normalized base states $|1\rangle, |2\rangle$ has the form

$$\begin{pmatrix} \langle 1|H|1\rangle & \langle 1|H|2\rangle \\ \langle 2|H|1\rangle & \langle 2|H|2\rangle \end{pmatrix} = \begin{pmatrix} 3a\cos 2\theta & a\sin 2\theta \\ a\sin 2\theta & a\cos 2\theta \end{pmatrix}$$

where a is real and positive. Find the energy eigenvalues E_+ and E_-, and express the corresponding normalized eigenstates $|+\rangle$ and $|-\rangle$ in terms of $|1\rangle$ and $|2\rangle$.

At time $t = 0$ the system is in state $|1\rangle$. Show that the probability that it will be found to be in state $|2\rangle$ at a later time t is

$$\sin^2 2\theta \sin^2(at/\hbar).$$

Discuss how a formalism of this kind can be used in the context of neutrino oscillations. How might the existence of neutrino oscillations explain the solar neutrino problem?

2.6 We saw in section 2.7 that the extreme weakness of gravity (for individual particles) was related to the very large energy scale $G_{\mathrm{N}}^{-1/2} = M_{\mathrm{P}}$, associated with Newton's constant. In an interesting speculation, it has been suggested (Arkani-Hamad *et al* 1998, 1999, Antoniadis *et al* 1998) that the weakness of gravity as observed in our (apparently) three-dimensional world could be due to the fact that gravity actually extends into additional 'compactified' dimensions (that is, dimensions which have the geometry of a circle, rather than of an infinite line). For the particles and forces of the Standard Model, however, such leakage into extra dimensions has to be confined to currently probed distances, which are of order M_{W}^{-1}.

(i) Consider Newtonian gravity in $(3 + d)$ spatial dimensions. Explain why you would expect that the analogue of (2.23) will have the form

$$V_{\mathrm{N},3+d}(r) = \frac{m_1 m_2 G_{\mathrm{N},3+d}}{r^{d+1}}. \tag{2.27}$$

[Think about how the '$1/r^2$' fall-off of the *force* is related to the *surface area* of a sphere in the case $d = 0$. Note that the formula works for $d = -2$! In what ways is the case $d = -1$ special?]

(ii) Show that $G_{\mathrm{N},3+d}$ has dimensions $(\text{mass})^{-(2+d)}$. This allows us to introduce the 'true' Planck scale—i.e. the one for the underlying theory in $3+d$ spatial dimensions—as $G_{\mathrm{N},3+d} = (M_{\mathrm{P},3+d})^{-(2+d)}$.

(iii) Now suppose that the form (2.27) only holds when the distance r between the masses is much smaller R, the size of the compactified dimensions. If the masses are placed at distances $r \gg R$, their gravitational flux cannot continue to penetrate into the extra dimensions, and the potential (2.27) should reduce to the familiar three-dimensional one, (2.23); so we must have

$$V_{\mathrm{N},3+d}(r \gg R) = \frac{m_1 m_2 G_{\mathrm{N},3+d}}{R^d}\frac{1}{r}. \tag{2.28}$$

Show that this implies that

$$M_{\mathrm{P}}^2 = M_{\mathrm{P},3+d}^2 (R M_{\mathrm{P},3+d})^d. \tag{2.29}$$

(iv) Suppose that $d = 2$ and $R \sim 1$ mm: what would $M_{\mathrm{P},3+d}$ be, in TeV? Suggest ways in which this theory might be tested experimentally. Taking $M_{\mathrm{P},3+d} \sim 1$ TeV, explore other possibilities for d and R.

3

ELECTROMAGNETISM AS A GAUGE THEORY

3.1 Introduction

The previous chapters introduced the basic ideas of the Standard Model of particle physics, in which quarks and leptons interact via the exchange of gauge field quanta. We must now look more closely into what is the main concern of this book—namely, the particular nature of these '*gauge theories*'.

One of the relevant forces—electromagnetism—has been well understood in its classical guise for many years. Over a century ago, Faraday, Maxwell and others developed the theory of electromagnetic interactions, culminating in Maxwell's paper of 1864 (Maxwell 1864). Today Maxwell's theory still stands—unlike Newton's 'classical mechanics' which was shown by Einstein to require modifications at relativistic speeds, approaching the speed of light. Moreover, Maxwell's electromagnetism, when suitably married with quantum mechanics, gives us '*quantum electrodynamics*' or QED. We shall see in chapter 11 that this theory is in truly remarkable agreement with experiment. As we have already indicated, the theories of the weak and strong forces included in the Standard Model are generalizations of QED, and promise to be as successful as that theory. The simplest of the three, QED, is therefore our paradigmatic theory.

From today's perspective, the crucial thing about electromagnetism is that it is a theory in which the *dynamics* (i.e. the behaviour of the forces) is intimately related to a *symmetry* principle. In the everyday world, a symmetry operation is something that can be done to an object that leaves the object looking the same after the operation as before. By extension, we may consider mathematical operations—or 'transformations'—applied to the objects in our theory such that the physical laws look the same after the operations as they did before. Such transformations are usually called *invariances* of the laws. Familiar examples are, for instance, the translation and rotation invariance of all fundamental laws: Newton's laws of motion remain valid whether or not we translate or rotate a system of interacting particles. But of course—precisely because they do apply to all laws, classical or quantum—these two invariances have no special connection with any particular force law. Instead, they constrain the form of the allowed laws to a considerable extent, but by no means uniquely determine them. Nevertheless, this line of argument leads one to speculate whether it might in fact be possible to impose further types of symmetry constraints so that the forms of the force laws *are* essentially determined. This would then be one possible answer to the question: why are the force laws the way they are? (Ultimately of course this only replaces one question by another!)

In this chapter we shall discuss electromagnetism from this point of view. This is not the historical route to the theory, but it is the one which generalizes to the other two interactions. This is why we believe it important to present the central ideas of this approach in the familiar context of electromagnetism at this early stage.

A distinction that is vital to the understanding of all these interactions is that between a *global* invariance and a *local* invariance. In a global invariance the same transformation is carried out at all space–time points: it has an 'everywhere simultaneously' character. In a local invariance different transformations are carried out at different individual space–time points. In general, as we shall see, a theory that is globally invariant will not be invariant under locally varying transformations. However, by introducing new force fields that interact with the original particles in the theory in a specific way, and which also transform in a particular way under the local transformations, a sort of local invariance can be restored. We will see all these things more clearly when we go into more detail, but the important conceptual point to be grasped is this: one may view these special force fields and their interactions as existing in order to to permit certain local invariances to be true. The particular local invariance relevant to electromagnetism is the well-known *gauge invariance* of Maxwell's equations: in the quantum form of the theory this property is directly related to an invariance under *local phase transformations* of the quantum fields. A generalized form of this phase invariance also underlies the theories of the weak and strong interactions. For this reason they are all known as 'gauge theories'.

A full understanding of gauge invariance in electrodynamics can only be reached via the formalism of quantum field theory, which is not easy to master—and the theory of quantum gauge fields is particularly tricky, as we shall see in chapter 7. Nevertheless, many of the crucial ideas can be perfectly adequately discussed within the more familiar framework of ordinary quantum mechanics, rather than quantum field theory, treating electromagnetism as a purely classical field. This is the programme followed in the rest of part 1. In the present chapter we shall discuss these ideas in the context of non-relativistic quantum mechanics; in the following one, we shall explore the generalization to relativistic quantum mechanics, for particles of spin-0 (via the Klein–Gordon equation) and spin-$\frac{1}{2}$ (via the Dirac equation). While containing substantial physics in their own right, these chapters constitute essential groundwork for the quantum field treatment in parts 2–4.

3.2 The Maxwell equations: current conservation

We begin by considering the basic laws of classical electromagnetism, the Maxwell equations. We use a system of units (Heaviside–Lorentz) which is convenient in particle physics (see appendix C). Before Maxwell's work these laws were

$$\nabla \cdot \boldsymbol{E} = \rho_{\text{em}} \qquad \text{(Gauss' law)} \tag{3.1}$$

$$\nabla \times \boldsymbol{E} = -\frac{\partial \boldsymbol{B}}{\partial t} \qquad \text{(Faraday–Lenz laws)} \tag{3.2}$$

$$\nabla \cdot \boldsymbol{B} = 0 \qquad \text{(no magnetic charges)} \tag{3.3}$$

and, for steady currents,

$$\nabla \times \boldsymbol{B} = \boldsymbol{j}_{\text{em}} \qquad \text{(Ampère's law).} \tag{3.4}$$

Here ρ_{em} is the charge density and $\boldsymbol{j}_{\text{em}}$ is the current density; these densities act as 'sources' for the $\boldsymbol{E}$ and $\boldsymbol{B}$ fields. Maxwell noticed that taking the divergence of this last equation leads to conflict with the continuity equation for electric charge

$$\frac{\partial \rho_{\text{em}}}{\partial t} + \nabla \cdot \boldsymbol{j}_{\text{em}} = 0. \tag{3.5}$$

Since

$$\nabla \cdot (\nabla \times \boldsymbol{B}) = 0 \tag{3.6}$$

from (3.4) there follows the result

$$\nabla \cdot \boldsymbol{j}_{\text{em}} = 0. \tag{3.7}$$

This can only be true in situations where the charge density is constant in time. For the general case, Maxwell modified Ampère's law to read

$$\nabla \times \boldsymbol{B} = \boldsymbol{j}_{\text{em}} + \frac{\partial \boldsymbol{E}}{\partial t} \tag{3.8}$$

which is now consistent with (3.5). Equations (3.1)–(3.3), together with (3.8), constitute Maxwell's equations in free space (apart from the sources).

The vitally important continuity equation (3.5) states that the rate of decrease of charge in any arbitrary volume Ω is due precisely and only to the flux of current out of its surface; that is, no net charge can be created or destroyed in Ω. Since Ω can be made as small as we please, this means that *electric charge must be locally conserved*: a process in which charge is created at one point and destroyed at a distant one is not allowed, despite the fact that it conserves the charge overall or 'globally'. The ultimate reason for this is that the global form of charge conservation would necessitate the instantaneous propagation of signals (such as 'now, create a positron over there'), and this conflicts with special relativity—a theory which, historically, flowered from the soil of electrodynamics. The extra term introduced by Maxwell—the 'electric displacement current'—owes its place in the dynamical equations to a local conservation requirement.

We remark at this point that we have just introduced another local/global distinction, similar to that discussed earlier in connection with invariances. In this case the distinction applies to a conservation law, but since invariances are related to conservation laws in both classical and quantum mechanics, we should perhaps not be too surprised by this. However, as with invariances, conservation laws—such as charge conservation in electromagnetism—play a central role in gauge theories in that they are closely related to the dynamics. The point is simply illustrated by asking how we could measure the charge of a newly created subatomic particle X. There are two conceptually different ways:

(i) We could arrange for X to be created in a reaction such as

$$\mathrm{A} + \mathrm{B} \to \mathrm{C} + \mathrm{D} + \mathrm{X}$$

where the charges of A, B, C and D are already known. In this case we can use *charge conservation* to determine the charge of X.

(ii) We could see how particle X responded to known electromagnetic fields. This uses *dynamics* to determine the charge of X.

Either way gives the same answer: it is the conserved charge which determines the particle's response to the field. By contrast, there are several other conservation laws that seem to hold in particle physics, such as lepton number and baryon number, that apparently have no dynamical counterpart (cf the remarks in section 2.6). To determine the baryon number of a newly produced particle, we have to use B conservation and tot up the total baryon number on either side of the reaction. As far as we know there is no baryonic force field.

Thus gauge theories are characterized by a close interrelation between *three* conceptual elements: symmetries, conservation laws and dynamics. In fact, it is now widely believed that the *only* exact quantum number conservation laws are those which have an associated gauge theory force field—see comment (i) in section 3.6. Thus one might suspect that baryon number is not absolutely conserved—as is indeed the case in proposed unified gauge theories of the strong, weak and electromagnetic interactions. In this discussion we have briefly touched on the connection between two pairs of these three elements: symmetries $\leftrightarrow$ dynamics; and conservation laws $\leftrightarrow$ dynamics. The precise way in which the remaining link is made—between the symmetry of electromagnetic gauge invariance and the conservation law of charge - is more technical. We will discuss this connection with the help of simple ideas from quantum field theory in chapter 7, section 7.4. For the present we continue with our study of the Maxwell equations and, in particular, of the gauge invariance they exhibit.

3.3 The Maxwell equations: Lorentz covariance and gauge invariance

In classical electromagnetism, and especially in quantum mechanics, it is convenient to introduce the vector potential $A_\mu(x)$ in place of the fields $\boldsymbol{E}$ and $\boldsymbol{B}$. We write:

$$\boldsymbol{B} = \nabla \times \boldsymbol{A} \tag{3.9}$$

$$\boldsymbol{E} = -\nabla V - \frac{\partial \boldsymbol{A}}{\partial t} \tag{3.10}$$

which defines the 3-vector potential $\boldsymbol{A}$ and the scalar potential V. With these definitions, equations (3.2) and (3.3) are then automatically satisfied.

The origin of gauge invariance in classical electromagnetism lies in the fact that the potentials $\boldsymbol{A}$ and V are not unique for given physical fields $\boldsymbol{E}$ and $\boldsymbol{B}$. The transformations that $\boldsymbol{A}$ and V may undergo while preserving $\boldsymbol{E}$ and $\boldsymbol{B}$ (and hence the Maxwell equations) unchanged are called gauge transformations, and the associated invariance of the Maxwell equations is called gauge invariance.

What are these transformations? Clearly $\boldsymbol{A}$ can be changed by

$$\boldsymbol{A} \rightarrow \boldsymbol{A}' = \boldsymbol{A} + \nabla\chi \tag{3.11}$$

where χ is an arbitrary function, with no change in $\boldsymbol{B}$ since $\nabla \times \nabla f = 0$, for any scalar function f. To preserve $\boldsymbol{E}$, V must then change simultaneously by

$$V \rightarrow V' = V - \frac{\partial \chi}{\partial t}. \tag{3.12}$$

These transformations can be combined into a single compact equation by introducing the 4-vector potential[1]:

$$A^\mu = (V, \boldsymbol{A}) \tag{3.13}$$

and noting (from problem 3.1) that the differential operators $(\partial/\partial t, -\nabla)$ form the components of a 4-vector operator ∂^μ. A gauge transformation is then specified by

$$\boxed{A^\mu \rightarrow A'^\mu = A^\mu - \partial^\mu \chi.} \tag{3.14}$$

The Maxwell equations can also be written in a manifestly *Lorentz covariant form* (see appendix D) using the 4-current j^μ_{em} given by

$$j^\mu_{\mathrm{em}} = (\rho_{\mathrm{em}}, \boldsymbol{j}_{\mathrm{em}}) \tag{3.15}$$

in terms of which the continuity equation takes the form (problem 3.1):

$$\partial_\mu j^\mu_{\mathrm{em}} = 0. \tag{3.16}$$

The Maxwell equations (3.1) and (3.8) then become (problem 3.2):

$$\partial_\mu F^{\mu\nu} = j^\nu_{\mathrm{em}} \tag{3.17}$$

where we have defined the field strength tensor:

$$F^{\mu\nu} \equiv \partial^\mu A^\nu - \partial^\nu A^\mu. \tag{3.18}$$

[1] See appendix D for relativistic notation and for an explanation of the very important concept of *covariance*, which we are about to invoke in the context of Lorentz transformations, and will use again in the next section in the context of gauge transformations; we shall also use it in other contexts in later chapters.

Under the gauge transformation

$$A^\mu \to A'^\mu = A^\mu - \partial^\mu \chi \tag{3.19}$$

$F^{\mu\nu}$ remains unchanged:

$$F^{\mu\nu} \to F'^{\mu\nu} = F^{\mu\nu} \tag{3.20}$$

so $F^{\mu\nu}$ is gauge invariant and so, therefore, are the Maxwell equations in the form (3.17). The 'Lorentz-covariant and gauge-invariant field equations' satisfied by A^μ then follow from equations (3.17) and (3.18):

$$\Box A^\nu - \partial^\nu(\partial_\mu A^\mu) = j^\nu_{\mathrm{em}}. \tag{3.21}$$

Since gauge transformations turn out to be of central importance in the quantum theory of electromagnetism, it would be nice to have some insight into why Maxwell's equations are gauge invariant. The all-important 'fourth' equation (3.8) was inferred by Maxwell from local charge conservation, as expressed by the continuity equation

$$\partial_\mu j^\mu_{\mathrm{em}} = 0. \tag{3.22}$$

The field equation

$$\partial_\mu F^{\mu\nu} = j^\mu_{\mathrm{em}} \tag{3.23}$$

then of course automatically embodies (3.22). The mathematical reason it does so is that $F^{\mu\nu}$ is a four-dimensional kind of 'curl'

$$F^{\mu\nu} \equiv \partial^\mu A^\nu - \partial^\nu A^\mu \tag{3.24}$$

which (as we have seen in (3.20)) is unchanged by a gauge transformation

$$A^\mu \to A'^\mu = A^\mu - \partial^\mu \chi. \tag{3.25}$$

Hence there is the suggestion that the gauge invariance is related in some way to charge conservation. However, the connection is not so simple. Wigner (1949) has given a simple argument to show that the principle that no physical quantity can depend on the absolute value of the electrostatic potential, when combined with energy conservation, implies the conservation of charge. Wigner's argument relates charge (and energy) conservation to an invariance under transformation of the electrostatic potential by a constant: charge conservation alone does not seem to require the more general space–time-dependent transformation of gauge invariance.

Changing the value of the electrostatic potential by a constant amount is an example of what we have called a *global* transformation (since the change in the potential is the same everywhere). Invariance under this global transformation is related to a conservation law: that of charge. But this global invariance is not sufficient to generate the full Maxwellian dynamics. However, as remarked by 't Hooft (1980), one can regard equations (3.11) and (3.12) as expressing the fact that the *local* change in the electrostatic potential V (the $\partial\chi/\partial t$ term in (3.12)) can be compensated—in the sense of leaving the Maxwell equations unchanged—by a corresponding local change in the magnetic vector potential $\boldsymbol{A}$. Thus by including magnetic effects, the global invariance under a change of V by a constant can be extended to a local invariance (which is a much more restrictive condition to satisfy). Hence there is a beginning of a suggestion that one might almost 'derive' the complete Maxwell equations, which unify electricity and magnetism, from the requirement that the theory be expressed in terms of potentials in such a way as to be invariant under local (gauge) transformations on those potentials. Certainly special relativity must play a role too: this also links electricity and magnetism, via the magnetic effects of charges as seen by an observer moving relative to them. If a 4-vector potential A^{μ} is postulated, and it is then demanded that the theory involve it only in a way which is insensitive to local changes of the form (3.14), one is led naturally to the idea that the physical fields enter only via the quantity $F^{\mu\nu}$, which is invariant under (3.14). From this, one might conjecture the field equation on grounds of Lorentz covariance.

It goes without saying that this is certainly not a 'proof' or 'derivation' of the Maxwell equations. Nevertheless, the idea that *dynamics* (in this case, the complete interconnection of electric and magnetic effects) may be intimately related to a local invariance requirement (in this case, electromagnetic gauge invariance) turns out to be a fruitful one. As indicated in section 3.1, it is generally the case that, when a certain global invariance is generalized to a local one, the existence of a new 'compensating' field is entailed, interacting in a specified way. The first example of dynamical theory 'derived' from a local invariance requirement seems to be the theory of Yang and Mills (1954) (see also Shaw 1955). Their work was extended by Utiyama (1956), who developed a general formalism for such compensating fields. As we have said, these types of dynamical theories, based on local invariance principles, are called gauge theories.

It is a remarkable fact that the interactions in the Standard Model of particle physics are of precisely this type. We have briefly discussed the Maxwell equations in this light, and we will continue with (quantum) electrodynamics in the following two sections. The two other fundamental interactions—the strong interaction between quarks and the weak interaction between quarks and leptons—also seem to be described by gauge theories (of essentially the Yang–Mills type), as we shall see in detail in the second volume of this book. A fourth example, but one which we shall not pursue in this book, is that of general relativity (the theory of gravitational interactions). Utiyama (1956) showed that this theory could be arrived at by generalizing the global (space–time independent) coordinate transformations of special relativity to local ones; as with electromagnetism, the more restrictive local invariance requirements entailed the existence of a new field—the gravitational one—with an (almost) prescribed form of interaction. Unfortunately, despite this 'gauge' property, no consistent quantum field theory of general relativity is known, as discussed earlier in section 2.7.

In order to proceed further, we must now discuss how such (gauge) ideas are incorporated into quantum mechanics.

3.4 Gauge invariance (and covariance) in quantum mechanics

The Lorentz force law for a non-relativistic particle of charge q moving with velocity $\boldsymbol{v}$ under the influence of both electric and magnetic fields is

$$\boldsymbol{F} = q\boldsymbol{E} + q\boldsymbol{v} \times \boldsymbol{B}. \tag{3.26}$$

It may be derived, via Hamilton's equations, from the classical Hamiltonian[2]

$$H = \frac{1}{2m}(\boldsymbol{p} - q\boldsymbol{A})^2 + qV. \tag{3.27}$$

The Schrödinger equation for such a particle in an electromagnetic field is

$$\left(\frac{1}{2m}(-\mathrm{i}\nabla - q\boldsymbol{A})^2 + qV\right)\psi(\boldsymbol{x},t) = \mathrm{i}\frac{\partial\psi(\boldsymbol{x},t)}{\partial t} \tag{3.28}$$

which is obtained from the classical Hamiltonian by the usual prescription, $\boldsymbol{p} \rightarrow -\mathrm{i}\nabla$, for Schrödinger's wave mechanics ($\hbar = 1$). Note the appearance of the operator combinations

$$\boxed{\begin{aligned} \boldsymbol{D} &\equiv \nabla - \mathrm{i}q\boldsymbol{A} \\ D^0 &\equiv \partial/\partial t + \mathrm{i}qV \end{aligned}} \tag{3.29}$$

in place of ∇ and $\partial/\partial t$, in going from the free-particle Schrödinger equation to the electromagnetic field case.

The solution of $\psi(\boldsymbol{x},t)$ of the Schrödinger equation (3.28) describes completely the state of the particle moving under the influence of the potentials V, $\boldsymbol{A}$. However, these potentials are not unique, as we have already seen: they can be changed by a gauge transformation

$$\boldsymbol{A} \rightarrow \boldsymbol{A}' = \boldsymbol{A} + \nabla\chi \tag{3.30}$$
$$V \rightarrow V' = V - \partial\chi/\partial t \tag{3.31}$$

and the Maxwell equations for the fields $\boldsymbol{E}$ and $\boldsymbol{B}$ will remain the same. This immediately raises a serious question: if we carry out such a change of potentials in equation (3.28), will the solution of the resulting equation describe the same physics as the solution of equation (3.28)? If it does, we shall be able to assume

[2] We set $\hbar = c = 1$ throughout (see appendix B).

the validity of Maxwell's theory for the quantum world; if not, some modification will be necessary, since the gauge symmetry possessed by the Maxwell equations will be violated in the quantum theory.

The answer to the question just posed is evidently negative, since it is clear that the same 'ψ' cannot possibly satisfy both (3.28) and the analogous equation with $(V, \boldsymbol{A})$ replaced by $(V', \boldsymbol{A}')$. Unlike Maxwell's equations, the Schrödinger equation is not gauge invariant. But we must remember that the wavefunction ψ is not a directly observable quantity, as the electromagnetic fields $\boldsymbol{E}$ and $\boldsymbol{B}$ are. Perhaps ψ does not need to remain unchanged (invariant) when the potentials are changed by a gauge transformation. In fact, in order to have any chance of 'describing the same physics' in terms of the gauge-transformed potentials, *we will have to allow ψ to change as well.* This is a crucial point: for quantum mechanics to be consistent with Maxwell's equations it is necessary for the gauge transformations (3.30) and (3.31) of the Maxwell potentials to be accompanied also by a transformation of the quantum-mechanical wavefunction, $\psi \to \psi'$, where ψ' satisfies the equation

$$\left(\frac{1}{2m}(-\mathrm{i}\nabla - q\boldsymbol{A}')^2 + qV'\right)\psi'(\boldsymbol{x}, t) = \mathrm{i}\frac{\partial\psi'(\boldsymbol{x}, t)}{\partial t}. \tag{3.32}$$

Note that the *form* of (3.32) is exactly the same as the form of (3.28)—it is this that will effectively ensure that both 'describe the same physics'. Readers of appendix D will expect to be told that—if we can find such a ψ'—we may then assert that (3.28) is *gauge covariant*, meaning that it maintains the same form under a gauge transformation. (The transformations relevant to this use of 'covariance' are gauge transformations.)

Since we know the relations (3.30) and (3.31) between $\boldsymbol{A}$, V and $\boldsymbol{A}'$, V', we can actually find what $\psi'(\boldsymbol{x}, t)$ must be in order that equation (3.32) be consistent with (3.28). We shall state the answer and then verify it; then we shall discuss the physical interpretation. The required $\psi'(\boldsymbol{x}, t)$ is

$$\psi'(\boldsymbol{x}, t) = \exp[\mathrm{i}q\chi(\boldsymbol{x}, t)]\psi(\boldsymbol{x}, t) \tag{3.33}$$

where χ is the same space–time-dependent function as appears in equations (3.30) and (3.31). To verify this we consider

$$\begin{aligned}(-\mathrm{i}\nabla - q\boldsymbol{A}')\psi' &= [-\mathrm{i}\nabla - q\boldsymbol{A} - q(\nabla\chi)][\exp(\mathrm{i}q\chi)\psi] \\ &= q(\nabla\chi)\exp(\mathrm{i}q\chi)\psi + \exp(\mathrm{i}q\chi)\cdot(-\mathrm{i}\nabla\psi) \\ &\quad + \exp(\mathrm{i}q\chi)\cdot(-q\boldsymbol{A}\psi) - q(\nabla\chi)\exp(\mathrm{i}q\chi)\psi. \end{aligned} \tag{3.34}$$

The first and the last terms cancel leaving the result:

$$(-\mathrm{i}\nabla - q\boldsymbol{A}')\psi' = \exp(\mathrm{i}q\chi)\cdot(-\mathrm{i}\nabla - q\boldsymbol{A})\psi \tag{3.35}$$

which may be written using equation (3.29) as:

$$(-\mathrm{i}\boldsymbol{D}'\psi') = \exp(\mathrm{i}q\chi)\cdot(-\mathrm{i}\boldsymbol{D}\psi). \tag{3.36}$$

Thus, although the space–time-dependent phase factor feels the action of the gradient operator ∇, it 'passes through' the combined operator $\boldsymbol{D}'$ and converts it into $\boldsymbol{D}$: in fact comparing the equations (3.33) and (3.36), we see that $\boldsymbol{D}'\psi'$ bears to $\boldsymbol{D}\psi$ exactly the same relation as ψ' bears to ψ. In just the same way we find (cf equation (3.29))

$$(\mathrm{i}D^{0\prime}\psi') = \exp(\mathrm{i}q\chi)\cdot(\mathrm{i}D^0\psi) \tag{3.37}$$

where we have used equation (3.31) for V'. Once again, $D^{0\prime}\psi'$ is simply related to $D^0\psi$. Repeating the operation which led to equation (3.36) we find

$$\begin{aligned}\frac{1}{2m}(-\mathrm{i}\boldsymbol{D}')^2\psi' &= \exp(\mathrm{i}q\chi)\cdot\frac{1}{2m}(-\mathrm{i}\boldsymbol{D})^2\psi\\ &= \exp(\mathrm{i}q\chi)\cdot\mathrm{i}D^0\psi \qquad \text{(using equation (3.28))}\\ &= \mathrm{i}D^{0\prime}\psi' \qquad \text{(using equation (3.29)).}\end{aligned} \tag{3.38}$$

Equation (3.38) is just (3.32) written in the D notation of equation (3.29), so we have verified that (3.33) is the correct relationship between ψ' and ψ to ensure consistency between equations (3.28) and (3.32). Precisely this consistency is summarized by the statement that (3.28) is gauge covariant.

Do ψ and ψ' describe the same physics, in fact? The answer is yes, but it is not quite trivial. It is certainly obvious that the probability densities $|\psi|^2$ and $|\psi'|^2$ are equal, since in fact ψ and ψ' in equation (3.33) are related by a *phase* transformation. However, we can be interested in other observables involving the derivative operators ∇ or $\partial/\partial t$—for example, the current, which is essentially $\psi^*(\nabla\psi) - (\nabla\psi)^*\psi$. It is easy to check that this current is *not* invariant under (3.33), because the phase $\chi(\boldsymbol{x}, t)$ is $\boldsymbol{x}$-dependent. But equations (3.36) and (3.37) show us what we must do to construct *gauge-invariant currents*: namely, we must replace ∇ by $\boldsymbol{D}$ (and in general also $\partial/\partial t$ by D^0) since then:

$$\psi^{*\prime}(\boldsymbol{D}'\psi') = \psi^*\exp(-\mathrm{i}q\chi)\cdot\exp(\mathrm{i}q\chi)\cdot(\boldsymbol{D}\psi) = \psi^*\boldsymbol{D}\psi \tag{3.39}$$

for example. Thus the identity of the physics described by ψ and ψ' is indeed ensured. Note, incidentally, that the *equality* between the first and last terms in (3.39) is indeed a statement of *(gauge) invariance*.

We summarize these important considerations by the statement that the gauge invariance of Maxwell equations re-emerges as a covariance in quantum mechanics provided we make the combined transformation

$$\boxed{\begin{aligned}&\boldsymbol{A}\to\boldsymbol{A}' = \boldsymbol{A}+\nabla\chi\\ &V\to V' = V-\partial\chi/\partial t\\ &\psi\to\psi' = \exp(\mathrm{i}q\chi)\psi\end{aligned}} \tag{3.40}$$

on the potential and on the wavefunction.

The Schrödinger equation is non-relativistic, but the Maxwell equations are of course fully relativistic. One might therefore suspect that the prescriptions discovered here are actually true relativistically as well, and this is indeed the case. We shall introduce the spin-0 and spin-$\frac{1}{2}$ relativistic equations in chapter 4. For the present we note that (3.29) can be written in manifestly Lorentz covariant form as

$$\boxed{D^\mu \equiv \partial^\mu + \mathrm{i}qA^\mu} \tag{3.41}$$

in terms of which (3.36) and (3.37) become

$$-\mathrm{i}D'^\mu\psi' = \exp(\mathrm{i}q\chi)\cdot(-\mathrm{i}D^\mu\psi). \tag{3.42}$$

It follows that any equation involving the operator ∂^μ can be made gauge invariant under the combined transformation

$$\begin{aligned} A^\mu &\rightarrow A'^\mu = A^\mu - \partial^\mu\chi \\ \psi &\rightarrow \psi' = \exp(\mathrm{i}q\chi)\psi \end{aligned}$$

if ∂^μ is replaced by D^μ. In fact, we seem to have a very simple prescription for obtaining the wave equation for a particle in the presence of an electromagnetic field from the corresponding *free particle* wave equation: make the replacement

$$\boxed{\partial^\mu \rightarrow D^\mu \equiv \partial^\mu + \mathrm{i}qA^\mu.} \tag{3.43}$$

In the following section this will be seen to be the basis of the so-called 'gauge principle' whereby, in accordance with the idea advanced in the previous sections, the form of the *interaction* is determined by the insistence on (local) gauge invariance.

One final remark: this new kind of derivative

$$D^\mu \equiv \partial^\mu + \mathrm{i}qA^\mu \tag{3.44}$$

turns out to be of fundamental importance—it will be the operator which generalizes from the (Abelian) phase symmetry of QED (see comment (iii) of section 3.6) to the (non-Abelian) phase symmetry of our weak and strong interaction theories. It is called the '*gauge covariant derivative*', the term being usually shortened to 'covariant derivative' in the present context. The geometrical significance of this term will be explained in volume 2.

3.5 The argument reversed: the gauge principle

In the preceding section, we took it as *known* that the Schrödinger equation, for example, for a charged particle in an electromagnetic field, has the form

$$\left[\frac{1}{2m}(-\mathrm{i}\boldsymbol{\nabla} - q\boldsymbol{A})^2 + qV\right]\psi = \mathrm{i}\partial\psi/\partial t. \tag{3.45}$$

We then checked its gauge invariance under the combined transformation

$$\begin{aligned} \boldsymbol{A} &\rightarrow \boldsymbol{A}' = \boldsymbol{A} + \nabla \chi \\ V &\rightarrow V' = V - \partial \chi / \partial t \\ \psi &\rightarrow \psi' = \exp(\mathrm{i} q \chi) \psi. \end{aligned} \tag{3.46}$$

We now want to reverse the argument: we shall start by demanding that our theory is invariant under the *space–time-dependent phase transformation*

$$\psi(\boldsymbol{x}, t) \rightarrow \psi'(\boldsymbol{x}, t) = \exp[\mathrm{i} q \chi(\boldsymbol{x}, t)] \psi(\boldsymbol{x}, t). \tag{3.47}$$

We shall demonstrate that such a phase invariance is not possible for a free theory, but rather requires an *interacting* theory involving a (4-vector) field whose interactions with the charged particle are precisely determined, and which undergoes the transformation

$$\boldsymbol{A} \rightarrow \boldsymbol{A}' = \boldsymbol{A} + \nabla \chi \tag{3.48}$$

$$V \rightarrow V' = V - \partial \chi / \partial t \tag{3.49}$$

when $\psi \rightarrow \psi'$. The demand of this type of phase invariance will have then dictated the form of the interaction—this is the basis of the *gauge principle*.

Before proceeding we note that the resulting equation—which will of course turn out to be (3.28)—will not strictly speaking be invariant under (3.47), but rather covariant (in the gauge sense), as we saw in the preceding section. Nevertheless, we shall in this section sometimes continue (slightly loosely) to speak of 'local phase invariance'. When we come to implement these ideas in quantum field theory in chapter 7 (section 7.4), using the Lagrangian formalism, we shall see that the relevant Lagrangians are indeed invariant under (3.47).

We therefore focus attention on the phase of the wavefunction. The absolute phase of a wavefunction in quantum mechanics cannot be measured; only relative phases are measurable, via some sort of interference experiment. A simple example is provided by the diffraction of particles by a two-slit system. Downstream from the slits, the wavefunction is a coherent superposition of two components, one originating from each slit: symbolically,

$$\psi = \psi_1 + \psi_2. \tag{3.50}$$

The probability distribution $|\psi|^2$ will then involve, in addition to the separate intensities $|\psi_1|^2$ and $|\psi_2|^2$, the *interference* term

$$2\,\mathrm{Re}(\psi_1^* \psi_2) = 2|\psi_1||\psi_2| \cos \delta \tag{3.51}$$

where δ $(= \delta_1 - \delta_2)$ is the *phase difference* between components ψ_1 and ψ_2. The familiar pattern of alternating intensity maxima and minima is then attributed to variation in the phase difference δ. Where the components are in phase, the interference is constructive and $|\psi|^2$ has a maximum; where they are out of phase, it is destructive and $|\psi|^2$ has a minimum. It is clear that if the individual phases δ_1 and δ_2 are each shifted by the same amount, there will be no observable consequences, since only the phase difference δ enters.

The situation in which the wavefunction can be changed in a certain way without leading to any observable effects is precisely what is entailed by a symmetry or invariance principle in quantum mechanics. In the case under discussion, the invariance is that of a constant overall change in phase. In performing calculations it is necessary to make some definite choice of phase; that is, to adopt a 'phase convention'. The invariance principle guarantees that any such choice, or convention, is equivalent to any other.

Invariance under a constant change in phase is an example of a *global* invariance, according to the terminology introduced in the previous section. We make this point quite explicit by writing out the transformation as

$$\boxed{\begin{array}{c}\psi \to \psi' = e^{i\alpha}\psi \\ \alpha = \text{constant}\end{array}} \quad \text{global phase invariance.} \qquad (3.52)$$

That α in (3.52) is a constant, the same for all space–time points, expresses the fact that once a phase convention (choice of α) has been made at one space–time point, the same must be adopted at all other points. Thus in the two-slit experiment we are not free to make a *local* chance of phase: for example, as discussed by 't Hooft (1980), inserting a half-wave plate behind just one of the slits will certainly have observable consequences.

There is a sense in which this may seem an unnatural state of affairs . Once a phase convention has been adopted at one space–time point, the same convention must be adopted at all other ones: the half-wave plate must extend instantaneously across all of space, or not at all. Following this line of thought, one might then be led to 'explore the possibility' of requiring invariance under *local* phase transformations: that is, independent choices of phase convention at each space–time point. By itself, the foregoing is not a compelling motivation for such step. However, as we pointed out in section 3.3, such a move from a global to a local invariance is apparently of crucial significance in classical electromagnetism and general relativity, and seems now to provide the key to an understanding of the other interactions in the Standard Model. Let us see, then, where the demand of '*local* phase invariance'

$$\boxed{\psi(\boldsymbol{x}, t) \to \psi'(\boldsymbol{x}, t) = \exp[i\alpha(\boldsymbol{x}, t)]\psi(\boldsymbol{x}, t)} \quad \text{local phase invariance} \quad (3.53)$$

leads us.

There is immediately a problem: this is *not* an invariance of the free-particle Schrödinger equation or of any free-particle relativistic wave equation! For example, if the original wavefunction $\psi(\boldsymbol{x}, t)$ satisfied the free-particle

Schrödinger equation

$$\frac{1}{2m}(-\mathrm{i}\nabla^2)\psi(\boldsymbol{x},t) = \mathrm{i}\partial\psi(\boldsymbol{x},t)/\partial t \tag{3.54}$$

then the wavefunction ψ', given by the local phase transformation, will not, since both ∇ and $\partial/\partial t$ now act on $\alpha(\boldsymbol{x},t)$ in the phase factor. Thus local phase invariance is not an invariance of the free-particle wave equation. If we wish to satisfy the demands of local phase invariance, we are obliged to modify the free-particle Schrödinger equation into something for which there is a local phase invariance—or rather, more accurately, a corresponding covariance. But this modified equation will no longer describe a free particle: in other words, the freedom to alter the phase of a charged particle's wavefunction locally is only possible if some kind of force field is introduced in which the particle moves. In more physical terms, the covariance will now be manifested in the inability to distinguish observationally between the effect of making a local change in phase convention and the effect of some new field in which the particle moves.

What kind of field will this be? In fact, we know immediately what the answer is, since the local phase transformation

$$\psi \to \psi' = \exp[\mathrm{i}\alpha(\boldsymbol{x},t)]\psi \tag{3.55}$$

with $\alpha = q\chi$ is just the phase transformation associated with electromagnetic gauge invariance! Thus we must modify the Schrödinger equation

$$\frac{1}{2m}(-\mathrm{i}\nabla)^2\psi = \mathrm{i}\partial\psi/\partial t \tag{3.56}$$

to

$$\frac{1}{2m}(-\mathrm{i}\nabla - q\boldsymbol{A})^2\psi = (\mathrm{i}\partial\psi/\partial t - qV)\psi \tag{3.57}$$

and satisfy the local phase invariance

$$\psi \to \psi' = \exp[\mathrm{i}\alpha(\boldsymbol{x},t)]\psi \tag{3.58}$$

by demanding that $\boldsymbol{A}$ and V transform by

$$\begin{aligned} \boldsymbol{A} &\to \boldsymbol{A}' = \boldsymbol{A} + q^{-1}\nabla\alpha \\ V &\to V' = V - q^{-1}\partial\alpha/\partial t \end{aligned} \tag{3.59}$$

when $\psi \to \psi'$. But the modified wave equation is of course precisely the Schrödinger equation describing the interaction of the charged particle with the electromagnetic field described by $\boldsymbol{A}$ and V.

In a Lorentz covariant treatment, $\boldsymbol{A}$ and V will be regarded as parts of a 4-vector A^{μ}, just as $-\nabla$ and $\partial/\partial t$ are parts of ∂^{μ} (see problem 3.1). Thus the presence of the vector field A^{μ}, interacting in a 'universal' prescribed way with any particle of charge q, is dictated by local phase invariance. A vector field such as A^μ, introduced to guarantee local phase invariance, is called a 'gauge field'. The principle that the interaction should be so dictated by the phase (or gauge) invariance is called the *gauge principle*: it allows us to write down the wave equation for the interaction directly from the free particle equation via the replacement (3.43)[3]. As before, the method clearly generalizes to the four-dimensional case.

3.6 Comments on the gauge principle in electromagnetism

Comment (i)

A properly sceptical reader may have detected an important sleight of hand in the previous discussion. Where exactly did the electromagnetic charge appear from? The trouble with our argument as so far presented is that we could have defined fields $\boldsymbol{A}$ and V so that they coupled equally to all particles—instead we smuggled in a factor q.

Actually we can do a bit better than this. We can use the fact that the electromagnetic charge is absolutely conserved to claim that there can be no quantum mechanical interference between states of different charge q. Hence different phase changes are allowed within each 'sector' of definite q:

$$\psi' = \exp(\mathrm{i}q\chi)\psi \tag{3.60}$$

let us say. When this becomes a local transformation, $\chi \to \chi(\boldsymbol{x}, t)$, we shall need to cancel a term $q\nabla\chi$, which will imply the presence of a '$-q\boldsymbol{A}$' term, as required. Note that such an argument is only possible for an *absolutely* conserved quantum number q—otherwise we cannot split up the states of the system into non-communicating sectors specified by different values of q. Reversing this line of reasoning, a conservation law such as baryon number conservation, with no related gauge field, would therefore now be suspected of not being absolutely conserved.

We still have not tied down why q is the electromagnetic charge and not some other absolutely conserved quantum number. A proper discussion of the reasons for identifying A^μ with the electromagnetic potential and q with the particle's charge will be given in chapter 7 with the help of quantum field theory.

Comment (ii)

Accepting these identifications, we note that the form of the interaction contains but one parameter, the electromagnetic charge q of the particle in question. It is the *same* whatever the type of particle with charge q, whether it be lepton, hadron,

[3] Actually the electromagnetic interaction is uniquely specified by this procedure only for particles of spin-0 or $\frac{1}{2}$. The spin-1 case will be discussed in volume 2.

nucleus, ion, atom, etc. Precisely this type of 'universality' is present in the weak couplings of quarks and leptons, as we shall see in volume 2. This strongly suggests that some form of gauge principle must be at work in generating weak interactions as well. The associated symmetry or conservation law is, however, of a very subtle kind. Incidentally, although all particles of a given charge q interact electromagnetically in a universal way, there is nothing at all in the preceding argument to indicate why, in nature, the charges of observed particles are all integer multiples of one basic charge.

Comment (iii)

Returning to comment (i), we may wish that we had not had to introduce the absolute conservation of charge as a separate axiom. As remarked earlier, at the end of section 3.2, we should like to relate that conservation law to the symmetry involved, namely invariance under (3.53). It is worth looking at the nature of this symmetry in a little more detail. It is not a symmetry which—as in the case of translation and rotation invariances for instance—involves changes in the space–time coordinates $\boldsymbol{x}$ and t. Instead, it operates on the *real and imaginary parts of the wavefunction*. Let us write

$$\psi = \psi_{\mathrm{R}} + \mathrm{i}\psi_{\mathrm{I}}. \tag{3.61}$$

Then

$$\psi' = \mathrm{e}^{\mathrm{i}\alpha}\psi = \psi'_{\mathrm{R}} + \mathrm{i}\psi'_{\mathrm{I}} \tag{3.62}$$

can be written as

$$\begin{aligned} \psi'_{\mathrm{R}} &= (\cos\alpha)\psi_{\mathrm{R}} - (\sin\alpha)\psi_{\mathrm{I}} \\ \psi'_{\mathrm{I}} &= (\sin\alpha)\psi_{\mathrm{R}} + \cos\alpha)\psi_{\mathrm{I}} \end{aligned} \tag{3.63}$$

from which we can see that it is indeed a kind of 'rotation', but in the ψ_{R}–ψ_{I} plane, whose 'coordinates' are the real and imaginary parts of the wavefunction. We call this plane an *internal* space and the associated symmetry an *internal symmetry*. Thus our phase invariance can be looked upon as a kind of internal space rotational invariance.

We can imagine doing two successive such transformations

$$\psi \to \psi' \to \psi'' \tag{3.64}$$

where

$$\psi'' = \mathrm{e}^{\mathrm{i}\beta}\psi' \tag{3.65}$$

and so

$$\psi'' = \mathrm{e}^{\mathrm{i}(\alpha+\beta)}\psi = \mathrm{e}^{\mathrm{i}\delta}\psi \tag{3.66}$$

with $\delta = \alpha + \beta$. This is a transformation of the same form as the original one. The set of all such transformations forms what mathematicians call a *group*, in this case U(1), meaning the group of all unitary one-dimensional matrices. A unitary matrix $\mathbf{U}$ is one such that

$$\mathbf{U}\mathbf{U}^{\dagger} = \mathbf{U}^{\dagger}\mathbf{U} = \mathbf{1} \tag{3.67}$$

where $\mathbf{1}$ is the identity matrix and † denotes the Hermitian conjugate. A one-dimensional matrix is of course a single number—in this case a complex number. Condition (3.67) limits this to being a simple phase: the set of phase factors of the form $\mathrm{e}^{\mathrm{i}\alpha}$, where α is any real number, form the elements of a U(1) group. These are just the factors that enter into our gauge (or phase) transformations for wavefunctions. Thus we say that the electromagnetic gauge group is U(1). We must remember, however, that it is a *local* U(1), meaning (cf (3.53)) that the phase parameters $\alpha, \beta, \ldots$ depend on the space–time point x.

The transformations of the U(1) group have the simple property that it does not matter in what order they are performed: referring to (3.64)–(3.66), we would have got the same final answer if we had done the β 'rotation' first and then the α one, instead of the other way around; this is because, of course,

$$\exp(\mathrm{i}\alpha)\cdot\exp(\mathrm{i}\beta) = \exp[\mathrm{i}(\alpha+\beta)] = \exp(\mathrm{i}\beta)\cdot\exp(\mathrm{i}\alpha). \tag{3.68}$$

This property remains true even in the 'local' case when α and β depend on x. Mathematicians call U(1) an *Abelian* group: different transformations commute. We shall see later (in volume 2) that the 'internal' symmetry spaces relevant to the strong and weak gauge invariances are not so simple. The 'rotations' in these cases are more like full three-dimensional rotations of real space, rather than the two-dimensional rotation of (3.63). We know that, in general, such real-space rotations do *not* commute, and the same will be true of the strong and weak rotations. Their gauge groups are called *non-Abelian*.

Once again, we shall have to wait until chapter 7 before understanding how the symmetry represented by (3.62) is really related to the conservation law of charge.

Comment (iv)

The attentive reader may have picked up one further loose end. The vector potential $\boldsymbol{A}$ is related to the magnetic field $\boldsymbol{B}$ by

$$\boldsymbol{B} = \nabla\times\boldsymbol{A}. \tag{3.69}$$

Thus if $\boldsymbol{A}$ has the special form

$$\boldsymbol{A} = \nabla f \tag{3.70}$$

$\boldsymbol{B}$ will vanish. The question we must answer, therefore, is: how do we know that the $\boldsymbol{A}$ field introduced by our gauge principle is not of the form (3.70), leading to a trivial theory ($\boldsymbol{B} = \mathbf{0}$)? The answer to this question will lead us on a very worthwhile detour.

The Schrödinger equation with $\boldsymbol{\nabla} f$ as the vector potential is

$$\frac{1}{2m}(-\mathrm{i}\boldsymbol{\nabla} - q\boldsymbol{\nabla} f)^2\psi = E\psi. \tag{3.71}$$

We can write the formal solution to this equation as

$$\psi = \exp\left(\mathrm{i}q\int_{-\infty}^{x} \boldsymbol{\nabla} f \cdot \mathrm{d}\boldsymbol{l}\right) \cdot \psi(f = 0) \tag{3.72}$$

which may be checked by using the fact that

$$\frac{\partial}{\partial a}\int^{a} f(t)\,\mathrm{d}t = f(a). \tag{3.73}$$

The notation $\psi(f = 0)$ means just the free-particle solution with $f = 0$; the line integral is taken along an arbitrary path ending in the point $\boldsymbol{x}$. But we have

$$\mathrm{d}f = \frac{\partial f}{\partial x}\,\mathrm{d}x + \frac{\partial f}{\partial y}\,\mathrm{d}y + \frac{\partial f}{\partial z}\,\mathrm{d}z \equiv \boldsymbol{\nabla} f \cdot \mathrm{d}\boldsymbol{l}. \tag{3.74}$$

Hence the integral can be done trivially and the solution becomes

$$\psi = \exp[\mathrm{i}q(f(\boldsymbol{x}) - f(-\infty))] \cdot \psi(f = 0). \tag{3.75}$$

We say that the phase factor introduced by the (in reality, field-free) vector potential $\boldsymbol{A} = \boldsymbol{\nabla} f$ is *integrable*: the effect of this particular $\boldsymbol{A}$ is merely to multiply the free-particle solution by an $\boldsymbol{x}$-dependent phase (apart from a trivial constant phase). Since this $\boldsymbol{A}$ should give no real electromagnetic effect, we must hope that such a change in the wavefunction is also somehow harmless. Indeed Dirac showed (Dirac 1981, pp 92–3) that such a phase factor corresponds merely to a redefinition of the momentum operator $\hat{\boldsymbol{p}}$. The essential point is that (in one dimension, say) $\hat{p}$ is defined ultimately by the commutator ($\hbar = 1$)

$$[\hat{x}, \hat{p}] = \mathrm{i}. \tag{3.76}$$

Certainly the familiar choice

$$\hat{p} = -\mathrm{i}\frac{\partial}{\partial x} \tag{3.77}$$

satisfies this commutation relation. But we can also add any function of x to $\hat{p}$, and this modified $\hat{p}$ will be still satisfactory since x commutes with any function of x. More detailed considerations by Dirac showed that this arbitrary function must actually have the form $\partial F/\partial x$, where F is arbitrary. Thus

$$\hat{p}' = -\mathrm{i}\frac{\partial}{\partial x} + \frac{\partial F}{\partial x} \tag{3.78}$$

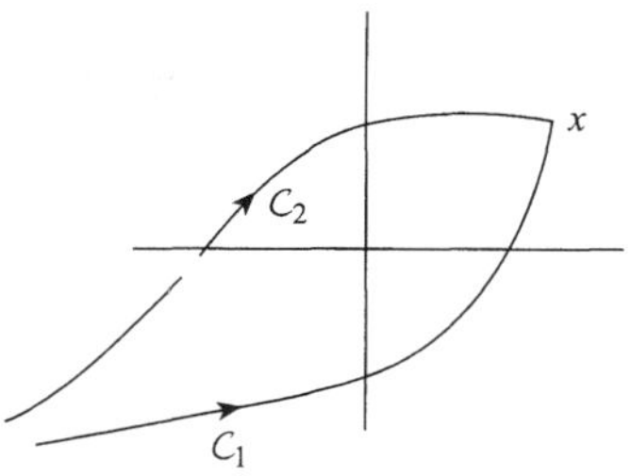

Figure 3.1. Two paths C_1 and C_2 (in two dimensions for simplicity) from $-\infty$ to the point x.

is an acceptable momentum operator. Consider then the quantum mechanics defined by the wavefunction $\psi(f = 0)$ and the momentum operator $\hat{p} = -\mathrm{i}\partial/\partial x$. Under the unitary transformation (cf (3.75))

$$\psi(f = 0) \rightarrow \mathrm{e}^{\mathrm{i}qf(x)}\psi(f = 0) \tag{3.79}$$

$\hat{p}$ will be transformed to

$$\hat{p} \rightarrow \mathrm{e}^{\mathrm{i}qf(x)}\hat{p}\mathrm{e}^{-\mathrm{i}qf(x)}. \tag{3.80}$$

But the right-hand side of this equation is just $\hat{p} - q\partial f/\partial x$ (problem 3.3), which is an equally acceptable momentum operator, identifying qf with the F of Dirac. Thus the case $\boldsymbol{A} = \boldsymbol{\nabla} f$ is indeed equivalent to the field-free case.

What of the physically interesting case in which $\boldsymbol{A}$ is *not* of the form $\boldsymbol{\nabla} f$? The equation is now

$$\frac{1}{2m}(-\mathrm{i}\boldsymbol{\nabla} - q\boldsymbol{A})^2\psi = E\psi \tag{3.81}$$

to which the solution is

$$\psi = \exp\left(\mathrm{i}q\int_{-\infty}^{\boldsymbol{x}} \boldsymbol{A}\cdot\mathrm{d}\boldsymbol{l}\right)\cdot\psi(\boldsymbol{A} = 0). \tag{3.82}$$

The line integral can now not be done so trivially: one says that the $\boldsymbol{A}$-field has produced a *non-integrable phase factor*. There is more to this terminology than the mere question of whether the integral is easy to do. The crucial point is that the integral now depends on the *path followed* in reaching the point $\boldsymbol{x}$, whereas the integrable phase factor in (3.72) depends only on the end-points of the integral, not on the path joining them.

Consider two paths C_1 and C_2 (figure 3.1) from $-\infty$ to the point $\boldsymbol{x}$. The difference in the two line integrals is the integral over a *closed* curve C, which can be evaluated by Stokes' theorem:

$$\int_{C_1}^{\boldsymbol{x}} \boldsymbol{A}\cdot\mathrm{d}\boldsymbol{l} - \int_{C_2}^{\boldsymbol{x}} \boldsymbol{A}\cdot\mathrm{d}\boldsymbol{l} = \oint_C \boldsymbol{A}\cdot\mathrm{d}\boldsymbol{l} = \iint_S \boldsymbol{\nabla}\times\boldsymbol{A}\cdot\mathrm{d}\boldsymbol{S} = \iint_S \boldsymbol{B}\cdot\mathrm{d}\boldsymbol{S} \tag{3.83}$$

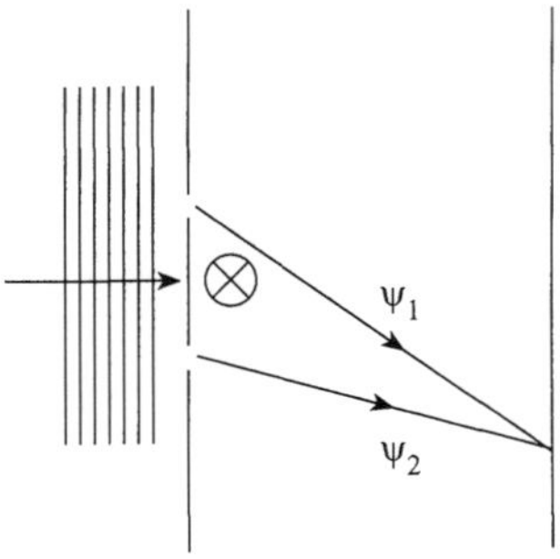

Figure 3.2. The Aharonov–Bohm effect.

where S is any surface spanning the curve $\mathcal{C}$. In this form we see that if $\boldsymbol{A} = \nabla f$, then indeed the line integrals over $\mathcal{C}_1$ and $\mathcal{C}_2$ are equal since $\nabla \times \nabla f = 0$, but if $\boldsymbol{B} = \nabla \times \boldsymbol{A}$ is not zero, the difference between the integrals is determined by the enclosed flux of $\boldsymbol{B}$.

This analysis turns out to imply the existence of a remarkable phenomenon—the Aharonov–Bohm effect, named after its discoverers (Aharonov and Bohm 1959). Suppose we go back to our two-slit experiment of section 3.5, only this time we imagine that a long thin solenoid is inserted between the slits, so that the components ψ_1 and ψ_2 of the split beam pass one on each side of the solenoid (figure 3.2). After passing round the solenoid, the beams are recombined, and the resulting interference pattern is observed downstream. At any point $\boldsymbol{x}$ of the pattern, the phase of the ψ_1 and ψ_2 components will be modified—relative to the $\boldsymbol{B} = \boldsymbol{0}$ case—by factors of the form (3.82). These factors depend on the respective paths, which are different for the two components ψ_1 and ψ_2. The phase difference between these components, which determines the interference pattern, will therefore involve the $\boldsymbol{B}$-dependent factor (3.83). Thus, even though the field $\boldsymbol{B}$ is essentially totally contained within the solenoid, and the beams themselves have passed through $\boldsymbol{B} = \boldsymbol{0}$ regions only, there is nevertheless an observable effect on the pattern provided $\boldsymbol{B} \neq \boldsymbol{0}$! This effect—a shift in the pattern as $\boldsymbol{B}$ varies—was first confirmed experimentally by Chambers (1960), soon after its prediction by Aharonov and Bohm. It was anticipated in work by Ehrenburg and Siday (1949); further references and discussion are contained in Berry (1984).

Comment (v)

In conclusion, we must emphasize that there is ultimately no compelling logic for the vital leap to a local phase invariance from a global one. The latter is, by itself, both necessary and sufficient in quantum field theory to guarantee local charge conservation. Nevertheless, the gauge principle—deriving interactions from the requirement of local phase invariance—provides a satisfying conceptual unification of the interactions present in the Standard Model. In volume 2 of this book we shall consider generalizations of the electromagnetic gauge principle. It will be important always to bear in mind that any attempt to base theories of non-electromagnetic interactions on some kind of gauge principle can only make sense if there is an exact symmetry involved. The reason for this will only become clear when we consider the *renormalizability* of QED in chapter 11.

Problems

3.1

(a) A Lorentz transformation in the x^1 direction is given by

$$\begin{aligned} t' &= \gamma(t - vx^1) \\ x^{1\prime} &= \gamma(-vt + x^1) \\ x^{2\prime} &= x^2, \qquad x^{3\prime} = x^3 \end{aligned}$$

where $\gamma = (1 - v^2)^{-1/2}$ and $c = 1$. Write down the inverse of this transformation (i.e. express (t, x^1) in terms of $(t', x^{1\prime})$), and use the 'chain rule' of partial differentiation to show that, under the Lorentz transformation, the two quantities $(\partial/\partial t, -\partial/\partial x^1)$ transform in the same way as (t, x^1).

[The general result is that the four-component quantity $(\partial/\partial t, -\partial/\partial x^1, -\partial/\partial x^2, -\partial/\partial x^3) \equiv (\partial/\partial t, -\boldsymbol{\nabla})$ transforms in the same way as (t, x^1, x^2, x^3). Four-component quantities transforming this way are said to be 'contravariant 4-vectors', and are written with an upper 4-vector index; thus $(\partial/\partial t, -\boldsymbol{\nabla}) \equiv \partial^\mu$. Upper indices can be lowered by using the metric tensor $g_{\mu\nu}$, see appendix D, which reverses the sign of the spatial components. Thus $\partial^\mu = (\partial/\partial t, \partial/\partial x_1, \partial/\partial x_2, \partial/\partial x_3)$. Similarly the four quantities $(\partial/\partial t, \boldsymbol{\nabla}) = (\partial/\partial t, \partial/\partial x^1, \partial/\partial x^2, \partial/\partial x^3)$ transform as $(t, -x^1, -x^2, -x^3)$ and are a 'covariant 4-vector', denoted by ∂_μ.]

(b) Check that equation (3.5) can be written as (3.16).

3.2 How many independent components does the field strength $F^{\mu\nu}$ have? Express each component in terms of electric and magnetic field components. Hence verify that equation (3.17) correctly reproduces both equations (3.1) and (3.8).

3.3 Verify the result

$$\mathrm{e}^{\mathrm{i}qf(x)}\hat{p}\mathrm{e}^{-\mathrm{i}qf(x)} = \hat{p} - q\frac{\partial f}{\partial x}.$$

4

RELATIVISTIC QUANTUM MECHANICS

It is clear that the non-relativistic Schrödinger equation is quite inadequate to analyse the results of experiments at energies far higher than the rest mass energies of the particles involved. Besides, the quarks and leptons have spin-$\frac{1}{2}$, a degree of freedom absent from the Schrödinger wavefunction. We therefore need two generalizations—from non-relativistic to relativistic for spin-0 particles, and from spin-0 to spin-$\frac{1}{2}$. The first step is to the Klein–Gordon equation (section 4.1), the second to the Dirac equation (section 4.2). Then after some further work on solutions of the Dirac equation (sections 4.3–4.5), we shall consider (section 4.6) some simple consequences of including the electromagnetic interaction via the gauge principle replacement (3.43).

4.1 The Klein–Gordon equation

The non-relativistic Schrödinger equation may be put into correspondence with the non-relativistic energy–momentum relation

$$E = \boldsymbol{p}^2/2m \tag{4.1}$$

by means of the operator replacements[1]

$$E \rightarrow \mathrm{i}\partial/\partial t \tag{4.2}$$

$$\boldsymbol{p} \rightarrow -\mathrm{i}\boldsymbol{\nabla} \tag{4.3}$$

these differential operators being understood to act on the Schrödinger wavefunction.

For a relativistic wave equation we must start with the correct relativistic energy–momentum relation. Energy and momentum appear as the 'time' and 'space' components of the momentum 4-vector

$$p^\mu = (E, \boldsymbol{p}) \tag{4.4}$$

which satisfy the mass-shell condition

$$p^2 = p_\mu p^\mu = E^2 - \boldsymbol{p}^2 = m^2. \tag{4.5}$$

[1] Recall $\hbar = c = 1$ throughout (see appendix B).

Since energy and momentum are merely different components of a 4-vector, an attempt to base a relativistic theory on the relation

$$E = +(\boldsymbol{p}^2 + m^2)^{1/2} \tag{4.6}$$

is unattractive, as well as having obvious difficulties in interpretation for the square root operator. Schrödinger, before settling for the less ambitious non-relativistic Schrödinger equation, and later Klein and Gordon, attempted to build relativistic quantum mechanics (RQM) from the squared relation

$$E^2 = \boldsymbol{p}^2 + m^2. \tag{4.7}$$

Using the operator replacements for E and $\boldsymbol{p}$ we are led to

$$-\partial^2\phi/\partial t^2 = (-\nabla^2 + m^2)\phi \tag{4.8}$$

which is the Klein–Gordon equation (KG equation). We consider the case of a one-component scalar wavefunction $\phi(\boldsymbol{x}, t)$: one expects this to be appropriate for the description of spin-0 bosons.

4.1.1 Solutions in coordinate space

In terms of the D'Alembertian operator

$$\square \equiv \partial_\mu \partial^\mu = \frac{\partial^2}{\partial t^2} - \nabla^2 \tag{4.9}$$

the KG equation reads:

$$(\square + m^2)\phi(\boldsymbol{x}, t) = 0. \tag{4.10}$$

Let us look for a plane-wave solution of the form

$$\phi(\boldsymbol{x}, t) = N\mathrm{e}^{-\mathrm{i}Et+\mathrm{i}\boldsymbol{p}\cdot\boldsymbol{x}} = N\mathrm{e}^{-\mathrm{i}p\cdot x} \tag{4.11}$$

where we have written the exponent in suggestive 4-vector scalar product notation

$$p \cdot x = p_\mu x^\mu = Et - \boldsymbol{p}\cdot\boldsymbol{x} \tag{4.12}$$

and N is a normalization factor which need not be decided upon here (see section 8.1.1). In order that this wavefunction be a solution of the KG equation, we find by direct substitution that E must be related to $\boldsymbol{p}$ by the condition

$$E^2 = \boldsymbol{p}^2 + m^2. \tag{4.13}$$

This looks harmless enough, but it actually implies that for a given 3-momentum $\boldsymbol{p}$ there are in fact *two* possible solutions for the energy, namely

$$E = \pm(\boldsymbol{p}^2 + m^2)^{1/2}. \tag{4.14}$$

As Schrödinger and others quickly found, it is not possible to ignore the negative solutions without obtaining inconsistencies. What then do these negative-energy solutions mean?

4.1.2 Probability current for the KG equation

In exactly the same way as for the non-relativistic Schrödinger equation, it is possible to derive a conservation law for a 'probability current' of the KG equation. We have

$$\frac{\partial^2 \phi}{\partial t^2} - \nabla^2 \phi + m^2 \phi = 0 \tag{4.15}$$

and by multiplying this equation by ϕ^*, and subtracting ϕ times the complex conjugate of equation (4.15), one obtains, after some manipulation (see problem 4.1), the result

$$\frac{\partial \rho}{\partial t} + \nabla \cdot \boldsymbol{j} = 0 \tag{4.16}$$

where

$$\rho = \mathrm{i}\left[\phi^* \frac{\partial \phi}{\partial t} - \left(\frac{\partial \phi^*}{\partial t}\right)\phi\right] \tag{4.17}$$

and

$$\boldsymbol{j} = \mathrm{i}^{-1}[\phi^* \nabla \phi - (\nabla \phi^*)\phi] \tag{4.18}$$

(the derivatives $(\partial_\mu \phi^*)$ act only within the bracket). In explicit 4-vector notation this conservation condition reads (cf problem 3.1 and equation (D.4) in appendix D)

$$\partial_\mu j^\mu = 0 \tag{4.19}$$

with

$$j^\mu \equiv (\rho, \boldsymbol{j}) = \mathrm{i}[\phi^* \partial^\mu \phi - (\partial^\mu \phi^*)\phi]. \tag{4.20}$$

Since ϕ of (4.11) is Lorentz invariant and ∂^μ is a contravariant 4-vector, equation (4.20) shows explicitly that j^μ is a contravariant 4-vector, as anticipated in the notation.

The spatial current $\boldsymbol{j}$ is identical in form to the Schrödinger current, but for the KG case the 'probability density' now contains time derivatives since the KG equation is second order in $\partial/\partial t$. This means that ρ is not constrained to be positive definite—so how can ρ represent a probability density? We can see this problem explicitly for the plane-wave solutions

$$\phi = N\mathrm{e}^{-\mathrm{i}Et+\mathrm{i}\boldsymbol{p}\cdot\boldsymbol{x}} \tag{4.21}$$

which give (problem 4.1)

$$\rho = 2|N|^2 E \tag{4.22}$$

and E can be positive or negative: that is, the sign of ρ is the sign of energy.

Historically, this problem of negative probabilities coupled with that of negative energies led to the abandonment of the KG equation. For the moment we will follow history, and turn to the Dirac equation. We shall see in section 4.5, however, how the negative-energy solutions of the KG equation do after all have a role to play, following Feynman's interpretation, in processes involving anti-particles. Later, in chapters 5–7, we shall see how this interpretation arises naturally within the formalism of quantum field theory.

4.2 The Dirac equation

In the case of the KG equation it is clear why the problem arose:

(a) In constructing a wave equation in close correspondence with the squared energy–momentum relation

$$E^2 = \boldsymbol{p}^2 + m^2$$

we immediately allowed negative-energy solutions.

(b) The KG equation has a $\partial^2/\partial t^2$ term: this leads to a continuity equation with a 'probability density' containing $\partial/\partial t$, and hence to negative probabilities.

Dirac approached these problems in his characteristically direct way. In order to obtain a positive-definite probability density $\rho \geq 0$, he required an equation linear in $\partial/\partial t$. Then, for relativistic covariance (see later), the equation must also be linear in $\boldsymbol{\nabla}$. He postulated the equation (Dirac 1928)

$$\begin{aligned} \mathrm{i}\frac{\partial\psi(\boldsymbol{x},t)}{\partial t} &= \left[-\mathrm{i}\left(\alpha_1\frac{\partial}{\partial x^1} + \alpha_2\frac{\partial}{\partial x^2} + \alpha_3\frac{\partial}{\partial x^3}\right) + \beta m\right]\psi(\boldsymbol{x},t) \\ &= (-\mathrm{i}\boldsymbol{\alpha}\cdot\boldsymbol{\nabla} + \beta m)\psi(\boldsymbol{x},t). \end{aligned} \tag{4.23}$$

What are the α's and β? To find the conditions on the α's and β, consider what we require of a relativistic wave equation:

(a) the correct relativistic relation between E and $\boldsymbol{p}$, namely

$$E = +(\boldsymbol{p}^2 + m^2)^{1/2}$$

(b) the equation should be covariant under Lorentz transformations.

We shall postpone discussion of (b) until section 4.4. To solve requirement (a), Dirac in fact demanded that his wavefunction ψ satisfy, in addition, a KG-type condition

$$-\partial^2\psi/\partial t^2 = (-\boldsymbol{\nabla}^2 + m^2)\psi. \tag{4.24}$$

We note with hindsight that we have once more opened the door to negative-energy solutions: Dirac's remarkable achievement was to turn this apparent defect into one of the triumphs of theoretical physics!

We can now derive conditions on $\boldsymbol{\alpha}$ and β. We have

$$\mathrm{i}\partial\psi/\partial t = (-\mathrm{i}\boldsymbol{\alpha}\cdot\boldsymbol{\nabla} + \beta m)\psi \tag{4.25}$$

and so, squaring the operator on both sides,

$$\left(\mathrm{i}\frac{\partial}{\partial t}\right)^2\psi = (-\mathrm{i}\boldsymbol{\alpha}\cdot\boldsymbol{\nabla} + \beta m)(-\mathrm{i}\boldsymbol{\alpha}\cdot\boldsymbol{\nabla} + \beta m)\psi$$

$$= -\sum_{i=1}^{3} \alpha_i^2 \frac{\partial^2 \psi}{(\partial x^i)^2} - \sum_{\substack{i,j=1 \\ i>j}}^{3} (\alpha_i \alpha_j + \alpha_j \alpha_i) \frac{\partial^2 \psi}{\partial x^i \partial x^j}$$

$$- \mathrm{i}m \sum_{i=1}^{3} (\alpha_i \beta + \beta \alpha_i) \frac{\partial \psi}{\partial x^i} + \beta^2 m^2 \psi. \tag{4.26}$$

But by our assumption that ψ also satisfies the KG condition, we must have

$$\left(\mathrm{i}\frac{\partial}{\partial t}\right)^2 \psi = -\sum_{i=1}^{3} \frac{\partial^2 \psi}{(\partial x^i)^2} + m^2 \psi. \tag{4.27}$$

It is thus evident that the α's and β cannot be ordinary, classical, commuting quantities. Instead they must satisfy the following *anti-commutation relations* in order to eliminate the unwanted terms on the right-hand side of equation (4.26):

$$\alpha_i \beta + \beta \alpha_i = 0 \qquad i = 1, 2, 3 \tag{4.28}$$

$$\alpha_i \alpha_j + \alpha_j \alpha_i = 0 \qquad i, j = 1, 2, 3;\ i \neq j. \tag{4.29}$$

In addition we require

$$\alpha_i^2 = \beta^2 = 1. \tag{4.30}$$

Dirac proposed that the α's and β should be interpreted as matrices, acting on a wavefunction which had several components arranged as a column vector. Anticipating somewhat the results of the next section, we would expect that, since each such component obeys the same wave equation, the physical states which they represent would have the same energy. This would mean that the different components represent some *degeneracy*, associated with a new degree of freedom.

The degree of freedom is, of course, *spin*—an entirely quantum mechanical angular momentum, analogous to (but not equivalent to) orbital angular momentum. Consider, for example, the wavefunctions for the 2p state in the simple non-relativistic theory of the hydrogen atom. There are three of them, all degenerate with energy given by the $n = 2$ Bohr energy. The three corresponding states all have orbital angular momentum quantum number l equal to 1; they differ in their values of the 'magnetic' quantum number m (i.e. the eigenvalue of the z-component of the orbital angular momentum operator $\hat{L}_z$). Specifically, these three wavefunctions have the form (omitting normalization constants) $(r \sin\theta \mathrm{e}^{\mathrm{i}\phi}, r \sin\theta \mathrm{e}^{-\mathrm{i}\phi}, r\cos\theta)\mathrm{e}^{-r/2r_{\mathrm{B}}}$, where r_{B} is the Bohr radius. Remembering the expressions for the Cartesian coordinates x, y and z in terms of the spherical polar coordinates r, θ and ϕ, we see that by a suitable linear combination (always allowed for degenerate states) we can write these wavefunctions as $(x, y, z) f(r)$, where again a normalization factor has been omitted. In this form it is plain that the multiplicity of the p-state wavefunctions can be interpreted in simple geometrical terms: they are effectively the components of a *vector* (multiplication by the scalar function $f(r)$ does not affect this).

The several components of the Dirac wavefunction together make up a similar, but quite distinct, object called a *spinor*. We shall have more to say about this in section 4.4. For the moment we continue with the problem of finding the matrices α_i and β to satisfy (4.28)–(4.30).

As problem 4.2 shows, the smallest possible dimension of the matrices for which the Dirac conditions can be satisfied is 4×4. One conventional choice of the α's and β is

$$\alpha_i = \begin{pmatrix} \mathbf{0} & \sigma_i \\ \sigma_i & \mathbf{0} \end{pmatrix} \qquad \beta = \begin{pmatrix} \mathbf{1} & \mathbf{0} \\ \mathbf{0} & -\mathbf{1} \end{pmatrix} \tag{4.31}$$

where we have written these 4×4 matrices in 2×2 'block diagonal' form, the σ_i's are the 2×2 *Pauli matrices*, $\mathbf{1}$ is the 2×2 unit matrix, and $\mathbf{0}$ is the 2×2 null matrix. The Pauli matrices (see appendix A) are defined by

$$\sigma_x = \begin{pmatrix} 0 & 1 \\ 1 & 0 \end{pmatrix} \qquad \sigma_y = \begin{pmatrix} 0 & -i \\ i & 0 \end{pmatrix} \qquad \sigma_z = \begin{pmatrix} 1 & 0 \\ 0 & -1 \end{pmatrix}. \tag{4.32}$$

Readers unfamiliar with the labour-saving 'block' form of (4.31) should verify, both by using the corresponding explicit 4×4 matrices, such as

$$\alpha_1 = \begin{pmatrix} 0 & 0 & 0 & 1 \\ 0 & 0 & 1 & 0 \\ 0 & 1 & 0 & 0 \\ 1 & 0 & 0 & 0 \end{pmatrix} \tag{4.33}$$

and so on, and by the block diagonal form, that this choice does indeed satisfy the required conditions. These are

$$\{\alpha_i, \beta\} = 0 \tag{4.34}$$

$$\{\alpha_i, \alpha_j\} = 2\delta_{ij}\mathbf{1} \tag{4.35}$$

$$\beta^2 = \mathbf{1} \tag{4.36}$$

where $\{\mathbf{A}, \mathbf{B}\}$ is the anti-commutator of two matrices, $\mathbf{AB} + \mathbf{BA}$, and $\mathbf{1}$ is here the 4×4 unit matrix.

At this point we can already begin to see that the extra multiplicity is very likely to have something to do with an angular momentum-like degree of freedom. In fact, if we define the spin matrices $\boldsymbol{S}$ by $\boldsymbol{S} = \frac{1}{2}\boldsymbol{\sigma}$ ($\hbar = 1$), we find from (4.32) that

$$[S_x, S_y] = iS_z \tag{4.37}$$

(with obvious cyclic permutations), which are precisely the commutation relations satisfied by the components $\hat{J}_x$, $\hat{J}_y$ and $\hat{J}_z$ of the angular momentum operator $\hat{\boldsymbol{J}}$ in quantum mechanics (see appendix A). Furthermore, the eigenvalues of S_z are $\pm\frac{1}{2}$, and of $\mathbf{S}^2$ are $s(s+1)$ with $s = \frac{1}{2}$. So these matrices undoubtedly represent quantum mechanical angular momentum operators, appropriate to a state with angular momentum quantum number $j = \frac{1}{2}$. This is precisely what 'spin' is. We will discuss this in more detail in section 4.3.

It is important to note that the choice (4.31) of α and β is not unique. In fact, all matrices related to these by any unitary 4×4 matrix $\mathbf{U}$ (which thus preserves the anti-commutation relations) are allowed:

$$\alpha_i' = \mathbf{U}\alpha_i\mathbf{U}^{-1} \tag{4.38}$$

$$\beta' = \mathbf{U}\beta\mathbf{U}^{-1}. \tag{4.39}$$

For the most part we shall use the standard representation (4.31), but in some cases it is convenient (but not necessary) to use other representations in which β is not diagonal. Another commonly used representation is introduced in section 4.4, and explored further in problem 4.15.

4.2.1 Free-particle solutions

Since the Dirac Hamiltonian now involves 4×4 matrices, it is clear that we must interpret the Dirac wavefunction ψ as a four-component column vector—the so-called Dirac spinor. Let us look at the explicit form of the free-particle solutions. As in the KG case, we look for solutions in which the space–time behaviour is of plane-wave form and put

$$\psi = \omega \mathrm{e}^{-\mathrm{i}p\cdot x} \tag{4.40}$$

where ω is a four-component spinor independent of x, and $\mathrm{e}^{-\mathrm{i}p\cdot x}$, with $p^\mu = (E, \boldsymbol{p})$, is the plane-wave solution corresponding to 4-momentum p^μ. We substitute this into the Dirac equation

$$\mathrm{i}\partial\psi/\partial t = (-\mathrm{i}\boldsymbol{\alpha}\cdot\boldsymbol{\nabla} + \beta m)\psi \tag{4.41}$$

using the explicit $\boldsymbol{\alpha}$ and β matrices. In order to use the 2×2 block form, it is conventional (and convenient) to split the spinor ω into two two-component spinors ϕ and χ:

$$\omega = \begin{pmatrix} \phi \\ \chi \end{pmatrix}. \tag{4.42}$$

We obtain the matrix equation (see problem 4.3)

$$E\begin{pmatrix} \phi \\ \chi \end{pmatrix} = \begin{pmatrix} m\mathbf{1} & \boldsymbol{\sigma}\cdot\boldsymbol{p} \\ \boldsymbol{\sigma}\cdot\boldsymbol{p} & -m\mathbf{1} \end{pmatrix}\begin{pmatrix} \phi \\ \chi \end{pmatrix} \tag{4.43}$$

representing two coupled equations for ϕ and χ:

$$(E - m)\phi = \boldsymbol{\sigma}\cdot\boldsymbol{p}\chi \tag{4.44}$$

and

$$(E + m)\chi = \boldsymbol{\sigma}\cdot\boldsymbol{p}\phi. \tag{4.45}$$

Solving for χ from (4.45), the general four-component spinor may be written (without worrying about normalization for the moment)

$$\omega = \begin{pmatrix} \phi \\ \dfrac{\boldsymbol{\sigma} \cdot \boldsymbol{p}}{E + m}\phi \end{pmatrix}. \tag{4.46}$$

What is the relation between E and $\boldsymbol{p}$ for this to be a solution of the Dirac equation? If we substitute χ from (4.45) into (4.44) and remember that (problem 4.4)

$$(\boldsymbol{\sigma} \cdot \boldsymbol{p})^2 = \boldsymbol{p}^2 \mathbf{1} \tag{4.47}$$

we find that

$$(E - m)(E + m)\phi = \boldsymbol{p}^2 \phi \tag{4.48}$$

for any ϕ. Hence we arrive at the same result as for the KG equation in that for a given value of $\boldsymbol{p}$, two values of E are allowed:

$$E = \pm(\boldsymbol{p}^2 + m^2)^{1/2} \tag{4.49}$$

i.e. positive *and* negative solutions are still admitted.

The Dirac equation does not therefore solve this problem. What about the probability current?

4.2.2 Probability current for the Dirac equation

Consider the following quantity which we denote (suggestively) by ρ:

$$\rho = \psi^\dagger(x)\psi(x). \tag{4.50}$$

Here $\psi^\dagger$ is the Hermitian conjugate row vector of the column vector ψ. In terms of components

$$\rho = (\psi_1^*, \psi_2^*, \psi_3^*, \psi_4^*) \begin{pmatrix} \psi_1 \\ \psi_2 \\ \psi_3 \\ \psi_4 \end{pmatrix} \tag{4.51}$$

so

$$\rho = \sum_{a=1}^{4} |\psi_a|^2 > 0 \tag{4.52}$$

and we see that ρ is a scalar density which is explicitly positive-definite. This is one property we require of a probability density: in addition, we require a conservation law, coming from the Dirac equation, and a corresponding probability current density. In fact (see problem 4.11) we can demonstrate, using the Dirac equation,

$$\mathrm{i}\partial\psi/\partial t = (-\mathrm{i}\boldsymbol{\alpha} \cdot \nabla + \beta m)\psi \tag{4.53}$$

and its Hermitian conjugate

$$-\mathrm{i}\partial\psi^\dagger = \psi^\dagger(+\mathrm{i}\boldsymbol{\alpha}\cdot\overleftarrow{\nabla} + \beta m) \tag{4.54}$$

that there is a conservation law of the required form

$$\partial\rho/\partial t + \nabla\cdot\boldsymbol{j} = 0. \tag{4.55}$$

The notation $\psi^\dagger\overleftarrow{\nabla}$ requires some comment: it is shorthand for three row matrices

$$\psi^\dagger\overleftarrow{\nabla}_x \equiv \partial\psi^\dagger/\partial x \qquad \text{etc}$$

(recall that $\psi^\dagger$ is a row matrix).

In equation (4.55), with ρ being given by (4.50), the probability current density $\boldsymbol{j}$ is

$$\boldsymbol{j}(x) = \psi^\dagger(x)\boldsymbol{\alpha}\psi(x) \tag{4.56}$$

representing a 3-vector with components

$$(\psi^\dagger\alpha_1\psi, \psi^\dagger\alpha_2\psi, \psi^\dagger\alpha_3\psi). \tag{4.57}$$

We therefore have a positive-definite ρ and an associated $\boldsymbol{j}$ satisfying the required conservation law (4.55), which, as usual, we can write in invariant form as $\partial_\mu j^\mu = 0$, where

$$j^\mu = (\rho, \boldsymbol{j}). \tag{4.58}$$

Thus j^μ is an acceptable probability current, unlike the current for the KG equation—as we might have anticipated.

The form of equation (4.55) implies that j^μ of (4.58) is a contravariant 4-vector (cf equation (D.4)), as we verified explicitly in the KG case. The corresponding verification is more difficult in the Dirac case, since the Dirac spinor ψ transforms non-trivially under Lorentz transformations, unlike the KG wavefunction ϕ. We shall come back to this problem in section 4.4.

We now turn to further discussion of the spin degree of freedom, postponing consideration of the negative-energy solutions until section 4.5.

4.3 Spin

Four-momentum is not the only physical property of a particle obeying the Dirac equation. We must now interpret the column vector (Dirac spinor) part, ω, of the solution (4.40). The particular properties of the $\boldsymbol{\sigma}$-matrices, appearing in the $\boldsymbol{\alpha}$-matrices, have already led us to think in terms of spin. A further indication that this is correct comes when we consider the explicit form of ω given in (4.46). In this equation the two-component spinor ϕ is completely arbitrary. It may be chosen in just two linearly independent ways, for example

$$\phi_\uparrow = \begin{pmatrix}1\\0\end{pmatrix} \qquad \phi_\downarrow = \begin{pmatrix}0\\1\end{pmatrix} \tag{4.59}$$

which (as the notation of course indicates) are in fact eigenvectors of $S_z = \frac{1}{2}\sigma_z$ with eigenvalues $\pm\frac{1}{2}$ ('up' and 'down' along the z-axis). Remember that, in quantum mechanics, linear combinations of wavefunctions can be formed using complex numbers as superposition coefficients, in general; so the most general ϕ can always be written as

$$\phi = \begin{pmatrix} a \\ b \end{pmatrix} = a\phi_\uparrow + b\phi_\downarrow \tag{4.60}$$

where a and b are complex numbers. Hence, there are precisely *two* linearly independent solutions, for a given 4-momentum, just as we would expect for a quantum system with $j = \frac{1}{2}$ (the multiplicity is $2j + 1$, in general).

In the rest frame of the particle ($\boldsymbol{p} = \mathbf{0}$) this interpretation is straightforward. In this case choosing (4.59) for the two independent ϕ's, the solutions (4.60) for $E = m$ reduce to

$$\underset{\text{(a)}}{\begin{pmatrix} 1 \\ 0 \\ 0 \\ 0 \end{pmatrix} \mathrm{e}^{-\mathrm{i}mt}} \quad \text{and} \quad \underset{\text{(b)}}{\begin{pmatrix} 0 \\ 1 \\ 0 \\ 0 \end{pmatrix} \mathrm{e}^{-\mathrm{i}mt}}. \tag{4.61}$$

Since we have degeneracy between these two solutions (both have $E = m$) there must be some operator which commutes with the energy operator, and whose eigenvalues would distinguish the solutions (4.61). In this case the energy operator is just βm (from (4.53) setting $-\mathrm{i}\nabla$ to zero, since $\boldsymbol{p} = \mathbf{0}$) and the required operator commuting with β is

$$\Sigma_z = \begin{pmatrix} \sigma_z & 0 \\ 0 & \sigma_z \end{pmatrix} \tag{4.62}$$

which has eigenvalues 1 (twice) and -1 (twice). Our rest-frame spinors appearing in (4.61) are indeed eigenstates of Σ_z, with eigenvalues ± 1 as can be easily verified.

Generalizing (4.62), we introduce the three matrices $\boldsymbol{\Sigma}$ where

$$\boldsymbol{\Sigma} = \begin{pmatrix} \boldsymbol{\sigma} & \mathbf{0} \\ \mathbf{0} & \boldsymbol{\sigma} \end{pmatrix}. \tag{4.63}$$

Then the operators $\frac{1}{2}\boldsymbol{\Sigma}$ are such that

$$[\tfrac{1}{2}\Sigma_x, \tfrac{1}{2}\Sigma_y] = \mathrm{i}\tfrac{1}{2}\Sigma_z \tag{4.64}$$

and $(\frac{1}{2}\boldsymbol{\Sigma})^2 = \frac{3}{4}\mathbf{I}$ where $\mathbf{I}$ is now the unit 4×4 matrix. These are just the properties expected of quantum-mechanical angular momentum operators (see appendix A) belonging to magnitude $j = \frac{1}{2}$ (we already know that the eigenvalues of $\frac{1}{2}\Sigma_z$ are

$\pm\frac{1}{2}$). So we can interpret $\frac{1}{2}\boldsymbol{\Sigma}$ as spin-$\frac{1}{2}$ operators appropriate to our rest-frame solutions; and—at least in the rest frame—we may say that the Dirac equation describes a particle of spin-$\frac{1}{2}$.

It seems reasonable to suppose that the magnitude of a spin of a particle could not be changed by doing a Lorentz transformation, as would be required in order to discuss the spin in a general frame with $\boldsymbol{p} \neq \boldsymbol{0}$. But $\frac{1}{2}\boldsymbol{\Sigma}$ is then no longer a suitable spin operator, since it fails to commute with the energy operator, which is now $(\boldsymbol{\alpha}\cdot\boldsymbol{p}+\beta m)$ from (4.53), for a plane-wave solution with momentum $\boldsymbol{p}$. Yet there are still just two independent states for a given 4-momentum as our explicit solution (4.46) shows: ϕ can still be chosen in only two linearly independent ways. Hence there must be some operator which does commute with $\boldsymbol{\alpha}\cdot\boldsymbol{p}+\beta m$, and whose eigenvalues can be used to distinguish the two states. Actually this condition is not enough to specify such an operator uniquely, and several choices are common. One of the most useful is the *helicity* operator $h(\boldsymbol{p})$ defined by

$$h(\boldsymbol{p}) = \begin{pmatrix} \dfrac{\boldsymbol{\sigma}\cdot\boldsymbol{p}}{|\boldsymbol{p}|} & \boldsymbol{0} \\ \boldsymbol{0} & \dfrac{\boldsymbol{\sigma}\cdot\boldsymbol{p}}{|\boldsymbol{p}|} \end{pmatrix} \tag{4.65}$$

which (see problem 4.5) does commute with $\boldsymbol{\alpha}\cdot\boldsymbol{p}+\beta m$. We can therefore choose our general $\boldsymbol{p} \neq \boldsymbol{0}$ states to be eigenstates of $h(\boldsymbol{p})$. These will be called 'helicity states': physically they are eigenstates of $\boldsymbol{\Sigma}$ resolved along the direction of $\boldsymbol{p}$.

Using (4.47) it is easy to see that the eigenvalues of $h(\boldsymbol{p})$ are $+1$ (twice) and -1 (twice). Our general four-component spinor (4.46) is therefore an eigenstate of $h(\boldsymbol{p})$ if

$$\begin{pmatrix} \dfrac{\boldsymbol{\sigma}\cdot\boldsymbol{p}}{|\boldsymbol{p}|} & \boldsymbol{0} \\ \boldsymbol{0} & \dfrac{\boldsymbol{\sigma}\cdot\boldsymbol{p}}{|\boldsymbol{p}|} \end{pmatrix} \begin{pmatrix} \phi \\ \dfrac{\boldsymbol{\sigma}\cdot\boldsymbol{p}}{E+m}\phi \end{pmatrix} = \pm \begin{pmatrix} \phi \\ \dfrac{\boldsymbol{\sigma}\cdot\boldsymbol{p}}{E+m}\phi \end{pmatrix}. \tag{4.66}$$

Taking the $+$ sign first, this will hold if

$$\frac{\boldsymbol{\sigma}\cdot\boldsymbol{p}}{|\boldsymbol{p}|}\phi_+ = \phi_+ \tag{4.67}$$

where the $+$ subscript has been added to indicate that this ϕ is a solution of (4.67). Such a ϕ_+ is called a two-component helicity spinor. The explicit form of ϕ_+ can be found by solving (4.67)—see problem 4.6. Similarly, the four-component spinor will be an eigenstate of $h(\boldsymbol{p})$ belonging to the eigenvalue -1 if it contains ϕ_- where

$$\frac{\boldsymbol{\sigma}\cdot\boldsymbol{p}}{|\boldsymbol{p}|}\phi_- = -\phi_-. \tag{4.68}$$

Again, these two choices ϕ_+ and ϕ_- are linearly independent.

After all this discussion about how the Dirac equation describes particles of spin-$\frac{1}{2}$, one might well ask why the Dirac spinor has *four* components and not two, as might have been expected from the $(2j+1)$-multiplicity, with $j = \frac{1}{2}$. We shall return to this interesting question at the end of section 4.4.

4.4 Lorentz transformation properties of spinors

In quantum mechanics, there is an intimate connection between the angular momentum associated with a wavefunction, and the way the wavefunction transforms under rotations of the coordinate system. To take a simple case, the 2p wavefunctions mentioned in section 4.2 correspond to $l = 1$ on the one hand and, on the other, to the components of a vector—indeed the most basic vector of all, the position vector $\boldsymbol{r} = (x, y, z)$ itself. Now if, for example, we decide to rotate the coordinate system by an angle α in a positive sense about the x-axis, then the position vector referred to the new system is $\boldsymbol{r}' = (x', y', z')$ where

$$\begin{pmatrix} x' \\ y' \\ z' \end{pmatrix} = \begin{pmatrix} 1 & 0 & 0 \\ 0 & \cos\alpha & \sin\alpha \\ 0 & -\sin\alpha & \cos\alpha \end{pmatrix} \begin{pmatrix} x \\ y \\ z \end{pmatrix}. \tag{4.69}$$

Thus the components in the new system transform into simple linear combinations of the components in the old system.

Very much the same thing happens in the case of spinor wavefunctions, except that they transform in a way different from—though closely related to—that of vectors. In the present section we shall discuss how this works for three-dimensional rotations of the spatial coordinate system, and explain how it generalizes to four-dimensional Lorentz transformations which include transformations of the time coordinate as well.

We begin by considering the (simpler) two-component helicity spinors ϕ_+ and ϕ_- defined by (4.67) and (4.68). Suppose we start by describing the states using a particular coordinate system with the z-axis aligned along the direction of $\boldsymbol{p}$, so that $\boldsymbol{p} = (0, 0, p)$. Then a normalized ϕ_+ is simply $\binom{1}{0}$ and a normalized ϕ_- is $\binom{0}{1}$. Now rotate the coordinate system by a positive angle α about the x-axis. Since $\boldsymbol{p}$ is a vector, it transforms just the same way as $\boldsymbol{r}$, namely (from (4.69)) to

$$\begin{aligned} p'_x &= p_x = 0 \\ p'_y &= \cos\alpha p_y + \sin\alpha p_z = p\sin\alpha \\ p'_z &= -\sin\alpha p_y + \cos\alpha p_z = p\cos\alpha. \end{aligned} \tag{4.70}$$

In this new coordinate system, the spinor ϕ'_+ has to satisfy

$$\frac{\boldsymbol{\sigma}\cdot\boldsymbol{p}'}{|\boldsymbol{p}'|}\phi'_+ = \phi'_+ \tag{4.71}$$

since it must still represent the helicity $+1$ state; the state is merely being described in a different coordinate system. In demanding (4.71), we are once more invoking a 'covariance' requirement (see appendix D), which we met before in the context of Lorentz transformations and Maxwell's equations (section 3.3), and of gauge transformations and Schrödinger's equation (section 3.4). This time the situation is rather similar to the Schrödinger case, where we used the (gauge) covariance requirement to find the form of the gauge-transformed wavefunction ψ'. Here, the transformations are the familiar rotations but—as with that ψ'—we shall use (4.71) to determine what the form of ϕ'_+ must be. (In the Maxwell case, we used the known transformation properties of the quantities involved to establish the covariance of the equations.)

It is straightforward to find ϕ'_+ from (4.71): from (4.70) we know $\boldsymbol{p}'$ in terms of $\boldsymbol{p}$, hence we can solve (4.71) explicitly for ϕ'_+. Remembering that $|\boldsymbol{p}'| = |\boldsymbol{p}|$, we see from (4.71) that we need to solve

$$\begin{pmatrix} \cos\alpha & -\mathrm{i}\sin\alpha \\ \mathrm{i}\sin\alpha & -\cos\alpha \end{pmatrix} \begin{pmatrix} a \\ b \end{pmatrix} = \begin{pmatrix} a \\ b \end{pmatrix} \tag{4.72}$$

where we have written

$$\phi'_+ = \begin{pmatrix} a \\ b \end{pmatrix}. \tag{4.73}$$

It is simple to verify (problem 4.7) that a normalized solution (i.e. $\phi'^{\dagger}_+\phi'_+ = 1$) of (4.72) is

$$\phi'_+ = \begin{pmatrix} \cos\alpha/2 \\ \mathrm{i}\sin\alpha/2 \end{pmatrix}. \tag{4.74}$$

The result (4.74) may be rewritten, in a form analogous to (4.69), as

$$\phi'_+ = \begin{pmatrix} \cos\alpha/2 & \mathrm{i}\sin\alpha/2 \\ \mathrm{i}\sin\alpha/2 & \cos\alpha/2 \end{pmatrix} \phi_+. \tag{4.75}$$

In a similar way, one can verify (problem 4.7) that a normalized negative helicity spinor in the rotated coordinate system, ϕ'_-, is given by

$$\phi'_- = \begin{pmatrix} \mathrm{i}\sin\alpha/2 \\ \cos\alpha/2 \end{pmatrix} = \begin{pmatrix} \cos\alpha/2 & \mathrm{i}\sin\alpha/2 \\ \mathrm{i}\sin\alpha/2 & \cos\alpha/2 \end{pmatrix} \phi_-. \tag{4.76}$$

These ϕ'_+ and ϕ'_- preserve the orthogonality of the original ϕ_+ and ϕ_- in the sense that

$$\phi'^{\dagger}_-\phi'_+ = 0. \tag{4.77}$$

From (4.75) and (4.76) we can therefore see that—just as in the (vector) $l = 1$ case—the components of the spinor wavefunctions in the rotated coordinate system are linear combinations of the corresponding components in the original system, as determined by a certain matrix. In the spinor case this matrix is 2×2 rather than 3×3 as in the vector case, since the wavefunctions (spinors) have just two components. The entries in the matrix again depend on the rotation angle, but in the spinor case—interestingly—it is the half-angle $\alpha/2$ that enters.

The results (4.75) and (4.76) can both be put into a suggestive compact form:

$$\phi' = \mathrm{e}^{\mathrm{i}\sigma_x \alpha/2}\phi \tag{4.78}$$

where the exponential of the matrix is understood as the infinite series

$$\mathrm{e}^{\mathrm{i}\sigma_x \alpha/2} = 1 + \mathrm{i}\sigma_x\alpha/2 + \tfrac{1}{2}(\mathrm{i}\sigma_x\alpha/2)^2 + \cdots. \tag{4.79}$$

Multiplying out the terms in (4.79) and remembering that $\sigma_x^2 = 1$, we see that

$$\mathrm{e}^{\mathrm{i}\sigma_x \alpha/2} = \cos\alpha/2 + \mathrm{i}\sigma_x \sin\alpha/2 = \begin{pmatrix} \cos\alpha/2 & \mathrm{i}\sin\alpha/2 \\ \mathrm{i}\sin\alpha/2 & \cos\alpha/2 \end{pmatrix} \tag{4.80}$$

from which (4.78) easily follows.

Equation (4.78) therefore summarizes very neatly how the two-component helicity spinors transform under this rotation of α about the x-axis. For an arbitrary rotation through an angle θ about an axis specified by the unit vector $\hat{\boldsymbol{n}}$, ϕ and ϕ' are related by

$$\phi' = \mathrm{e}^{\mathrm{i}\boldsymbol{\sigma}\cdot\hat{\boldsymbol{n}}\theta/2}\phi \equiv \mathbf{U}\phi \tag{4.81}$$

where the matrix $\mathbf{U}$ is a unitary matrix satisfying $\mathbf{U}^\dagger\mathbf{U} = \mathbf{U}\mathbf{U}^\dagger = \mathbf{1}$. This property guarantees the preservation of the normalization of ϕ under rotations: $\phi^\dagger\phi = \phi'^\dagger\phi'$. Note that $\mathbf{U}$ is determined by the quantities $\hat{\boldsymbol{n}}$ and θ which specify the rotation in question. Although we have approached the determination of ϕ' by considering the particular case of helicity eigenstates, the transformation (4.81) holds for *any* two-component spinor. We may therefore add *spinors* to our tally (invariants, vectors, tensors) of 'things with well-defined transformation character under rotations'.

Now we have to consider the four-component Dirac spinor ω, equation (4.46). If, for definiteness, we again consider the 'helicity' choice $\phi = \phi_\pm$, ω becomes just

$$\omega_\pm = \begin{pmatrix} \phi_\pm \\ \pm\left(\dfrac{|\boldsymbol{p}|}{E+m}\right)\phi_\pm \end{pmatrix} \tag{4.82}$$

from which we see at once that the lower two components have—as far as their spinor nature is concerned—exactly the same rotational transformation properties as the top two (namely as $\phi_\pm$), since the factor multiplying $\phi_\pm$ in the lower half of (4.82) is a rotational invariant. Thus the generalization of (4.81) to the four-component case is

$$\omega' = \mathrm{e}^{\mathrm{i}\boldsymbol{\Sigma}\cdot\hat{\boldsymbol{n}}\theta/2}\omega \tag{4.83}$$

with $\boldsymbol{\Sigma}$ as in (4.63), showing that ω' transforms by a 4×4 unitary matrix. Once again, (4.83) is true for any choice of ϕ. This is an important result since, as we shall immediately show, it implies that the Dirac probability density is *invariant* under rotations, as we would surely hope. Placing the solution (4.40) into $\rho = \psi^\dagger \psi$, we see that $\rho = \omega^\dagger \omega$, since the plane-wave factor cancels out. Thus

$$\begin{aligned}\rho' = \omega'^\dagger \omega' &= \omega^\dagger \exp(-\mathrm{i}\boldsymbol{\Sigma} \cdot \hat{\boldsymbol{n}}\theta/2) \exp(\mathrm{i}\boldsymbol{\Sigma} \cdot \hat{\boldsymbol{n}}\theta/2)\omega \\ &= \omega^\dagger \omega = \rho \end{aligned} \tag{4.84}$$

establishing the invariance of ρ under rotations.

More interestingly, we can proceed to examine the properties of the probability current density $\boldsymbol{j}$ of (4.56) under rotations. Of course, it should behave as a vector, and this is checked in problem 4.8.

Finally we come to the behaviour of the Dirac spinor under Lorentz transformations which mix t with the spatial coordinates. As in the case of spatial rotations, we shall approach this by considering a particular transformation and then giving the generalization. The idea is very simple. We know the form of the spinor in the particle's rest frame (equation (4.61)); and we also know it in a frame where the particle has momentum $\boldsymbol{p}$ (equation (4.46)). But these two frames are related by a Lorentz transformation—hence we can deduce the corresponding transformation that takes the rest frame spinor to the moving frame one[2].

Let us call S the frame in which the particle is at rest, with $E = m$ and $\boldsymbol{p} = \boldsymbol{0}$, and let S' be the frame in which the particle has momentum $\boldsymbol{p}' = (p'_x, 0, 0)$ and energy E'. Thus the required Lorentz transformation is one in which S' has velocity $(-v'_x, 0, 0)$ relative to S, where $v'_x = p'_x/E'$. It is convenient to give the general form of this transformation in terms of the transforms of the 4-vector $(E, \boldsymbol{p})$, rather than $(t, \boldsymbol{x})$, namely

$$E' = \cosh\vartheta\, E + \sinh\vartheta\, p_x \tag{4.85}$$

$$p'_x = \sinh\vartheta\, E + \cosh\vartheta\, p_x \tag{4.86}$$

where $\tanh\vartheta = v'_x = p'_x/E$ and $E' = (p'^2_x + m^2)^{1/2}$. In the case we are considering, (4.85) and (4.86) just reconstruct $E' = \gamma' m$ and $p'_x = \gamma' m v'_x$, where $\gamma' = (1 - v'^2_x)^{-1/2}$. Now consider the forms of the spinors in S and S'. Let us take the simple case in which the two-component 'ϕ' in (4.60) is just

$$\phi = \begin{pmatrix} 1 \\ 0 \end{pmatrix}. \tag{4.87}$$

Then the spinor in S is

$$\omega = \begin{pmatrix} 1 \\ 0 \\ 0 \\ 0 \end{pmatrix} \tag{4.88}$$

[2] Readers who may be wondering why we do not make use of another covariance argument, this time for the full four-component Dirac equation under Lorentz transformations, may turn to problem 4.15.

as in (4.61(a)), and in S' is

$$\omega' = N \begin{pmatrix} \begin{pmatrix} 1 \\ 0 \end{pmatrix} \\ \dfrac{p'_x}{E+m} \begin{pmatrix} 0 \\ 1 \end{pmatrix} \end{pmatrix} \tag{4.89}$$

from (4.46), using

$$\sigma_x \begin{pmatrix} 1 \\ 0 \end{pmatrix} = \begin{pmatrix} 0 \\ 1 \end{pmatrix}.$$

Note that we have included a normalization constant N, for a reason which will soon be clear.

We need to understand how (4.88) and (4.89) are related. First, note that the transformations (4.85) and (4.86) are like rotations through a purely imaginary angle ($\vartheta \rightarrow \mathrm{i}\theta$, $\cosh \mathrm{i}\theta = \cos\theta$ and $\sinh \mathrm{i}\theta = \mathrm{i}\sin\theta$). Second, it is clear from (4.88) and (4.89) that the upper two-component spinor ϕ and the lower two-component spinor χ must transform differently under such 'velocity' transformations (even though they behaved the same under rotations), because E and $\boldsymbol{p}$ enter differently in these coupled equations. Thus we expect that, as compared to the rotation case (4.78), we shall see (a) a purely imaginary angle and (b) a matrix that couples the ϕ piece to the χ piece. The answer is

$$\omega' = \mathrm{e}^{\alpha_x \vartheta/2}\omega \tag{4.90}$$

as we shall now verify. Expanding out the exponential as in (4.79), we find

$$\mathrm{e}^{\alpha_x \vartheta/2} = \cosh \vartheta/2 + \alpha_x \sinh \vartheta/2. \tag{4.91}$$

A small amount of algebra reveals that

$$\begin{aligned} \cosh \vartheta/2 &= \left(\frac{E'+m}{2m}\right)^{1/2} \\ \sinh \vartheta/2 &= \frac{p'_x}{[2m(E'+m)]^{1/2}}. \end{aligned} \tag{4.92}$$

It is then simple to verify (problem 4.9) that (4.88) and (4.89) are indeed related by (4.90), where the normalization constant is given by

$$N = \left(\frac{E'+m}{2m}\right)^{1/2}. \tag{4.93}$$

So now *four-component spinors* and their transformation character under Lorentz transformations can be added to our tally.

The appearance of this particular N is not an accident. First note that, in contrast to (4.83), the transformation (4.90) is *not* a unitary one; indeed if we denote the matrix $\exp(\alpha_x \vartheta/2)$ by B_x, we have $B_x^\dagger = B_x$ rather that $B_x^\dagger = B_x^{-1}$. So B_x does not preserve normalization in the sense that $\omega^\dagger \omega \neq \omega'^\dagger \omega'$. But actually we would not want it to! We already know from section 4.2.2 that $\psi^\dagger \psi$ (which equals $\omega^\dagger \omega$ for these plane-wave solutions) ought to transform under Lorentz transformations as the fourth component ρ of the 4-vector $j^\mu = (\rho, \boldsymbol{j})$. We shall check this very shortly, but assuming it to be true, $\omega^\dagger \omega$ cannot be invariant under B_x. In fact, using (4.89) with N given by (4.93) we find

$$\omega'^\dagger \omega' = \frac{E'}{m} \tag{4.94}$$

which indeed shows the transformation character of the fourth component of a 4-vector. Equation (4.94) reduces to $\omega^\dagger \omega = 1$ in the rest frame, consistently with (4.88). (As we shall see later on, it is often more convenient to adopt a slightly different normalization, in which $\omega'^\dagger \omega' = 2E'$; this avoids trouble in the $m \to 0$ limit.) Incidentally, the quantity that *is* invariant under (4.90) is $\bar{\psi}\psi$ where $\bar{\psi} = \psi^\dagger \beta$ (see problem 4.10).

It is now simple to check that $\psi^\dagger \psi$ transforms as the fourth component of a 4-vector under the transformation (4.90). We have

$$\begin{aligned} \rho = \psi^\dagger \psi \to \psi'^\dagger \psi' &= \psi^\dagger \mathrm{e}^{\alpha_x \vartheta/2} \mathrm{e}^{\alpha_x \vartheta/2} \phi \\ &= \psi^\dagger \mathrm{e}^{\alpha_x \vartheta} \psi \\ &= \psi^\dagger (\cosh \vartheta + \alpha_x \sinh \vartheta) \psi, \end{aligned}$$

i.e.

$$\rho' = \cosh \vartheta \rho + \sinh \vartheta j_x$$

exactly as in (4.85), with $j_x = \psi^\dagger \alpha_x \psi$ as required. The reader may verify that $\boldsymbol{j} = \psi^\dagger \boldsymbol{\alpha} \psi$ transforms correctly as the spatial part of a 4-vector, under (4.90).

At this point it is beginning to be clear that a more 'covariant-looking' (see appendix D) notation would be very desirable. In the case of the KG probability current, the 4-vector index μ was clearly visible in the expression on the right-hand side of (4.20), but there is nothing similar in the Dirac case so far. In problem 4.11 the four 'γ matrices' are introduced, defined by $\gamma^\mu = (\gamma^0, \boldsymbol{\gamma})$ with $\gamma^0 = \beta$ and $\boldsymbol{\gamma} = \beta \boldsymbol{\alpha}$, together with the quantity $\bar{\psi} \equiv \psi^\dagger \gamma^0$, in terms of which the Dirac ρ of (4.50) and $\boldsymbol{j}$ of (4.56) can be written as $\bar{\psi}(x)\gamma^0 \psi(x)$ and $\bar{\psi}(x)\boldsymbol{\gamma}\psi(x)$ respectively. The complete Dirac 4-current is then (see also problem 4.11)

$$j^\mu = \bar{\psi}(x) \gamma^\mu \psi(x). \tag{4.95}$$

In this form the index 'μ' is *manifest*, but of course this is hardly a proof that j^μ is indeed a 4-vector: the reader will have a chance to check this important result in problem 8.4.

The generalization of (4.90) is

$$\omega' = \exp\left(\frac{\boldsymbol{\alpha} \cdot \boldsymbol{v}}{|\boldsymbol{v}|} \vartheta/2 \right) \omega \tag{4.96}$$

for a transformation from S to S' which is moving with velocity $-\boldsymbol{v}$ with respect to S, where $\tanh\vartheta = |\boldsymbol{v}|$. Equation (4.96) should be compared with (4.83).

Before leaving this section, we return to the question raised at the end of section 4.3: why does the Dirac spinor need four-components to describe a spin-$\frac{1}{2}$ particle? The answer has to do with another type of coordinate transformation—namely *parity*. This is most easily seen by employing an alternative representation of the $\boldsymbol{\alpha}$ and β matrices in which

$$\boldsymbol{\alpha} = \begin{pmatrix} \boldsymbol{\sigma} & \mathbf{0} \\ \mathbf{0} & -\boldsymbol{\sigma} \end{pmatrix} \qquad \beta = \begin{pmatrix} \mathbf{0} & \mathbf{1} \\ \mathbf{1} & \mathbf{0} \end{pmatrix}. \tag{4.97}$$

In this representation (4.44) and (4.45) become, instead,

$$E\phi = \boldsymbol{\sigma}\cdot\boldsymbol{p}\phi + m\chi \tag{4.98}$$
$$E\chi = -\boldsymbol{\sigma}\cdot\boldsymbol{p}\chi + m\phi. \tag{4.99}$$

Under parity, $\boldsymbol{p} \to -\boldsymbol{p}$. Thus in the parity-transformed frame (4.98) and (4.99) are

$$E\phi' = -\boldsymbol{\sigma}\cdot\boldsymbol{p}\phi' + m\chi' \tag{4.100}$$
$$E\chi' = \boldsymbol{\sigma}\cdot\boldsymbol{p}\chi' + m\phi' \tag{4.101}$$

from which, comparing with (4.98) and (4.99), we can deduce $\phi' = \chi$ and $\chi' = \phi$. Thus the two-component spinors ϕ and χ are interchanged (in this representation) under parity and in a relativistic theory we would not be able to represent the effect of parity satisfactorily with only one of them.

4.5 The negative-energy solutions

In this section we shall first look more closely at the form of both the positive- and negative-energy solutions of the Dirac equation, and we shall then concentrate on the physical interpretation of the negative-energy solutions of both the Dirac and the KG equations.

It will be convenient, from now on, to reserve the symbol 'E' for the *positive* square root in (4.49): $E = +(\boldsymbol{p}^2 + m^2)$. The general 4-momentum in the plane-wave solution (4.40) will be denoted by $p^\mu = (p^0, \boldsymbol{p})$ where p^0 may be either positive or negative. With this notation equation (4.43) becomes

$$p^0\begin{pmatrix} \phi \\ \chi \end{pmatrix} = \begin{pmatrix} m\mathbf{1} & \boldsymbol{\sigma}\cdot\boldsymbol{p} \\ \boldsymbol{\sigma}\cdot\boldsymbol{p} & -m\mathbf{1} \end{pmatrix}\begin{pmatrix} \phi \\ \chi \end{pmatrix}. \tag{4.102}$$

in our original representation for $\boldsymbol{\alpha}$ and β.

4.5.1 Positive-energy spinors

For these

$$p^0 = +(\boldsymbol{p}^2 + m^2)^{1/2} \equiv E > 0. \tag{4.103}$$

We eliminate χ and obtain positive-energy spinors in the form

$$\omega^{1,2} = N \begin{pmatrix} \phi^{1,2} \\ \dfrac{\boldsymbol{\sigma}\cdot\boldsymbol{p}}{E+m}\phi^{1,2} \end{pmatrix}. \tag{4.104}$$

We shall now choose N so that for these positive-energy solutions $\omega^\dagger\omega = 2E$. In this case the spinors will be denoted by $u(p, s)$, where (problem 4.12)

$$u(p,s) = (E+m)^{1/2} \begin{pmatrix} \phi^s \\ \dfrac{\boldsymbol{\sigma}\cdot\boldsymbol{p}}{E+m}\phi^s \end{pmatrix} \qquad s = 1, 2 \tag{4.105}$$

and s labels the spin degree of freedom in some suitable way (e.g. the helicity eigenvalues). The complete plane-wave solution ψ for such a positive 4-momentum state is then

$$\psi = u(p,s)\mathrm{e}^{-\mathrm{i}p_+\cdot x} \tag{4.106}$$

with $p_+^\mu = (E, \boldsymbol{p})$.

4.5.2 Negative-energy spinors

Now we look for spinors appropriate to the solution

$$p^0 = -(\boldsymbol{p}^2 + m^2)^{1/2} \equiv -E < 0 \tag{4.107}$$

(E is always defined to be positive). Consider first what are appropriate solutions at rest. We have now

$$p^0 = -m \qquad \boldsymbol{p} = \boldsymbol{0} \tag{4.108}$$

and

$$-m\begin{pmatrix} \phi \\ \chi \end{pmatrix} = \begin{pmatrix} m\mathbf{1} & \mathbf{0} \\ \mathbf{0} & -m\mathbf{1} \end{pmatrix}\begin{pmatrix} \phi \\ \chi \end{pmatrix} \tag{4.109}$$

leading to

$$\phi = 0. \tag{4.110}$$

Thus the two independent negative-energy solutions at rest are just

$$\omega(p^0 = -m, s) = \begin{pmatrix} 0 \\ \chi^s \end{pmatrix}. \tag{4.111}$$

The solution for finite momentum $+\boldsymbol{p}$, i.e. for 4-momentum $(-E, \boldsymbol{p})$, is then

$$\omega(p^0 = -E, \boldsymbol{p}, s) = \begin{pmatrix} \dfrac{-\boldsymbol{\sigma}\cdot\boldsymbol{p}}{E+m}\chi^s \\ \chi^s \end{pmatrix}. \tag{4.112}$$

However, it is clearly much more in keeping with relativity if, in addition to changing the sign of E, we also change the sign of $\boldsymbol{p}$ and consider solutions corresponding to negative 4-momentum $(-E, -\boldsymbol{p}) = -p_+^\mu$. We therefore define

$$\omega(p^0 = -E, -\boldsymbol{p}, s) \equiv \omega^{3,4} = N \begin{pmatrix} \dfrac{\boldsymbol{\sigma} \cdot \boldsymbol{p}}{E+m} \chi^{1,2} \\ \chi^{1,2} \end{pmatrix}. \tag{4.113}$$

Adopting the same N as in (4.105) implies the same normalization ($\omega^\dagger \omega = 2E$) for (4.113) as in (4.105); in this case the spinors are called $v(p, s)$ where (problem 4.12)

$$v(p, s) = (E+m)^{1/2} \begin{pmatrix} \dfrac{\boldsymbol{\sigma} \cdot \mathbf{p}}{E+m} \chi^s \\ \chi^s \end{pmatrix} \qquad s = 1, 2. \tag{4.114}$$

(There is a small subtlety in the choice of χ^1 and χ^2 which we will come to shortly.) The solution ψ for such negative 4-momentum states is then

$$\psi = v(p, s)\mathrm{e}^{-\mathrm{i}(-p_+)\cdot x} = v(p, s)\mathrm{e}^{\mathrm{i}p_+\cdot x}. \tag{4.115}$$

4.5.3 Dirac's interpretation of the negative-energy solutions of the Dirac equation

The physical interpretation of the positive-energy solution (4.106) is straightforward, in terms of the ρ and $\boldsymbol{j}$ given in section 4.2.2. They describe spin-$\frac{1}{2}$ particles with 4-momentum $(E, \boldsymbol{p})$ and spin appropriate to the choice of ϕ^s; ρ and the energy p^0 are both positive.

Unfortunately ρ is also positive for the *negative*-energy solutions (4.115), so we cannot eliminate them on that account. This means that for a free Dirac particle (e.g. an electron) the available positive- and negative-energy levels are as shown in figure 4.1. This, in turn, implies that a particle with initially positive energy can 'cascade down' through the negative-energy levels, without limit; in this case no stable positive-energy state would exist!

In order to prevent positive-energy electrons making transitions to the lower, negative-energy states, Dirac postulated that the normal 'empty', or 'vacuum', state—that with no positive-energy electrons present—is such that all the negative-energy states are filled with electrons. The Pauli exclusion principle then forbids any positive-energy electrons from falling into these lower energy levels. The 'vacuum' now has infinite negative charge and energy, but since all observations represent *finite* fluctuations in energy and charge with respect to this vacuum, this leads to an acceptable theory. For example, if one negative-energy electron is absent from the Dirac sea, we have a 'hole' relative to the normal vacuum:

$$\text{energy of 'hole'} = -(E_{\text{neg}}) \to \text{positive energy}$$
$$\text{charge of 'hole'} = -(q_{\text{e}}) \to \text{positive charge.}$$

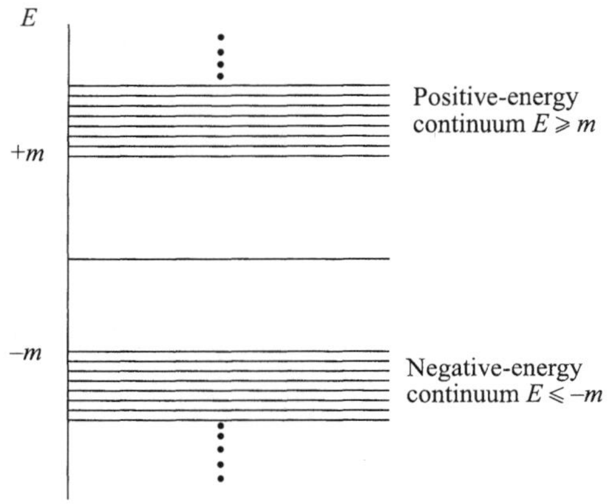

Figure 4.1. Energy levels for Dirac particle.

Thus the *absence* of a negative-energy electron is equivalent to the *presence* of a positive-energy positively charged version of the electron, that is a positron. In the same way, the absence of a 'spin-up' negative-energy electron is equivalent to the presence of a 'spin-down' positive-energy positron. This last point is the reason for the subtlety in the choice of χ^s mentioned after (4.114): we choose

$$\chi^1 = \begin{pmatrix} 0 \\ 1 \end{pmatrix} \qquad \chi^2 = \begin{pmatrix} 1 \\ 0 \end{pmatrix} \tag{4.116}$$

the opposite way round from the choice for the positive-energy spinors (4.105).

Dirac's brilliant re-interpretation of (unfilled) negative-energy solutions in terms of anti-particles is one of the triumphs of theoretical physics[3]: Carl Anderson received the Nobel Prize for his discovery of the positron in 1932 (Anderson 1932).

In this way it proved possible to obtain sensible results from the Dirac equation and its negative-energy solutions. It is clear, however, that the theory is no longer really a 'single-particle' theory, since we can excite electrons from the infinite 'sea' of filled negative-energy states that constitute the normal 'empty state'. For example, if we excite one negative-energy electron to a positive-energy state, we have in the final state a positive-energy electron plus a positive-energy positron 'hole' in the vacuum: this corresponds physically to the process of e^+e^- pair creation. Thus this way of dealing with the negative-energy problem for fermions leads us directly to the need for a quantum field theory. The appropriate formalism will be presented later in section 7.2.

[3] At that time, this was not universally recognized. For example, Pauli (1933) wrote: 'Dirac has tried to identify holes with anti-electrons. . . we do not believe that this explanation can be seriously considered.'

4.5.4 Feynman's interpretation of the negative-energy solutions of the KG and Dirac equations

It is clear that despite its brilliant success for spin-$\frac{1}{2}$ particles, Dirac's interpretation cannot be applied to spin-0 particles, since bosons are not subject to the exclusion principle. Besides, spin-0 particles also have their corresponding anti-particles (e.g. π^+ and π^-), and so do spin-1 particles (W^+ and W^-, for instance). A consistent picture for both bosons and fermions does emerge from quantum field theory, as we shall see in chapters 5–7, which is perhaps one of the strongest reasons for mastering it. Nevertheless, it is useful to have an alternative, non-field-theoretic, interpretation of the negative-energy solutions which works for both bosons and fermions. Such an interpretation is due to Feynman: in essence, the idea is that the negative 4-momentum solutions will be used to describe anti-particles, for both bosons and fermions.

We begin with bosons—for example pions, which for the present purposes we take to be simple spin-0 particles whose wavefunctions obey the KG equation. We decide by convention that the π^+ is the 'particle'. We will then have

$$\text{positive 4-momentum } \pi^+ \text{ solutions: } N\mathrm{e}^{-\mathrm{i}p\cdot x} \tag{4.117}$$
$$\text{negative 4-momentum } \pi^+ \text{ solutions: } N\mathrm{e}^{\mathrm{i}p\cdot x}. \tag{4.118}$$

where $p^\mu = [(m^2 + \boldsymbol{p}^2)^{1/2}, \boldsymbol{p}]$. Consider first the electromagnetic current for a physical (positive-energy) π^+. We shall assume (and justify in chapter 8) that it is given by the probability current for a positive-energy solution multiplied by the charge $Q(= +e)$:

$$j^\mu_{\text{em}}(\pi^+) = (+e) \times (\text{probability current for positive energy } \pi^+) \tag{4.119}$$
$$= (+e)2|N|^2[(m^2 + \boldsymbol{p}^2)^{1/2}, \boldsymbol{p}] \tag{4.120}$$

using (4.20) and (4.117) (see problem 4.1). What about the current for the π^-? For physical π^- particles of positive energy $(m^2 + \boldsymbol{p}^2)^{1/2}$ and momentum $\boldsymbol{p}$ we expect

$$j^\mu_{\text{em}}(\pi^-) = (-e)2|N|^2[(m^2 + \boldsymbol{p}^2)^{1/2}, \boldsymbol{p}] \tag{4.121}$$

by simply changing the sign of the charge in (4.120). But it is evident that (4.121) may be written as

$$j^\mu_{\text{em}}(\pi^-) = (+e)2|N|^2[-(m^2 + \boldsymbol{p}^2)^{1/2}, -\boldsymbol{p}] \tag{4.122}$$

which is just $j^\mu_{\text{em}}(\pi^+)$ with *negative* 4-momentum. This suggests some equivalence between anti-particle solutions with positive 4-momentum and particle solutions with negative 4-momentum.

Can we push this equivalence further? Consider what happens when a system A absorbs a π^+ with positive 4-momentum p: its charge increases by $+e$, and its 4-momentum increases by p. Now suppose that A emits a physical π^- with 4-momentum q, where the energy q^0 is positive. Then the charge of A

Figure 4.2. Coulomb scattering of a π^- by a static charge Ze illustrating the Feynman interpretation of negative 4-momentum states.

will increase by $+e$, and its 4-momentum will decrease by q. Now this increase in the charge of A could equally well be caused by the absorption of a π^+—and indeed we can make the effect (as far as A is concerned) of the π^- emission process fully equivalent to a π^+ absorption process if we say that the equivalent absorbed π^+ has negative 4-momentum, $-q$; in particular the equivalent absorbed π^+ has negative energy $-q^0$. In this way, we view the emission of a physical 'anti-particle' π^- with positive 4-momentum q as equivalent to the absorption of a 'particle' π^+ with (unphysical) negative 4-momentum $-q$. Similar reasoning will apply to the absorption of a π^- of positive 4-momentum, which is equivalent to the emission of a π^+ of negative 4-momentum. Thus we are led to the following hypothesis (due to Feynman):

> *The emission (absorption) of an anti-particle of 4-momentum p^μ is physically equivalent to the absorption (emission) of a particle of 4-momentum $-p^\mu$.*

In other words the unphysical negative 4-momentum solutions of the 'particle' equation do have a role to play: they can be used to describe physical processes involving positive 4-momentum anti-particles, if we reverse the role of 'entry' and 'exit' states.

The idea is illustrated in figure 4.2, for the case of Coulomb scattering of a π^- particle by a static charge Ze, which will be discussed later in section 8.1.3. By convention we are taking π^- to be the anti-particle. In the physical process of figure 4.2(*a*) the incoming physical anti-particle π^- has 4-momentum p_i, and the final π^- has 4-momentum p_f: both E_i and E_f are, of course, positive. Figure 4.2(*b*) shows how the amplitude for the process can be calculated using π^+ solutions with negative 4-momentum. The initial state π^- of 4-momentum p_i becomes a final state π^+ with 4-momentum $-p_i$, and similarly the final state π^- of 4-momentum p_f becomes an initial state π^+ of 4-momentum $-p_f$. Note that in this and similar figures, the sense of the arrows always indicates the 'flow' of 4-momentum, positive 4-momentum corresponding to forward flow.

It is clear that the basic physical idea here is not limited to bosons. But there is a difference between the KG and Dirac cases in that the Dirac equation was explicitly designed to yield a probability density (and probability current density) which was independent of the sign of the energy:

$$\rho = \psi^\dagger \psi \qquad \boldsymbol{j} = \psi^\dagger \boldsymbol{\alpha} \psi. \tag{4.123}$$

Thus for any solutions of the form

$$\psi = \omega \phi(\boldsymbol{x}, t) \tag{4.124}$$

we have

$$\rho = \omega^\dagger \omega |\phi(\boldsymbol{x}, t)|^2 \tag{4.125}$$

and

$$\boldsymbol{j} = \omega^\dagger \boldsymbol{\alpha} \omega |\phi(\boldsymbol{x}, t)|^2 \tag{4.126}$$

and $\rho \geq 0$ always. We nevertheless want to set up a correspondence so that positive-energy solutions describe *electrons* (taken to be the 'particle', by convention, in this case) and negative-energy solutions describe *positrons*, if we reverse the sense of incoming and outgoing waves. For the KG case this was straightforward, since the probability current was proportional to the 4-momentum:

$$j^\mu(\mathrm{KG}) \sim p^\mu. \tag{4.127}$$

We were therefore able to set up the correspondence for the electromagnetic current of π^+ and π^-:

$$\pi^+ : \quad j^\mu_{\mathrm{em}} \sim e p^\mu \qquad \text{positive energy } \pi^+ \tag{4.128}$$

$$\pi^- : \quad j^\mu_{\mathrm{em}} \sim (-e) p^\mu \qquad \text{positive energy } \pi^- \tag{4.129}$$

$$\equiv (+e)(-p^\mu) \qquad \text{negative energy } \pi^+. \tag{4.130}$$

This simple connection does not hold for the Dirac case since $\rho \geq 0$ for both signs of the energy. It is still possible to set up the correspondence, but now an extra minus sign must be inserted 'by hand' whenever we have a negative-energy fermion in the final state. We shall make use of this rule in section 8.2.4. We therefore state the Feynman hypothesis for fermions:

The invariant amplitude for the emission (absorption) of an anti-fermion of 4-momentum p^μ and spin projection s_z in the rest frame is equal to the amplitude (minus the amplitude) for the absorption (emission) of a fermion of 4-momentum $-p^\mu$ and spin projection $-s_z$ in the rest frame.

As we shall see in chapters 5–7, the Feynman interpretation of the negative-energy solutions is naturally embodied in the field theory formalism.

4.6 Inclusion of electromagnetic interactions via the gauge principle: the Dirac prediction of $g = 2$ for the electron

Having set up the relativistic spin-0 and spin-$\frac{1}{2}$ free-particle wave equations, we are now in a position to use the machinery developed in chapter 3, in order to include electromagnetic interactions. All we have to do is make the replacement

$$\partial^\mu \to D^\mu \equiv \partial^\mu + \mathrm{i}qA^\mu \tag{4.131}$$

for a particle of charge q. For the spin-0 KG equation (4.10) we obtain, after some rearrangement (problem 4.13),

$$\begin{aligned}(\Box + m^2)\phi &= -\mathrm{i}q(\partial_\mu A^\mu + A^\mu\partial_\mu)\phi + q^2A^2\phi && (4.132)\\ &= -\hat{V}_{\mathrm{KG}}\phi. && (4.133)\end{aligned}$$

Note that the potential $\hat{V}_{\mathrm{KG}}$ contains the differential operator ∂_μ; the sign of $\hat{V}_{\mathrm{KG}}$ is a convention chosen so as to maintain the same relative sign between $\boldsymbol{\nabla}^2$ and $\hat{V}$ as in the Schrödinger equation—for example that in (A.5).

For the Dirac equation the replacement (4.131) leads to

$$\mathrm{i}\frac{\partial\psi}{\partial t} = [\boldsymbol{\alpha}\cdot(-\mathrm{i}\boldsymbol{\nabla} - q\boldsymbol{A}) + \beta m + qA^0]\psi \tag{4.134}$$

where $A^\mu = (A^0, \boldsymbol{A})$. The potential due to A^μ is therefore $\hat{V}_{\mathrm{D}} = qA^0\mathbf{1} - q\boldsymbol{\alpha}\cdot\boldsymbol{A}$, which is a 4×4 matrix acting on the Dirac spinor.

The non-relativistic limit of (4.134) is of great importance, both physically and historically. It was, of course, first obtained by Dirac; and it provided, in 1928, a sensational explanation of why the g-factor of the electron had the value $g = 2$, which was then the empirical value, without any theoretical basis.

By way of background, recall from appendix A that the Schrödinger equation for a non-relativistic spinless particle of charge q in a magnetic field $\boldsymbol{B}$ described by a vector potential $\boldsymbol{A}$ such that $\boldsymbol{B} = \boldsymbol{\nabla}\times\boldsymbol{A}$ is

$$-\frac{1}{2m}\boldsymbol{\nabla}^2\psi - \frac{q}{2m}\boldsymbol{B}\cdot\hat{\boldsymbol{L}}\psi + \frac{q^2}{2m}\boldsymbol{A}^2\psi = \mathrm{i}\frac{\partial\psi}{\partial t}. \tag{4.135}$$

Taking $\boldsymbol{B}$ along the z-axis, the $\boldsymbol{B}\cdot\hat{\boldsymbol{L}}$ term will cause the usual splitting (into states of different magnetic quantum number) of the $(2l+1)$-fold degeneracy associated with a state of definite l. In particular, though, there should be no splitting of the hydrogen ground state which has $l = 0$. But experimentally splitting into two levels is observed, indicating a two-fold degeneracy and thus (see earlier) a $j = \frac{1}{2}$-like degree of freedom.

Uhlenbeck and Goudsmit (1925) suggested that the doubling of the hydrogen ground state could be explained if the electron were given an additional quantum number corresponding to an angular-momentum-like observable, having magnitude $j = \frac{1}{2}$. The operators $\boldsymbol{S} = \frac{1}{2}\boldsymbol{\sigma}$ which we have already met serve to represent such a *spin* angular momentum. If the contribution to the energy operator of the particle due to its spin $\boldsymbol{S}$ enters into the effective Schrödinger equation in exactly the same way as that due to its orbital angular momentum, then we would expect an additional term on the left-hand side of (4.135) of the form

$$-\frac{q}{2m}\boldsymbol{B}\cdot\boldsymbol{S}. \tag{4.136}$$

The corresponding wavefunction must now have two (spinor) components, acted on by the 2×2 matrices in $\boldsymbol{S}$.

The energy difference between the two levels with eigenvalues $S_z = \pm\frac{1}{2}$ would then be $qB/2m$ in magnitude. Experimentally the splitting was found to be just twice this value. Thus empirically the term (4.136) was modified to

$$-g\frac{q}{2m}\boldsymbol{B}\cdot\boldsymbol{S} \tag{4.137}$$

where g is the 'gyromagnetic ratio' of the particle, with $g \approx 2$. Let us now see how Dirac deduced the term (4.137), with the precise value $g = 2$, from his equation.

To achieve a non-relativistic limit, we expect that we have somehow to reduce the four-component Dirac equation to one involving just two components, since the desired term (4.137) is only a 2×2 matrix. Looking at the explicit form (4.104) for the free-particle positive-energy solutions, we see that the lower two components are of order v (i.e. v/c with $c = 1$) times the upper two. This suggests that, to get a non-relativistic limit, we should regard the lower two components of ψ as being small (at least in the specific representation we are using for $\boldsymbol{\alpha}$ and β). However, since (4.134) includes the A^μ-field, this will have to be demonstrated (see (4.144)). Also, if we write the total energy operator as $m + \hat{H}_1$, we expect $\hat{H}_1$ to be the non-relativistic energy operator.

We let

$$\psi = \begin{pmatrix} \Psi \\ \Phi \end{pmatrix} \tag{4.138}$$

where Ψ and Φ are not free-particle solutions, and they carry the space–time dependence as well as the spinor character (each has two components). We set

$$\hat{H}_1 = \boldsymbol{\alpha}\cdot(-\mathrm{i}\boldsymbol{\nabla} - q\boldsymbol{A}) + \beta m + eA^0 - m \tag{4.139}$$

where a 4×4 unit matrix multiplying the last two terms is understood. Then

$$\hat{H}_1\begin{pmatrix} \Psi \\ \Phi \end{pmatrix} = \begin{pmatrix} \boldsymbol{0} & \boldsymbol{\sigma}\cdot(-\mathrm{i}\boldsymbol{\nabla} - q\boldsymbol{A}) \\ \boldsymbol{\sigma}\cdot(-\mathrm{i}\boldsymbol{\nabla} - q\boldsymbol{A}) & \boldsymbol{0} \end{pmatrix}\begin{pmatrix} \Psi \\ \Phi \end{pmatrix}$$
$$-2m\begin{pmatrix} 0 \\ \Phi \end{pmatrix} - eA^0\begin{pmatrix} \Psi \\ \Phi \end{pmatrix}. \tag{4.140}$$

Multiplying out (4.140), we obtain

$$\hat{H}_1\Psi = \boldsymbol{\sigma}\cdot(-\mathrm{i}\nabla - q\boldsymbol{A})\Phi - eA^0\Psi \tag{4.141}$$
$$\hat{H}_1\Phi = \boldsymbol{\sigma}\cdot(-\mathrm{i}\nabla - q\boldsymbol{A})\Psi - eA^0\Phi - 2m\Phi. \tag{4.142}$$

From (4.142), we obtain

$$(\hat{H}_1 + eA^0 + 2m)\Phi = \boldsymbol{\sigma}\cdot(-\mathrm{i}\nabla - q\boldsymbol{A})\psi. \tag{4.143}$$

So, if $\hat{H}_1$ (or rather any matrix element of it) is $\ll m$ and if A^0 is positive or, if negative, much less in magnitude than m/e, we can deduce

$$\Phi \sim (\text{velocity}) \times \Psi \tag{4.144}$$

as in the free case, provided that the magnetic energy $\sim \boldsymbol{\sigma}\cdot\boldsymbol{A}$ is not of order m. Further, if $\hat{H}_1 \ll m$ and the conditions on the fields are met, we can drop $\hat{H}_1$ and eA^0 on the left-hand side of (4.143), as a first approximation, so that

$$\Phi \approx \frac{\boldsymbol{\sigma}\cdot(-\mathrm{i}\nabla - q\boldsymbol{A})}{2m}\Psi. \tag{4.145}$$

Hence, in (4.141),

$$\hat{H}_1\Psi \approx \frac{1}{2m}\{\boldsymbol{\sigma}\cdot(-\mathrm{i}\nabla - q\boldsymbol{A})\}^2\Psi - eA^0\Psi. \tag{4.146}$$

The right-hand side of (4.146) should therefore be the non-relativistic energy operator for a spin-$\frac{1}{2}$ particle of charge q and mass m in a field A^μ.

Consider then the case $A^0 = 0$ which is sufficient for the discussion of g. We need to evaluate

$$\{\boldsymbol{\sigma}\cdot(-\mathrm{i}\nabla - q\boldsymbol{A})\}^2\Psi. \tag{4.147}$$

This requires care, because although it is true that (for example) $(\boldsymbol{\sigma}\cdot\boldsymbol{p})^2 = \boldsymbol{p}^2$ if $\boldsymbol{p} = (p_x, p_y, p_z)$ are ordinary numbers which commute with each other, the components of '$-\mathrm{i}\nabla - q\boldsymbol{A}$' do *not* commute due to the presence of the differential operator ∇, and the fact that $\boldsymbol{A}$ depends on $\boldsymbol{r}$. In problem 4.14 it is shown that

$$\{\boldsymbol{\sigma}\cdot(-\mathrm{i}\nabla - q\boldsymbol{A})\}^2\Psi = (-\mathrm{i}\nabla - q\boldsymbol{A})^2\Psi - q\boldsymbol{\sigma}\cdot\boldsymbol{B}\Psi. \tag{4.148}$$

The first term on the right-hand side of (4.148) when inserted into (4.146), gives precisely the spin-0 non-relativistic Hamiltonian appearing on the left-hand side of (4.135) (see appendix A), while the second term in (4.148) yields exactly (4.137) with $g = 2$, recalling that $\boldsymbol{S} = \frac{1}{2}\boldsymbol{\sigma}$. Thus the non-relativistic reduction of the Dirac equation leads to the prediction $g = 2$ for a spin-$\frac{1}{2}$ particle.

In actual fact, the measured g-factor of the electron (and muon) is slightly greater than this value: $g_{\mathrm{exp}} = 2(1 + a)$. The 'anomaly' a, which is of order 10^{-3} in size, is measured with quite extraordinary precision (see section 11.7) for both the e^- and e^+. This small correction can also be computed with equally extraordinary accuracy, using the full theory of QED, as we shall briefly explain in chapter 11. The agreement between theory and experiment is phenomenal and is one example of such agreement exhibited by our 'paradigm theory'.

It may be worth noting that spin-$\frac{1}{2}$ hadrons, such as the proton, have g-factors very different from the Dirac prediction. This is because they are, as we know, composite objects and are thus (in this respect) more like atoms in nuclei than 'elementary particles'.

Problems

4.1

(a) In natural units $\hbar = c = 1$ and with $2m = 1$, the Schrödinger equation may be written as

$$-\nabla^2\psi + V\psi - \mathrm{i}\partial\psi/\partial t = 0.$$

Multiply this equation from the left by ψ^* and multiply the complex conjugate of this equation by ψ (assume V is real). Subtract the two equations and show that your answer may be written in the form of a continuity equation

$$\partial\rho/\partial t + \nabla\cdot\boldsymbol{j} = 0$$

where $\rho = \psi^*\psi$ and $\boldsymbol{j} = \mathrm{i}^{-1}[\psi^*(\nabla\psi) - (\nabla\psi^*)\psi]$.

(b) Perform the same operations for the Klein–Gordon equation and derive the corresponding 'probability' density current. Show also that for a free-particle solution

$$\phi = N\mathrm{e}^{-\mathrm{i}p\cdot x}$$

with $p^\mu = (E, \boldsymbol{p})$, the probability current $j^\mu = (\rho, \boldsymbol{j})$ is proportional to p^μ.

4.2 Prove the following properties of the matrices α_i and β:

(i) α_i and β $(i = 1, 2, 3)$ are all Hermitian [*hint*: what is the Hamiltonian?].
(ii) $\mathrm{Tr}\,\alpha_i = \mathrm{Tr}\,\beta = 0$ where 'Tr' means the trace, i.e. the sum of the diagonal elements [*hint*: use $\mathrm{Tr}(\boldsymbol{AB}) = \mathrm{Tr}(\boldsymbol{BA})$ for any matrices $\boldsymbol{A}$ and $\boldsymbol{B}$—and prove this too!].
(iii) The eigenvalues of α_i and β are ± 1 [*hint*: square α_i and β].
(iv) The dimensionality of α_i and β is even [*hint*: the trace of a matrix is equal to the sum of its eigenvalues].

4.3 For free-particle solutions of the Dirac equation

$$\psi = \omega\mathrm{e}^{-\mathrm{i}p\cdot x}$$

the four-component spinor ω may be written in terms of the two-component spinors

$$\omega = \begin{pmatrix} \phi \\ \chi \end{pmatrix}.$$

From the Dirac equation for ψ

$$\mathrm{i}\partial\psi/\partial t = (-\mathrm{i}\boldsymbol{\alpha} \cdot \nabla + \beta m)\psi$$

using the explicit forms for the Dirac matrices

$$\boldsymbol{\alpha} = \begin{pmatrix} \mathbf{0} & \boldsymbol{\sigma} \\ \boldsymbol{\sigma} & \mathbf{0} \end{pmatrix} \qquad \beta = \begin{pmatrix} \mathbf{1} & \mathbf{0} \\ \mathbf{0} & -\mathbf{1} \end{pmatrix}$$

show that ϕ and χ satisfy the coupled equations

$$(E - m)\phi = \boldsymbol{\sigma} \cdot \boldsymbol{p}\chi$$
$$(E + m)\chi = \boldsymbol{\sigma} \cdot \boldsymbol{p}\phi$$

where $p^\mu = (E, \boldsymbol{p})$.

4.4

(a) Using the explicit forms for the 2×2 Pauli matrices, verify the commutation (square brackets) and anti-commutation (braces) relation [note the summation convention for repeated indices: $\epsilon_{ijk} \equiv \sum_{k=1}^{3} \epsilon_{ijk}\sigma_k$]:

$$[\sigma_i, \sigma_j] = 2\mathrm{i}\epsilon_{ijk}\sigma_k \qquad \{\sigma_i, \sigma_j\} = 2\delta_{ij}\mathbf{1}$$

where ϵ_{ijk} is the usual antisymmetric tensor

$$\epsilon_{ijk} = \begin{cases} +1 & \text{for an even permutation of 1, 2, 3} \\ -1 & \text{for an odd permutation of 1, 2, 3} \\ 0 & \text{if two or more indices are the same,} \end{cases}$$

δ_{ij} is the usual Kronecker delta, and $\mathbf{1}$ is the 2×2 matrix. Hence show that

$$\sigma_i\sigma_j = \delta_{ij}\mathbf{1} + \mathrm{i}\epsilon_{ijk}\sigma_k.$$

(b) Use this last identity to prove the result

$$(\boldsymbol{\sigma} \cdot \boldsymbol{a})(\boldsymbol{\sigma} \cdot \boldsymbol{b}) = \boldsymbol{a} \cdot \boldsymbol{b}\mathbf{1} + \mathrm{i}\boldsymbol{\sigma} \cdot \boldsymbol{a} \times \boldsymbol{b}.$$

Using the explicit 2×2 form for

$$\boldsymbol{\sigma} \cdot \boldsymbol{p} = \begin{pmatrix} p_z & p_x - ip_y \\ p_x + ip_y & -p_z \end{pmatrix}$$

show that

$$(\boldsymbol{\sigma} \cdot \boldsymbol{p})^2 = \boldsymbol{p}^2\mathbf{1}.$$

4.5 Check that $h(\boldsymbol{p})$ as given by (4.65) does commute with $\boldsymbol{\alpha} \cdot \boldsymbol{p} + \beta m$, the momentum–space free Dirac Hamiltonian.

4.6 Let ϕ be an arbitrary two-component spinor, and let $\hat{\boldsymbol{u}}$ be a unit vector.

(i) Show that $\frac{1}{2}(1 + \boldsymbol{\sigma} \cdot \hat{\boldsymbol{u}})\phi$ is an eigenstate of $\boldsymbol{\sigma} \cdot \hat{\boldsymbol{u}}$ with eigenvalue $+1$. The operator $\frac{1}{2}(1 + \boldsymbol{\sigma} \cdot \hat{\boldsymbol{u}})$ is called a projector operator for the $\boldsymbol{\sigma} \cdot \hat{\boldsymbol{u}} = +1$ eigenstate since when acting on any ϕ this is what it 'projects out'. Write down a similar operator which projects out the $\boldsymbol{\sigma} \cdot \hat{\boldsymbol{u}} = -1$ eigenstate.

(ii) Construct two two-component spinors ϕ_+ and ϕ_- which are eigenstates of $\boldsymbol{\sigma} \cdot \hat{\boldsymbol{u}}$ belonging to eigenvalues ± 1, and normalized to $\phi_r^\dagger \phi_s = \delta_{rs}$ for $(r, s) = (+, -)$, for the case $\hat{\boldsymbol{u}} = (\sin\theta \cos\phi, \sin\theta \sin\phi, \cos\theta)$ [*hint*: take the arbitrary $\phi = \binom{1}{0}$].

4.7 Verify equations (4.74), (4.75) and (4.76).

4.8 Check that the probability current $\boldsymbol{j} = \omega^\dagger \boldsymbol{\alpha} \omega$ (for a plane-wave solution) transforms correctly under the x-axis rotation specified by (4.69) [*hint*: using $\omega' = \exp(\mathrm{i}\Sigma_x \alpha/2)\omega$ and (4.69) you need to show that $j'_x = \omega'^\dagger \alpha_x \omega' = j_x$, that $j'_y = \omega'^\dagger \alpha_y \omega' = \cos\alpha j_y + \sin\alpha j_z$, and that $j'_z = \omega'^\dagger \alpha_z \omega' = -\sin\alpha j_y + \cos\alpha j_z$].

4.9 Verify that the transformation law (4.90) does indeed generate ω' of (4.89) from ω of (4.88), with $N = ((E' + m)/2m)^{1/2}$.

4.10 Show that if we define $\bar{\psi} = \psi^\dagger \beta$, then the quantity $\bar{\psi}\psi$ is invariant under the transformation (4.83) and under the transformation (4.90).

4.11

(a) Verify explicitly that the matrices $\boldsymbol{\alpha}$ and β of problem 4.3 satisfy the Dirac anti-commutation relations. Defining the four 'γ matrices'

$$\gamma^\mu = (\gamma^0, \boldsymbol{\gamma})$$

where $\gamma^0 = \beta$ and $\boldsymbol{\gamma} = \beta\boldsymbol{\alpha}$, show that the Dirac equation can be written in the form $(\mathrm{i}\gamma^\mu \partial_\mu - m)\psi = 0$. Find the anti-commutation relations of the γ matrices.

(b) Define the conjugate spinor

$$\bar{\psi}(x) = \psi^\dagger(x)\gamma^0$$

and use the previous result to find the equation satisfied by $\bar{\psi}$ in γ matrix notation.

(c) The Dirac probability current may be written as

$$j^\mu = \bar{\psi}(x)\gamma^\mu \psi(x).$$

Show that it satisfies the conservation law

$$\partial_\mu j^\mu = 0.$$

4.12 Positive-energy spinors $u(p, s)$ are defined by

$$u(p,s) = (E+m)^{1/2} \begin{pmatrix} \phi^s \\ \dfrac{\boldsymbol{\sigma} \cdot \boldsymbol{p}}{E+m}\phi^s \end{pmatrix} \qquad s = 1, 2.$$

Verify that these satisfy $u^\dagger u = 2E$.

In a similar way, negative-energy spinors $v(p, s)$ are defined by

$$v(p,s) = (E+m)^{1/2} \begin{pmatrix} \dfrac{\boldsymbol{\sigma} \cdot \boldsymbol{p}}{E+m}\chi^s \\ \chi^s \end{pmatrix} \qquad s = 1, 2.$$

Verify that $v^\dagger v = 2E$.

4.13 Using the KG equation together with the replacement $\partial^\mu \to \partial^\mu + \mathrm{i}qA^\mu$, find the form of the potential $\hat{V}_{\mathrm{KG}}$ in the corresponding equation

$$(\Box + m^2)\phi = -\hat{V}_{\mathrm{KG}}\phi$$

in terms of A^μ.

4.14 Evaluate

$$\{\boldsymbol{\sigma} \cdot (-\mathrm{i}\nabla - q\boldsymbol{A})\}^2\psi$$

by following the subsequent steps (or doing it your own way):

(i) Multiply the operator by itself to get

$$\{(\boldsymbol{\sigma} - \mathrm{i}\nabla)^2 + \mathrm{i}q(\boldsymbol{\sigma} \cdot \nabla)(\boldsymbol{\sigma} \cdot \boldsymbol{A}) + \mathrm{i}q(\boldsymbol{\sigma} \cdot \boldsymbol{A})(\boldsymbol{\sigma} \cdot \nabla) + q^2(\boldsymbol{\sigma} \cdot \boldsymbol{A})^2\}\psi.$$

The first and last terms are, respectively, $-\nabla^2$ and $q^2\boldsymbol{A}^2$ where the 2×2 unit matrix $\mathbf{1}$ is understood. The second and third terms are $\mathrm{i}q(\boldsymbol{\sigma} \cdot \nabla)(\boldsymbol{\sigma} \cdot \boldsymbol{A}\psi)$ and $\mathrm{i}q(\boldsymbol{\sigma} \cdot \boldsymbol{A})(\boldsymbol{\sigma} \cdot \nabla\psi)$. These may be simplified using the identity of problem 4.4(b), but we must be careful to treat ∇ correctly as a differential operator.

(ii) Show that $(\boldsymbol{\sigma} \cdot \nabla)(\boldsymbol{\sigma} \cdot \boldsymbol{A})\psi = \nabla \cdot (\boldsymbol{A}\psi) + \mathrm{i}\boldsymbol{\sigma} \cdot \{\nabla \times (\boldsymbol{A}\psi)\}$. Now use $\nabla \times (\boldsymbol{A}\psi) = (\nabla \times \boldsymbol{A})\psi - \boldsymbol{A} \times \nabla\psi$ to simplify the last term.

(iii) Similarly, show that $(\boldsymbol{\sigma} \cdot \boldsymbol{A})(\boldsymbol{\sigma} \cdot \nabla)\psi = \boldsymbol{A} \cdot \nabla\psi + \mathrm{i}\boldsymbol{\sigma} \cdot (\boldsymbol{A} \times \nabla\psi)$.

(iv) Hence verify (4.148).

4.15 A possible wave equation for a free massless spin-$\frac{1}{2}$ particle is

$$\mathrm{i}\frac{\partial\psi}{\partial t} = \boldsymbol{\sigma} \cdot -\mathrm{i}\nabla\psi$$

since, squaring the operators on both sides, we obtain

$$-\frac{\partial^2\psi}{\partial t^2} = -\nabla^2\psi \qquad (4.149)$$

using problem 4.4(b), and (4.149) is a KG equation for the case $m = 0$. Inserting a plane-wave solution of the form

$$\psi \sim \mathrm{e}^{-\mathrm{i}Et+\mathrm{i}\boldsymbol{p}\cdot\boldsymbol{x}}u,$$

the two-component spinor u satisfies

$$Eu = \boldsymbol{\sigma}\cdot\boldsymbol{p}u.$$

But also $E = |\boldsymbol{p}|$ for a massless particle. Hence for such solutions u is, in fact, a helicity eigenfunction as in (4.67), with eigenvalue +1. In the present context, we shall denote it by u_{R}, and the corresponding ψ by ψ_{R}.

(i) Explain why helicity is a Lorentz-invariant property for a massless spin-$\frac{1}{2}$ particle but not for a massive one.

(ii) Consider the Lorentz transformation properties of u_{R} satisfying

$$Eu_{\mathrm{R}} = \boldsymbol{\sigma}\cdot\boldsymbol{p}u_{\mathrm{R}}. \tag{4.150}$$

Let $\boldsymbol{p} = (p, 0, 0)$ and consider the particular Lorentz transformation of (4.85) and (4.86), for an *infinitesimal* velocity v'_x (to make the algebra easier), which we denote by ϵ_x:

$$E' = E + \epsilon_x p_x \tag{4.151}$$

$$p'_x = \epsilon_x E + p_x \tag{4.152}$$

together with $p'_y = p_y$, $p'_z = p_z$. In (4.151) and (4.152) we have used $\cosh\vartheta \approx 1$ and $\sinh\vartheta \approx \epsilon_x$. In the transformed system we must have

$$E'u'_{\mathrm{R}} = \boldsymbol{\sigma}\cdot\boldsymbol{p}'u'_{\mathrm{R}}. \tag{4.153}$$

Show that (4.150)–(4.153) are consistent if

$$u'_{\mathrm{R}} = (1 + \tfrac{1}{2}\epsilon_x\sigma_x)u_{\mathrm{R}}. \tag{4.154}$$

For a three-dimensional Lorentz transformation parametrized by $\boldsymbol{\epsilon} = (\epsilon_x, \epsilon_y, \epsilon_z)$, (4.154) generalizes to

$$u'_{\mathrm{R}} = (1 + \tfrac{1}{2}\boldsymbol{\epsilon}\cdot\boldsymbol{\sigma})u_{\mathrm{R}} \tag{4.155}$$

which may be compared with the infinitesimal form of (4.96). What is the *finite* version of (4.155)?

(iii) Another possible massless spin-$\frac{1}{2}$ wave equation is

$$\mathrm{i}\frac{\partial\psi_{\mathrm{L}}}{\partial t} = \boldsymbol{\sigma}\cdot\mathrm{i}\boldsymbol{\nabla}\psi_{\mathrm{L}},$$

since on squaring the operators this also leads to the massless KG equation (4.149). Plane-wave solutions of the form

$$\psi_{\mathrm{L}} \sim \mathrm{e}^{-\mathrm{i}Et+\mathrm{i}\boldsymbol{p}\cdot\boldsymbol{x}}u_{\mathrm{L}}$$

lead to

$$Eu_{\mathrm{L}} = |\boldsymbol{p}|u_{\mathrm{L}} = -\boldsymbol{\sigma}\cdot\boldsymbol{p}u_{\mathrm{L}}$$

showing that u_{L} is an eigenfunction of helicity with eigenvalue -1. Show that under the infinitesimal Lorentz transformation (4.151) and (4.152),

$$u'_{\mathrm{L}} = (1 - \tfrac{1}{2}\epsilon_x\sigma_x)u_{\mathrm{L}}. \tag{4.156}$$

This generalizes to

$$u'_{\mathrm{L}} = (1 - \tfrac{1}{2}\boldsymbol{\epsilon}\cdot\boldsymbol{\sigma})u_{\mathrm{L}} \tag{4.157}$$

as in (4.155).

(iv) Consider the Dirac equation using the representation (4.97) for $\boldsymbol{\alpha}$ and β. Referring to (4.98) and (4.99) we see that when $m \to 0$, the upper two components ϕ of the Dirac wavefunction may be identified with u_{R}, and the lower two χ with u_{L} (in this representation). Discuss why the mass 'couples the states of different helicity'.

(v) Show that the Dirac spinor components, in the form (4.98) and (4.99) with $\phi \to u_{\mathrm{R}}$ and $\chi \to u_{\mathrm{L}}$, transform consistently under the Lorentz transformation (4.151) and (4.152), with u'_{R} and u'_{L} given by (4.154) and (4.156).

PART 2

INTRODUCTION TO QUANTUM FIELD THEORY

It was a wonderful world my father told me about.

You might wonder what he got out of it all. I went to MIT. I went to Princeton. I went home and he said, 'Now you've got a science education. I have always wanted to know something that I have never understood; and so, my son, I want you to explain it to me.' I said yes.

He said, 'I understand that they say that light is emitted from an atom when it goes from one state to another, from an excited state to a state of lower energy.'

I said 'That's right.'

'And light is a kind of particle, a photon I think they call it.'

'Yes.'

'So if the photon comes out of the atom when it goes from the excited to the lower state, the photon must have been in the atom in the excited state.'

I said, 'Well, no.'

He said, 'Well, how do you look at it so you can think of a particle photon coming out without it having been in there in the excited state?'

I thought a few minutes, and I said, 'I'm sorry; I don't know. I can't explain it to you.'

He was very disappointed after all these years and years trying to teach me something, that it came out with such poor results.

R P Feynman, *The Physics Teacher*, vol 7, No 6, September 1969

All the fifty years of conscious brooding have brought me no closer to the answer to the question, 'What are light quanta?' Of course today every rascal thinks he knows the answer, but he is deluding himself.

A Einstein (1951)

Quoted in 'Einstein's research on the nature of light'

E Wolf (1979), *Optic News*, vol 5, No 1, page 39.

I never satisfy myself until I can make a mechanical model of a thing. If I can make a mechanical model I can understand it. As long as I cannot make a mechanical model all the way through I cannot understand; and that is why I cannot get the electromagnetic theory.

Sir William Thomson, Lord Kelvin, 1884 *Notes of Lectures on Molecular Dynamics and the Wave Theory of Light delivered at the Johns Hopkins University, Baltimore*, stenographic report by A S Hathaway (Baltimore: Johns Hopkins University) Lecture XX, pp 270–1.

5

QUANTUM FIELD THEORY I: THE FREE SCALAR FIELD

In this chapter we shall give an elementary introduction to quantum field theory, which is the established 'language' of the Standard Model of particle physics. Even so long after Maxwell's theory of the (classical) electromagnetic field, the concept of a 'disembodied' field is not an easy one; and we are going to have to add the complications of quantum mechanics to it. In such a situation, it is helpful to have some physical model in mind. For most of us, as for Lord Kelvin, this still means a mechanical model. Thus in the following two sections we begin by considering a mechanical model for a quantum field. At the end, we shall—like Maxwell—throw away the 'mechanism' and have simply quantum field theory. Section 5.1 describes this programme qualitatively; section 5.2 presents a more complete formalism, for the simple case of a field whose quanta are massless, and move in only one spatial dimension. The appropriate generalizations for massive quanta in three dimensions are given in section 5.3.

5.1 The quantum field: (i) descriptive

Mechanical systems are usefully characterized by the number of *degrees of freedom* they possess: thus a one-dimensional pendulum has one degree of freedom, two coupled one-dimensional pendulums have two degrees of freedom—which may be taken to be their angular displacements, for example. A scalar field $\phi(x, t)$ corresponds to a system with an infinite number of degrees of freedom, since at each continuously varying point x an independent 'displacement' $\phi(x, t)$, which also varies with time, has to be determined. Thus quantum field theory involves two major mathematical steps: the description of continuous systems (fields) which have infinitely many degrees of freedom, and the application of *quantum* theory to such systems. These two aspects are clearly separable. It is certainly easier to begin by considering systems with a discrete—but possibly very large—number of degrees of freedom, for example a solid. We shall treat such systems first classically and then quantum mechanically. Then, returning to the classical case, we shall allow the number of degrees of freedom to become infinite, so that the system corresponds to a classical field. Finally, we shall apply quantum mechanics directly to fields.

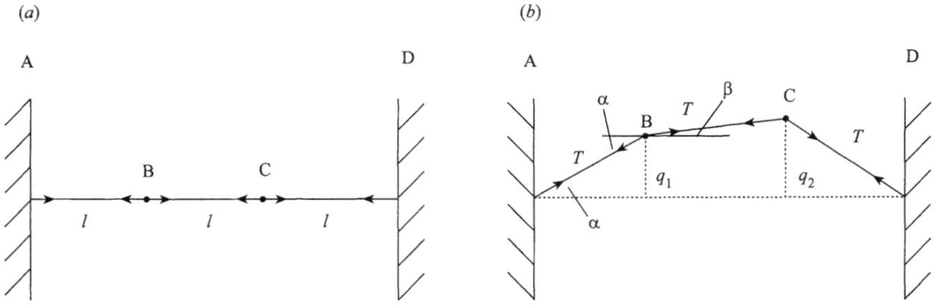

Figure 5.1. A vibrating system with two degrees of freedom: (*a*) two mass points at rest, with the strings under tension; (*b*) a small transverse displacement.

We begin by considering a rather small solid—one that has only two atoms free to move. The atoms, each of mass m, are connected by a string, and each is connected to a fixed support by a similar string (figure 5.1(*a*)); all the strings are under tension T. We consider *small transverse vibrations* of the atoms (figure 5.1(*b*)), and we call $q_r(t)$ ($r = 1, 2$) the transverse displacements. We are interested in the total energy E of the system. According to classical mechanics, this is equal to the sum of the kinetic energies $\frac{1}{2}m\dot{q}_r^2$ of each atom, together with a potential energy V which can be calculated as follows. Referring to figure 5.1(*b*), when atom 1 is displaced by q_1, it experiences a restoring force

$$F_1 = T\sin\alpha - T\sin\beta \tag{5.1}$$

assuming a constant tension T along the string. For small displacements q_1 and q_2 (i.e. $q_{1,2} \ll l$) we have

$$\begin{aligned} \sin\alpha &= q_1/(l^2 + q_1^2)^{1/2} \approx q_1/l \\ \sin\beta &= (q_2 - q_1)/[l^2 + (q_2 - q_1)^2]^{1/2} \approx (q_2 - q_1)/l \end{aligned} \tag{5.2}$$

where terms of order $(q_{1,2}/l)^3$ and higher have been neglected. Thus the restoring force on particle 1 is, in this approximation,

$$F_1 = k(2q_1 - q_2) \tag{5.3}$$

with $k = T/l$. Similarly, the restoring force on particle 2 is

$$F_2 = k(2q_2 - q_1) \tag{5.4}$$

and the equations of motion are

$$m\ddot{q}_1 = -k(2q_1 - q_2) \tag{5.5}$$

$$m\ddot{q}_2 = -k(2q_2 - q_1). \tag{5.6}$$

The potential energy is then determined (up to an irrelevant constant) by the requirement that (5.5) and (5.6) are of the form

$$m\ddot{q}_1 = -\partial V/\partial q_1 \tag{5.7}$$

$$m\ddot{q}_2 = -\partial V/\partial q_2. \tag{5.8}$$

Thus we deduce that

$$V = k(q_1^2 + q_2^2 - q_1 q_2). \tag{5.9}$$

Equations (5.5) and (5.6) form a pair of *linear*, *coupled* differential equations. Each of the italicized words is important. By 'linear', is meant that only the first power of q_1 and q_2 and their time derivatives appear in the equations of motion; terms such as q_1^2, $q_1 q_2$, $\dot{q}_1^2$, q_1^3 and so on would render the equations of motion 'nonlinear'. This linear/nonlinear distinction is a crucial one in dynamics. Most importantly, the solutions of linear differential equations may be added together with constant coefficients ('linearly superposed') to make new valid solutions of the equations. In contrast, solutions of nonlinear differential equations—besides being very hard to find!—cannot be linearly superposed to get new solutions. In addition, nonlinear dynamical equations may typically lead to *chaotic motion*.

The notion of linearity/nonlinearity carries over also into the equations of motion for fields. In this context, an equation for a field $\phi(x, t)$ is said to be linear if ϕ and its space—or time—derivatives appear only to the first power. As we shall see, this is true for Maxwell's equations for the electromagnetic field and it is, of course, the mathematical reason behind all the physics of such things as interference and diffraction, which may be understood precisely in terms of superposition of solutions of these equations. Likewise the equations of quantum mechanics (e.g. Schrödinger's equation) are all linear in this sense, consistent with the principle of superposition in quantum mechanics.

It is clear, then, that in looking at simple mechanical models as a guide to the field systems in which we will ultimately be interested, we should consider ones in which the equations of motion are linear. In the present case, this is true, but only because we have made the approximation that q_1 and q_2 are small (compared to l). Referring to equation (5.2), we can immediately see that if we had kept the full expression for $\sin\alpha$ and $\sin\beta$, the resulting equations of motion would have been highly nonlinear. A similar 'small displacement' approximation has to be made in determining the familiar wave equation, describing waves on continuous strings, for example (see (5.30) later). Most significantly, however, quantum mechanics is believed to be a linear theory *without* any approximation.

The appearance of only linear terms in q_1 and q_2 in the equations of motion implies, via (5.7) and (5.8), that the potential energy can only involve quadratic powers of the q's, i.e. q_1^2, q_2^2 and $q_1 q_2$, as in (5.9). Once again, had we used the general expression for the potential energy in a stretched string as 'tension×extension' we would have obtained an expression containing all powers of the q's via such terms as $\{[l^2 + q_1^2]^{1/2} - l\}$.

We turn now to the *coupled* aspect of (5.5) and (5.6). By this we mean that the right-hand side of the q_1 equation depends on q_2 as well as q_1, and similarly for the q_2 equation. This 'mathematical' coupling has its origin in the term $-kq_1q_2$ in V, which corresponds to the 'physical' coupling of the string BC connecting the two atoms. If this coupling were absent, equations (5.5) and (5.6) would describe two independent (uncoupled) harmonic oscillators, each of frequency $(2k/m)^{1/2}$. When we consider the addition of more and more particles (see later) we certainly do not want them to vibrate independently, otherwise we would not be able to get wave-like displacements propagating through the system. So we need to retain at least this minimal kind of 'quadratic' coupling.

With the coupling, the solutions of (5.5) and (5.6) are not quite so obvious. However, a simple step makes the equations much easier. Suppose we add the two equations so as to obtain

$$m(\ddot{q}_1 + \ddot{q}_2) = -k(q_1 + q_2) \tag{5.10}$$

and subtract them to obtain

$$m(\ddot{q}_1 - \ddot{q}_2) = -3k(q_1 - q_2). \tag{5.11}$$

A remarkable thing has happened: the two *combinations* $q_1 + q_2$ and $q_1 - q_2$ of the original coordinates satisfy *uncoupled* equations—which are of course very easy to solve. The combination $q_1 + q_2$ oscillates with frequency $\omega_1 = (k/m)^{1/2}$, while $q_1 - q_2$ oscillates with frequency $\omega_2 = (3k/m)^{1/2}$.

Let us introduce

$$Q_1 = (q_1 + q_2)/\sqrt{2} \qquad Q_2 = (q_1 - q_2)/\sqrt{2} \tag{5.12}$$

(the $\sqrt{2}$'s are for later convenience). Then the solutions of (5.10) and (5.11) are:

$$Q_1(t) = A\cos\omega_1 t + B\sin\omega_1 t \tag{5.13}$$

$$Q_2(t) = C\cos\omega_2 t + D\sin\omega_2 t. \tag{5.14}$$

Suppose that the initial conditions are such that

$$q_1(0) = q_2(0) = a \qquad \dot{q}_1(0) = \dot{q}_2(0) = 0 \tag{5.15}$$

i.e. the atoms are released from rest, at equal transverse displacements a. In terms of the Q_r's, the conditions (5.15) are

$$\left.\begin{aligned} Q_2(0) = \dot{Q}_2(0) = 0 \\ Q_1(0) = \sqrt{2}a \qquad \dot{Q}_1(0) = 0. \end{aligned}\right. \tag{5.16}$$

Thus from (5.13) and (5.14) we find that the complete solution, for these initial conditions, is

$$Q_1(t) = \sqrt{2}a\cos\omega_1 t \tag{5.17}$$

$$Q_2(t) = 0. \tag{5.18}$$

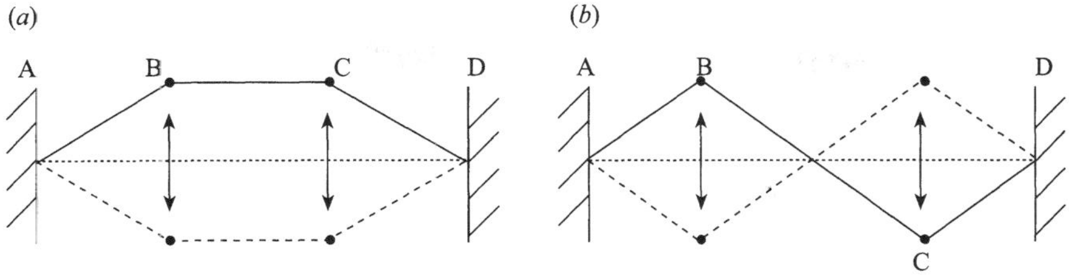

Figure 5.2. Motion in the two normal modes: (*a*) frequency ω_1; (*b*) frequency ω_2.

We see from (5.18) that the motion is such that $q_1 = q_2$ throughout, and from (5.17) that the system vibrates with a single definite frequency ω_1. A form of motion in which the system as a whole moves with a definite frequency is called a '*normal mode*' or simply a 'mode' for short. Figure 5.2(*a*) shows two 'snapshot' configurations of our two-atom system when it is oscillating in the mode characterized by $q_1 = q_2$. In this mode, only $Q_1(t)$ changes; $Q_2(t)$ is always zero. Another mode also exists in which $q_1 = -q_2$ at all times: here $Q_1(t)$ is zero and $Q_2(t)$ oscillates with frequency ω_2. Figure 5.2(*b*) shows two snapshots of the atoms when they are vibrating in this second mode. The coordinate combinations Q_1, Q_2, in terms of which this 'single frequency motion' occurs, are called 'normal mode coordinates' or '*normal coordinates*' for short.

In general, the initial conditions will not be such that the motion is a pure mode; both $Q_1(t)$ and $Q_2(t)$ will be non-zero. From (5.12) we have

$$q_1(t) = [Q_1(t) + Q_2(t)]/\sqrt{2} \tag{5.19}$$

and

$$q_2(t) = [Q_1(t) - Q_2(t)]/\sqrt{2} \tag{5.20}$$

so that q_1 and q_2 are expressed as a sum of two terms oscillating with frequencies ω_1 and ω_2. We say the system is in 'a superposition of modes'. Nevertheless, the mode idea is still very important as regards the total energy of the system, as we shall now see. The kinetic energy can be written in terms of the mode coordinates Q_r as

$$T = \tfrac{1}{2}m\dot{Q}_1^2 + \tfrac{1}{2}m\dot{Q}_2^2 \tag{5.21}$$

while the potential energy becomes

$$V = \tfrac{1}{2}m\omega_1^2 Q_1^2 + \tfrac{1}{2}m\omega_2^2 Q_2^2 \equiv V(Q_1, Q_2). \tag{5.22}$$

The total energy is therefore

$$E = [\tfrac{1}{2}m\dot{Q}_1^2 + \tfrac{1}{2}m\dot{Q}_2^2] + [\tfrac{1}{2}m\omega_1^2 Q_1^2 + \tfrac{1}{2}m\omega_2^2 Q_2^2]. \tag{5.23}$$

This equation shows that, when written in terms of the normal coordinates, the total energy contains *no* couplings terms of the form $Q_1 Q_2$; indeed, the energy has the remarkable form of a simple sum of two independent uncoupled oscillators, one with characteristic frequency ω_1, the other with frequency ω_2. The energy (5.23) has exactly the form appropriate to a system of two *non-interacting* 'things', each executing simple harmonic motion: the 'things' are actually the two modes. Modes do not interact, whereas the original atoms do! Of course, this decoupling in the expression for the total energy is reflected in the decoupling of the equations of motion for the Q variables:

$$m\ddot{Q}_r = -\frac{\partial V(Q_1, Q_2)}{\partial Q_r} \qquad r = 1, 2. \tag{5.24}$$

It is most important to realize that the modes are non-interacting by virtue of the fact that we ignored higher than quadratic terms in $V(q_1, q_2)$. Although the simple change of variables $(q_1, q_2) \rightarrow (Q_1, Q_2)$ of (5.12) does remove the $q_1 q_2$ coupling, this would not be the case if, say, cubic terms in V were to be considered. Such higher order 'anharmonic' corrections would produce couplings between the modes—indeed, this will be the basis of the quantum field theory description of particle interactions (see the following chapter)!

The system under discussion had just two degrees of freedom. We began by describing it in terms of the obvious degree of freedom, the physical displacements of the two atoms q_1 and q_2. But we have learned that it is very illuminating to describe it in terms of the normal coordinate combinations Q_1 and Q_2. The normal coordinates are really the relevant degrees of freedom. Of course, for just two particles, the choice between the q_r's and the Q_r's may seem rather academic; but the important point—and the reason for going through these simple manipulations in detail—is that the basic idea of the normal mode, and of normal coordinates, generalizes immediately to the much less trivial N-atom problem (and also to the field problem). For N atoms there are (for one-dimensional displacements) N degrees of freedom, and if we take them to be the actual atomic displacements, the total energy will be

$$E = \sum_{r=1}^{N} \tfrac{1}{2} m\dot{q}_r^2 + V(q_1, \ldots, q_r) \tag{5.25}$$

which includes all the couplings between atoms. We assume, as before, that the q_r's are small enough so that only quadratic terms need to be kept in V (a constant is as usual irrelevant, and the linear terms vanish if the q_r's are the displacements from equilibrium). In this case, the equations of motion will be linear. By a linear transformation of the form (generalizing (5.12))

$$Q_r = \sum_{s=1}^{N} a_{rs} q_r \tag{5.26}$$

it is possible to write E as a sum of N separate terms, just as in (5.23):

$$E = \sum_{r=1}^{N} [\tfrac{1}{2}m\dot{Q}_r^2 + \tfrac{1}{2}m\omega_r^2 Q_r^2]. \tag{5.27}$$

The Q_r's are the normal coordinates and the ω_r's are the normal frequencies, and there are N of them. If only one of the Q_r's is non-zero, the N atoms are moving in a single mode. The fact that the total energy in (5.27) is a *sum* of N single-mode energies allows us to say that our N-atom solid *behaves as if it consisted of N separate and free harmonic oscillators*—which, however, are *not* to be identified with the coordinates of the original atoms. Once again, and now much more crucially, it is the *mode coordinates* that are the relevant degrees of freedom rather than those of the original particles.

The second stage in our programme is to treat such systems quantum mechanically, as we should certainly have to for a real solid. It is still true that—if the potential energy is a quadratic function of the displacements—the transformation (5.26) allows us to write the total energy as a sum of N mode energies, each of which has the form of a harmonic oscillator. Now, however, these oscillators obey the laws of quantum mechanics, so that each mode oscillator exists only in certain definite states, whose energy eigenvalues are quantized . For each mode of frequency ω_r, the allowed energy values are

$$\epsilon_r = (n_r + \tfrac{1}{2})\hbar\omega_r \tag{5.28}$$

where n_r is a positive integer or zero. This is in sharp contrast to the classical case, of course, in which arbitrary values are allowed for the oscillator energies. The total energy eigenvalue then has the form

$$E = \sum_{r=1}^{N} (n_r + \tfrac{1}{2})\hbar\omega_r. \tag{5.29}$$

The frequencies ω_r are determined by the interatomic forces and are common to both the classical and quantum descriptions; in quantum theory, though, *the states of definite energy of the vibrating N-body system are characterized by the values of a set of integers* $(n_1, n_2, \ldots, n_N)$*, which determine the energies of each mode oscillator.*

For each mode oscillator, $\hbar\omega_r$ measures the quantum of vibrational energy; the energy of an allowed mode state is determined uniquely by the number n_r of such quanta of energy in the state. We now make a profound reinterpretation of this result (first given, almost *en passant* by Born, Heisenberg and Jordan (Born *et al* 1926) in one of the earliest papers on quantum mechanics). We forget about the original N degrees of freedom $q_1, q_2, \ldots, q_N$ and the original N 'atoms', which indeed are only remembered in (5.29) via the fact that there are N different mode frequencies ω_r. Instead we concentrate on the *quanta* and treat *them* as 'things' which really determine the behaviour of our quantum system. We say that 'in a state with energy $(n_r + \frac{1}{2})\hbar\omega_r$ there are n_r quanta present'. For the state characterized by $(n_1, n_2, \ldots, n_N)$ there are n_1 quanta of mode 1 (frequency ω_1), n_2 of mode 2, ... and n_N of mode N. Note particularly that although the number of modes N is fixed, the values of the n_r's are unrestricted, except insofar as the total energy is fixed. Thus we are moving from a 'fixed number' picture (N degrees of freedom) to a 'variable number' picture (the n_r's restricted only by the total energy constraint (5.29)). In the case of a real solid, these quanta of vibrational energy are called *phonons*. We summarize the point we have reached by the important statement that *a phonon is an elementary quantum of vibrational excitation*.

Now we take one step backward in order, afterwards, to take two steps forward. We return to the classical mechanical model with N harmonically interacting degrees of freedom. It is possible to imagine increasing the number N to infinity, and decreasing the interatomic spacing a to zero, in such a way that the product Na stays finite, say $Na = L$. We then have a classical *continuous* system—for example a string of length L. (We stay in one dimension for simplicity.) The transverse vibrations of this string are now described by a *field* $\phi(x, t)$, where at each point x of the string $\phi(x, t)$ measures the displacement from equilibrium, at the time t, of a small element of string around the point x. Thus we have passed from a system described by a discrete number of degrees of freedom, $q_r(t)$ or $Q_r(t)$, to one described by a continuous degree of freedom, the displacement field $\phi(x, t)$. The discrete suffix r has become the continuous argument x—and to prepare for later abstraction, we have denoted the displacement by $\phi(x, t)$ rather than, say, $q(x, t)$.

In the continuous problem the analogue of the small-displacement assumption, which limited the potential energy in the discrete case to quadratic powers, implies that $\phi(x, t)$ obeys the wave equation

$$\frac{1}{c^2}\frac{\partial^2\phi(x,t)}{\partial t^2} = \frac{\partial^2\phi(x,t)}{\partial x^2} \tag{5.30}$$

where c is the wave propagation velocity. Note that (5.30) is linear, but only by virtue of having made the small-displacement assumption. Again, we consider first the classical treatment of this system. Our aim is to find, for this continuous field problem, the analogue of the normal coordinates—or in physical terms, the *modes* of vibration—which were so helpful in the discrete case. Fortunately, the string's modes are very familiar. By imposing suitable boundary conditions at each end of the string, we determine the allowed wavelengths of waves travelling along the string. Suppose, for simplicity, that the string is stretched between $x = 0$ and $x = L$. This constrains $\phi(x, t)$ to vanish at these end points. A suitable form for $\phi(x, t)$ which does this is

$$\phi_r(x,t) = A_r(t)\sin\left(\frac{r\pi x}{L}\right) \tag{5.31}$$

Figure 5.3. String motion in two normal modes: (*a*) $r = 1$ in equation (5.31); (*b*) $r = 2$.

where $r = 1, 2, 3, \ldots$, which expresses the fact that an exact number of half-wavelengths must fit onto the interval $(0, L)$. Inserting (5.31) into (5.30), we find

$$\ddot{A}_r = -\omega_r^2 A_r \tag{5.32}$$

where

$$\omega_r^2 = r^2\pi^2c^2/L^2. \tag{5.33}$$

Thus the amplitude $A_r(t)$ of the particular waveform (5.31) executes simple harmonic motion with frequency ω_r. Each motion of the string which has a definite wavelength also has a definite frequency; it is therefore precisely a mode. Figure 5.3(*a*) shows two snapshots of the string when it is oscillating in the mode for which $r = 1$, and figure 5.3(*b*) shows the same for the mode $r = 2$; these may be compared with figures 5.2(*a*) and (*b*). Just as in the discrete case, the general motion of the string is a superposition of modes

$$\phi(x, t) = \sum_{r=1}^{\infty} A_r(t) \sin\left(\frac{r\pi x}{L}\right) \tag{5.34}$$

in short, a Fourier series!

We must now examine the total energy of the vibrating string, which we expect to be greatly simplified by the use of the mode concept. The total energy is the continuous analogue of the discrete summation in (5.25), namely the integral

$$E = \int_0^L \left[\frac{1}{2}\rho\left(\frac{\partial\phi}{\partial t}\right)^2 + \frac{1}{2}\rho c^2\left(\frac{\partial\phi}{\partial x}\right)^2\right] \mathrm{d}x \tag{5.35}$$

where the first term is the kinetic energy and the second is the potential energy (ρ is the mass per unit length of the string, assumed constant). As noted earlier, the potential energy term arises from an approximation which limits it to the quadratic power. To relate this to the earlier discrete case, note that the derivative may be regarded as $[\phi(x + \delta x) - \phi(x)]/\delta x$ as $\delta x \to 0$, so that the square of the derivative involves the 'nearest neighbour coupling' $\phi(x + \delta x)\phi(x)$, analogous to the q_1q_2 term in (5.9).

Inserting (5.34) into (5.35), and using the orthonormality of the sine functions on the interval $(0, L)$, one obtains (problem 5.1) the crucial result

$$E = (L/2)\sum_{r=1}^{\infty}[\tfrac{1}{2}\rho\dot{A}_r^2 + \tfrac{1}{2}\rho\omega_r^2A_r^2]. \tag{5.36}$$

Indeed, just as in the discrete case, the total energy of the string can be written as a sum of individual mode energies. We note that *the Fourier amplitude A_r acts as a normal coordinate*. Comparing (5.36) with (5.27), we see that the string behaves exactly like a system of independent uncoupled oscillators, the only difference being that now there are an infinite number of them, corresponding to the infinite number of degrees of freedom in the continuous field $\phi(x, t)$. The normal coordinates $A_r(t)$ are, for many purposes, a much more relevant set of degrees of freedom than the original displacements $\phi(x, t)$.

The final step is to apply quantum mechanics to this classical field system. Once again, the total energy is equivalent to that of a sum of (infinitely many) mode oscillators, each of which has to be quantized. The total energy eigenvalue has the form (5.29), except that now the sum extends to infinity:

$$E = \sum_{r=1}^{\infty} (n_r + \tfrac{1}{2})\hbar\omega_r. \tag{5.37}$$

The excited states of the quantized field $\phi(x, t)$ are characterized by saying how many phonons of each frequency are present; the ground state has no phonons at all. We remark that as $L \to \infty$, the mode sum in (5.36) or (5.37) will be replaced by an integral over a continuous frequency variable.

We have now completed, in outline, the programme introduced earlier, ending up with the quantization of a 'mechanical' system. All of the foregoing, it must be clearly emphasized, is absolutely basic to modern solid state physics. The essential idea—quantizing independent modes—can be applied to an enormous variety of 'oscillations'. In all cases the crucial concept is the elementary excitation—the mode quantum. Thus we have plasmons (quanta of plasma oscillations), magnons (magnetic oscillations), ..., as well as phonons (vibrational oscillations). All this is securely anchored in the physics of many-body systems.

Now we come to the use of these ideas as an *analogy*, to help us understand the (presumably non-mechanical) quantum fields with which we shall actually be concerned in this book—for example the electromagnetic field. Consider a region of space containing electromagnetic fields. These fields obey (a three-dimensional version of) the wave equation (5.30), with c now standing for the speed of light. By imposing suitable boundary conditions, the total electromagnetic energy in any region of space can be written as a sum of mode energies. Each mode has the form of an oscillator, whose amplitude is (see (5.31)) the Fourier component of the wave, for a given wavelength. These oscillators are all quantized. Their quanta are called photons. Thus, *a photon is an elementary quantum of excitation of the electromagnetic field.*

So far the only kind of 'particle' we have in our relativistic quantum field theoretic world is the photon. What about the electron, say? Well, recalling Feynman again, 'There is one lucky break, however—electrons behave just like light'. In other words, we shall also regard an electron as an elementary quantum of excitation of an 'electron field'. What is 'waving' to supply the vibrations for this electron field? We do not answer this question just as we did not for the photon. We *postulate* a relativistic quantum field for the electron which obeys some suitable wave equation—in this case, for non-interacting electrons, the Dirac equation. The field is expanded as a sum of Fourier components, as with the electromagnetic field. Each component behaves as an independent oscillator degree of freedom (and there are, of course, an infinite number of them); the quanta of these oscillators are electrons.

Actually this, though correctly expressing the basic idea, omits one crucial factor, which makes it almost fraudulently oversimplified. There is of course one very big difference between photons and electrons. The former are *bosons* and the latter are *fermions*; photons have spin angular momentum of one (in unit of $\hbar$), electrons of one-half. It is very difficult, if not downright impossible, to construct any mechanical model at all which has fermionic excitations. Phonons have spin-1, in fact, corresponding to the three states of polarization of the corresponding vibrational waves. But 'phonons' carrying spin-$\frac{1}{2}$ are hard to come by. No matter, you may say, Maxwell has weaned us away from jelly, so we shall be grown up and boldly postulate the electron field as a basic *thing*.

Certainly this is what we do. But we also know that fermionic particles, like electrons, have to obey an exclusion principle: no two identical fermions can have the same quantum numbers. In chapter 7, we shall learn how the idea sketched here must be modified for fields whose quanta are fermions.

5.2 The quantum field: (ii) Lagrange–Hamilton formulation

5.2.1 The action principle: Lagrangian particle mechanics

We must now make the foregoing qualitative picture more mathematically precise. It is clear that we would like a formalism capable of treating, within a single overall framework, the mechanics of both fields and particles, in both classical and quantum aspects. Remarkably enough, such a framework does exist (and was developed long before quantum field theory): Hamilton's principle of *least action*, with the action defined in terms of a *Lagrangian*. We strongly recommend the reader with no prior acquaintance with this profound approach to physical laws to read chapter 19 of volume 2 of Feynman's *Lectures on Physics* (Feynman 1964).

The least action approach differs radically from the more familiar one which can be conveniently be called 'Newtonian'. Consider the simplest case, that of classical particle mechanics. In the Newtonian approach, equations of motion are postulated which involve forces as the essential physical input; from these, the trajectories of the particle can be calculated. In the least action approach, equations of motion are *not* postulated as basic, and the primacy of forces yields to that of *potentials*. The path by which a particle actually travels is determined by the postulate (or principle) that it has to follow that particular path, out of infinitely

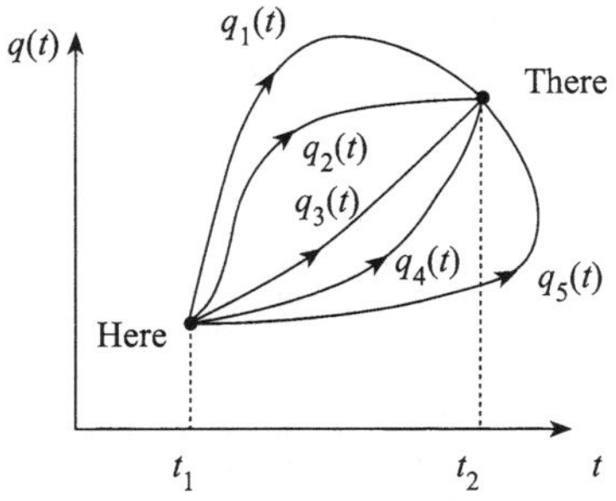

Figure 5.4. Possible space–time trajectories from ‘here’ ($q(t_1)$) to ‘there’ ($q(t_2)$).

many possible ones, for which a certain quantity—the *action*—is minimized. The action S is defined by

$$S = \int_{t_1}^{t_2} L(q(t), \dot{q}(t))\, \mathrm{d}t \tag{5.38}$$

where $q(t)$ is the position of the particle as a function of time, $\dot{q}(t)$ is its velocity and the all-important function L is the Lagrangian. Given L as an explicit function of the variables $q(t)$ and $\dot{q}(t)$, we can imagine evaluating S for all sorts of possible $q(t)$'s starting at time t_1 and ending at time t_2. We can draw these different *possible trajectories* on a q versus t diagram as in figure 5.4. For each path we evaluate S: the *actual* path is the one for which S is smallest, by hypothesis.

But what is L? In simple cases (as we shall verify later) L is just $T - V$, the difference of kinetic and potential energies. Thus for a single particle in a potential V

$$L = \tfrac{1}{2}m\dot{x}^2 - V(x). \tag{5.39}$$

Knowing $V(x)$, we can try and put the ‘action principle’ into action. However, how can we set about finding which trajectory minimizes S? It is quite interesting to play with some simple specific examples and actually calculate S for several ‘fictitious’ trajectories—i.e. ones that we know from the Newtonian approach are *not* followed by the particle—and try and get a feeling for what the actual trajectory that minimizes S might be like (of course it is the Newtonian one—see problem 5.2). But clearly this is not a practical answer to the general problem of finding the $q(t)$ that minimizes S. Actually, we can solve this problem by calculus.

Our problem is something like the familiar one of finding the point t_0 at which a certain function $f(t)$ has a stationary value. In the present case, however, the function S is not a simple function of t—rather it is a function of the entire set of points $q(t)$. It is a *function of the function* $q(t)$, or a ‘*functional*’ of $q(t)$. We want to know what particular ‘$q_c(t)$’ minimizes S.

By analogy with the single-variable case, we consider a small variation $\delta q(t)$ in the path from $q(t_1)$ to $q(t_2)$. At the minimum, the change δS corresponding to the change δq must vanish. This change in the action is given by

$$\delta S = \int_{t_1}^{t_2} \left(\frac{\partial L}{\partial q(t)} \delta q(t) + \frac{\partial L}{\partial \dot{q}(t)} \delta \dot{q}(t) \right) \mathrm{d}t. \tag{5.40}$$

Using $\delta \dot{q}(t) = \mathrm{d}(\delta q(t))/\mathrm{d}t$ and integrating the second term by parts yields

$$\delta S = \int_{t_1}^{t_2} \delta q(t) \left[\frac{\partial L}{\partial q(t)} - \frac{\mathrm{d}}{\mathrm{d}t} \frac{\partial L}{\partial \dot{q}(t)} \right] \mathrm{d}t + \left[\frac{\partial L}{\partial \dot{q}(t)} \delta q(t) \right]_{t_1}^{t_2}. \tag{5.41}$$

Since we are considering variations of path in which all trajectories start at t_1 and end at t_2, $\delta q(t_1) = \delta q(t_2) = 0$. So the condition that S be stationary is

$$\delta S = \int_{t_1}^{t_2} \delta q(t) \left[\frac{\partial L}{\partial q(t)} - \frac{\mathrm{d}}{\mathrm{d}t} \frac{\partial L}{\partial \dot{q}(t)} \right] \mathrm{d}t = 0. \tag{5.42}$$

Since this must be true for *arbitrary* $\delta q(t)$, we must have

$$\boxed{\frac{\partial L}{\partial q(t)} - \frac{\mathrm{d}}{\mathrm{d}t} \frac{\partial L}{\partial \dot{q}(t)} = 0.} \tag{5.43}$$

This is the celebrated Euler–Lagrange equation of motion. Its solution gives the '$q_c(t)$' which the particle actually follows.

We can see how this works for the simple case (5.39) where q is the coordinate x. We have immediately

$$\partial L/\partial \dot{x} = m\dot{x} = p \tag{5.44}$$

and

$$\partial L/\partial x = -\partial V/\partial x = F \tag{5.45}$$

where p and F are, respectively, the momentum and the force of the Newtonian approach. The Euler–Lagrange equation then reads

$$F = \mathrm{d}p/\mathrm{d}t \tag{5.46}$$

precisely the Newtonian equation of motion. For the special case of a harmonic oscillator (obviously fundamental for the quantum field idea, as section 5.1 should have made clear), we have

$$L = \tfrac{1}{2}m\dot{x}^2 - \tfrac{1}{2}m\omega^2 x^2 \tag{5.47}$$

which can be immediately generalized to N independent oscillators (see section 5.1) via

$$L = \sum_{r=1}^{N} \left(\tfrac{1}{2}m\dot{Q}_r^2 - \tfrac{1}{2}m\omega_r^2 Q_r^2\right). \tag{5.48}$$

For many dynamical systems, the Lagrangian has the form '$T - V$' indicated in (5.47) and (5.48). We now consider the passage to quantum mechanics.

5.2.2 Quantum mechanics à la Heisenberg–Lagrange–Hamilton

It seems likely that a particularly direct correspondence between the quantum and the classical cases will be obtained if we use the Heisenberg formulation (or 'picture') of quantum mechanics (see appendix I). In the Schrödinger picture, the dynamical variables such as position x are independent of time, and the time dependence is carried by the wavefunction. Thus we seem to have nothing like the $q(t)$'s. However, one can always do a unitary transformation to the Heisenberg picture, in which the wavefunction is fixed and the dynamical variables change with time. This is what we want in order to parallel the Lagrangian quantities $q(t)$. But of course there is one fundamental difference between quantum mechanics and classical mechanics: in the former, the dynamical variables are operators which in general do not commute. In particular, the fundamental commutator states that ($\hbar = 1$)

$$[\hat{q}(t), \hat{p}(t)] = \mathrm{i} \tag{5.49}$$

where $\hat{}$ indicates the operator character of the quantity. Here $\hat{p}$ is defined by the generalization of (5.44):

$$\hat{p} = \partial \hat{L} / \partial \dot{\hat{q}}. \tag{5.50}$$

In this formulation of quantum mechanics we do not have the Schrödinger-type equation of motion. Instead we have the Heisenberg equation of motion

$$\dot{\hat{A}} = -\mathrm{i}[\hat{A}, \hat{H}] \tag{5.51}$$

where the *Hamiltonian* operator $\hat{H}$ is defined in terms of the Lagrangian operator $\hat{L}$ by

$$\hat{H} = \hat{p}\dot{\hat{q}} - \hat{L} \tag{5.52}$$

and $\hat{A}$ is any dynamical observable. For example, in the oscillator case

$$\hat{L} = \tfrac{1}{2} m \dot{\hat{q}}^2 - \tfrac{1}{2} m \omega^2 \hat{q}^2 \tag{5.53}$$

$$\hat{p} = m \dot{\hat{q}} \tag{5.54}$$

and

$$\hat{H} = \frac{1}{2m} \hat{p}^2 + \frac{1}{2} m \omega^2 \hat{q}^2 \tag{5.55}$$

which is the total energy operator. Note that $\hat{p}$, obtained from the Lagrangian using (5.50), had better be consistent with the Heisenberg equation of motion for the operator $\hat{A} = \hat{q}$. The Heisenberg equation of motion for $\hat{A} = \hat{p}$ leads to

$$\dot{\hat{p}} = -m \omega^2 \hat{q} \tag{5.56}$$

which is an operator form of Newton's law for the harmonic oscillator. Using the expression for $\hat{p}$ (5.54), we find

$$\ddot{\hat{q}} = -\omega^2 \hat{q}. \tag{5.57}$$

Now, although this looks like the familiar classical equation of motion for the position of the oscillator—and recovering it from the Lagrangian formalism is encouraging—we must be very careful to appreciate that this is an equation stating how an *operator* evolves with time. Where the quantum particle will actually be *found* is an entirely different matter. By sandwiching (5.57) between wavefunctions, we can at once see that the *average* position of the particle will follow the classical trajectory (remember that wavefunctions are independent of time in the Heisenberg formulation). But *fluctuations* about this trajectory will certainly occur: a quantum particle does not follow a ray-like classical trajectory. Come to think of it, neither does a photon!

In the original formulations of quantum theory, such fluctuations were generally taken to imply that the very notion of a 'path' was no longer a useful one. However, just as the differential equations satisfied by operators in the Heisenberg picture are quantum generalizations of Newtonian mechanics, so there is an analogous quantum generalization of the 'path-contribution to the action' approach to classical mechanics. The idea was first hinted at by Dirac (1933, 1981, section 32), but it was Feynman who worked it out completely. The book by Feynman and Hibbs (1965) presents a characteristically fascinating discussion—here we only wish to indicate the central idea. We ask: how does a particle get from the point $q(t_1)$ at time t_1 to the point $q(t_2)$ at t_2? Referring back to figure 5.4, in the classical case we imagined (infinitely) many possible paths $q_i(t)$, of which, however, only *one* was the actual path followed, namely the one we called $q_c(t)$ which minimized the action integral (5.38) as a functional of $q(t)$. In the quantum case, however, we previously noted that a particle will no longer follow any definite path, because of quantum fluctuations. But rather than, as a consequence, throwing away the whole idea of a path, Feynman's insight was to appreciate that the 'opposite' viewpoint is also possible: since unique paths are forbidden in quantum theory, we should in principle include *all possible* paths! In other words, we take all the trajectories on figure 5.4 as physically possible (together with all the other infinitely many ways of accomplishing the trip).

However, surely not all paths are equally *likely*: after all, we must presumably recover the classical trajectory as $\hbar \to 0$, in some sense. Thus we must find an appropriate weighting for the paths. Feynman's recipe is beautifully simple: weight each path by the factor

$$\mathrm{e}^{\mathrm{i}S/\hbar} \tag{5.58}$$

where S is the action for that particular path. At first sight this is a rather strange proposal, since all paths—even the classical one—are weighted by a quantity which is of unit modulus. But of course contributions of the form (5.58) from all the paths have to be added coherently—just as we superposed the amplitudes in the 'two-slit' discussion in section 3.5. What distinguishes the classical path $q_c(t)$ is that it makes S stationary under small changes of path: thus in its vicinity paths have a strong tendency to add up constructively, while far from it the phase factors will tend to produce cancellations. The amount a quantum particle can 'stray' from the classical path depends on the magnitude of the corresponding action relative to $\hbar$, the quantum of action: the scale of coherence is set by $\hbar$.

In summary, then, the quantum mechanical amplitude to go from $q(t_1)$ to $q(t_2)$ is proportional to

$$\sum_{\text{all paths } q(t)} \exp\left(\frac{\mathrm{i}}{\hbar}\int_{t_1}^{t_2} L(q(t), \dot{q}(t))\,\mathrm{d}t\right). \tag{5.59}$$

There is an evident generalization to quantum field theory. We shall not, however, make use of the 'path integral' approach to quantum field theory in this volume. Its use was, in fact, decisive in obtaining the Feynman rules for non-Abelian gauge theories; and it is the only approach suitable for numerical studies of quantum field theories (how can operators be simulated numerically?). Nevertheless, for a first introduction to quantum field theory, there is still much to be said for the traditional approach based on 'quantizing the modes', and this is the path we shall follow in the rest of this volume. Not the least of its advantages is that it contains the intuitively powerful 'calculus' of creation and annihilation operators, as we now describe.

5.2.3 Interlude: the quantum oscillator

As we saw in section 5.1, we need to know the energy spectrum and associated states of a quantum harmonic oscillator. This is a standard problem, but there is one particular way of solving it—the 'operator' approach due to Dirac (1981, chapter 6)—that is so crucial to all subsequent development that we include a discussion here in the body of the text.

For the oscillator Hamiltonian

$$\hat{H} = \frac{1}{2m}\hat{p}^2 + \frac{1}{2}m\omega^2\hat{q}^2 \tag{5.60}$$

if $\hat{p}$ and $\hat{q}$ were not operators, we could attempt to factorize the Hamiltonian in the form '$(q + \mathrm{i}p)(q - \mathrm{i}p)$' (apart from the factors of $2m$ and ω). In the quantum case, in which $\hat{p}$ and $\hat{q}$ do not commute, it still turns out to be very helpful to introduce such combinations. If we define the operator

$$\hat{a} = \frac{1}{\sqrt{2}}\left(\sqrt{m\omega}\hat{q} + \frac{\mathrm{i}}{\sqrt{m\omega}}\hat{p}\right) \tag{5.61}$$

and its Hermitian conjugate

$$\hat{a}^\dagger = \frac{1}{\sqrt{2}}\left(\sqrt{m\omega}\hat{q} - \frac{\mathrm{i}}{\sqrt{m\omega}}\hat{p}\right) \tag{5.62}$$

the Hamiltonian may be written as (see problem 5.4)

$$\hat{H} = \tfrac{1}{2}(\hat{a}^\dagger\hat{a} + \hat{a}\hat{a}^\dagger)\omega = (\hat{a}^\dagger\hat{a} + \tfrac{1}{2})\omega. \tag{5.63}$$

The second form for $\hat{H}$ may be obtained from the first using the commutation relation between $\hat{a}$ and $\hat{a}^\dagger$

$$[\hat{a}, \hat{a}^\dagger] = 1 \tag{5.64}$$

derived using the fundamental commutator between $\hat{p}$ and $\hat{q}$. Using this basic commutator (5.64) and our expression for $\hat{H}$, (5.63), one can prove the relations (see problem 5.4)

$$\begin{aligned} [\hat{H}, \hat{a}] &= -\omega\hat{a} \\ [\hat{H}, \hat{a}^\dagger] &= \omega\hat{a}^\dagger. \end{aligned} \tag{5.65}$$

Consider now a state $|n\rangle$ which is an eigenstate of $\hat{H}$ with energy E_n:

$$\hat{H}|n\rangle = E_n|n\rangle. \tag{5.66}$$

Using this definition and the commutators (5.65), we can calculate the energy of the states $(\hat{a}^\dagger|n\rangle)$ and $(\hat{a}|n\rangle)$. We find

$$\hat{H}(\hat{a}^\dagger|n\rangle) = (E_n + \omega)(\hat{a}^\dagger|n\rangle) \tag{5.67}$$

$$\hat{H}(\hat{a}|n\rangle) = (E_n - \omega)(\hat{a}|n\rangle). \tag{5.68}$$

Thus the operators $\hat{a}$ and $\hat{a}^\dagger$ respectively raise and lower the energy of $|n\rangle$ by one unit of ω ($\hbar = 1$). Now since $\hat{H} \sim \hat{p}^2 + \hat{q}^2$ with $\hat{p}$ and $\hat{q}$ Hermitian, we can prove that $\langle\psi|\hat{H}|\psi\rangle$ is positive-definite for any state $|\psi\rangle$. Thus the operator $\hat{a}$ cannot lower the energy indefinitely: there must exist a lowest state $|0\rangle$ such that

$$\boxed{\hat{a}|0\rangle = 0.} \tag{5.69}$$

This defines the lowest-energy state of the system; its energy is

$$\hat{H}|0\rangle = \tfrac{1}{2}\omega|0\rangle \tag{5.70}$$

the 'zero-point energy' of the quantum oscillator. The first excited state is

$$|1\rangle = \hat{a}^\dagger|0\rangle \tag{5.71}$$

with energy $(1 + \frac{1}{2})\omega$. The nth state has energy $(n + \frac{1}{2})\omega$ and is proportional to $(\hat{a}^\dagger)^n|0\rangle$. To obtain a normalization

$$\langle n|n\rangle = 1 \tag{5.72}$$

the correct normalization factor can be shown to be (problem 5.4)

$$|n\rangle = \frac{1}{\sqrt{n!}}(\hat{a}^\dagger)^n|0\rangle. \tag{5.73}$$

Returning to the eigenvalue equation for $\hat{H}$, we have arrived at the result

$$\hat{H}|n\rangle = (\hat{a}^\dagger\hat{a} + \tfrac{1}{2})\omega|n\rangle = (n + \tfrac{1}{2})\omega|n\rangle \tag{5.74}$$

so that the state $|n\rangle$ defined by (5.73) is an eigenstate of the *number operator* $\hat{n} = \hat{a}^\dagger \hat{a}$, with integer eigenvalue n:

$$\hat{n}|n\rangle = n|n\rangle. \tag{5.75}$$

It is straightforward to generalize all the foregoing to a system whose Lagrangian is a sum of N independent oscillators, as in (5.48):

$$\hat{L} = \sum_{r=1}^{N} (\tfrac{1}{2}m\dot{\hat{q}}_r^2 - \tfrac{1}{2}m\omega_r^2\hat{q}_r^2). \tag{5.76}$$

The required generalization of the basic commutation relations (5.49) is

$$\begin{aligned} [\hat{q}_r, \hat{p}_s] &= \mathrm{i}\delta_{rs} \\ [\hat{q}_r, \hat{q}_s] &= [\hat{p}_r, \hat{p}_s] = 0 \end{aligned} \tag{5.77}$$

since the different oscillators labelled by the index r or s are all independent. The Hamiltonian is (cf (5.27))

$$\hat{H} = \sum_{r=1}^{N} [(1/2m)\hat{p}_r^2 + \tfrac{1}{2}m\omega_r^2\hat{q}_r^2] \tag{5.78}$$

$$= \sum_{r=1}^{N} (\hat{a}_r^\dagger \hat{a}_r + \tfrac{1}{2})\omega_r \tag{5.79}$$

with $\hat{a}_r$ and $\hat{a}_r^\dagger$ defined via the analogues of (5.61) and (5.62). Since the eigenvalues of *each* number operator $\hat{n}_r = \hat{a}_r^\dagger \hat{a}_r$ are n_r, by the previous results, the eigenvalues of $\hat{H}$ indeed have the form (5.29),

$$E = \sum_{r=1}^{N} (n_r + \tfrac{1}{2})\omega_r. \tag{5.80}$$

The corresponding eigenstates are products $|n_1\rangle|n_2\rangle\cdots|n_N\rangle$ of N individual oscillator eigenstates, where $|n_r\rangle$ contains n_r quanta of excitation, of frequency ω_r; the product state is usually abbreviated to $|n_1, n_2, \ldots, n_N\rangle$. In the ground state of the system, each individual oscillator is unexcited: this state is $|0, 0, \ldots, 0\rangle$, which is abbreviated to $|0\rangle$, where it is understood that

$$\hat{a}_r|0\rangle = 0 \qquad \text{for all } r. \tag{5.81}$$

The operators $\hat{a}_r^\dagger$ *create oscillator quanta*; the operators $\hat{a}_r$ *destroy oscillator quanta*.

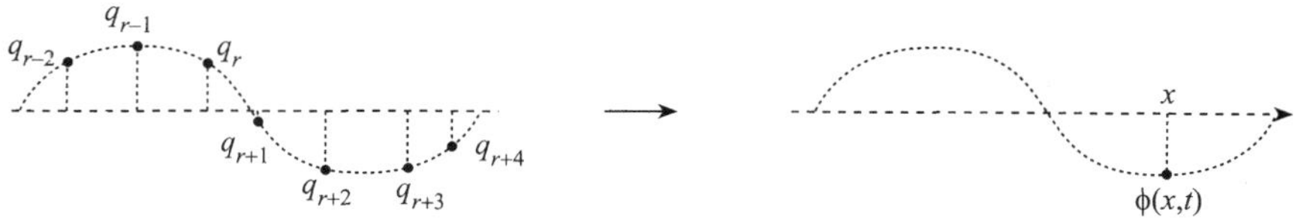

Figure 5.5. The passage from a large number of discrete degrees of freedom (mass points) to a continuous degree of freedom (field).

5.2.4 Lagrange–Hamilton classical field mechanics

We now consider how to use the Lagrange–Hamilton approach for a *field*, starting again with the classical case and limiting ourselves to one dimension to start with.

As explained in the previous section, we shall have in mind the $N \to \infty$ limit of the N degrees of freedom case

$$\{q_r(t); r = 1, 2, \ldots, N\} \underset{N\to\infty}{\longrightarrow} \phi(x, t) \tag{5.82}$$

where x is now a continuous variable labelling the displacement of the 'string' (to picture a concrete system, see figure 5.5). At each point x we have an independent degree of freedom $\phi(x, t)$—thus the *field* system has a 'continuous infinity' of degrees of freedom. We now formulate everything in terms of a Lagrangian density $\mathcal{L}$:

$$S = \int \mathrm{d}t\, L \tag{5.83}$$

where (in one dimension)

$$L = \int \mathrm{d}x\, \mathcal{L}. \tag{5.84}$$

Equation (5.82) suggests that ϕ has dimension of [length], and since in the discrete case $L = T - V$, $\mathcal{L}$ has dimension [energy/length]. (In general $\mathcal{L}$ has dimension [energy/volume].)

A new feature arises because ϕ is now a continuous function of x, so that $\mathcal{L}$ can depend on $\partial\phi/\partial x$ as well as on ϕ and $\dot{\phi} = \partial\phi/\partial t$: $\mathcal{L} = \mathcal{L}(\phi, \partial\phi/\partial x, \dot{\phi})$.

As before, we postulate the same fundamental principle

$$\delta S = 0 \tag{5.85}$$

meaning that the dynamics of the field ϕ is governed by minimizing S. This time the total variation is given by

$$\delta S = \int \mathrm{d}t \int \left[\frac{\partial\mathcal{L}}{\partial\phi}\delta\phi + \frac{\partial\mathcal{L}}{\partial(\partial\phi/\partial x)}\delta\left(\frac{\partial\phi}{\partial x}\right) + \frac{\partial\mathcal{L}}{\partial\dot{\phi}}\delta\dot{\phi}\right] \mathrm{d}x. \tag{5.86}$$

Integrating the $\delta\dot{\phi}$ by parts in t, and the $\delta(\partial\phi/\partial x)$ by parts in x, and discarding the resulting 'surface' terms, we obtain

$$\delta S = \int \mathrm{d}t \int \mathrm{d}x\, \delta\phi \left[\frac{\partial\mathcal{L}}{\partial\phi} - \frac{\partial}{\partial x}\left(\frac{\partial\mathcal{L}}{\partial(\partial\phi/\partial x)}\right) - \frac{\partial}{\partial t}\left(\frac{\partial\mathcal{L}}{\partial\dot{\phi}}\right)\right]. \tag{5.87}$$

Since $\delta\phi$ is an *arbitrary* function, the requirement $\delta S = 0$ yelds the Euler–Lagrange *field* equation

$$\frac{\partial \mathcal{L}}{\partial \phi} - \frac{\partial}{\partial x}\left(\frac{\partial \mathcal{L}}{\partial(\partial\phi/\partial x)}\right) - \frac{\partial}{\partial t}\left(\frac{\partial \mathcal{L}}{\partial \dot{\phi}}\right) = 0. \tag{5.88}$$

The generalization to three dimensions is

$$\boxed{\frac{\partial \mathcal{L}}{\partial \phi} - \nabla\cdot\left(\frac{\partial \mathcal{L}}{\partial(\nabla\phi)}\right) - \frac{\partial}{\partial t}\left(\frac{\partial \mathcal{L}}{\partial \dot{\phi}}\right) = 0.} \tag{5.89}$$

As an example, consider

$$\mathcal{L}_\rho = \frac{1}{2}\rho\left(\frac{\partial\phi}{\partial t}\right)^2 - \frac{1}{2}\rho c^2\left(\frac{\partial\phi}{\partial x}\right)^2 \tag{5.90}$$

where the factor ρ (mass density) and c (a velocity) have been introduced to get the dimension of $\mathcal{L}$ right. Inserting this into the Euler–Lagrangian field equation (5.88), we obtain

$$\frac{\partial^2\phi}{\partial x^2} - \frac{1}{c^2}\frac{\partial^2\phi}{\partial t^2} = 0 \tag{5.91}$$

which is precisely the wave equation (5.30), now obtained via the Euler–Lagrange field equations. Note that the Lagrange density $\mathcal{L}$ has the expected form (cf (5.48)) of 'kinetic energy density minus potential energy density'.

For the final step—the passage to quantum mechanics for a field system—we shall be interested in the Hamiltonian (total energy) of the system, just as we were for the discrete case. Though we shall not actually *use* the Hamiltonian in the classical field case, we shall introduce it here, generalizing it to the quantum theory in the following section. We recall that Hamiltonian mechanics is formulated in terms of coordinate variables ('q') and momentum variables ('p'), rather than the q and $\dot{q}$ of Lagrangian mechanics. In the continuum (field) case, the Hamiltonian H is written as the integral of a density $\mathcal{H}$ (we remain in one dimension)

$$H = \int \mathrm{d}x\, \mathcal{H} \tag{5.92}$$

while the coordinates $q_r(t)$ become the 'coordinate field' $\phi(x, t)$. The question is what is the corresponding 'momentum field'?

The answer to this is provided by a continuum version of the generalized momentum derived from the Lagrangian approach (cf equation (5.44))

$$p = \partial L/\partial\dot{q}. \tag{5.93}$$

We define a 'momentum field' $\pi(x, t)$—technically called the 'momentum canonically conjugate to ϕ'—by

$$\pi(x, t) = \partial\mathcal{L}/\partial\dot{\phi}(x, t) \tag{5.94}$$

where $\mathcal{L}$ is now the Lagrangian density. Note that π has dimensions of a momentum density. In the classical particle mechanics case we define the Hamiltonian by

$$H(p, q) = p\dot{q} - L. \tag{5.95}$$

Here we define a Hamiltonian density $\mathcal{H}$ by

$$\mathcal{H}(\phi, \pi) = \pi(x, t)\dot{\phi}(x, t) - \mathcal{L}. \tag{5.96}$$

Let us see how this works for the one-dimensional string with $\mathcal{L}$ given by

$$\mathcal{L}_\rho = \frac{1}{2}\rho\left(\frac{\partial\phi}{\partial t}\right)^2 - \frac{1}{2}\rho c^2\left(\frac{\partial\phi}{\partial x}\right)^2. \tag{5.97}$$

We have

$$\pi(x, t) = \rho\partial\phi/\partial t \tag{5.98}$$

and

$$\begin{aligned}\mathcal{H}_\rho &= \frac{1}{\rho}\pi^2 - \frac{1}{2}\left[\frac{1}{\rho}\pi^2 - \rho c^2\left(\frac{\partial\phi}{\partial x}\right)^2\right] \\ &= \frac{1}{2}\left[\frac{1}{\rho}\pi^2 + \rho c^2\left(\frac{\partial\phi}{\partial x}\right)^2\right]\end{aligned} \tag{5.99}$$

so that

$$H_\rho = \int\left[\frac{1}{2\rho}\pi^2(x, t) + \frac{1}{2}\rho c^2\left(\frac{\partial\phi(x, t)}{\partial x}\right)^2\right]\mathrm{d}x. \tag{5.100}$$

This has exactly the form we expect (see (5.35)), thus verifying the plausibility of the above prescription.

5.2.5 Heisenberg–Lagrange–Hamilton quantum field mechanics

We must now *quantize* the classical field formalism: *a priori*, it is not at all obvious how to set about this task. Indeed, the fact that this formalism indicates a definite answer to this problem is one reason for developing the quite elaborate Lagrange–Hamilton field formalism in the first place. (Another is that it is very well suited to the analysis of symmetries, a crucial aspect of gauge theories—see chapter 7.) We have introduced the 'coordinate-like' field $\phi(x, t)$ and also the 'momentum-like' field $\pi(x, t)$. To pass to the quantized version of the field theory, we mimic the procedure followed in the discrete case and promote both the quantities ϕ and π to operators $\hat{\phi}$ and $\hat{\pi}$, in the Heisenberg picture. As usual, the distinctive feature of quantum theory is the non-commutativity of certain basic quantities in the theory—for example, the fundamental commutator ($\hbar = 1$)

$$[\hat{q}_r(t), \hat{p}_s(t)] = \mathrm{i}\delta_{rs} \tag{5.101}$$

of the discrete case. Thus we expect that the operators $\hat{\phi}$ and $\hat{\pi}$ will obey some commutation relation which is a continuum generalization of (5.101). The commutator will be of the form $[\hat{\phi}(x,t), \hat{\pi}(y,t)]$, since—recalling figure 5.5—the discrete index r or s becomes the continuous variable x or y; we also note that (5.101) is between operators at equal times. The continuum generalization of the δ_{rs} symbol is the Dirac δ function, $\delta(x-y)$, with the properties

$$\int_{-\infty}^{\infty} \delta(x)\,\mathrm{d}x = 1 \tag{5.102}$$

$$\int_{-\infty}^{\infty} \delta(x-y) f(x)\,\mathrm{d}x = f(y) \tag{5.103}$$

for all reasonable functions f (see appendix E). Thus the fundamental commutator of quantum field theory is taken to be

$$\boxed{[\hat{\phi}(x,t), \hat{\pi}(y,t)] = \mathrm{i}\delta(x-y)} \tag{5.104}$$

in the one-dimensional case, with obvious generalization to the three-dimensional case via the symbol $\delta^3(\boldsymbol{x}-\boldsymbol{y})$. Remembering that we have set $\hbar = 1$, it is straightforward to check that the dimensions are consistent on both sides. Variables $\hat{\phi}$ and $\hat{\pi}$ obeying such a commutation relation are said to be 'conjugate' to each other.

What about the commutator of two $\hat{\phi}$'s or two $\hat{\pi}$'s? In the discrete case, two different $\hat{q}$'s (in the Heisenberg picture) will commute at equal times, $[\hat{q}_r(t), \hat{q}_s(t)] = 0$, and so will two different $\hat{p}$'s. We therefore expect to supplement (5.104) with

$$[\hat{\phi}(x,t), \hat{\phi}(y,t)] = [\hat{\pi}(x,t), \hat{\pi}(y,t)] = 0. \tag{5.105}$$

Let us now proceed to explore the effect of these fundamental commutator assumptions, for the case of the Lagrangian which yielded the wave equation via the Euler–Lagrange equations, namely

$$\hat{\mathcal{L}}_\rho = \frac{1}{2}\rho\left(\frac{\partial\hat{\phi}}{\partial t}\right)^2 - \frac{1}{2}\rho c^2\left(\frac{\partial\hat{\phi}}{\partial x}\right)^2. \tag{5.106}$$

If we remove ρ, and set $c = 1$, we obtain

$$\hat{\mathcal{L}} = \frac{1}{2}\left(\frac{\partial\hat{\phi}}{\partial t}\right)^2 - \frac{1}{2}\left(\frac{\partial\hat{\phi}}{\partial x}\right)^2 \tag{5.107}$$

for which the Euler–Lagrangian equation yields the field equation

$$\frac{\partial^2\hat{\phi}}{\partial t^2} - \frac{\partial^2\hat{\phi}}{\partial x^2} = 0. \tag{5.108}$$

We can think of (5.108) as a highly simplified (spin-0, one-dimensional) version of the wave equation satisfied by the electromagnetic potentials. We may guess, then, that the associated quanta are massless, as we shall soon confirm.

The Lagrangian (5.107) is our prototype quantum field Lagrangian. Applying the quantized version of (5.94) we then have

$$\hat{\pi}(x,t) = \frac{\partial \mathcal{L}}{\partial \dot{\hat{\phi}}(x,t)} = \dot{\hat{\phi}}(x,t) \tag{5.109}$$

and the Hamiltonian density is

$$\hat{\mathcal{H}} = \hat{\pi}\dot{\hat{\phi}} - \hat{\mathcal{L}} = \frac{1}{2}\hat{\pi}^2 + \frac{1}{2}\left(\frac{\partial\hat{\phi}}{\partial x}\right)^2. \tag{5.110}$$

The total Hamiltonian is

$$\hat{H} = \int \hat{\mathcal{H}}\,\mathrm{d}x = \int \frac{1}{2}\left[\hat{\pi}^2 + \left(\frac{\partial\hat{\phi}}{\partial x}\right)^2\right]\mathrm{d}x. \tag{5.111}$$

In this form it is not immediately clear how to find the eigenvalues and eigenstates of $\hat{H}$. However, it is exactly at this point that all our preliminary work on *normal modes* comes into its own. If we can write the Hamiltonian as some kind of sum over independent oscillators—i.e. modes—we shall know how to proceed. But we learned that for the classical 'string' field considered in section 5.1, the mode expansion was simply a Fourier expansion. In the present case we do not particularly want to consider a field defined on a finite interval $(0, L)$ (though sometimes one does 'put the field in a box' in this way); we shall effectively let $L \to \infty$, so that the Fourier series is replaced by a Fourier integral, and standing waves are replaced by travelling waves. For the classical field obeying the wave equation (5.30) there are plane-wave solutions

$$\phi(x,t) \propto \mathrm{e}^{\mathrm{i}kx-\mathrm{i}\omega t} \tag{5.112}$$

where $(c = 1)$

$$\omega = k \tag{5.113}$$

which is just the dispersion relation of light *in vacuo*. The general field may be Fourier expanded in terms of these solutions:

$$\phi(x,t) = \int_{-\infty}^{\infty} \frac{\mathrm{d}k}{2\pi\sqrt{2\omega}}[a(k)\mathrm{e}^{\mathrm{i}kx-\mathrm{i}\omega t} + a^*(k)\mathrm{e}^{-\mathrm{i}kx+\mathrm{i}\omega t}] \tag{5.114}$$

where we have required ϕ to be real. (The rather fussy factors $(2\pi\sqrt{2\omega})^{-1}$ are purely conventional, and determine the normalization of the expansion coefficients a, a^* and $\hat{a}$, $\hat{a}^\dagger$ later; in turn, the latter enter into the definition, and normalization, of the states—see (5.130)). Similarly, the 'momentum field' $\pi = \dot{\phi}$ is expanded as

$$\pi = \int_{-\infty}^{\infty} \frac{\mathrm{d}k}{2\pi\sqrt{2\omega}}(-\mathrm{i}\omega)[a(k)\mathrm{e}^{\mathrm{i}kx-\mathrm{i}\omega t} - a^*(k)\mathrm{e}^{-\mathrm{i}kx+\mathrm{i}\omega t}]. \tag{5.115}$$

We quantize these mode expressions by promoting $\phi \to \hat{\phi}$, $\pi \to \hat{\pi}$ and assuming the commutator (5.104). Thus we write

$$\hat{\phi} = \int_{-\infty}^{\infty} \frac{\mathrm{d}k}{2\pi\sqrt{2\omega}}[\hat{a}(k)\mathrm{e}^{\mathrm{i}kx-\mathrm{i}\omega t} + \hat{a}^\dagger(k)\mathrm{e}^{-\mathrm{i}kx+\mathrm{i}\omega t}] \tag{5.116}$$

and similarly for $\hat{\pi}$. The commutator (5.104) now *determines* the commutators of the *mode operators* $\hat{a}$ and $\hat{a}^\dagger$:

$$\begin{aligned}[][\hat{a}(k), \hat{a}^\dagger(k')] &= 2\pi\delta(k-k') \\ [\hat{a}(k), \hat{a}(k')] &= [\hat{a}^\dagger(k), \hat{a}^\dagger(k')] = 0\end{aligned} \tag{5.117}$$

as shown in problem 5.6. These are the desired continuum analogues of the *discrete* oscillator commutation relations

$$\begin{aligned}[][\hat{a}_r, \hat{a}_s^\dagger] &= \delta_{rs} \\ [\hat{a}_r, \hat{a}_s] &= [\hat{a}_r^\dagger, \hat{a}_s^\dagger] = 0.\end{aligned} \tag{5.118}$$

The precise factor in front of the δ-function in (5.117) depends on the normalization choice made in the expansion of $\hat{\phi}$, (5.116). Problem 5.6 also shows that the commutation relations (5.117) lead to (5.105) as expected.

The form of the $\hat{a}$, $\hat{a}^\dagger$ commutation relations (5.117) already suggests that the $\hat{a}(k)$ and $\hat{a}^\dagger(k)$ operators are precisely the single-quantum destruction and creation operators for the continuum problem. To verify this interpretation and find the eigenvalues of $\hat{H}$, we now insert the expansion for $\hat{\phi}$ and $\hat{\pi}$ into $\hat{H}$. One finds the remarkable result (problem 5.7)

$$\hat{H} = \int_{-\infty}^{\infty} \frac{\mathrm{d}k}{2\pi}\left\{\frac{1}{2}[\hat{a}^\dagger(k)\hat{a}(k) + \hat{a}(k)\hat{a}^\dagger(k)]\omega\right\}. \tag{5.119}$$

Comparing this with the single-oscillator result

$$\hat{H} = \tfrac{1}{2}(\hat{a}^\dagger\hat{a} + \hat{a}\hat{a}^\dagger)\omega \tag{5.120}$$

shows that, as anticipated in section 5.1, each classical mode of the 'string' field can be quantized, and behaves like a separate oscillator coordinate, with its own frequency $\omega = k$. The operator $\hat{a}^\dagger(k)$ creates, and $\hat{a}(k)$ destroys, a quantum of the k mode. The factor $(2\pi)^{-1}$ in $\hat{H}$ arises from our normalization choice.

We note that in the field operator $\hat{\phi}$ of (5.116), those terms which destroy quanta go with the factor $\mathrm{e}^{-\mathrm{i}\omega t}$, while those which create quanta go with $e^{+i\omega t}$. This choice is deliberate and is consistent with the 'absorption' and 'emission' factors $e^{\pm i\omega t}$ of ordinary time-dependent perturbation theory in quantum mechanics (cf equation (A.33) of appendix A).

What is the mass of these quanta? We know that their frequency ω is related to their wavenumber k by (5.113), which—restoring $\hbar$'s and c's—can be regarded as equivalent to $\hbar\omega = \hbar ck$, or $E = cp$, where we use the Einstein and de Broglie relations. This is precisely the E–p relation appropriate to a *massless* particle, as expected.

What is the energy spectrum? We expect the ground state to be determined by the continuum analogue of

$$\hat{a}_r|0\rangle = 0 \qquad \text{for all } r; \tag{5.121}$$

namely

$$\hat{a}(k)|0\rangle = 0 \qquad \text{for all } k. \tag{5.122}$$

However, there is a problem with this. If we allow the Hamiltonian of (5.119) to act on $|0\rangle$ the result is not (as we would expect) zero, because of the $\hat{a}(k)\hat{a}^\dagger(k)$ term (the other term does give zero by (5.122)). In the *single* oscillator case, we rewrote $\hat{a}\hat{a}^\dagger$ in terms of $\hat{a}^\dagger\hat{a}$ by using the commutation relation (5.64), and this led to the 'zero-point energy', $\frac{1}{2}\omega$, of the oscillator ground state. Adopting the same strategy here, we write $\hat{H}$ of (5.119) as

$$\hat{H} = \int \frac{\mathrm{d}k}{2\pi}\hat{a}^\dagger(k)\hat{a}(k)\omega + \int \frac{\mathrm{d}k}{2\pi}\frac{1}{2}[\hat{a}(k), \hat{a}^\dagger(k)]\omega. \tag{5.123}$$

Now consider $\hat{H}|0\rangle$: we see from the definition of the vacuum (5.122) that the first term will give zero as expected—but the second term is infinite, since the commutation relation (5.117) produces the infinite quantity '$\delta(0)$' as $k \to k'$; moreover, the k integral diverges.

This term is obviously the continuum analogue of the zero-point energy $\frac{1}{2}\omega$—but because there are infinitely many oscillators, it is infinite. The conventional ploy is to argue that only energy *differences*, relative to a conveniently defined ground state, really matter—so that we may discard the infinite constant in (5.123). Then the ground state $|0\rangle$ has energy zero, by definition, and the eigenvalues of $\hat{H}$ are of the form

$$\int \frac{\mathrm{d}k}{2\pi} n(k)\omega \tag{5.124}$$

where $n(k)$ is the number of quanta (counted by the number operator $\hat{a}^\dagger(k)\hat{a}(k)$) of energy $\omega = k$. For each definite k, and hence ω, the spectrum is like that of the simple harmonic oscillator. The process of going from (5.119) to (5.123) *without* the second term is called 'normally ordering' the $\hat{a}$ and $\hat{a}^\dagger$ operators: in a 'normally ordered' expression, all $\hat{a}^\dagger$'s are to the left of all $\hat{a}$'s, with the result that the vacuum value of such expressions is by definition zero.

It has to be admitted that the argument that only energy differences matter is false as far as gravity is concerned, which couples to all sources of energy. It would ultimately be desirable to have theories in which the vacuum energy came out finite from the start (as actually happens in 'supersymmetric' field theories—see for example Weinberg (1995), p 325); see also comment (3).

We proceed on to the excited states. Any desired state in which excitation quanta are present can be formed by the appropriate application of $\hat{a}^{\dagger}(k)$ operators to the ground state $|0\rangle$. For example, a two-quantum state containing one quantum of momentum k_1 and another of momentum k_2 may be written (cf (5.73))

$$|k_1, k_2\rangle \propto \hat{a}^{\dagger}(k_1)\hat{a}^{\dagger}(k_2)|0\rangle. \tag{5.125}$$

A general state will contain an arbitrary number of quanta.

Once again, and this time more formally, we have completed the programme outlined in section 5.1, ending up with the 'quantization' of a classical field $\phi(x, t)$, as exemplified in the basic expression (5.116), together with the interpretation of the operators $\hat{a}(k)$ and $\hat{a}^{\dagger}(k)$ as destruction and creation operators for mode quanta. We have, at least implicitly, still retained up to this point the 'mechanical model'—thus our field $\hat{\phi}(x, t)$ could describe the quantized vibrations of a (one-dimensional) solid in the Debye (continuum) approximation. The quanta are then phonons, and $|0\rangle$ is the ground state of the solid. We now throw away the mechanical props and embrace the unadorned quantum field theory! We do not ask *what* is waving, we simply postulate a field—such as ϕ—and quantize it. *Its quanta of excitation are what we call particles*—for example, photons in the electromagnetic case.

We end this long section with some further remarks about the formalism, and the physical interpretation of our quantum field $\hat{\phi}$.

Comment (1)

The alert reader, who has studied appendix I, may be worried about the following (possible) consistency problem. The fields $\hat{\phi}$ and $\hat{\pi}$ are Heisenberg picture operators, and obey the equations of motion

$$\dot{\hat{\phi}}(x, t) = -\mathrm{i}[\hat{\phi}(x, t), \hat{H}] \tag{5.126}$$

$$\dot{\hat{\pi}}(x, t) = -\mathrm{i}[\hat{\pi}(x, t), \hat{H}] \tag{5.127}$$

where $\hat{H}$ is given by (5.119). It is a good exercise to check (problem 5.8(a)) that (5.126) yields just the expected relation $\dot{\hat{\phi}}(x, t) = \hat{\pi}(x, t)$ (cf (5.109)). Thus (5.127) becomes

$$\ddot{\hat{\phi}}(x, t) = -\mathrm{i}[\hat{\pi}(x, t), \hat{H}]. \tag{5.128}$$

However, we have assumed in our work here that $\hat{\phi}$ obeyed the wave equation (cf.(5.108))

$$\ddot{\hat{\phi}} = \frac{\partial^2}{\partial x^2}\hat{\phi}(x, t) \tag{5.129}$$

as a consequence of the quantized version of the Euler–Lagrange equation (5.88). Thus the right-hand sides of (5.128) and (5.129) need to be the same, for consistency—and they are: see problem 5.8(b). Thus—at least in this case—the Heisenberg operator equations of motion are consistent with the Euler–Lagrange equations.

Comment (2)

Following on from this, we may note that this formalism encompasses both the wave and the particle aspects of matter and radiation. The former is evident from the plane-wave expansion functions in the expansion of $\hat{\phi}$, (5.116), which in turn originate from the fact that $\hat{\phi}$ obeys the wave equation (5.108). The latter follows from the discrete nature of the energy spectrum and the associated operators $\hat{a}$, $\hat{a}^\dagger$ which refer to individual quanta i.e. *particles*.

Comment (3)

Next, we may ask: what is the meaning of the ground state $|0\rangle$ for a quantum field? It is undoubtedly the state with $n(k) = 0$ for all k, i.e. the state with no quanta in it—and hence no *particles* in it, on our new interpretation. It is therefore the vacuum! As we shall see later, this understanding of the vacuum as the ground state of a field system is fundamental to much of modern particle physics—for example, to quark confinement and to the generation of mass for the weak vector bosons. Note that although we discarded the overall (infinite) constant in $\hat{H}$, differences in zero-point energies *can* be detected; for example, in the Casimir effect (Casimir 1948, Kitchener and Prosser 1957, Sparnaay 1958, Lamoreaux 1997, 1998). These and other aspects of the quantum field theory vacuum are discussed in Aitchison (1985).

Comment (4)

Consider the two-particle state (5.125): $|k_1, k_2\rangle \propto \hat{a}^\dagger(k_1)\hat{a}^\dagger(k_2)|0\rangle$. Since the $\hat{a}^\dagger$ operators commute, (5.117), this state is symmetric under the interchange $k_1 \leftrightarrow k_2$. This is an inevitable feature of the formalism as so far developed—there is no possible way of distinguishing one quantum of energy from another, and we expect the two-quantum state to be indifferent to the order in which the quanta are put in it. However, this has an important implication for the *particle* interpretation: since the state is symmetric under interchange of the particle labels k_1 and k_2, it must describe identical *bosons*. How the formalism is modified in order to describe the antisymmetric states required for two fermionic quanta will be discussed in section 7.2.

Comment (5)

Finally, the reader may well wonder how to connect the quantum field theory formalism to ordinary 'wavefunction' quantum mechanics. The ability to see this connection will be important in subsequent chapters and it is indeed quite simple. Suppose we form a state containing one quantum of the $\hat{\phi}$ field, with momentum k':

$$|k'\rangle = N\hat{a}^\dagger(k')|0\rangle \tag{5.130}$$

where N is a normalization constant. Now consider the amplitude $\langle 0|\hat{\phi}(x,t)|k'\rangle$. We expand this out as

$$\langle 0|\hat{\phi}(x,t)|k'\rangle = \langle 0|\int \frac{\mathrm{d}k}{2\pi\sqrt{2\omega}}[\hat{a}(k)\mathrm{e}^{\mathrm{i}kx-\mathrm{i}\omega t} + \hat{a}^\dagger(k)\mathrm{e}^{-\mathrm{i}kx+\mathrm{i}\omega t}]N\hat{a}^\dagger(k')|0\rangle. \tag{5.131}$$

The '$\hat{a}^\dagger\hat{a}^\dagger$' term will give zero since $\langle 0|\hat{a}^\dagger = 0$. For the other term we use the commutation relation (5.117) to write it as

$$\langle 0|\int \frac{\mathrm{d}k}{2\pi\sqrt{2\omega}}[\hat{a}^\dagger(k')\hat{a}(k) + 2\pi\delta(k-k')]\mathrm{e}^{\mathrm{i}kx-\mathrm{i}\omega t}|0\rangle = N\frac{\mathrm{e}^{\mathrm{i}k'x-\mathrm{i}\omega' t}}{\sqrt{2\omega'}} \tag{5.132}$$

using the vacuum condition once again, and integrating over the δ function using the property (5.103) which sets $k = k'$ and hence $\omega = \omega'$. The vacuum is normalized to unity, $\langle 0|0\rangle = 1$. The normalization constant N can be adjusted according to the desired convention for the normalization of the states and wavefunctions. The result is just the plane-wave *wavefunction* for a particle in the state $|k'\rangle$! Thus we discover that the vacuum to one-particle matrix elements of the field operators are just the familiar wavefunctions of single-particle quantum mechanics. In this connection we can explain some common terminology. The path to quantum field theory that we have followed is sometimes called 'second quantization'—ordinary single-particle quantum mechanics being the first-quantized version of the theory.

5.3 Generalizations: four dimensions, relativity and mass

In the previous section we have shown how quantum mechanics may be married to field theory, but we have considered only one spatial dimension, for simplicity. Now we must generalize to three and incorporate the demands of relativity. This is very easy to do in the Lagrangian approach, for the scalar field $\phi(\boldsymbol{x}, t)$. 'Scalar' means that the field has only one independent component at each point $(\boldsymbol{x}, t)$—unlike the electromagnetic field, for instance, for which the analogous quantity has four components, making up a 4-vector field $A^\mu(\boldsymbol{x}, t) = (A_0(\boldsymbol{x}, t), \boldsymbol{A}(\boldsymbol{x}, t))$ (see chapter 7). In the quantum case, a one-component field (or wavefunction) is appropriate for spin-0 particles.

As we saw in (5.89), the three-dimensional Euler–Lagrange equations are

$$\frac{\partial \mathcal{L}}{\partial \phi} - \nabla \cdot \frac{\partial \mathcal{L}}{\partial (\nabla \phi)} - \frac{\partial}{\partial t}\left(\frac{\partial \mathcal{L}}{\partial \dot{\phi}}\right) = 0 \tag{5.133}$$

which may immediately be rewritten in relativistically invariant form

$$\frac{\partial \mathcal{L}}{\partial \phi} - \partial_\mu \left(\frac{\partial \mathcal{L}}{\partial (\partial_\mu \phi)}\right) = 0 \tag{5.134}$$

where $\partial_\mu = \partial/\partial x^\mu$. Similarly, the action

$$S = \int \mathrm{d}t \int \mathrm{d}^3\boldsymbol{x}\, \mathcal{L} = \int \mathrm{d}^4x\, \mathcal{L} \tag{5.135}$$

will be relativistically invariant if $\mathcal{L}$ is, since the volume element d^4x is invariant. Thus, to construct a relativistic field theory, we have to construct an invariant density $\mathcal{L}$ and use the already given covariant Euler–Lagrange equation. Thus our previous string Lagrangian

$$\mathcal{L}_\rho = \frac{1}{2}\rho\left(\frac{\partial \phi}{\partial t}\right)^2 - \frac{1}{2}\rho c^2\left(\frac{\partial \phi}{\partial x}\right)^2 \tag{5.136}$$

with $\rho = c = 1$ generalizes to

$$\mathcal{L} = \tfrac{1}{2}\partial_\mu \phi \partial^\mu \phi \tag{5.137}$$

and produces the invariant wave equation

$$\partial_\mu \partial^\mu \phi = \left(\frac{\partial^2}{\partial t^2} - \nabla^2\right)\phi = 0. \tag{5.138}$$

All of this goes through just the same when the fields are quantized.

This invariant Lagrangian describes a field whose quanta are massless. To find the Lagrangian for the case of massive quanta, we need to find the Lagrangian that gives us the Klein–Gordon equation (see section 4.1)

$$(\Box + m^2)\phi(\boldsymbol{x}, t) = 0 \tag{5.139}$$

via the Euler–Lagrangian equations.

The answer is a simple generalization of (5.137):

$$\mathcal{L}_{\mathrm{KG}} = \tfrac{1}{2}\partial_\mu \phi \partial^\mu \phi - \tfrac{1}{2}m^2\phi^2. \tag{5.140}$$

The plane-wave solutions of the field equation—now the KG equation—have frequencies (or energies) given by

$$\omega^2 = \boldsymbol{k}^2 + m^2 \tag{5.141}$$

which is the correct energy–momentum relation for a massive particle.

How do we quantize this field theory? The four-dimensional analogue of the Fourier expansion of the field ϕ takes the form

$$\hat{\phi}(x) = \int_{-\infty}^{\infty} \frac{\mathrm{d}^3\boldsymbol{k}}{(2\pi)^3\sqrt{2\omega}}[\hat{a}(k)\mathrm{e}^{-\mathrm{i}k\cdot x} + \hat{a}^\dagger(k)\mathrm{e}^{\mathrm{i}k\cdot x}] \tag{5.142}$$

with a similar expansion for the 'conjugate momentum' $\hat{\pi} = \dot{\hat{\phi}}$:

$$\hat{\pi}(x) = \int_{-\infty}^{\infty} \frac{\mathrm{d}^3\boldsymbol{k}}{(2\pi)^3\sqrt{2\omega}}(-\mathrm{i}\omega)[\hat{a}(k)\mathrm{e}^{-\mathrm{i}k\cdot x} - \hat{a}^\dagger(k)\mathrm{e}^{\mathrm{i}k\cdot x}]. \tag{5.143}$$

Here $k \cdot x$ is the four-dimensional dot product $k \cdot x = \omega t - \boldsymbol{k} \cdot \boldsymbol{x}$, and $\omega = +(\boldsymbol{k}^2 + m^2)^{1/2}$. The Hamiltonian is found to be

$$\hat{H}_{\mathrm{KG}} = \int \mathrm{d}^3\boldsymbol{x}\hat{\mathcal{H}}_{\mathrm{KG}} = \int_{-\infty}^{\infty} \mathrm{d}^3\boldsymbol{x}\tfrac{1}{2}[\hat{\pi}^2 + \nabla\hat{\phi}\cdot\nabla\hat{\phi} + m^2\hat{\phi}^2] \tag{5.144}$$

and this can be expressed in terms of the $\hat{a}$'s and the $\hat{a}^\dagger$'s using the expansion for $\hat{\phi}$ and $\hat{\pi}$ and the commutator

$$[\hat{a}(k), \hat{a}^\dagger(k')] = (2\pi)^3\delta^3(\boldsymbol{k} - \boldsymbol{k}') \tag{5.145}$$

with all others vanishing. The result is, as expected,

$$\hat{H}_{\mathrm{KG}} = \frac{1}{2}\int \frac{\mathrm{d}^3\boldsymbol{k}}{(2\pi)^3}[\hat{a}^\dagger(k)\hat{a}(k) + \hat{a}(k)\hat{a}^\dagger(k)]\omega \tag{5.146}$$

and, normally ordering as usual, we arrive at

$$\hat{H}_{\mathrm{KG}} = \int \frac{\mathrm{d}^3\boldsymbol{k}}{(2\pi)^3}\hat{a}^\dagger(k)\hat{a}(k)\omega. \tag{5.147}$$

This supports the physical interpretation of the mode operators $\hat{a}^\dagger$ and $\hat{a}$ as creation and destruction operators for quanta of the field $\hat{\phi}$ as before, except that now the energy–momentum relation for these particles is the relativistic one, for particles of mass m.

Since $\hat{\phi}$ is real ($\hat{\phi} = \hat{\phi}^\dagger$) and has no spin degrees of freedom, it is called a real scalar field. Only field quanta of one type enter—those created by $\hat{a}^\dagger$ and destroyed by $\hat{a}$. Thus $\hat{\phi}$ would correspond physically to a case where there was a unique particle state of a given mass m—for example the π^0 field. Actually, of course, we would not want to describe the π^0 in any fundamental sense in terms of such a field, since we know it is not a point-like object ('ϕ' is defined only at the single space–time point $(\boldsymbol{x}, t)$). The question of whether true 'elementary' scalar fields exist in nature is an interesting one: in the Standard Model, as we shall eventually see in volume 2, the Higgs field is a scalar field (though it contains several components with different charge). It remains to be seen if this field—and the associated quantum, the Higgs boson—does exist, and if so whether it is elementary or composite.

We have learned how to describe free relativistic spinless particles of finite mass as the quanta of a relativistic quantum field. We now need to understand *interactions* in quantum field theory.

Problems

5.1 Verify equation (5.36).

5.2 Consider one-dimensional motion under gravity so that $V(x) = -mgx$ in (5.39). Evaluate S of (5.38) for $t_1 = 0$, $t_2 = t_0$, for three possible trajectories:

(a) $x(t) = at$,
(b) $x(t) = \frac{1}{2}gt^2$ (the Newtonian result) and
(c) $x(t) = bt^3$

where the constants a and b are to be chosen so that all the trajectories end at the same point $x(t_0)$.

5.3

(a) Use (5.49) and (5.55) to verify that

$$\hat{p} = m\dot{\hat{q}}$$

is consistent with the Heisenberg equation of motion for $\hat{A} = \hat{q}$.

(b) By similar methods verify that

$$\dot{\hat{p}} = -m\omega^2\hat{q}.$$

5.4

(a) Rewrite the Hamiltonian $\hat{H}$ of (5.53) in terms of the operators $\hat{a}$ and $\hat{a}^\dagger$.
(b) Evaluate the commutator between $\hat{a}$ and $\hat{a}^\dagger$ and use this result together with your expression for $\hat{H}$ from part (a) to verify equation (5.65).
(c) Verify that for $|n\rangle$ given by equation (5.73) the normalization condition

$$\langle n|n\rangle = 1$$

is satisfied.

(d) Verify (5.75) directly using the commutation relation (5.64).

5.5 Treating ψ and ψ^* as independent classical fields, show that the Lagrangian density

$$\mathcal{L} = \mathrm{i}\psi^*\dot{\psi} - (1/2m)\nabla\psi^* \cdot \nabla\psi$$

gives the Schrödinger equation for ψ and ψ^* correctly.

5.6

(a) Verify that the commutation relations for $\hat{a}(k)$ and $\hat{a}^{\dagger}(k)$ (equations (5.117)) are consistent with the equal time commutation relation between $\hat{\phi}$ and $\hat{\pi}$ (equation (5.104)), and with (5.105).

(b) Consider the *unequal time* commutator $D(x_1, x_2) \equiv [\hat{\phi}(\boldsymbol{x}_1, t_1), \hat{\phi}(\boldsymbol{x}_2, t_2)]$, where $\hat{\phi}$ is a massive KG field in three dimensions. Show that

$$D(x_1, x_2) = \int \frac{\mathrm{d}^3\boldsymbol{k}}{(2\pi)^3 2E}[\mathrm{e}^{-\mathrm{i}k\cdot(x_1-x_2)} - \mathrm{e}^{\mathrm{i}k\cdot(x_1-x_2)}] \tag{5.148}$$

where $k \cdot (x_1 - x_2) = E(t_1 - t_2) - \boldsymbol{k} \cdot (\boldsymbol{x}_1 - \boldsymbol{x}_2)$, and $E = (\boldsymbol{k}^2 + m^2)^{1/2}$. Note that D is not an operator, and that it depends only on the difference of coordinates $x_1 - x_2$, consistent with translation invariance. Show that $D(x_1, x_2)$ vanishes for $t_1 = t_2$. Explain why the right-hand side of (5.148) is Lorentz invariant (see the exercise in appendix E), and use this fact to show that $D(x_1, x_2)$ vanishes for all *space-like* separations $(x_1 - x_2)^2 < 0$. Discuss the significance of this result—or see the discussion in section 6.3.2!

5.7 Insert the plane-wave expansions for the operators $\hat{\phi}$ and $\hat{\pi}$ into the equation for $\hat{H}$, (5.111), and verify equation (5.119). [*Hint*: note that ω is defined to be always positive, so that (5.113) should strictly be written $\omega = |k|$.]

5.8

(a) Use (5.104) and (5.111) to verify that $\hat{\pi}(x, t) = \dot{\hat{\phi}}(x, t)$ is consistent with the Heisenberg equation of motion for $\hat{\phi}(x, t)$. [*Hint*: write the integral in (5.111) as over y, not x!]

(b) Similarly, verify the consistency of (5.127) and (5.108).

6

QUANTUM FIELD THEORY II: INTERACTING SCALAR FIELDS

6.1 Interactions in quantum field theory: qualitative introduction

In the previous chapter we considered only free—i.e. non-interacting—quantum fields. The fact that they are non-interacting is evident in a number of ways. The mode expansions (5.116) and (5.142), are written in terms of the (free) plane-wave solutions of the associated wave equations. Also the Hamiltonians turned out to be just the sum of individual oscillator Hamiltonians for each mode frequency, as in (5.119) or (5.146). The energies of the quanta add up—they are non-interacting quanta. Finally, since the Hamiltonians are just sums of number operators

$$\hat{n}(k) = \hat{a}^{\dagger}(k)\hat{a}(k) \tag{6.1}$$

it is obvious that each such operator commutes with the Hamiltonian and is therefore a constant of the motion. Thus two waves, each with one excitation quantum, travelling towards each other will pass smoothly through each other and emerge unscathed on the other side—they will not interact at all.

How can we get the mode quanta to interact? If we return to our discussion of classical mechanical systems in section 5.1, we see that the crucial step in arriving at the 'sum over oscillators' form for the energy was the assumption that the potential energy was quadratic in the small displacements q_r. We expect that 'modes will interact' when we go *beyond this harmonic approximation*. The same is true in the continuous (wave or field) case. In the derivation of the appropriate wave equation you will find that somewhere an approximation like $\tan\phi \approx \phi$ or $\sin\phi \approx \phi$ is made. This linearizes the equation, and solutions to linear equations can be linearly superposed to make new solutions. If we retain higher powers of ϕ, such as ϕ^3, the resulting nonlinear equation has solutions that cannot be obtained by superposing two independent solutions. Thus two waves travelling towards each other will not just pass smoothly through each other: various forms of interaction and distortion of the original waveforms will occur.

What happens when we quantize such anharmonic systems? To gain some idea of the new features that emerge, consider just one 'anharmonic oscillator' with Hamiltonian

$$\hat{H} = (1/2m)\hat{p}^2 + \tfrac{1}{2}m\omega^2\hat{q}^2 + \lambda\hat{q}^3. \tag{6.2}$$

In terms of the $\hat{a}$ and $\hat{a}^\dagger$ combinations this becomes

$$\hat{H} = \frac{1}{2}(\hat{a}^\dagger\hat{a} + \hat{a}\hat{a}^\dagger)\omega + \frac{\lambda}{(2m\omega)^{3/2}}(\hat{a} + \hat{a}^\dagger)^3 \tag{6.3}$$

$$\equiv \hat{H}_0 + \lambda\hat{H}' \tag{6.4}$$

where $\hat{H}_0$ is our previous free oscillator Hamiltonian. The algebraic tricks we used to find the spectrum of $\hat{H}_0$ do *not* work for this new $\hat{H}$ because of the addition of the $\hat{H}'$ interaction term. In particular, although $\hat{H}_0$ commutes with the number operator $\hat{a}^\dagger\hat{a}$, $\hat{H}'$ does not. Therefore, whatever the eigenstates of $\hat{H}$ are, they will not in general have a definite number of '$\hat{H}_0$ quanta'. In fact, we cannot find an exact algebraic solution to this new eigenvalue problem, and we must resort to *perturbation theory* or to numerical methods.

The perturbative solution to this problem treats $\lambda\hat{H}'$ as a perturbation and expands the true eigenstates of $\hat{H}$ in terms of the eigenstates of $\hat{H}_0$:

$$|\bar{r}\rangle = \sum_n c_{rn}|n\rangle. \tag{6.5}$$

From this expansion we see that, as expected, the true eigenstates $|\bar{r}\rangle$ will 'contain different numbers of $\hat{H}_0$ quanta': $|c_{rn}|^2$ is the probability of finding n '$\hat{H}_0$ quanta' in the state $|\bar{r}\rangle$. Perturbation theory now proceeds by expanding the coefficients c_{rn} and exact energy eigenvalues $\bar{E}_r$ as power series in the strength λ of the perturbation. For example, the exact energy eigenvalue has the expansion

$$\bar{E}_r = E_r^{(0)} + \lambda E_r^{(1)} + \lambda^2 E_r^{(2)} + \cdots \tag{6.6}$$

where

$$\hat{H}_0|r\rangle = E_r^{(0)}|r\rangle \tag{6.7}$$

and

$$E_r^{(1)} = \langle r|\hat{H}'|r\rangle \tag{6.8}$$

$$E_r^{(2)} = \sum_{s\neq r} \frac{\langle r|\hat{H}'|s\rangle\langle s|\hat{H}'|r\rangle}{E_r^{(0)} - E_s^{(0)}}. \tag{6.9}$$

To evaluate the second-order shift in energy, we therefore need to consider matrix elements of the form

$$\langle s|(\hat{a} + \hat{a}^\dagger)^3|r\rangle. \tag{6.10}$$

Keeping careful track of the order of the $\hat{a}$ and $\hat{a}^\dagger$ operators, we can evaluate these matrix elements and find, in this case, that there are non-zero matrix elements for states $\langle s| = \langle r+3|$, $\langle r+1|$, $\langle r-1|$ and $\langle r-3|$.

What about the quantum mechanics of two coupled nonlinear oscillators? In the same way, the general state is assumed to be a superposition

$$|\bar{r}\rangle = \sum_{n_1,n_2} c_{r,n_1n_2}|n_1\rangle|n_2\rangle \tag{6.11}$$

of states of arbitrary numbers of quanta of the unperturbed oscillator Hamiltonians $\hat{H}_{0(1)}$ and $\hat{H}_{0(2)}$. States of the unperturbed system contain definite numbers n_1 and n_2, say, of the '1' and '2' quanta. Perturbation calculations of the interacting system will involve matrix elements connecting such $|n_1\rangle|n_2\rangle$ states to states $|n'_1\rangle|n'_2\rangle$ with different numbers of these quanta.

All this can be summarized by the remark that the typical feature of quantized interacting modes is that we need to consider processes in which the numbers of the different mode quanta are not constants of the motion. This is, of course, exactly what happens when we have collisions between high-energy particles. When far apart the particles, definite in number, are indeed free and are just the mode quanta of some quantized fields. But, when they interact, we must expect to see changes in the numbers of quanta, and can envisage processes in which the number of quanta which emerge finally as free particles is different from the number that originally collided. From the quantum mechanical examples we have discussed, we expect that these interactions will be produced by terms like $\hat{\phi}^3$ or $\hat{\phi}^4$, since the free—'harmonic'—case has $\hat{\phi}^2$, analogous to $\hat{q}^2$ in the quantum mechanics example. Such terms arise in the solid state phonon application precisely from anharmonic corrections involving the atomic displacements. These terms lead to non-trivial phonon–phonon scattering, the treatment of which forms the basis of the quantum theory of thermal resistivity of insulators. In the quantum field theory case, when we have generalized the formalism to fermions and photons, the nonlinear interaction terms will produce e^+e^- scattering, $q\bar{q}$ annihilation and so on. As in the quantum mechanical case, the basic calculational method will be perturbation theory.

As remarked earlier, the trouble with all these 'real-life' cases is that they involve significant complications due to spin; the corresponding fields then have several components, with attendant complexity in the solutions of the associated free-particle wave equations (Maxwell, Dirac). So in this chapter we shall seek to explain the essence of *the perturbative approach to quantum field dynamics*—which we take to be essentially the Feynman graph version of Yukawa's exchange mechanism—in the context of simple models involving only scalar fields; Maxwell (vector) and Dirac (spinor) fields will be introduced in the following chapter. The route we follow to the 'Feynman rules' is the one first given (with remarkable clarity) by Dyson (1949a), which rapidly became the standard formulation.

Before proceeding it may be worth emphasizing that in introducing a 'non-harmonic' term such as $\hat{\phi}^3$ and thus departing from linearity in that sense, we are in no way affecting the basic linearity of state vector superposition in quantum mechanics (cf (6.11)), which continues to hold.

6.2 Perturbation theory for interacting fields: the Dyson expansion of the S-matrix

> On the third day of the journey a remarkable thing happened; going into a sort of semi-stupor as one does after 48 hours of bus-riding, I began to think very hard about physics, and particularly about the rival radiation theories of Schwinger and Feynman. Gradually my thoughts grew more coherent, and before I knew where I was, I had solved the problem that had been in the back of my mind all this year, which was to prove the equivalence of the two theories.

From a letter from F. J. Dyson to his parents, 18 September 1948, as quoted in Schweber (1994), p 505.

For definiteness, let us consider the Lagrangian

$$\hat{\mathcal{L}} = \tfrac{1}{2}\partial_\mu\hat{\phi}\partial^\mu\hat{\phi} - \tfrac{1}{2}m^2\hat{\phi}^2 - \lambda\hat{\phi}^3 \equiv \hat{\mathcal{L}}_{\mathrm{KG}} - \lambda\hat{\phi}^3 \tag{6.12}$$

with $\lambda > 0$. Equation (6.12) is like '$\hat{\mathcal{L}} = \hat{T} - \hat{V}$' where $\hat{V} = \frac{1}{2}(\nabla\hat{\phi})^2 + \frac{1}{2}m^2\hat{\phi}^2 + \lambda\hat{\phi}^3$ is the 'potential'. Though simple, this Lagrangian is unfortunately not physically sensible. The classical particle analogue potential would have the form $V(q) = \frac{1}{2}\omega q^2 + \lambda q^3$. If we sketch $V(q)$ as a function of q we see that, for small λ, it retains the shape of an oscillator well near $q = 0$, but for q sufficiently large and negative it will 'turn over', tending ultimately to $-\infty$ as $q \to -\infty$. Classically we expect to be able to set up a successful perturbation theory for oscillations about the equilibrium position $q = 0$, provided that the amplitude of the oscillations is not so large as to carry the particle over the 'lip' of the potential; in the latter case, the particle will escape to $q = -\infty$, invalidating a perturbative approach. In the quantum mechanical case the same potential $V(q)$ is more problematical, since the particle can *tunnel* through the barrier separating it from the region where $V \to -\infty$. This means that the ground state will not be stable. An analogous disease affects the quantum field case—the supposed vacuum state will be unstable, and indeed the energy will not be positive-definite.

Nevertheless, as the reader may already have surmised, and we shall confirm later in this chapter, the 'ϕ-cubed' interaction is precisely of the form relevant to Yukawa's exchange mechanism. As we have seen in the previous section, such an interaction will typically give rise to matrix elements between one-quantum and two-quantum states, for example, exactly like the basic Yukawa emission and absorption process. In fact, all that is necessary to make the $\hat{\phi}^3$-type interaction physical is to let it describe, not the 'self-coupling' of a single field, but the 'interactive coupling' of at least two different fields. For example, we may have two scalar fields with quanta 'A' and 'B', and an interaction between them of the form $\lambda\hat{\phi}_{\mathrm{A}}^2\hat{\phi}_{\mathrm{B}}$. This will allow processes such as $\mathrm{A} \leftrightarrow \mathrm{A} + \mathrm{B}$. Or we may have three such fields, and an interaction $\lambda\hat{\phi}_{\mathrm{A}}\hat{\phi}_{\mathrm{B}}\hat{\phi}_{\mathrm{C}}$, allowing $\mathrm{A} \leftrightarrow \mathrm{B} + \mathrm{C}$ and similar transitions. In these cases the problems with the $\hat{\phi}^3$ self-interaction do not arise. (Incidentally those problems can be eliminated by the addition of a suitable higher-power term, for instance $g\hat{\phi}^4$.) In later sections we shall be considering the 'ABC' model specifically, but for the present it will be simpler to continue with the single field $\hat{\phi}$ and the self-interaction $\lambda\hat{\phi}^3$, as described by the Lagrangian (6.12). The associated Hamiltonian is

$$\hat{H} = \hat{H}_{\mathrm{KG}} + \hat{H}' \tag{6.13}$$

where (as is usual in perturbation theory) we have separated the Hamiltonian into a part we can handle exactly, which is the free Klein–Gordon Hamiltonian

$$\hat{H}_{\mathrm{KG}} = \int \mathrm{d}^3\boldsymbol{x}\, \hat{\mathcal{H}}_{\mathrm{KG}} = \tfrac{1}{2}\int \mathrm{d}^3\boldsymbol{x}\, [\hat{\pi}^2 + (\nabla\hat{\phi})^2 + m^2\hat{\phi}^2] \tag{6.14}$$

and the part we shall treat perturbatively

$$\hat{H}' = \int \mathrm{d}^3\boldsymbol{x}\, \hat{\mathcal{H}}' = \lambda \int \mathrm{d}^3\boldsymbol{x}\, \hat{\phi}^3. \tag{6.15}$$

6.2.1 The interaction picture

We begin with a crucial formal step. In our introduction to quantum field theory in the previous chapter, we worked in the Heisenberg picture (HP). There, however, we only dealt with free (non-interacting) fields. The time dependence of the operators as given by the mode expansion (5.142) is that generated by the free KG Hamiltonian (6.14) via the Heisenberg equations of motion (see problem 5.8). But as soon as we include the interaction term $\hat{H}'$, we cannot make progress in the HP, since we do not then know the time dependence of the operators—which is generated by the full Hamiltonian $\hat{H} = \hat{H}_{\mathrm{KG}} + \hat{H}'$.

Instead, we might consider using the Schrödinger picture (SP) in which the states change with time according to

$$\hat{H}|\psi(t)\rangle = \mathrm{i}\frac{\mathrm{d}}{\mathrm{d}t}|\psi(t)\rangle \tag{6.16}$$

and the operators are time-independent (see appendix I). Note that although (6.16) is a 'Schrödinger picture' equation, there is nothing non-relativistic about it: on the contrary, $\hat{H}$ is the relevant relativistic Hamiltonian. In this approach, the field operators appearing in the density $\hat{\mathcal{H}}$ are all evaluated at a fixed time, say $t = 0$ by convention, which is the time at which the Schrödinger and Heisenberg pictures coincide. At this fixed time, mode expansions of the form (5.142) with $t = 0$ are certainly possible, since the basis functions form a complete set.

One problem with this formulation, however, is that it is not going to be manifestly 'Lorentz invariant' (or covariant), because a particular time ($t = 0$) has been singled out. In the end, physical quantities should come out correct, but it is much more convenient to have everything looking nice and consistent with relativity as we go along. This is one of the reasons for choosing to work in yet a third 'picture', an ingenious kind of half-way-house between the other two, called the 'interaction picture' (IP). We shall see other good reasons shortly.

In the HP, all the time dependence is carried by the operators and none by the state, while in the SP it is exactly the other way around. In the IP, both states and operators are time-dependent but in a way that is well adapted to perturbation theory, especially in quantum field theory. The operators have a time dependence generated by the free Hamiltonian $\hat{H}_0$, say, and so a 'free-particle' mode expansion like (5.142) survives intact (here $\hat{H}_0 = \hat{H}_{\mathrm{KG}}$). The states have a time dependence generated by the interaction $\hat{H}'$. Thus as $\hat{H}' \to 0$ we return to the free-particle HP.

The way this works formally is as follows. In terms of the time-independent SP operator $\hat{A}$ (cf appendix I), we define the corresponding IP operator $\hat{A}_{\mathrm{I}}(t)$ by

$$\hat{A}_{\mathrm{I}}(t) = \mathrm{e}^{\mathrm{i}\hat{H}_0 t}\hat{A}\mathrm{e}^{-\mathrm{i}\hat{H}_0 t}. \tag{6.17}$$

This is just like the definition of the HP operator $\hat{A}(t)$ in appendix I, except that $\hat{H}_0$ appears instead of the full $\hat{H}$. It follows that the time dependence of $\hat{A}_{\mathrm{I}}(t)$ is given by (I.8) with $\hat{H} \to \hat{H}_0$:

$$\frac{\mathrm{d}\hat{A}_{\mathrm{I}}(t)}{\mathrm{d}t} = -\mathrm{i}[\hat{A}_{\mathrm{I}}(t), \hat{H}_0]. \tag{6.18}$$

Equation (6.18) can also, of course, be derived by carefully differentiating (6.17). Thus—as mentioned already—the time dependence of $\hat{A}_{\mathrm{I}}(t)$ is generated by the free part of the Hamiltonian, by construction.

As applied to our model theory (6.12), then, our field $\hat{\phi}$ will now be specified as being in the IP, $\hat{\phi}_{\mathrm{I}}(\boldsymbol{x}, t)$. What about the field canonically conjugate to $\hat{\phi}_{\mathrm{I}}(t)$, in the case when the interaction is included? In the HP, as long as the interaction does not contain time derivatives, as is the case here, the field canonically conjugate to the interacting field remains the same as the free-field case:

$$\hat{\pi}(\boldsymbol{x}, t) = \frac{\partial\hat{\mathcal{L}}}{\partial\dot{\hat{\phi}}(\boldsymbol{x}, t)} = \frac{\partial\hat{\mathcal{L}}_{\mathrm{KG}}}{\partial\dot{\hat{\phi}}(\boldsymbol{x}, t)} = \dot{\hat{\phi}}(\boldsymbol{x}, t) \tag{6.19}$$

so that we continue to adopt the equal-time commutation relation

$$[\hat{\phi}(\boldsymbol{x}, t), \hat{\pi}(\boldsymbol{y}, t)] = \mathrm{i}\delta^3(\boldsymbol{x} - \boldsymbol{y}) \tag{6.20}$$

for the Heisenberg fields. But the IP fields are related to the HP fields by a unitary transformation $\hat{U}$, as we can see by combining (6.17) with (I.7):

$$\begin{aligned}\hat{A}_{\mathrm{I}}(t) &= \mathrm{e}^{\mathrm{i}\hat{H}_0 t}\mathrm{e}^{-\mathrm{i}\hat{H}t}\hat{A}(t)\mathrm{e}^{\mathrm{i}\hat{H}t}\mathrm{e}^{-\mathrm{i}\hat{H}_0 t} \\ &= \hat{U}\hat{A}(t)\hat{U}^{-1}\end{aligned} \tag{6.21}$$

where $\hat{U} = \mathrm{e}^{\mathrm{i}\hat{H}_0 t}\mathrm{e}^{-\mathrm{i}\hat{H}t}$, and it is easy to check that $\hat{U}\hat{U}^\dagger = \hat{U}^\dagger\hat{U} = \hat{I}$. So taking equation (6.20) and pre-multiplying by $\hat{U}$ and post-multiplying by $\hat{U}^{-1}$ on both sides, we obtain

$$[\hat{\phi}_{\mathrm{I}}(\boldsymbol{x}, t), \hat{\pi}_{\mathrm{I}}(\boldsymbol{y}, t)] = \mathrm{i}\delta^3(\boldsymbol{x} - \boldsymbol{y}) \tag{6.22}$$

showing that, *in the interacting case*, the IP fields $\hat{\phi}_{\mathrm{I}}$ and $\hat{\pi}_{\mathrm{I}}$ obey the free field commutation relation. Thus in the IP case the interacting fields obey the same equations of motion and the same commutation relations as the free-field operators. It follows that the mode expansion (5.142), and the commutation relations (5.145) for the mode creation and annihilation operators, can be taken straight over for the IP operators.

We now turn to the states in the IP. To preserve consistency between the matrix elements in the Schrödinger and interaction pictures (cf the step from (I.6) to (I.7)) we define the corresponding IP state vector by

$$|\psi(t)\rangle_{\mathrm{I}} = \mathrm{e}^{\mathrm{i}\hat{H}_0 t}|\psi(t)\rangle \tag{6.23}$$

in terms of the SP state $|\psi(t)\rangle$. We now use (6.23) to find the equation of motion of $|\psi(t)\rangle_{\mathrm{I}}$. We have

$$\begin{aligned}
\mathrm{i}\frac{\mathrm{d}}{\mathrm{d}t}|\psi(t)\rangle_{\mathrm{I}} &= \mathrm{e}^{\mathrm{i}\hat{H}_0 t}\left\{-\hat{H}_0|\psi(t)\rangle + \mathrm{i}\frac{\mathrm{d}}{\mathrm{d}t}|\psi(t)\rangle\right\} \\
&= \mathrm{e}^{\mathrm{i}\hat{H}_0 t}\{-\hat{H}_0|\psi(t)\rangle + (\hat{H}_0 + \hat{H}')|\psi(t)\rangle\} \\
&= \mathrm{e}^{\mathrm{i}\hat{H}_0 t}\hat{H}'|\psi(t)\rangle \\
&= \mathrm{e}^{\mathrm{i}\hat{H}_0 t}\hat{H}'\mathrm{e}^{-\mathrm{i}\hat{H}_0 t}|\psi(t)\rangle_{\mathrm{I}}
\end{aligned} \tag{6.24}$$

or

$$\boxed{\mathrm{i}\frac{\mathrm{d}}{\mathrm{d}t}|\psi(t)\rangle_{\mathrm{I}} = \hat{H}'_{\mathrm{I}}(t)|\psi(t)\rangle_{\mathrm{I}}} \tag{6.25}$$

where

$$\hat{H}'_{\mathrm{I}} = \mathrm{e}^{\mathrm{i}\hat{H}_0 t}\hat{H}'\mathrm{e}^{-\mathrm{i}\hat{H}_0 t} \tag{6.26}$$

is the interaction Hamiltonian *in the interaction picture*. The italicised words are important: they mean that all operators in $\hat{H}'_{\mathrm{I}}$ have the (known) free-field time dependence, which would not be the case for $\hat{H}'$ in the HP. Thus, as mentioned earlier, the states in the IP have a time dependence generated by the interaction Hamiltonian, and this derivation has shown us that it is, in fact, the interaction Hamiltonian in the IP which is the appropriate generator of time change in this picture.

Equation (6.25) is a slightly simplified form of the Tomonaga–Schwinger equation, which formed the starting point of the approach to QED followed by Schwinger (Schwinger 1948b, 1949a, b) and independently by Tomonaga and his group (Tomonaga 1946, Koba *et al* 1947a, b, Kanesawa and Tomonaga 1948a, b, Koba and Tomonaga 1948, Koba and Takeda 1948, 1949).

6.2.2 The S-matrix and the Dyson expansion

We now start the job of applying the IP formalism to scattering and decay processes in quantum field theory, treated in perturbation theory; for this, following Dyson (1949a, b), the crucial quantity is the *scattering matrix*, or S-matrix for short, which we now introduce. A scattering process may plausibly be described in the following terms. At a time $t \to -\infty$, long before any interaction has occurred, we expect the effect of $\hat{H}'_{\mathrm{I}}$ to be negligible so that, from (6.25), $|\psi(-\infty)\rangle_{\mathrm{I}}$ will be a constant state vector $|\mathrm{i}\rangle$, which is in fact an eigenstate of $\hat{H}_0$. Thus $|\mathrm{i}\rangle$ will contain a certain number of non-interacting particles with definite momenta, and $|\psi(-\infty)\rangle_{\mathrm{I}} = |\mathrm{i}\rangle$. As time evolves, the particles approach each other and may scatter, leading in the distant future (at $t \to \infty$) to another constant state $|\psi(\infty)\rangle_{\mathrm{I}}$ containing non-interacting particles. Note that $|\psi(\infty)\rangle_{\mathrm{I}}$ will in general contain many different components, each with (in principle) different numbers and types of particle; these different components in $|\psi(\infty)\rangle_{\mathrm{I}}$ will be denoted by $|\mathrm{f}\rangle$. The $\hat{S}$-*operator* is now defined via

$$|\psi(\infty)\rangle_{\mathrm{I}} = \hat{S}|\psi(-\infty)\rangle_{\mathrm{I}} = \hat{S}|\mathrm{i}\rangle. \tag{6.27}$$

A particular S-*matrix element* is then the amplitude for finding a particular final state $|\mathrm{f}\rangle$ in $|\psi(\infty)\rangle_{\mathrm{I}}$:

$$\langle\mathrm{f}|\psi(\infty)\rangle_{\mathrm{I}} = \langle\mathrm{f}|\hat{S}|\mathrm{i}\rangle \equiv S_{\mathrm{fi}}. \tag{6.28}$$

Thus we may write

$$|\psi(\infty)\rangle_{\mathrm{I}} = \sum_{\mathrm{f}} |\mathrm{f}\rangle\langle\mathrm{f}|\psi(\infty)\rangle_{\mathrm{I}} = \sum_{\mathrm{f}} S_{\mathrm{fi}}|\mathrm{f}\rangle. \tag{6.29}$$

It is clear that it is these S-matrix elements S_{fi} that we need to calculate, and the associated probabilities $|S_{\mathrm{fi}}|^2$.

Before proceeding we note an important property of $\hat{S}$. Assuming that $|\psi(\infty)\rangle_{\mathrm{I}}$ and $|\mathrm{i}\rangle$ are both normalized, we have

$$1 = {}_{\mathrm{I}}\langle\psi(\infty)|\psi(\infty)\rangle_{\mathrm{I}} = \langle\mathrm{i}|\hat{S}^{\dagger}\hat{S}|\mathrm{i}\rangle = \langle\mathrm{i}|\mathrm{i}\rangle \tag{6.30}$$

implying that $\hat{S}$ is *unitary*: $\hat{S}^{\dagger}\hat{S} = \hat{I}$. Taking matrix elements of this gives us the result

$$\sum_{\mathrm{k}} S^{*}_{\mathrm{kf}} S_{\mathrm{ki}} = \delta_{\mathrm{fi}}. \tag{6.31}$$

Putting $\mathrm{i} = \mathrm{f}$ in (6.31) yields $\sum_{\mathrm{k}} |S_{\mathrm{ki}}|^2 = 1$, which confirms that the expansion coefficients in (6.29) must obey the usual condition that the sum of all the partial probabilities must add up to 1. Note, however, that in the present case the states involved may contain different numbers of particles.

We set up a perturbation-theory approach to calculating $\hat{S}$ as follows. Integrating (6.25) subject to the condition at $t \to -\infty$ yields

$$|\psi(t)\rangle_{\mathrm{I}} = |\mathrm{i}\rangle - \mathrm{i}\int_{-\infty}^{t} \hat{H}'_{\mathrm{I}}(t')|\psi(t')\rangle_{\mathrm{I}}\,\mathrm{d}t'. \tag{6.32}$$

This is an integral equation in which the unknown $|\psi(t)\rangle_{\rm I}$ is buried under the integral on the right-hand side, rather similar to the one we encounter in non-relativistic scattering theory (equation (H.12) of appendix H). As in that case, we solve it iteratively. If $\hat{H}'_{\rm I}$ is neglected altogether, then the solution is

$$|\psi(t)\rangle_{\rm I}^{(0)} = |{\rm i}\rangle. \tag{6.33}$$

To get the first order in $\hat{H}'_{\rm I}$ correction to this, insert (6.33) in place of $|\psi(t')\rangle_{\rm I}$ on the right-hand side of (6.32) to obtain

$$|\psi(t)\rangle_{\rm I}^{(1)} = |{\rm i}\rangle + \int_{-\infty}^{t} (-{\rm i}\hat{H}'_{\rm I}(t_1)){\rm d}t_1\,|{\rm i}\rangle \tag{6.34}$$

recalling that $|{\rm i}\rangle$ is a constant state vector. Putting this back into (6.32) yields $|\psi(t)\rangle$ correct to second order in $\hat{H}'_{\rm I}$:

$$\begin{aligned}|\psi(t)\rangle_{\rm I}^{(2)} = \bigg\{1 &+ \int_{-\infty}^{t} (-{\rm i}\hat{H}'_{\rm I}(t_1))\,{\rm d}t_1 \\ &+ \int_{-\infty}^{t} {\rm d}t_1 \int_{-\infty}^{t_1} {\rm d}t_2\,(-{\rm i}\hat{H}'_{\rm I}(t_1))(-{\rm i}\hat{H}'_{\rm I}(t_2))\bigg\}|{\rm i}\rangle\end{aligned} \tag{6.35}$$

which is as far as we intend to go. Letting $t \to \infty$ then gives us our *perturbative series for the $\hat{S}$-operator*:

$$\hat{S} = 1 + \int_{-\infty}^{\infty} (-{\rm i}\hat{H}'_{\rm I}(t_1))\,{\rm d}t_1 + \int_{-\infty}^{\infty} {\rm d}t_1 \int_{-\infty}^{t_1} {\rm d}t_2\,(-{\rm i}\hat{H}'_{\rm I}(t_1))(-{\rm i}\hat{H}'_{\rm I}(t_2)) + \cdots \tag{6.36}$$

the dots indicating the higher-order terms, which are in fact summarized by the full formula

$$\hat{S} = \sum_{n=0}^{\infty}(-{\rm i})^n \int_{-\infty}^{\infty} {\rm d}t_1 \int_{-\infty}^{t_1} {\rm d}t_2 \cdots \int_{-\infty}^{t_{n-1}} {\rm d}t_n\, \hat{H}'_{\rm I}(t_1)\hat{H}'_{\rm I}(t_2)\ldots\hat{H}'_{\rm I}(t_n). \tag{6.37}$$

We could immediately start getting to work with (6.37), but there is one more useful technical adjustment to make. Remembering that

$$\hat{H}'_{\rm I}(t) = \int \hat{\mathcal{H}}'_{\rm I}(\boldsymbol{x}, t)\,{\rm d}^3\boldsymbol{x} \tag{6.38}$$

we can write the second term of (6.36) as

$$\iint_{t_1>t_2} {\rm d}^4x_1\,{\rm d}^4x_2\,(-{\rm i}\hat{\mathcal{H}}'_{\rm I}(x_1))(-{\rm i}\hat{\mathcal{H}}'_{\rm I}(x_2)) \tag{6.39}$$

which looks much more symmetrical in $\boldsymbol{x} - t$. However, there is still an awkward asymmetry between the $\boldsymbol{x}$-integrals and the t-integrals because of the $t_1 > t_2$

condition. The t-integrals can be converted to run from $-\infty$ to ∞ without constraint, like the $\boldsymbol{x}$ ones, by a clever trick. Note that the *ordering* of the operators $\hat{\mathcal{H}}'_{\mathrm{I}}$ is significant (since they will contain non-commuting bits), and that it is actually given by the order of their time arguments, 'earlier' operators appearing to the right of 'later' ones. This feature must be preserved, obviously, when we let the t-integrals run over the full infinite domain. We can arrange for this by introducing the time-ordering symbol T, which is defined by

$$\begin{aligned} T(\hat{\mathcal{H}}'_{\mathrm{I}}(x_1)\hat{\mathcal{H}}'_{\mathrm{I}}(x_2)) &= \hat{\mathcal{H}}'_{\mathrm{I}}(x_1)\hat{\mathcal{H}}'_{\mathrm{I}}(x_2) \qquad \text{for } t_1 > t_2 \\ &= \hat{\mathcal{H}}'_{\mathrm{I}}(x_2)\hat{\mathcal{H}}'_{\mathrm{I}}(x_1) \qquad \text{for } t_1 < t_2 \end{aligned} \tag{6.40}$$

and similarly for more products, and for arbitrary operators. Then (see problem 6.1) (6.39) can be written as

$$\frac{1}{2}\iint \mathrm{d}^4x_1\,\mathrm{d}^4x_2\, T[(-\mathrm{i}\hat{\mathcal{H}}'_{\mathrm{I}}(x_1))(-\mathrm{i}\hat{\mathcal{H}}'_{\mathrm{I}}(x_2))] \tag{6.41}$$

where the integrals are now unrestricted. Applying a similar analysis to the general term gives us the *Dyson expansion of the $\hat{S}$ operator*:

$$\boxed{\hat{S} = \sum_{n=0}^{\infty}\frac{(-\mathrm{i})^n}{n!}\int\cdots\int \mathrm{d}^4x_1\,\mathrm{d}^4x_2\ldots\mathrm{d}^4x_n\, T\{\hat{\mathcal{H}}'_{\mathrm{I}}(x_1)\hat{\mathcal{H}}'_{\mathrm{I}}(x_2)\cdots\hat{\mathcal{H}}'_{\mathrm{I}}(x_n)\}.} \tag{6.42}$$

This fundamental formula provides the bridge leading from the Tomonaga–Schwinger equation (6.25) to the Feynman amplitudes (Feynman 1949a, b), as we shall see in detail in section 6.3.2 for the 'ABC' case.

6.3 Applications to the 'ABC' theory

As previously explained, the simple self-interacting $\hat{\phi}^3$ theory is not respectable. Following Griffiths (1987) we shall instead apply the foregoing covariant perturbation theory to a hypothetical world consisting of three distinct types of scalar particles A, B and C, with masses m_{A}, m_{B}, m_{C}. Each is described by a real scalar field which, if free, would obey the appropriate KG equation; the interaction term is $g\hat{\phi}_{\mathrm{A}}\hat{\phi}_{\mathrm{B}}\hat{\phi}_{\mathrm{C}}$. We shall from now on omit the IP subscript 'I', since all operators are taken to be in the IP. Thus the Hamiltonian is

$$\hat{H} = \hat{H}_0 + \hat{H}' \tag{6.43}$$

where

$$\hat{H}_0 = \frac{1}{2}\sum_{i=\mathrm{A,B,C}}\int [\hat{\pi}_i{}^2 + (\nabla\hat{\phi}_i)^2 + m_i^2\hat{\phi}_i^2]\,\mathrm{d}^3\boldsymbol{x} \tag{6.44}$$

and

$$\hat{H}' = g\int \mathrm{d}^3\boldsymbol{x}\,\hat{\phi}_{\mathrm{A}}\hat{\phi}_{\mathrm{B}}\hat{\phi}_{\mathrm{C}} \equiv \int \mathrm{d}^3\boldsymbol{x}\,\hat{\mathcal{H}}'. \tag{6.45}$$

Each field $\hat{\phi}_i$, (i = A, B, C) has a mode expansion of the form (5.142), and associated creation and annihilation operators $\hat{a}_i^\dagger$ and $\hat{a}_i$ which obey the commutation relations

$$[\hat{a}_i(k), \hat{a}_j^\dagger(k')] = (2\pi)^3\delta^3(\boldsymbol{k}-\boldsymbol{k}')\delta_{ij} \qquad i, j = \mathrm{A, B, C}. \tag{6.46}$$

The new feature in (6.43) is that operators associated with distinct particles commute. In a similar way, we also have $[\hat{a}_i, \hat{a}_j] = [\hat{a}_i^\dagger, \hat{a}_j^\dagger] = 0$.

6.3.1 The decay C → A + B

As our first application of (6.42), we shall calculate the decay rate (or resonance width) for the decay C → A+B, to lowest order in g. Admittedly this is not yet a realistic, physical, example; even so, the basic steps in the calculation are common to more complicated physical examples, such as $\mathrm{W}^- \to \mathrm{e}^- + \bar{\nu}_\mathrm{e}$.

We suppose that the initial state $|\mathrm{i}\rangle$ consists of one C particle with 4-momentum p_C, and that the final state in which we are interested is that with one A and one B particle present, with 4-momenta p_A and p_B respectively. We want to calculate the matrix element

$$S_{\mathrm{fi}} = \langle p_\mathrm{A}, p_\mathrm{B}|\hat{S}|p_\mathrm{C}\rangle \tag{6.47}$$

to lowest order in g. (Note that the '1' term in (6.36) cannot contribute here because the initial and final states are plainly orthogonal.) This means that we need to evaluate the amplitude

$$\mathcal{A}_{\mathrm{fi}}^{(1)} = -\mathrm{i}g\langle p_\mathrm{A}, p_\mathrm{B}|\int \mathrm{d}^4x\, \hat{\phi}_\mathrm{A}(x)\hat{\phi}_\mathrm{B}(x)\hat{\phi}_\mathrm{C}(x)|p_\mathrm{C}\rangle. \tag{6.48}$$

To proceed we need to decide on the normalization of our states $|p_i\rangle$. We will define (for i = A, B, C)

$$|p_i\rangle = \sqrt{2E_i}\hat{a}_i^\dagger(p_i)|0\rangle \tag{6.49}$$

where $E_i = \sqrt{m_i^2 + \boldsymbol{p}_i^2}$, so that (using (6.46))

$$\langle p_i'|p_i\rangle = 2E_i(2\pi)^3\delta^3(\boldsymbol{p}_i' - \boldsymbol{p}_i). \tag{6.50}$$

The quantity $E_i\delta^3(\boldsymbol{p}_i' - \boldsymbol{p}_i)$ is Lorentz invariant. Note that the completeness relation for such states reads

$$\int \frac{\mathrm{d}^3\boldsymbol{p}_i}{(2\pi^3)}\frac{1}{2E_i}|p_i\rangle\langle p_i| = 1 \tag{6.51}$$

where the '1' on the right-hand side means the identity in the subspace of such one-particle states, and zero for all other states. The normalization choice (6.49) corresponds (see comment (5) in section 5.2.5) to a wavefunction normalization of $2E_i$ particles per unit volume.

Consider now just the $\hat{\phi}_{\mathrm{C}}(x)|p_{\mathrm{C}}\rangle$ piece of (6.48). This is

$$\int \frac{\mathrm{d}^3\boldsymbol{k}}{(2\pi)^3}\frac{1}{\sqrt{2E_k}}[\hat{a}_{\mathrm{C}}(k)\mathrm{e}^{-\mathrm{i}k\cdot x}+\hat{a}^\dagger_{\mathrm{C}}(k)\mathrm{e}^{\mathrm{i}k\cdot x}]\sqrt{2E_{\mathrm{C}}}\hat{a}^\dagger_{\mathrm{C}}(p_{\mathrm{C}})|0\rangle \tag{6.52}$$

where $k=(E_k,\boldsymbol{k})$ and $E_k=\sqrt{\boldsymbol{k}^2+m_{\mathrm{C}}^2}$. The term with two $\hat{a}^\dagger_{\mathrm{C}}$'s will give zero when bracketed with a final state containing no C particles. In the other term, we use (6.46) together with $\hat{a}_{\mathrm{C}}(k)|0\rangle=0$ to reduce (6.52) to

$$\int \frac{\mathrm{d}^3\boldsymbol{k}}{(2\pi^3)}\frac{1}{\sqrt{2E_k}}(2\pi)^3\delta^3(\boldsymbol{p}_{\mathrm{C}}-\boldsymbol{k})\sqrt{2E_{\mathrm{C}}}\mathrm{e}^{-\mathrm{i}k\cdot x}|0\rangle=\mathrm{e}^{-\mathrm{i}p_{\mathrm{C}}\cdot x}|0\rangle \tag{6.53}$$

where $p_{\mathrm{C}}=(\sqrt{\boldsymbol{p}_{\mathrm{C}}^2+m_{\mathrm{C}}^2},\boldsymbol{p}_{\mathrm{C}})$. In exactly the same way we find that, when bracketed with an initial state containing no A's or B's,

$$\langle p_{\mathrm{A}},p_{\mathrm{B}}|\hat{\phi}_{\mathrm{A}}(x)\hat{\phi}_{\mathrm{B}}(x)=\langle 0|e^{\mathrm{i}p_{\mathrm{A}}\cdot x}\mathrm{e}^{\mathrm{i}p_{\mathrm{B}}\cdot x}. \tag{6.54}$$

Hence the amplitude (6.48) becomes just

$$\mathcal{A}^{(1)}_{\mathrm{fi}}=-\mathrm{i}g\int \mathrm{d}^4x\mathrm{e}^{\mathrm{i}(p_{\mathrm{A}}+p_{\mathrm{B}}-p_{\mathrm{C}})\cdot x}=-\mathrm{i}g(2\pi)^4\delta^4(p_{\mathrm{A}}+p_{\mathrm{B}}-p_{\mathrm{C}}). \tag{6.55}$$

Unsurprisingly, but reassuringly, we have discovered that the amplitude vanishes unless the 4-momentum is conserved via the δ-function condition: $p_{\mathrm{C}}=p_{\mathrm{A}}+p_{\mathrm{B}}$.

It is clear that such a transition will not occur unless $m_{\mathrm{C}}>m_{\mathrm{A}}+m_{\mathrm{B}}$ (in the rest frame of the C, we need $m_{\mathrm{C}}=\sqrt{m_{\mathrm{A}}^2+\boldsymbol{p}^2}+\sqrt{m_{\mathrm{B}}^2+\boldsymbol{p}^2}$), so let us assume this to be the case. We would now like to calculate the rate for the decay $\mathrm{C}\to\mathrm{A}+\mathrm{B}$. To do this, we shall adopt a plausible generalization of the ordinary procedure followed in quantum mechanical time-dependent perturbation theory (the reader may wish to consult section H.3 of appendix H at this point, to see a non-relativistic analogue). The first problem is that the transition probability $|\mathcal{A}^{(1)}_{\mathrm{fi}}|^2$ apparently involves the square of the four-dimensional δ-function. This is bad news, since (to take a simple case) $\delta(x-a)\delta(x-a)=\delta(x-a)\delta(0)$ and $\delta(0)$ is infinite. In our case we have a four-fold infinity. This trouble has arisen because we have been using plane-wave solutions of our wave equation, and these notoriously lead to such problems. A proper procedure would set the whole thing up using wave packets, as is done, for instance, in Peskin and Schroeder (1995), section 4.5. An easier remedy is to adopt 'box normalization', in which we imagine that space has the finite volume V, and the interaction is turned on only for a time T. Then '$(2\pi)^4\delta^4(0)$' is effectively 'VT' (see Weinberg (1995, section 3.4)). Dividing this factor out, the transition rate per unit volume is then

$$\dot{P}_{\mathrm{fi}}=|\mathcal{A}^{(1)}_{\mathrm{fi}}|^2/VT=(2\pi)^4\delta^4(p_{\mathrm{A}}+p_{\mathrm{B}}-p_{\mathrm{C}})|\mathcal{M}_{\mathrm{fi}}|^2 \tag{6.56}$$

where (cf (6.55))

$$\mathcal{A}^{(1)}_{\mathrm{fi}}=(2\pi)^4\delta^4(p_{\mathrm{A}}+p_{\mathrm{B}}-p_{\mathrm{C}})\mathrm{i}\mathcal{M}_{\mathrm{fi}} \tag{6.57}$$

so that the *invariant amplitude* $\mathrm{i}\mathcal{M}_{\mathrm{fi}}$ is just $-\mathrm{i}g$, in this case.

Equation (6.56) is the probability per unit time for a transition to one specific final state $|\mathrm{f}\rangle$. But in the present case (and in all similar ones with at least two particles in the final state), the A + B final states form a continuum, and to get the total rate Γ we need to integrate $\dot{P}_{\mathrm{fi}}$ over all the continuum of final states, consistent with energy–momentum conservation. The corresponding differential decay rate $\mathrm{d}\Gamma$ is defined by $\mathrm{d}\Gamma = \dot{P}_{\mathrm{fi}}\mathrm{d}N_{\mathrm{f}}$ where $\mathrm{d}N_{\mathrm{f}}$ is the number of final states, per particle, lying in a momentum space volume $\mathrm{d}^3 p_{\mathrm{A}}\mathrm{d}^3 p_{\mathrm{B}}$ about p_{A} and p_{B}. For the normalization (6.49), this number is

$$\mathrm{d}N_{\mathrm{f}} = \frac{\mathrm{d}^3 \boldsymbol{p}_{\mathrm{A}}}{(2\pi)^3 2E_{\mathrm{A}}} \frac{\mathrm{d}^3 \boldsymbol{p}_{\mathrm{B}}}{(2\pi)^3 2E_{\mathrm{B}}}. \tag{6.58}$$

Finally, to get a normalization-independent quantity we must divide by the number of decaying particles per unit volume, which is $2E_{\mathrm{C}}$. Thus our final formula for the decay rate is

$$\Gamma = \int \mathrm{d}\Gamma = \frac{1}{2E_{\mathrm{C}}}(2\pi)^4 \int \delta^4(p_{\mathrm{A}} + p_{\mathrm{B}} - p_{\mathrm{C}})|\mathcal{M}_{\mathrm{fi}}|^2 \frac{\mathrm{d}^3 \boldsymbol{p}_{\mathrm{A}}}{(2\pi)^3 2E_{\mathrm{A}}} \frac{\mathrm{d}^3 \boldsymbol{p}_{\mathrm{B}}}{(2\pi)^3 2E_{\mathrm{B}}}. \tag{6.59}$$

Note that the '$\mathrm{d}^3\boldsymbol{p}/2E$' factors are Lorentz invariant (see the exercise in appendix E) and so are all the other terms in (6.59) except E_{C}, which contributes the correct Lorentz-transformation character for a rate (i.e. rate $\propto 1/\gamma$).

We now calculate the total rate Γ in the rest frame of the decaying C particle. In this case, the 3-momentum part of the δ^4 gives $\boldsymbol{p}_{\mathrm{A}} + \boldsymbol{p}_{\mathrm{B}} = \mathbf{0}$, so $\boldsymbol{p}_{\mathrm{A}} = \boldsymbol{p} = -\boldsymbol{p}_{\mathrm{B}}$, and the energy part becomes $\delta(E - m_{\mathrm{C}})$ where

$$E = \sqrt{m_{\mathrm{A}}^2 + \boldsymbol{p}^2} + \sqrt{m_{\mathrm{B}} + \boldsymbol{p}^2} = E_{\mathrm{A}} + E_{\mathrm{B}}. \tag{6.60}$$

So the total rate is

$$\Gamma = \frac{1}{2m_{\mathrm{C}}} \frac{g^2}{(2\pi)^2} \int \frac{\mathrm{d}^3 \boldsymbol{p}}{4E_{\mathrm{A}}E_{\mathrm{B}}} \delta(E - m_{\mathrm{C}}). \tag{6.61}$$

Differentiating (6.60) we find

$$\mathrm{d}E = \left(\frac{|\boldsymbol{p}|}{E_{\mathrm{A}}} + \frac{|\boldsymbol{p}|}{E_{\mathrm{B}}}\right) \mathrm{d}|\boldsymbol{p}| = \frac{|\boldsymbol{p}|E}{E_{\mathrm{A}}E_{\mathrm{B}}} \mathrm{d}|\boldsymbol{p}|. \tag{6.62}$$

Thus we may write

$$\mathrm{d}^3 \boldsymbol{p} = 4\pi|\boldsymbol{p}|^2 \mathrm{d}|\boldsymbol{p}| = 4\pi|\boldsymbol{p}| \frac{E_{\mathrm{A}}E_{\mathrm{B}}}{E} \mathrm{d}E \tag{6.63}$$

and use the energy δ-function in (6.61) to do the $\mathrm{d}E$ integral yielding finally

$$\Gamma = \frac{g^2}{8\pi} \frac{|\boldsymbol{p}|}{m_{\mathrm{C}}^2}. \tag{6.64}$$

The quantity $|\boldsymbol{p}|$ is actually determined from (6.60) now with $E = m_{\mathrm{C}}$; after some algebra, we find (problem 6.2)

$$|\boldsymbol{p}| = [m_{\mathrm{A}}^4 + m_{\mathrm{B}}^4 + m_{\mathrm{C}}^4 - 2m_{\mathrm{A}}^2 m_{\mathrm{B}}^2 - 2m_{\mathrm{B}}^2 m_{\mathrm{C}}^2 - 2m_{\mathrm{C}}^2 m_{\mathrm{A}}^2]^{1/2}/2m_{\mathrm{C}}. \tag{6.65}$$

Equation (6.64) is the result of an 'almost real life' calculation and a number of comments are in order. First, consider the question of dimensions. In our units $\hbar = c = 1$, Γ as an inverse time should have the dimensions of a mass (see appendix B), which can also be understood if we think of Γ as the width of an unstable resonance state. This requires 'g' to have the dimensions of a mass, i.e. $g \sim M$ in these units. Going back to our Hamiltonian (6.44) and (6.45), which must also have dimensions of a mass, we see from (6.44) that the scalar fields $\hat{\phi}_i \sim M$ (using $\mathrm{d}^3\boldsymbol{x} \sim M^{-3}$), and hence from (6.45) $g \sim M$ as required. It turns out that the dimensionality of the coupling constants (such as g) is of great significance in quantum field theory. In QED, the analogous quantity is the charge e, and this is dimensionless in our units ($\alpha = e^2/4\pi = 1/137$, see appendix C). However, we saw in (2.20) that Fermi's 'four-fermion' coupling constant G had dimensions $\sim M^{-2}$, while Yukawa's 'g_{s}' and 'g'' (see figure 2.4) were both dimensionless. In fact, as we shall explain in section 11.8, the dimensionality of a theory's coupling constant is an important guide as to whether the infinities generally present in the theory can be controlled by *renormalization* (see chapter 10) or not: in particular, theories in which the coupling constant has negative mass dimensions, such as the 'four-fermion' theory, are not renormalizable. Theories with dimensionless coupling constants, such as QED, are generally renormalizable, though not invariably so. Theories whose coupling constants have positive mass dimension, as in the ABC model, are 'super-renormalizable', meaning (roughly) that they have fewer basic divergences than ordinary renormalizable theories (see section 11.8).

In the present case, let us say that the mass of the decaying particle m_{C}, 'sets the scale' for g, so that we write $g = \tilde{g}m_{\mathrm{C}}$ and then

$$\Gamma = \frac{\tilde{g}^2}{8\pi}|\boldsymbol{p}| \tag{6.66}$$

where $\tilde{g}$ is dimensionless. Equation (6.66) shows us nicely that Γ is simply proportional to the energy release in the decay, as determined by $|\boldsymbol{p}|$ (one often says that Γ is determined 'by the available phase space'). If m_{C} is exactly equal to $m_{\mathrm{A}} + m_{\mathrm{B}}$, then $|\boldsymbol{p}|$ vanishes and so does Γ. At the opposite extreme, if m_{A} and m_{B} are negligible compared to m_{C}, we would have

$$\Gamma = \frac{\tilde{g}^2}{16\pi}m_{\mathrm{C}}. \tag{6.67}$$

Equation (6.67) shows that, even if $\tilde{g}^2/16\pi$ is small ($\sim 1/137$ say) Γ can still be surprisingly large if m_{C} is, as in $\mathrm{W}^- \to \mathrm{e}^- + \bar{\nu}_{\mathrm{e}}$ for example.

6.3.2 A + B → A + B scattering: the amplitudes

We now consider the two-particle → two-particle process

$$\mathrm{A} + \mathrm{B} \to \mathrm{A} + \mathrm{B} \tag{6.68}$$

in which the initial 4-momenta are p_{A}, p_{B} and the final 4-momenta are p'_{A}, p'_{B} so that $p_{\mathrm{A}} + p_{\mathrm{B}} = p'_{\mathrm{A}} + p'_{\mathrm{B}}$. Our main task is to calculate the matrix element $\langle p'_{\mathrm{A}}, p'_{\mathrm{B}}|\hat{S}|p_{\mathrm{A}}, p_{\mathrm{B}}\rangle$ to lowest non-trivial order in g. The result will be the derivation of our first 'Feynman rules' for amplitudes in perturbative quantum field theory.

The first term in the $\hat{S}$-operator expansion (6.42) is '1', which does not involve g at all. Nevertheless, it is a useful exercise to evaluate and understand this contribution (which in the present case does not vanish), namely

$$\langle 0|\hat{a}_{\mathrm{A}}(p'_{\mathrm{A}})\hat{a}_{\mathrm{B}}(p'_{\mathrm{B}})\hat{a}^{\dagger}_{\mathrm{A}}(p_{\mathrm{A}})\hat{a}^{\dagger}_{\mathrm{B}}(p_{\mathrm{B}})|0\rangle(16E_{\mathrm{A}}E_{\mathrm{B}}E'_{\mathrm{A}}E'_{\mathrm{B}})^{1/2}. \tag{6.69}$$

We shall have to evaluate many such *vacuum expectation values* (vev) of products of $\hat{a}^{\dagger}$'s and $\hat{a}$'s. The general strategy is to commute the $\hat{a}^{\dagger}$'s to the left, and the $\hat{a}$'s to the right, and then make use of the facts

$$\langle 0|\hat{a}^{\dagger}_{i} = \hat{a}_{i}|0\rangle = 0 \tag{6.70}$$

for any $i = \mathrm{A, B, C}$. Thus, remembering that all 'A' operators commute with all 'B' ones, the vev in (6.69) is equal to

$$\begin{aligned}
&\langle 0|\hat{a}_{\mathrm{A}}(p'_{\mathrm{A}})\hat{a}^{\dagger}_{\mathrm{A}}(p_{\mathrm{A}})\{(2\pi)^3\delta^3(\boldsymbol{p}_{\mathrm{B}} - \boldsymbol{p}'_{\mathrm{B}}) + \hat{a}^{\dagger}_{\mathrm{B}}(p_{\mathrm{B}})\hat{a}_{\mathrm{B}}(p'_{\mathrm{B}})\}|0\rangle \\
&\quad = \langle 0|\{(2\pi)^3\delta^3(\boldsymbol{p}_{\mathrm{A}} - \boldsymbol{p}'_{\mathrm{A}}) + \hat{a}^{\dagger}_{\mathrm{A}}(p_{\mathrm{A}})\hat{a}_{\mathrm{A}}(p'_{\mathrm{A}})\}(2\pi)^3\delta^3(\boldsymbol{p}_{\mathrm{B}} - \boldsymbol{p}'_{\mathrm{B}})|0\rangle \\
&\quad = (2\pi)^3\delta^3(\boldsymbol{p}_{\mathrm{A}} - \boldsymbol{p}'_{\mathrm{A}})(2\pi)^3\delta^3(\boldsymbol{p}_{\mathrm{B}} - \boldsymbol{p}'_{\mathrm{B}}).
\end{aligned} \tag{6.71}$$

The δ-functions enforce $E_{\mathrm{A}} = E'_{\mathrm{A}}$ and $E_{\mathrm{B}} = E'_{\mathrm{B}}$ so that (6.69) becomes

$$2E_{\mathrm{A}}(2\pi)^3\delta^3(\boldsymbol{p}_{\mathrm{A}} - \boldsymbol{p}'_{\mathrm{A}})2E_{\mathrm{B}}(2\pi)^3\delta^3(\boldsymbol{p}_{\mathrm{B}} - \boldsymbol{p}'_{\mathrm{B}}) \tag{6.72}$$

a result which just expresses the normalization of the states, and the fact that, with no 'g' entering, the particles have not interacted at all, but have continued on their separate ways, quite unperturbed ($\boldsymbol{p}_{\mathrm{A}} = \boldsymbol{p}'_{\mathrm{A}}$, $\boldsymbol{p}_{\mathrm{B}} = \boldsymbol{p}'_{\mathrm{B}}$). This contribution can be represented diagrammatically as figure 6.1.

Next, consider the term of order g, which we used in C → A + B. This is

$$-\mathrm{i}g\int \mathrm{d}^4x\, \langle p'_{\mathrm{A}}, p'_{\mathrm{B}}|\hat{\phi}_{\mathrm{A}}(x)\hat{\phi}_{\mathrm{B}}(x)\hat{\phi}_{\mathrm{C}}(x)|p_{\mathrm{A}}, p_{\mathrm{B}}\rangle. \tag{6.73}$$

We have to remember, now, that all the $\hat{\phi}_i$ operators are in the interaction picture and are therefore represented by standard mode expansions involving the *free* creation and annihilation operators $\hat{a}^{\dagger}_{i}$ and $\hat{a}_{i}$, i.e. the same ones used in defining

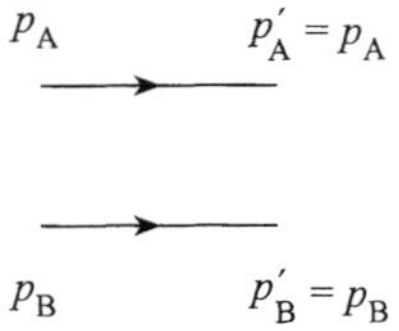

Figure 6.1. The order g^0 term in the perturbative expansion: the two particles do not interact.

the initial and final state vectors. It is then obvious that (6.73) must vanish, since no C-particle exists in either the initial or final state, and $\langle 0|\hat{\phi}_C|0\rangle = 0$.

So we move on to the term of order g^2, which will provide the real meat of this chapter. This term is

$$\frac{(-ig)^2}{2}\iint d^4x_1\, d^4x_2\, \langle 0|\hat{a}_A(p'_A)\hat{a}_B(p'_B) \\ \times T\{\hat{\phi}_A(x_1)\hat{\phi}_B(x_1)\hat{\phi}_C(x_1)\hat{\phi}_A(x_2)\hat{\phi}_B(x_2)\hat{\phi}_C(x_2)\} \\ \times \hat{a}_A^\dagger(p_A)\hat{a}_B^\dagger(p_B)|0\rangle(16E_A E_B E'_A E'_B)^{1/2}. \qquad (6.74)$$

The vev here involves the product of *ten* operators, so it will pay us to pause and think how such things may be efficiently evaluated.

Consider the case of just four operators

$$\langle 0|\hat{A}\hat{B}\hat{C}\hat{D}|0\rangle \qquad (6.75)$$

where each of $\hat{A}$, $\hat{B}$, $\hat{C}$, $\hat{D}$ is an $\hat{a}_i$, an $\hat{a}_i^\dagger$ or a linear combination of these. Let $\hat{A}$ have the generic form $\hat{A} = \hat{a} + \hat{a}^\dagger$. Then (using $\langle 0|a^\dagger = a|0\rangle = 0$)

$$\begin{aligned}\langle 0|\hat{A}\hat{B}\hat{C}\hat{D}|0\rangle &= \langle 0|\hat{a}\hat{B}\hat{C}\hat{D}|0\rangle \\ &= \langle 0|[\hat{a}, \hat{B}\hat{C}\hat{D}]|0\rangle.\end{aligned} \qquad (6.76)$$

Now it is an algebraic identity that

$$[\hat{a}, \hat{B}\hat{C}\hat{D}] = [\hat{a}, \hat{B}]\hat{C}\hat{D} + \hat{B}[\hat{a}, \hat{C}]\hat{D} + \hat{B}\hat{C}[\hat{a}, \hat{D}]. \qquad (6.77)$$

Hence

$$\langle 0|\hat{A}\hat{B}\hat{C}\hat{D}|0\rangle = [\hat{a}, \hat{B}]\langle 0|\hat{C}\hat{D}|0\rangle + [\hat{a}, \hat{C}]\langle 0|\hat{B}\hat{D}|0\rangle + [\hat{a}, \hat{D}]\langle 0|\hat{B}\hat{C}|0\rangle, \quad (6.78)$$

remembering that all the commutators—if non-vanishing—are just ordinary numbers (see (6.46)). We can rewrite (6.78) in more suggestive form by noting that

$$[\hat{a}, \hat{B}] = \langle 0|[\hat{a}, \hat{B}]|0\rangle = \langle 0|\hat{a}\hat{B}|0\rangle = \langle 0|\hat{A}\hat{B}|0\rangle. \qquad (6.79)$$

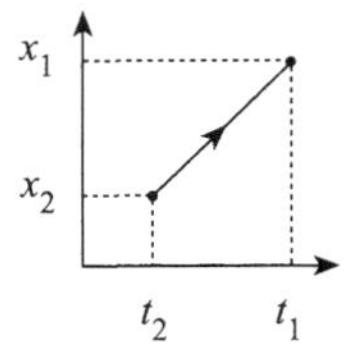

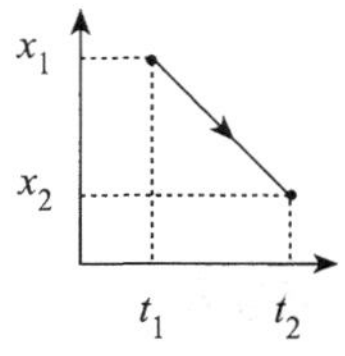

Figure 6.2. C-quantum propagating (*a*) for $t_1 > t_2$ (from x_2 to x_1) and (*b*) $t_1 < t_2$ (from x_1 to x_2).

Thus the vev of a product of four operators is just the sum of the products of all the possible pairwise 'contractions' (the name given to the vev of the product of two fields):

$$\langle 0|\hat{A}\hat{B}\hat{C}\hat{D}|0\rangle = \langle 0|\hat{A}\hat{B}|0\rangle\langle 0|\hat{C}\hat{D}|0\rangle + \langle 0|\hat{A}\hat{C}|0\rangle\langle 0|\hat{B}\hat{D}|0\rangle + \langle 0|\hat{A}\hat{D}|0\rangle\langle 0|\hat{B}\hat{C}|0\rangle. \tag{6.80}$$

This result *generalizes* to the vev of the product of any number of operators; there is also a similar result for the vev of time-ordered products of operators, which is known as Wick's theorem (Wick 1950), and is indispensable for a general discussion of quantum field perturbation theory.

Consider then the application of (6.80), as generalized to ten operators, to the vev in (6.74). The only kind of non-vanishing contractions are of the form $\langle 0|\hat{a}_i\hat{a}_i^\dagger|0\rangle$. As far as the C-operators are concerned, then, we can immediately conclude that the only surviving contraction is

$$\langle 0|T(\hat{\phi}_{\mathrm{C}}(x_1)\hat{\phi}_{\mathrm{C}}(x_2))|0\rangle. \tag{6.81}$$

This quantity is, in fact, of fundamental importance: it is called the *Feynman propagator* (in coordinate space) for the spin-0 C-particle. We shall derive the mathematical formula for it in due course, but for the moment let us understand its physical significance. Each of the $\hat{\phi}_{\mathrm{C}}$'s in (6.81) can create or destroy C-quanta, but for the vev to be non-zero anything created in the 'initial' state must be destroyed in the 'final' one. Which of the times t_1 and t_2 is initial or final is determined by the T-ordering symbol: for $t_1 > t_2$, a C-quantum is created at x_2 and destroyed at x_1, while for $t_1 < t_2$ a C-quantum is created at x_1 and destroyed at x_2. Thus the amplitude (6.81) may be represented pictorially as in figure 6.2, where time increases to the right, and the vertical axis is a one-dimensional version of three-dimensional space. It seems reasonable, indeed, to call this object the 'propagator', since it clearly has to do with a quantum propagating between two space–time points.

We might now worry that this explicit time-ordering seems to introduce a Lorentz non-invariant element into the calculation, ultimately threatening the

Lorentz invariance of the $\hat{S}$-operator (6.42). The reason that this is in fact not the case exposes an important property of quantum field theory. If the two points x_1 and x_2 are separated by a time-like interval (i.e. $(x_1 - x_2)^2 > 0$), then the time-ordering is Lorentz invariant; this is because no proper Lorentz transformation can alter the time-ordering of time-like separated events (here, the events are the creation/annihilation of particles/anti-particles at x_1 and x_2). By 'proper' is meant a transformation that does not reverse the sense of time; the behaviour of the theory under time-reversal is a different question altogether, not studied further here. The fact that time-ordering is invariant for time-like separated events is what guarantees that we cannot influence our past, only our future. But what if the events are space-like separated, $(x_1 - x_2)^2 < 0$? We know that the scalar fields $\hat{\phi}_i(x_1)$ and $\hat{\phi}_i(x_2)$ commute for equal times: remarkably, one can show (problem 5.6(b)) that they *also* commute for $(x_1 - x_2)^2 < 0$; so in this sector of $x_1 - x_2$ space the time-ordering symbol is irrelevant. Thus, contrary to appearances, the T-product vev is Lorentz invariant. For the same reason, the $\hat{S}$ operator of (6.42) is also Lorentz invariant: see, for example, Weinberg (1995, section 3.5).

The property

$$[\hat{\phi}_i(x_1), \hat{\phi}_i(x_2)] = 0 \qquad \text{for } (x_1 - x_2)^2 < 0 \tag{6.82}$$

has an important physical interpretation. In quantum mechanics, if operators representing physical observables commute with each other, then measurements of either observable can be performed without interfering with each other; the observables are said to be 'compatible'. This is just what we would want for measurements done at two points which are space-like separated—no signal with speed less than or equal to light can connect them, and so we would expect them to be non-interfering.

More mathematically, the amplitude (6.81) is in fact a *Green function* for the KG operator $(\Box + m_{\mathrm{C}}^2)$! (see appendix G, and problem 6.3). That is to say,

$$(\Box_{x_1} + m_{\mathrm{C}}^2)\langle 0|T(\hat{\phi}_{\mathrm{C}}(x_1)\hat{\phi}_{\mathrm{C}}(x_2))|0\rangle = -\mathrm{i}\delta^4(x_1 - x_2). \tag{6.83}$$

Actually, problem 6.3 shows that (6.83) is true even when the $\langle 0|$ and $|0\rangle$ are removed, i.e. the operator quantity $T(\hat{\phi}_{\mathrm{C}}(x_1)\hat{\phi}_{\mathrm{C}}(x_2))$ is itself a KG Green function. The work of appendices G and H indicates the central importance of such Green functions in scattering theory, so we need not be surprised to find such a thing appearing here.

Now let us figure out what are all the surviving terms in the vev in (6.74). As far as contractions involving $\hat{a}_{\mathrm{A}}(p'_{\mathrm{A}})$ are concerned, we have only three non-zero possibilities:

$$\langle 0|\hat{a}_{\mathrm{A}}(p'_{\mathrm{A}})\hat{a}^{\dagger}_{\mathrm{A}}(p_{\mathrm{A}})|0\rangle \qquad \langle 0|\hat{a}_{\mathrm{A}}(p'_{\mathrm{A}})\hat{\phi}_{\mathrm{A}}(x_1)|0\rangle \qquad \langle 0|\hat{a}_{\mathrm{A}}(p'_{\mathrm{A}})\hat{\phi}_{\mathrm{A}}(x_2)|0\rangle. \tag{6.84}$$

There are similar possibilities for $\hat{a}_{\mathrm{A}}^{\dagger}(p_{\mathrm{A}})$, $\hat{a}_{\mathrm{B}}(p'_{\mathrm{B}})$ and $\hat{a}_{\mathrm{B}}^{\dagger}(p_{\mathrm{B}})$. The upshot is that we have only the following pairings to consider:

$$\begin{aligned}&\langle 0|\hat{a}_{\mathrm{A}}(p'_{\mathrm{A}})\hat{a}_{\mathrm{A}}^{\dagger}(p_{\mathrm{A}})|0\rangle\langle 0|\hat{a}_{\mathrm{B}}(p'_{\mathrm{B}})\hat{a}_{\mathrm{B}}^{\dagger}(p_{\mathrm{B}})|0\rangle\\&\quad\times\langle 0|T(\hat{\phi}_{\mathrm{A}}(x_1)\hat{\phi}_{\mathrm{A}}(x_2))|0\rangle\langle 0|T(\hat{\phi}_{\mathrm{B}}(x_1)\hat{\phi}_{\mathrm{B}}(x_2))|0\rangle\langle 0|T(\hat{\phi}_{\mathrm{C}}(x_1)\hat{\phi}_{\mathrm{C}}(x_2))|0\rangle;\end{aligned}\tag{6.85}$$

$$\begin{aligned}&\langle 0|\hat{a}_{\mathrm{A}}(p'_{\mathrm{A}})\hat{a}_{\mathrm{A}}^{\dagger}(p_{\mathrm{A}})|0\rangle\langle 0|\hat{a}_{\mathrm{B}}(p'_{\mathrm{B}})\hat{\phi}_{\mathrm{B}}(x_1)|0\rangle\\&\quad\times\langle 0|\hat{\phi}_{\mathrm{B}}(x_2)\hat{a}_{\mathrm{B}}^{\dagger}(p_{\mathrm{B}})|0\rangle\langle 0|T(\hat{\phi}_{\mathrm{C}}(x_1)\hat{\phi}_{\mathrm{C}}(x_2))|0\rangle\langle 0|T(\hat{\phi}_{\mathrm{A}}(x_1)\hat{\phi}_{\mathrm{A}}(x_2))|0\rangle\\&\quad+x_1\leftrightarrow x_2;\end{aligned}\tag{6.86}$$

$$\begin{aligned}&\langle 0|\hat{a}_{\mathrm{B}}(p'_{\mathrm{B}})\hat{a}_{\mathrm{B}}^{\dagger}(p_{\mathrm{B}})|0\rangle\langle 0|\hat{a}_{\mathrm{A}}(p'_{\mathrm{A}})\hat{\phi}_{\mathrm{A}}(x_1)|0\rangle\\&\quad\times\langle 0|\hat{\phi}_{\mathrm{A}}(x_2)\hat{a}_{\mathrm{A}}^{\dagger}(p_{\mathrm{A}})|0\rangle\langle 0|T(\hat{\phi}_{\mathrm{C}}(x_1)\hat{\phi}_{\mathrm{C}}(x_2))|0\rangle\langle 0|T(\hat{\phi}_{\mathrm{B}}(x_1)\hat{\phi}_{\mathrm{B}}(x_2))|0\rangle\\&\quad+x_1\leftrightarrow x_2;\end{aligned}\tag{6.87}$$

$$\begin{aligned}&\langle 0|\hat{a}_{\mathrm{A}}(p'_{\mathrm{A}})\hat{\phi}_{\mathrm{A}}(x_1)|0\rangle\langle 0|\hat{\phi}_{\mathrm{A}}(x_2)\hat{a}_{\mathrm{A}}^{\dagger}(p_{\mathrm{A}})|0\rangle\langle 0|\hat{a}_{\mathrm{B}}(p'_{\mathrm{B}})\hat{\phi}_{\mathrm{B}}(x_1)|0\rangle\\&\quad\times\langle 0|\hat{\phi}_{\mathrm{B}}(x_2)\hat{a}_{\mathrm{B}}^{\dagger}(p_{\mathrm{B}})|0\rangle\langle 0|T(\hat{\phi}_{\mathrm{C}}(x_1)\hat{\phi}_{\mathrm{C}}(x_2))|0\rangle\\&\quad+x_1\leftrightarrow x_2;\end{aligned}\tag{6.88}$$

$$\begin{aligned}&\langle 0|\hat{a}_{\mathrm{A}}(p'_{\mathrm{A}})\hat{\phi}_{\mathrm{A}}(x_1)|0\rangle\langle 0|\hat{\phi}_{\mathrm{A}}(x_2)\hat{a}_{\mathrm{A}}^{\dagger}(p_{\mathrm{A}})|0\rangle\langle 0|\hat{a}_{\mathrm{B}}(p'_{\mathrm{B}})\hat{\phi}_{\mathrm{B}}(x_2)|0\rangle\\&\quad\times\langle 0|\hat{\phi}_{\mathrm{B}}(x_1)\hat{a}_{\mathrm{B}}^{\dagger}(p_{\mathrm{B}})|0\rangle\langle 0|T(\hat{\phi}_{\mathrm{C}}(x_1)\hat{\phi}_{\mathrm{C}}(x_2))|0\rangle\\&\quad+x_1\leftrightarrow x_2.\end{aligned}\tag{6.89}$$

We already know that quantities like $\langle 0|\hat{a}(p'_{\mathrm{A}})\hat{a}_{\mathrm{A}}^{\dagger}(p_{\mathrm{A}})|0\rangle$ yield something proportional to $\delta^3(\boldsymbol{p}_{\mathrm{A}}-\boldsymbol{p}'_{\mathrm{A}})$ and correspond to the initial A-particle going 'straight through'. The other factors in (6.85) which are new are quantities like $\langle 0|\hat{a}_{\mathrm{A}}(p'_{\mathrm{A}})\hat{\phi}_{\mathrm{A}}(x_1)|0\rangle$, which has the value (problem 6.4)

$$\langle 0|\hat{a}_{\mathrm{A}}(p'_{\mathrm{A}})\hat{\phi}_{\mathrm{A}}(x_1)|0\rangle=\frac{1}{\sqrt{2E_{\mathrm{A}}}}\mathrm{e}^{\mathrm{i}p'_{\mathrm{A}}\cdot x_1}\tag{6.90}$$

which is proportional (depending on the adopted normalization) to the wavefunction for an outgoing A-particle with 4-momentum p'_{A}.

We are now in a position to give a diagrammatic interpretation of all of (6.85)–(6.89). In these diagrams, we shall *not* (as we did in figure 6.2) draw two separately time-ordered pieces for each propagator. We shall not indicate the time-ordering at all and we shall understand that *both* time-orderings are always included in each propagator line. Term (6.85) then has the structure shown in figure 6.3(a); term (6.86) that shown in figure 6.3(b); term (6.87) that in figure 6.3(c); term (6.88) that in figure 6.3(d); and term (6.89) that in figure 6.3(e). We recognize in figure 6.3(e) the long-awaited Yukawa exchange process, which we shall shortly analyse in full - but the formalism has yielded much else besides! We shall come back to figures 6.3(a), (b) and (c) in section 6.3.5; for the moment we note that these processes do not represent true interactions between

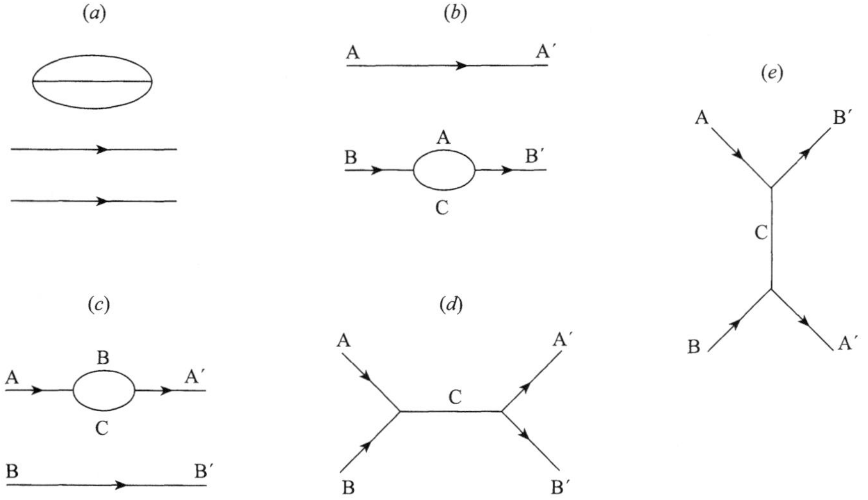

Figure 6.3. Graphical representation of (6.85)–(6.89): (*a*) (6.85); (*b*) (6.86); (*c*) (6.87); (*d*) (6.88); (*e*) (6.89).

the particles, since at least one goes through unscattered in each case. So we shall concentrate on figures 6.3(*d*) and (*e*), and *derive the Feynman rules for them.*

First, consider figure 6.3(*e*), corresponding to the contraction (6.89). When this is inserted into (6.74), the two terms in which x_1 and x_2 are interchanged give identical results (interchanging x_1 and x_2 in the integral), so the contribution we are discussing is

$$(-\mathrm{i}g)^2 \iint \mathrm{d}^4x_1 \mathrm{d}^4x_2 \mathrm{e}^{\mathrm{i}(p'_{\mathrm{A}}-p_{\mathrm{B}})\cdot x_1} \mathrm{e}^{\mathrm{i}(p'_{\mathrm{B}}-p_{\mathrm{A}})\cdot x_2} \langle 0|T(\hat{\phi}_{\mathrm{C}}(x_1)\hat{\phi}_{\mathrm{C}}(x_2))|0\rangle. \quad (6.91)$$

We must now turn our attention, as promised, to the propagator of (6.81), $\langle 0|T(\hat{\phi}_{\mathrm{C}}(x_1)\hat{\phi}_{\mathrm{C}}(x_2))|0\rangle$. Inserting the mode expansion (6.52) for each of $\hat{\phi}_{\mathrm{C}}(x_1)$ and $\hat{\phi}_{\mathrm{C}}(x_2)$, and using the commutation relations (6.46) and the vacuum conditions (6.70) we find (problem 6.5)

$$\langle 0|T(\hat{\phi}_{\mathrm{C}}(x_1)\hat{\phi}_{\mathrm{C}}(x_2))|0\rangle = \int \frac{\mathrm{d}^3\boldsymbol{k}}{(2\pi)^3 2\omega_k}[\theta(t_1-t_2)\mathrm{e}^{-\mathrm{i}\omega_k(t_1-t_2)+\mathrm{i}\boldsymbol{k}\cdot(\boldsymbol{x}_1-\boldsymbol{x}_2)} + \theta(t_2-t_1)\mathrm{e}^{-\mathrm{i}\omega_k(t_2-t_1)+\mathrm{i}\boldsymbol{k}\cdot(\boldsymbol{x}_2-\boldsymbol{x}_1)}] \quad (6.92)$$

where $\omega_k = (\boldsymbol{k}^2 + m_{\mathrm{C}}^2)^{1/2}$. This expression is very ‘uncovariant looking’, due to the presence of the θ-functions with time arguments. But the earlier discussion, after (6.81), has assured us that the left-hand side of (6.92) must be Lorentz invariant, and—by a clever trick—it is possible to recast the right-hand side in *manifestly* invariant form. We introduce an integral representation of the θ-

function via

$$\theta(t) = \mathrm{i}\int_{-\infty}^{\infty} \frac{\mathrm{d}z}{2\pi}\frac{\mathrm{e}^{-\mathrm{i}zt}}{z+\mathrm{i}\epsilon} \tag{6.93}$$

where ϵ is an infinitesimally small positive quantity (see appendix F). Multiplying (6.93) by $\mathrm{e}^{-\mathrm{i}\omega_k t}$ and changing z to $z + \omega_k$ in the integral we have

$$\theta(t)\mathrm{e}^{-\mathrm{i}\omega_k t} = \mathrm{i}\int_{-\infty}^{\infty} \frac{\mathrm{d}z}{2\pi}\frac{\mathrm{e}^{-\mathrm{i}zt}}{z-(\omega_k-\mathrm{i}\epsilon)}. \tag{6.94}$$

Putting (6.94) into (6.92) then yields

$$\langle 0|T(\hat{\phi}_{\mathrm{C}}(x_1)\hat{\phi}_{\mathrm{C}}(x_2))|0\rangle = \mathrm{i}\int \frac{\mathrm{d}^3\boldsymbol{k}\mathrm{d}z}{(2\pi)^4 2\omega_k}\left\{\frac{\mathrm{e}^{-\mathrm{i}z(t_1-t_2)+\mathrm{i}\boldsymbol{k}\cdot(\boldsymbol{x}_1-\boldsymbol{x}_2)}}{z-(\omega_k-\mathrm{i}\epsilon)} + \frac{\mathrm{e}^{\mathrm{i}z(t_1-t_2)-\mathrm{i}\boldsymbol{k}\cdot(\boldsymbol{x}_1-\boldsymbol{x}_2)}}{z-(\omega_k-\mathrm{i}\epsilon)}\right\}. \tag{6.95}$$

The exponentials and the volume element demand a more symmetrical notation: let us write $k_0 = z$ so that $(k_0 = z, \boldsymbol{k})$ form the components of a 4-vector k[1]. *Note very carefully, however, that* k_0 *is not* $(\boldsymbol{k}^2+m_{\mathrm{C}}^2)^{1/2}$! The variable k_0 is unrestricted, whereas it is ω_k that equals $(\boldsymbol{k}^2 + m_{\mathrm{C}}^2)^{1/2}$. With this change of notation, (6.95) becomes

$$\langle 0|T(\hat{\phi}_{\mathrm{C}}(x_1)\hat{\phi}_{\mathrm{C}}(x_2))|0\rangle = \int \frac{\mathrm{d}^4k}{(2\pi)^4}\frac{\mathrm{i}}{2\omega_k}\left\{\frac{\mathrm{e}^{-\mathrm{i}k\cdot(x_1-x_2)}}{k_0-(\omega_k-\mathrm{i}\epsilon)} + \frac{\mathrm{e}^{\mathrm{i}k\cdot(x_1-x_2)}}{k_0-(\omega_k-\mathrm{i}\epsilon)}\right\}. \tag{6.96}$$

Changing $k \to -k$ ($k_0 \to -k_0$, $\boldsymbol{k} \to -\boldsymbol{k}$) in the second term in (6.96), we finally have

$$\begin{aligned}\langle 0|T(\hat{\phi}_{\mathrm{C}}(x_1)\hat{\phi}_{\mathrm{C}}(x_2))|0\rangle &= \int \frac{\mathrm{d}^4k}{(2\pi)^4}\mathrm{e}^{-\mathrm{i}k\cdot(x_1-x_2)}\frac{\mathrm{i}}{2\omega_k}\left\{\frac{1}{k_0-(\omega_k-\mathrm{i}\epsilon)} - \frac{1}{k_0+\omega_k-\mathrm{i}\epsilon}\right\}\\ &= \int \frac{\mathrm{d}^4k}{(2\pi)^4}\mathrm{e}^{-\mathrm{i}k\cdot(x_1-x_2)}\frac{\mathrm{i}}{k_0^2-(\omega_k-\mathrm{i}\epsilon)^2},\end{aligned} \tag{6.97}$$

or

$$\boxed{\langle 0|T(\hat{\phi}_{\mathrm{C}}(x_1)\hat{\phi}_{\mathrm{C}}(x_2))|0\rangle = \int \frac{\mathrm{d}^4k}{(2\pi)^4}\mathrm{e}^{-\mathrm{i}k\cdot(x_1-x_2)}\frac{\mathrm{i}}{k_0^2-\boldsymbol{k}^2-m_{\mathrm{C}}^2+\mathrm{i}\epsilon}} \tag{6.98}$$

where in the last step we have used $\omega_k^2 = \boldsymbol{k}^2 + m_{\mathrm{C}}^2$ and written '$\mathrm{i}\epsilon$' for '$2\mathrm{i}\epsilon$' since what matters is just the sign of the small imaginary part (note that ω_k is defined

[1] We know that the left-hand side of (6.95) is Lorentz invariant, and that $(t_1 - t_2, \boldsymbol{x}_1 - \boldsymbol{x}_2)$ form the components of a 4-vector. The quantities $(k_0 = z, \boldsymbol{k})$ must also form the components of a 4-vector, in order for the exponentials in (6.95) to be invariant.

as the positive square root). In this final form, the Lorentz invariance of the scalar propagator is indeed manifest.

We shall have more to say about this propagator (Green function) in section 6.3.3. For the moment we simply note two points: first, it is the Fourier transform of $\mathrm{i}/k^2 - m_{\mathrm{C}}^2 + \mathrm{i}\epsilon$, as stated in appendix G, where $k^2 = k_0^2 - \boldsymbol{k}^2$; and second, it is a function of the coordinate difference $x_1 - x_2$, as it has to be since we do not expect physics to depend on the choice of origin. This second point gives us a clue as to how best to perform the $x_1 - x_2$ integral in (6.91). Let us introduce the new variables $x = x_1 - x_2$, $X = (x_1 + x_2)/2$. Then (problem 6.6) (6.91) reduces to

$$(-\mathrm{i}g)^2(2\pi)^4\delta^4(p_{\mathrm{A}} + p_{\mathrm{B}} - p'_{\mathrm{A}} - p'_{\mathrm{B}}) \int \mathrm{d}^4x\, \mathrm{e}^{\mathrm{i}q\cdot x} \int \frac{\mathrm{d}^4k}{(2\pi)^4} \mathrm{e}^{-\mathrm{i}k\cdot x} \frac{\mathrm{i}}{k^2 - m_{\mathrm{C}}^2 + \mathrm{i}\epsilon} \tag{6.99}$$

$$= (-\mathrm{i}g)^2(2\pi)^4\delta^4(p_{\mathrm{A}} + p_{\mathrm{B}} - p'_{\mathrm{A}} - p'_{\mathrm{B}}) \frac{\mathrm{i}}{q^2 - m_{\mathrm{C}}^2 + \mathrm{i}\epsilon} \tag{6.100}$$

where $q = p_{\mathrm{A}} - p'_{\mathrm{B}} = p'_{\mathrm{A}} - p_{\mathrm{B}}$ is the *4-momentum transfer* carried by the exchanged C-quantum in figure 6.4, and we have used the four-dimensional version of (E.26). We associate this single expression, which includes the two coordinate space processes of figure 6.2, with the *single momentum–space Feynman diagram* of figure 6.4. The arrows refer merely to the flow of 4-momentum, which is conserved at each 'vertex' (i.e. meeting of three lines). Thus although the arrow on the exchanged C-line is drawn as indicated, this has nothing to do with any presumed order of emission/absorption of the exchanged quantum. It cannot do so, after all, since in this diagram the states all have definite 4-momentum and hence are totally delocalized in space–time; equivalently, we recall from (6.91) that the amplitude in fact involves integrals over *all* space–time.

A similar analysis (problem 6.7) shows that the contribution of the contractions (6.88) to the S-matrix element (6.74) is

$$(-\mathrm{i}g)^2(2\pi)^4\delta^4(p_{\mathrm{A}} + p_{\mathrm{B}} - p'_{\mathrm{A}} - p'_{\mathrm{B}}) \frac{\mathrm{i}}{(p_{\mathrm{A}} + p_{\mathrm{B}})^2 - m_{\mathrm{C}}^2 + \mathrm{i}\epsilon} \tag{6.101}$$

which is represented by the momentum–space Feynman diagram of figure 6.5.

At this point we may start to write down the *Feynman rules for the* ABC *theory*, which enable us to associate a precise mathematical expression for an amplitude with a Feynman diagram such as figure 6.4 or figure 6.5. It is clear that we will always have a factor $(2\pi)^4\delta^4(p_{\mathrm{A}} + p_{\mathrm{B}} - p'_{\mathrm{A}} - p'_{\mathrm{B}})$ for all 'connected' diagrams, following from the flow of the conserved 4-momentum through the diagrams. It is conventional to extract this factor, and to define the invariant amplitude $\mathcal{M}_{\mathrm{fi}}$ via

$$S_{\mathrm{fi}} = \delta_{\mathrm{fi}} + \mathrm{i}(2\pi)^4\delta^4(p_{\mathrm{f}} - p_{\mathrm{i}})\mathcal{M}_{\mathrm{fi}} \tag{6.102}$$

in general (cf (6.57)). The rules reconstruct the invariant amplitude $\mathrm{i}\mathcal{M}_{\mathrm{fi}}$ corresponding to a given diagram, and for the present case they are:

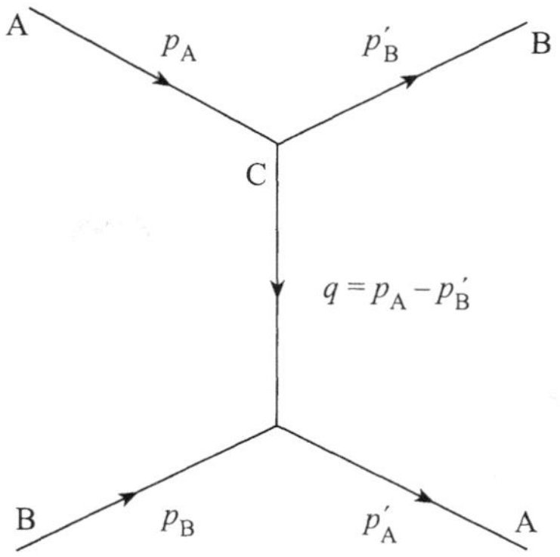

Figure 6.4. Momentum–space Feynman diagram corresponding to the $O(g^2)$ amplitude of (6.100).

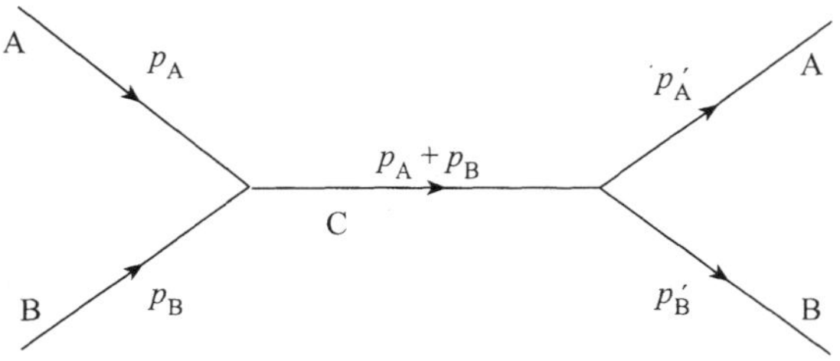

Figure 6.5. Momentum–space Feynman diagram corresponding to the $O(g^2)$ amplitude of (6.101).

(i) At each vertex, a factor $-ig$.
(ii) For each internal line, a factor

$$\frac{i}{q_i - m_i^2 + i\epsilon} \tag{6.103}$$

where i = A, B or C and q_i is the 4-momentum carried by that line. The factor (6.103) is the *Feynman propagator in momentum space*, for the scalar particle 'i'.

Of course, it is no big deal to give a set of rules which will just reconstruct (6.100) and (6.101). The real power of the 'rules' is that they work for all diagrams we can draw by joining together vertices and propagators (except that we have not yet explained what to do if more than one particle appears 'internally' between two vertices, as in figures 6.3(a)–(c): see section 6.3.5).

6.3.3 A + B → A + B scattering: the Yukawa exchange mechanism, *s* and *u* channel processes

Referring back to section 2.3, equation (2.17), we see that the amplitude for the exchange process of figure 6.4 indeed has the form suggested there, namely $\sim g^2/(q^2 - m_C^2)$ if C is exchanged. We have seen how, in the static limit, this may be interpreted as a Yukawa interaction of range $\hbar/m_C c$ between the particles A and B, treated in the Born approximation. Expression (6.100), then, provides us with the correct relativistic formula for this Yukawa mechanism.

There is more to be said about this fundamental amplitude (6.100), which is essentially the C propagator in momentum space. While it is always true that $p_i^2 = m_i^2$ for a free particle of 4-momentum p_i and rest mass m_i, it is not the case that $q^2 = m_C^2$ in (6.100). We emphasized after (6.95) that the variable k_0 introduced there was not equal to $(\boldsymbol{k}^2 + m_C^2)^{1/2}$, and the result of the step (6.99) → (6.100) was to replace k_0 by q_0 and $\boldsymbol{k}$ by $\boldsymbol{q}$, so that $q_0 \neq (\boldsymbol{q}^2 + m_C^2)^{1/2}$, i.e. $q^2 = q_0^2 - \boldsymbol{q}^2 \neq m_C^2$. So the exchanged quantum in figure 6.4 does not satisfy the 'mass-shell condition' $p_i^2 = m_i^2$; it is said to be 'off-mass shell' or 'virtual' (see also problem 6.8). It is quite a different entity from a free quantum. Indeed, as we saw in more elementary physical terms in section 2.4.2, it has a fleeting existence, as sanctioned by the uncertainty relation.

It is convenient, at this point, to introduce some kinematic variables which will appear often in following chapters. These are the 'Mandelstam variables' (Mandelstam 1958, 1959)

$$s = (p_A + p_B)^2 \qquad t = (p_A - p_A')^2 \qquad u = (p_A - p_B')^2. \tag{6.104}$$

They are clearly relativistically invariant. In terms of these variables the amplitude (6.100) is essentially $\sim 1/(u - m_C^2 + \mathrm{i}\epsilon)$, and the amplitude (6.101) is $\sim 1/(s - m_C^2 + \mathrm{i}\epsilon)$. The first is said to be a 'u-channel process', the second an 's-channel process'. Amplitudes of the form $(t - m^2)^{-1}$ or $(u - m^2)^{-1}$ are basically one-quantum exchange (i.e. 'force') processes, while those of the form $(s - m_C^2)^{-1}$ have a rather different interpretation, as we now discuss.

Let us first ask: can $s = (p_A + p_B)^2$ ever equal m_C^2 in (6.101)? Since s is invariant, we can evaluate it in any frame we like, for example the centre-of-momentum (CM) frame in which

$$(p_A + p_B)^2 = (E_A + E_B)^2 \tag{6.105}$$

with $E_A = (m_A^2 + \boldsymbol{p}^2)^{1/2}$, $E_B = (m_B^2 + \boldsymbol{p}^2)^{1/2}$. It is then clear that if $m_C < m_A + m_B$ the condition $(p_A + p_B)^2 = m_C^2$ can never be satisfied, and the internal quantum in figure 6.5 is always virtual (note that $p_A + p_B$ is the 4-momentum of the C-quantum). Depending on the details of the theory with which we are dealing, such an s-channel process can have different interpretations. In QED, for example, in the process $e^+ + e^- \to e^+ + e^-$ we could have a virtual

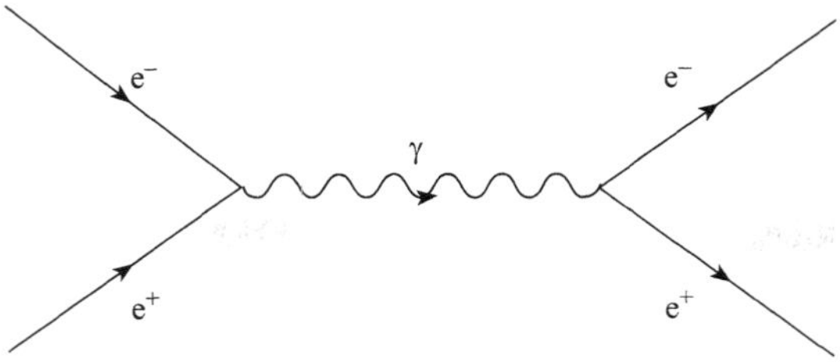

Figure 6.6. $O(e^2)$ contribution to $e^+e^- \rightarrow e^+e^-$ via annihilation to (and re-emission from) a virtual γ state.

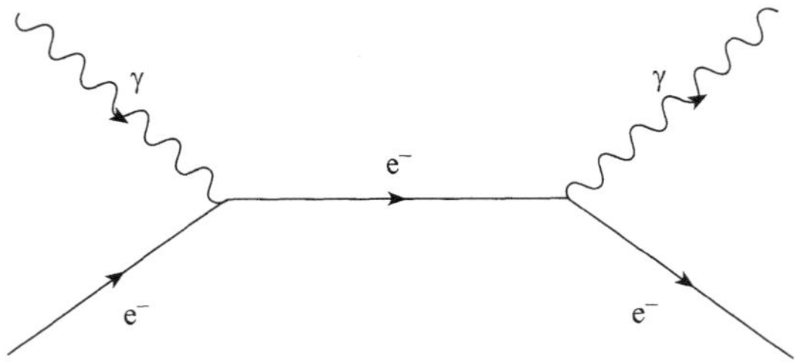

Figure 6.7. $O(e^2)$ contribution to $\gamma e^- \rightarrow \gamma e^-$ via absorption to (and re-emission from) a virtual e^- state.

γ s-channel process as shown in figure 6.6. This would be called an 'annihilation process' for obvious reasons. In the process $\gamma + e^- \rightarrow \gamma + e^-$, however, we could have figure 6.7 which would be interpreted as an absorption and re-emission process (i.e. of a photon).

However, if $m_C > m_A + m_B$, then we can indeed satisfy $(p_A + p_B)^2 = m_C^2$, and so (remembering that ϵ is infinitesimal) we seem to have an infinite result when s (the square of the CM energy) hits the value m_C^2. In fact, this is not the case. If $m_C > m_A + m_B$, the C-particle is unstable against decay to A+B, as we saw in section 6.3.1. The s-channel process must then be interpreted as the formation of a resonance, i.e. of the transitory and decaying state consisting of the single C-particle. Such a process would be described non-relativistically by a Breit–Wigner amplitude of the form

$$\mathcal{M} \propto 1/(E - E_R + i\Gamma/2) \tag{6.106}$$

which produces a peak in $|\mathcal{M}|^2$ centred at $E = E_R$ and full width Γ at half-height; Γ is, in fact, precisely the width calculated in section 6.3.1. The relativistic generalization of (6.106) is

$$\mathcal{M} \propto \frac{1}{s - M^2 + \mathrm{i}M\Gamma} \tag{6.107}$$

where M is the mass of the unstable particle. Thus in the present case the prescription for avoiding the infinity in our amplitude is to replace the infinitesimal '$\mathrm{i}\epsilon$' in (6.101) by the finite quantity $\mathrm{i}m_{\mathrm{C}}\Gamma$, with Γ as calculated in section 6.3.1. We shall see examples of such s-channel resonances in section 9.5.

6.3.4 A + B → A + B scattering: the differential cross section

We complete this exercise in the 'ABC' theory by showing how to calculate the cross section for A+B→ A+B scattering in terms of the invariant amplitude $\mathcal{M}_{\mathrm{fi}}$ of (6.102). The discussion will closely parallel the calculation of the decay rate Γ in section 6.3.1.

As in (6.56), the transition rate per unit volume, in this case, is

$$\dot{P}_{\mathrm{fi}} = (2\pi)^4\delta^4(p_{\mathrm{A}} + p_{\mathrm{B}} - p'_{\mathrm{A}} - p'_{\mathrm{B}})|\mathcal{M}_{\mathrm{fi}}|^2. \tag{6.108}$$

In order to obtain a quantity which may be compared from experiment to experiment, we must remove the dependence of the transition rate on the incident flux of particles and on the number of target particles per unit volume. Now the flux of beam particles ('A' ones, let us say) incident on a stationary target is just the number of particles per unit area reaching the target in unit time which, with our normalization of '$2E$ particles per unit volume', is just

$$|\boldsymbol{v}|2E_{\mathrm{A}} \tag{6.109}$$

where $\boldsymbol{v}$ is the velocity of the incident A in the rest frame of the target B. The number of target particles per unit volume is $2E_{\mathrm{B}}$ ($= 2m_{\mathrm{B}}$ for B at rest, of course).

We must also include the 'density of final states' factors, as in (6.59). Putting all this together, the total cross section σ is given in terms of the differential cross section $\mathrm{d}\sigma$ by

$$\begin{aligned}\sigma = \int \mathrm{d}\sigma &= \frac{1}{2E_{\mathrm{B}}2E_{\mathrm{A}}|\boldsymbol{v}|}(2\pi)^4 \int \delta^4(p_{\mathrm{A}} + p_{\mathrm{B}} - p'_{\mathrm{A}} - p'_{\mathrm{B}}) \\ &\quad \times |\mathcal{M}_{\mathrm{fi}}|^2 \frac{\mathrm{d}^3\boldsymbol{p}'_{\mathrm{A}}}{(2\pi)^3 2E'_{\mathrm{A}}} \frac{\mathrm{d}^3\boldsymbol{p}'_{\mathrm{B}}}{(2\pi)^3 2E'_{\mathrm{B}}} \\ &\equiv \frac{1}{4E_{\mathrm{A}}E_{\mathrm{B}}|\boldsymbol{v}|}\int |\mathcal{M}_{\mathrm{fi}}|^2 \mathrm{dLips}(s; p'_{\mathrm{A}}, p'_{\mathrm{B}}),\end{aligned} \tag{6.110}$$

where we have introduced the *Lorentz invariant phase space* $\mathrm{dLips}(s; p'_{\mathrm{A}}, p'_{\mathrm{B}})$ defined by

$$\boxed{\mathrm{dLips}(s; p'_{\mathrm{A}}, p'_{\mathrm{B}}) = \frac{1}{(4\pi)^2}\delta^4(p_{\mathrm{A}} + p_{\mathrm{B}} - p'_{\mathrm{A}} - p'_{\mathrm{B}})\frac{\mathrm{d}^3\boldsymbol{p}'_{\mathrm{A}}}{E'_{\mathrm{A}}}\frac{\mathrm{d}^3\boldsymbol{p}'_{\mathrm{B}}}{E'_{\mathrm{B}}}.} \tag{6.111}$$

We can write the flux factor for collinear collisions in invariant form using the relation (easily verified in a particular frame (problem 6.9))

$$E_{\mathrm{A}}E_{\mathrm{B}}|\boldsymbol{v}| = [(p_{\mathrm{A}}\cdot p_{\mathrm{B}})^2 - m_{\mathrm{A}}^2 m_{\mathrm{B}}^2]^{1/2}. \tag{6.112}$$

Everything in (6.110) is now written in invariant form.

It is a useful exercise to evaluate $\int \mathrm{d}\sigma$ in a given frame, and the simplest one is the centre-of-momentum (CM) frame defined by

$$\boldsymbol{p}_{\mathrm{A}} + \boldsymbol{p}_{\mathrm{B}} = \boldsymbol{p}'_{\mathrm{A}} + \boldsymbol{p}'_{\mathrm{B}} = \mathbf{0}. \tag{6.113}$$

However, before specializing to this frame, it is convenient to simplify our expression for dLips. Using the 3-momentum part of the δ-function in (6.110), we can eliminate the integral over $\mathrm{d}^3\boldsymbol{p}'_{\mathrm{B}}$:

$$\int \frac{\mathrm{d}^3\boldsymbol{p}'_{\mathrm{B}}}{E'_{\mathrm{B}}}\delta^4(p_{\mathrm{A}} + p_{\mathrm{B}} - p'_{\mathrm{A}} - p'_{\mathrm{B}}) = \frac{1}{E'_{\mathrm{B}}}\delta(E_{\mathrm{A}} + E_{\mathrm{B}} - E'_{\mathrm{A}} - E'_{\mathrm{B}}), \tag{6.114}$$

remembering also that now $\boldsymbol{p}'_{\mathrm{B}}$ has to be replaced by $\boldsymbol{p}_{\mathrm{A}} + \boldsymbol{p}_{\mathrm{B}} - \boldsymbol{p}'_{\mathrm{A}}$ in $\mathcal{M}_{\mathrm{fi}}$. On the right-hand side of (6.114), $\boldsymbol{p}'_{\mathrm{B}}$ and E'_{B} are no longer independent variables but are determined by the conditions

$$\boldsymbol{p}'_{\mathrm{B}} = \boldsymbol{p}_{\mathrm{A}} + \boldsymbol{p}_{\mathrm{B}} - \boldsymbol{p}'_{\mathrm{A}} \qquad E'_{\mathrm{B}} = (m_{\mathrm{B}}^2 + \boldsymbol{p}'^2_{\mathrm{B}})^{1/2}. \tag{6.115}$$

Next, convert $\mathrm{d}^3\boldsymbol{p}'_{\mathrm{A}}$ to angular variables

$$\mathrm{d}^3\boldsymbol{p}'_{\mathrm{A}} = \boldsymbol{p}'^2_{\mathrm{A}}\,\mathrm{d}|\boldsymbol{p}'_{\mathrm{A}}|\,\mathrm{d}\Omega. \tag{6.116}$$

The energy E'_{A} is given by

$$E'_{\mathrm{A}} = (m_{\mathrm{A}}^2 + \boldsymbol{p}'^2_{\mathrm{A}})^{1/2} \tag{6.117}$$

so that

$$E'_{\mathrm{A}}\,\mathrm{d}E'_{\mathrm{A}} = |\boldsymbol{p}'_{\mathrm{A}}|\,\mathrm{d}|\boldsymbol{p}'_{\mathrm{A}}|. \tag{6.118}$$

With all these changes we arrive at the result (valid in any frame)

$$\mathrm{dLips}(s;\, p'_{\mathrm{A}}, p'_{\mathrm{B}}) = \frac{1}{(4\pi)^2}\frac{|\boldsymbol{p}'_{\mathrm{A}}|\mathrm{d}E'_{\mathrm{A}}}{E'_{\mathrm{B}}}\,\mathrm{d}\Omega\,\delta(E_{\mathrm{A}} + E_{\mathrm{B}} - E'_{\mathrm{A}} - E'_{\mathrm{B}}). \tag{6.119}$$

We now specialize to the CM frame for which $\boldsymbol{p}_{\mathrm{A}} = \boldsymbol{p} = -\boldsymbol{p}_{\mathrm{B}}$, $\boldsymbol{p}'_{\mathrm{A}} = \boldsymbol{p}' = -\boldsymbol{p}'_{\mathrm{B}}$, and

$$E'_{\mathrm{A}} = (m_{\mathrm{A}}^2 + \boldsymbol{p}'^2)^{1/2} \qquad E'_{\mathrm{B}} = (m_{\mathrm{B}}^2 + \boldsymbol{p}'^2)^{1/2} \tag{6.120}$$

so that

$$E'_{\mathrm{A}}\,\mathrm{d}E'_{\mathrm{A}} = |\boldsymbol{p}'|\,\mathrm{d}|\boldsymbol{p}'| = E'_{\mathrm{B}}\,\mathrm{d}E'_{\mathrm{B}}. \tag{6.121}$$

Introduce the variable $W' = E'_{\mathrm{A}} + E'_{\mathrm{B}}$ (note that W' is only constrained to equal the total energy $W = E_{\mathrm{A}} + E_{\mathrm{B}}$ after the integral over the energy-conserving δ-function has been performed). Then (as in (6.62))

$$\mathrm{d}W' = \mathrm{d}E'_{\mathrm{A}} + \mathrm{d}E'_{\mathrm{B}} = \frac{W'|\boldsymbol{p}'|\,\mathrm{d}|\boldsymbol{p}'|}{E'_{\mathrm{A}}E'_{\mathrm{B}}} = \frac{W'}{E'_{\mathrm{B}}}\,\mathrm{d}E'_{\mathrm{A}} \tag{6.122}$$

where we have used (6.121) in each of the last two steps. Thus the factor

$$|\boldsymbol{p}'_{\mathrm{A}}|\frac{\mathrm{d}E'_{\mathrm{A}}}{E'_{\mathrm{B}}}\delta(E_{\mathrm{A}}+E_{\mathrm{B}}-E'_{\mathrm{A}}-E'_{\mathrm{B}}) \tag{6.123}$$

becomes

$$|\boldsymbol{p}'|\frac{\mathrm{d}W'}{W'}\delta(W-W') \tag{6.124}$$

which reduces to

$$|\boldsymbol{p}|/W$$

after integrating over W', since the energy-conservation relation forces $|\boldsymbol{p}'| = |\boldsymbol{p}|$. We arrive at the important result

$$\boxed{\mathrm{dLips}(s;\, p'_{\mathrm{A}},\, p'_{\mathrm{B}}) = \frac{1}{(4\pi)^2}\frac{|\boldsymbol{p}|}{W}\,\mathrm{d}\Omega} \tag{6.125}$$

for the two-body phase space in the CM frame.

The last piece in the puzzle is the evaluation of the flux factor (6.111) in the CM frame. In the CM we have

$$p_{\mathrm{A}}\cdot p_{\mathrm{B}} = (E_{\mathrm{A}},\, \boldsymbol{p})\cdot(E_{\mathrm{B}},\, -\boldsymbol{p}) \tag{6.126}$$

$$= E_{\mathrm{A}}E_{\mathrm{B}} + \boldsymbol{p}^2 \tag{6.127}$$

and a straightforward calculation shows that

$$(p_{\mathrm{A}}\cdot p_{\mathrm{B}})^2 - m_{\mathrm{A}}^2 m_{\mathrm{B}}^2 = \boldsymbol{p}^2 W^2.$$

Hence we finally have

$$\int \mathrm{d}\sigma = \frac{1}{4|\boldsymbol{p}|W}\frac{1}{(4\pi)^2}\frac{|\boldsymbol{p}|}{W}\int |\mathcal{M}_{\mathrm{fi}}|^2\,\mathrm{d}\Omega \tag{6.128}$$

and the *CM differential cross section* is

$$\boxed{\left.\frac{\mathrm{d}\sigma}{\mathrm{d}\Omega}\right|_{\mathrm{CM}} = \frac{1}{(8\pi W)^2}|\mathcal{M}_{\mathrm{fi}}|^2.} \tag{6.129}$$

6.3.5 A + B → A + B scattering: loose ends

We must now return to the amplitudes represented by figures 6.3(*a*)–(*c*), which we set aside earlier. Consider first figure 6.3(*b*). Here the A-particle has continued through without interacting, while the B-particle has made a virtual transition to the 'A + C' state, and then this state has reverted to the original B-state. So this is in the nature of a correction to the 'no-scattering' piece shown in figure 6.1, and does not contribute to $\mathcal{M}_{\mathrm{fi}}$. However, such a virtual transition B → A + C → B

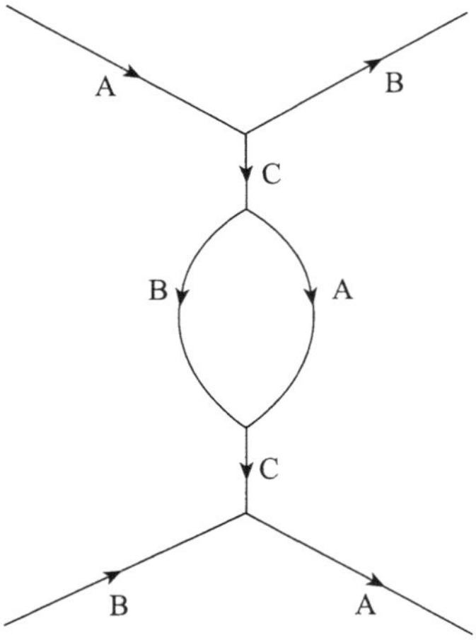

Figure 6.8. $O(g^4)$ contribution to the process $\mathrm{A}+\mathrm{B} \rightarrow \mathrm{A}+\mathrm{B}$, in which a virtual transition $\mathrm{C} \rightarrow \mathrm{A}+\mathrm{B} \rightarrow \mathrm{C}$ occurs in the C propagator.

does represent a modification of the properties of the original single B state, due to its interactions with other fields as specified in H_{I}'. We can easily imagine how, at order g^4, an amplitude will occur in which such a virtual process is inserted into the C propagator in figure 6.4 so as to arrive at figure 6.8, from which it is plausible that such emission and reabsorption processes by the same particle effectively modify the propagator for this particle. This, in turn, suggests that part, at least, of their effect will be to modify the mass of the affected particle, so as to change it from the original value specified in the Lagrangian. We may think of this physically as being associated, in some way, with a particle's carrying with it a 'cloud' of virtual particles, with which it is continually interacting; this will affect its mass, much as the mass of an electron in a solid becomes an 'effective' mass due to the various interactions experienced by the electron inside the solid.

We shall postpone the evaluation of amplitudes such as those represented by figures 6.3(b) and (c) to chapter 10. However, we note here just one feature: 4-momentum conservation applied at each vertex in figure 6.3(b) does not determine the individual 4-momenta of the intermediate A and C particles, only the sum of their 4-momenta, which is equal to p_{B} (and this is equal to p_{B}' also, so indeed no scattering has occurred). It is plausible that, if an internal 4-momentum in a diagram is undetermined in terms of the external (fixed) 4-momenta of the physical process, then that undetermined 4-momentum should be integrated over. This is the case, as can be verified straightforwardly by evaluating the amplitude (6.86), for example, as we evaluated (6.89); a similar calculation will be gone through in detail in chapter 10, section 10.1.1. The corresponding Feynman rule is

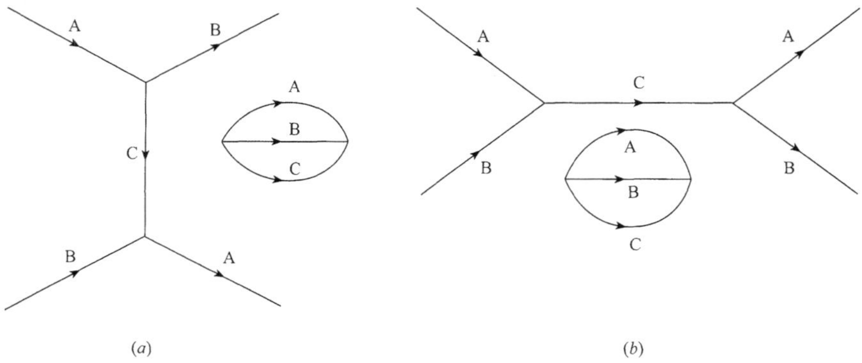

Figure 6.9. $O(g^4)$ disconnected diagrams in $\mathrm{A}+\mathrm{B} \to \mathrm{A}+\mathrm{B}$.

(iii) For each internal 4-momentum k which is not fixed by 4-momentum conservation, carry out the integration $\int \mathrm{d}^4k/(2\pi)^4$. One such integration with respect to an internal 4-momentum occurs for each closed loop.

If we apply this new rule to figure 6.3(b), we find that we need to evaluate the integral

$$\int \frac{\mathrm{d}^4 k}{(2\pi)^4} \frac{\mathrm{i}}{(k^2 - m_{\mathrm{A}}^2)} \frac{\mathrm{i}}{((p_{\mathrm{B}} - k)^2 - m_{\mathrm{C}}^2)} \tag{6.130}$$

which, by simple counting of powers of k in numerator and denominator, is logarithmically divergent. Thus we learn that, almost before we have started quantum field theory in earnest, we seem to have run into a serious problem, which is going to affect all higher-order processes containing loops. The procedure whereby these infinities are tamed is called renormalization, and we shall return to it in chapter 10.

Finally, what about figure 6.3(a)? In this case nothing at all has occurred to either of the scattering particles, and instead a virtual trio of $\mathrm{A}+\mathrm{B}+\mathrm{C}$ has appeared from the vacuum, and then disappeared back again. Such processes are called, obviously enough, vacuum diagrams. This particular one is in fact only (another) correction to figure 6.1, and it makes no contribution to $\mathcal{M}_{\mathrm{fi}}$. But as with figure 6.8, at $O(g^4)$ we can imagine such a vacuum process appearing 'alongside' figure 6.4 or figure 6.5, as in figures 6.9(a) and (b). These are called 'disconnected diagrams' and—since in them A and B have certainly interacted—they will contribute to $\mathcal{M}_{\mathrm{fi}}$ (note that they are in this respect quite different from the 'straight through' diagrams of figures 6.3(b) and (c)). However, it turns out, rather remarkably, that their effect is exactly compensated by another effect we have glossed over—namely the fact that the vacuum $|0\rangle$ we have used in our S-matrix elements is plainly the *unperturbed* vacuum (or ground state), whereas surely the introduction of interactions will perturb it. A careful analysis of this (Peskin and Schroeder 1995, section 7.2) shows that $\mathcal{M}_{\mathrm{fi}}$ is to be calculated from only the *connected* Feynman diagrams.

In this chapter we have seen how the Feynman rules for scattering and decay amplitudes in a simple scalar theory are derived, and also how cross sections and decay rates are calculated. A Yukawa (u-channel) exchange process has been found, in its covariant form, and the analogous s-channel process, together with a hint of the complications which arise when loops are considered, at higher order in g. Unfortunately, however, none of this applies directly to any real physical process, since we do not know of any physical 'scalar ABC' interaction. Rather, the interactions in the Standard Model are all gauge interactions similar to electrodynamics (with the exception of the Higgs sector, which has both cubic and quartic scalar interactions). The mediating quanta of these gauge interactions have spin-1, not zero; furthermore, the matter fields (again apart from the Higgs field) have spin-$\frac{1}{2}$. It is time to begin discussing the complications of spin and the particular form of dynamics associated with the 'gauge principle'.

Problems

6.1 Show that, for a quantum field $\hat{f}(t)$ (suppressing the space coordinates),

$$\int_{-\infty}^{\infty} \mathrm{d}t_1 \int_{-\infty}^{t_1} \mathrm{d}t_2\, \hat{f}(t_1)\hat{f}(t_2) = \tfrac{1}{2}\int_{-\infty}^{\infty} \mathrm{d}t_1 \int_{-\infty}^{\infty} \mathrm{d}t_2\, T(\hat{f}(t_1)\hat{f}(t_2))$$

where

$$\begin{aligned} T(\hat{f}(t_1)\hat{f}(t_2)) &= \hat{f}(t_1)\hat{f}(t_2) \qquad \text{for } t_1 > t_2 \\ &= \hat{f}(t_2)\hat{f}(t_1) \qquad \text{for } t_2 > t_1. \end{aligned}$$

6.2 Verify equation (6.65).

6.3 Let $\hat{\phi}(x,t)$ be a real scalar KG field in one space dimension, satisfying

$$(\Box_x + m^2)\hat{\phi}(x,t) \equiv \left(\frac{\partial^2}{\partial t^2} - \frac{\partial^2}{\partial x^2} + m^2\right)\hat{\phi}(x,t) = 0.$$

(i) Explain why

$$\begin{aligned} T(\hat{\phi}(x_1,t_1)\hat{\phi}(x_2,t_2)) &= \theta(t_1 - t_2)\hat{\phi}(x_1,t_1)\hat{\phi}(x_2,t_2) \\ &\quad + \theta(t_2 - t_1)\hat{\phi}(x_2,t_2)\hat{\phi}(x_1,t_1) \end{aligned}$$

(see equation (E.47) for a definition of the θ-function).

(ii) Using equation (E.46), show that

$$\frac{\mathrm{d}}{\mathrm{d}x}\theta(x-a) = \delta(x-a).$$

(iii) Using the result of (ii) with appropriate changes of variable, and equation (5.105), show that

$$\frac{\partial}{\partial t_1}\{T(\hat{\phi}(x_1, t_1)\hat{\phi}(x_2, t_2))\}$$
$$= \theta(t_1 - t_2)\dot{\hat{\phi}}(x_1, t_1)\hat{\phi}(x_2, t_2) + \theta(t_2 - t_1)\hat{\phi}(x_2, t_2)\dot{\hat{\phi}}(x_1, t_1).$$

(iv) Using (5.109) and (5.104) show that

$$\frac{\partial^2}{\partial t_1{}^2}\{T(\hat{\phi}(x_1, t_1)\hat{\phi}(x_2, t_2))\} = -\mathrm{i}\delta(x_1 - x_2)\delta(t_1 - t_2) + T(\ddot{\hat{\phi}}(x_1, t_1)\hat{\phi}(x_2, t_2))$$

and hence show that

$$\left(\frac{\partial^2}{\partial t_1{}^2} - \frac{\partial^2}{\partial x_1{}^2} + m^2\right) T(\hat{\phi}(x_1, t_1)\hat{\phi}(x_2, t_2)) = -\mathrm{i}\delta(x_1 - x_2)\delta(t_1 - t_2).$$

This shows that $T(\hat{\phi}(x_1, t_1)\hat{\phi}(x_2, t_2))$ is a Green function (see appendix G, equation (G.25)—the i is included here conventionally) for the KG operator

$$\frac{\partial^2}{\partial t_1{}^2} - \frac{\partial^2}{\partial x_1{}^2} + m^2.$$

The four-dimensional generalization is immediate.

6.4 Verify (6.90).

6.5 Verify (6.92).

6.6 Verify (6.99) and (6.100).

6.7 Show that the contribution of the contractions (6.88) to the S-matrix element (6.74) is given by (6.101).

6.8 Consider the case of equal masses $m_{\mathrm{A}} = m_{\mathrm{B}} = m_{\mathrm{C}}$. Evaluate u of (6.104) in the CM frame (compare section 2.6), and show that $u \leq 0$, so that u can never equal m_{C}^2 in (6.100). (This result is generally true for such single particle 'exchange' processes.)

6.9 Verify (6.112).

7

QUANTUM FIELD THEORY III: COMPLEX SCALAR FIELDS, DIRAC AND MAXWELL FIELDS; INTRODUCTION OF ELECTROMAGNETIC INTERACTIONS

In the previous two chapters we have introduced the formalism of relativistic quantum field theory for the case of free real scalar fields obeying the Klein–Gordon (KG) equation of section 4.1, extended it to describe interactions between such quantum fields and shown how the Feynman rules for a simple Yukawa-like theory are derived. It is now time to return to the unfortunately rather more complicated real world of quarks and leptons interacting via gauge fields—in particular electromagnetism. For this, several generalizations of the formalism of chapter 5 are necessary.

First, a glance back at chapter 3 will remind the reader that the electromagnetic interaction has everything to do with the *phase* of wavefunctions, and hence presumably of their quantum field generalizations: fields which are real must be electromagnetically neutral. Indeed, as noted very briefly in section 5.3, the quanta of a real scalar field are their own anti-particles; for a given mass, there is only one type of particle being created or destroyed. However, physical particles and anti-particles have identical masses (e.g. e^- and e^+), and it is actually a deep result of quantum field theory that this is so (see the end of section 7.1). In this case for a given mass m, there will have to be two distinct field degrees of freedom, one of which corresponds somehow to the 'particle', the other to the 'anti-particle'. This suggests that we will need a complex field if we want to distinguish particle from anti-particle, even in the absence of electromagnetism (for example, the (K^0, $\bar{K}^0$) pair). Such a distinction will have to be made in terms of some conserved quantum number (or numbers), having opposite values for 'particle' and 'anti-particle'. This conserved quantum number must be associated with some symmetry. Now, referring again to chapter 3, we recall that electromagnetism is associated with invariance under *local* U(1) phase transformations. Even in the absence of electromagnetism, however, a theory with complex fields can exhibit a *global* U(1) phase invariance. As we shall show in section 7.1, such a symmetry indeed leads to the existence of a conserved quantum number, in terms of which we can distinguish the particle and anti-particle parts of a complex scalar field.

In section 7.2 we generalize the complex scalar field to the complex spinor (Dirac) field, suitable for charged spin-$\frac{1}{2}$ particles. Again we find an analogous conserved quantum number, associated with a global U(1) phase invariance of the Lagrangian, which serves to distinguish particle from anti-particle. Central to the satisfactory physical interpretation of the Dirac field will be the requirement that it must be quantized with *anti-commutation* relations—the famous 'spin-statistics' connection.

The electromagnetic field must then be quantized, and section 7.3 describes the considerable difficulties this poses. With all this in place, we can easily introduce (section 7.4) electromagnetic interactions via the 'gauge principle' of chapter 3. The resulting Lagrangians and Feynman rules will be applied to simple processes in the following chapter.

7.1 The complex scalar field: global U(1) phase invariance, particles and anti-particles

Consider a Lagrangian for two free fields $\hat{\phi}_1$ and $\hat{\phi}_2$ having the same mass M:

$$\hat{\mathcal{L}} = \tfrac{1}{2}\partial_\mu\hat{\phi}_1\partial^\mu\hat{\phi}_1 - \tfrac{1}{2}M^2\hat{\phi}_1^2 + \tfrac{1}{2}\partial_\mu\hat{\phi}_2\partial^\mu\hat{\phi}_2 - \tfrac{1}{2}M^2\hat{\phi}_2^2. \tag{7.1}$$

We shall see how this is appropriate to a 'particle–anti-particle' situation.

In general 'particle' and 'anti-particle' are distinguished by having opposite values of one or more conserved additive quantum numbers. Since these quantum numbers are conserved, the operators corresponding to them commute with the Hamiltonian and are constant in time (in the Heisenberg formulation—see equation (5.51)); such operators are called *symmetry operators* and will be increasingly important in later chapters. For the present we consider the simplest case in which 'particle' and 'anti-particle' are distinguished by having opposite eigenvalues of just one symmetry operator. This situation is already realized in the simple Lagrangian of (7.1). The symmetry involved is just this: $\hat{\mathcal{L}}$ of (7.1) is left unchanged (is *invariant*) if $\hat{\phi}_1$ and $\hat{\phi}_2$ are replaced by $\hat{\phi}_1'$ and $\hat{\phi}_2'$, where (cf (3.63))

$$\begin{aligned}\hat{\phi}_1' &= (\cos\alpha)\hat{\phi}_1 - (\sin\alpha)\hat{\phi}_2\\ \hat{\phi}_2' &= (\sin\alpha)\hat{\phi}_1 + (\cos\alpha)\hat{\phi}_2\end{aligned} \tag{7.2}$$

where α is a real parameter. This is like a rotation of coordinates about the z-axis of ordinary space, but of course it mixes *field* degrees of freedom, not spatial coordinates. The symmetry transformation of (7.2) is sometimes called an 'O(2) transformation', referring to the two-dimensional rotation group O(2). We can easily check the invariance of $\hat{\mathcal{L}}$, i.e.

$$\hat{\mathcal{L}}(\hat{\phi}_1', \hat{\phi}_2') = \hat{\mathcal{L}}(\hat{\phi}_1, \hat{\phi}_2) \tag{7.3}$$

see problem 7.1.

Now let us see what is the conservation law associated with this symmetry. It is simpler (and sufficient) to consider an infinitesimal rotation characterized by the infinitesimal parameter ϵ, for which $\cos\epsilon \approx 1$ and $\sin\epsilon \approx \epsilon$ so that (7.2) becomes

$$\begin{aligned}\hat{\phi}_1' &= \hat{\phi}_1 - \epsilon\hat{\phi}_2 \\ \hat{\phi}_2' &= \hat{\phi}_2 + \epsilon\hat{\phi}_1\end{aligned} \tag{7.4}$$

and we can define changes $\delta\hat{\phi}_i$ by

$$\begin{aligned}\delta\hat{\phi}_1 &\equiv \hat{\phi}_1' - \hat{\phi}_1 = -\epsilon\hat{\phi}_2 \\ \delta\hat{\phi}_2 &\equiv \hat{\phi}_2' - \hat{\phi}_2 = +\epsilon\hat{\phi}_1.\end{aligned} \tag{7.5}$$

Under this transformation $\hat{\mathcal{L}}$ is invariant, and so $\delta\hat{\mathcal{L}} = 0$. But $\hat{\mathcal{L}}$ is an explicit function of $\hat{\phi}_1$, $\hat{\phi}_2$, $\partial_\mu\hat{\phi}_1$ and $\partial_\mu\hat{\phi}_2$. Thus we can write

$$0 = \delta\hat{\mathcal{L}} = \frac{\partial\hat{\mathcal{L}}}{\partial(\partial_\mu\hat{\phi}_1)}\delta(\partial_\mu\hat{\phi}_1) + \frac{\partial\hat{\mathcal{L}}}{\partial(\partial_\mu\hat{\phi}_2)}\delta(\partial_\mu\hat{\phi}_2) + \frac{\partial\hat{\mathcal{L}}}{\partial\hat{\phi}_1}\delta\hat{\phi}_1 + \frac{\partial\hat{\mathcal{L}}}{\partial\hat{\phi}_2}\delta\hat{\phi}_2. \tag{7.6}$$

This is a bit like the manipulations leading up to the derivation of the Euler–Lagrange equations in section 5.2.4, but now the changes $\delta\hat{\phi}_i$ ($i \equiv 1, 2$) have nothing to do with space–time trajectories—they mix up the two fields. However, we can *use* the equations of motion for $\hat{\phi}_1$ and $\hat{\phi}_2$ to rewrite $\delta\hat{\mathcal{L}}$ as

$$\begin{aligned}0 = &\frac{\partial\hat{\mathcal{L}}}{\partial(\partial_\mu\hat{\phi}_1)}\delta(\partial_\mu\hat{\phi}_1) + \frac{\partial\hat{\mathcal{L}}}{\partial(\partial_\mu\hat{\phi}_2)}\delta(\partial_\mu\hat{\phi}_2) \\ &+ \left[\partial_\mu\left(\frac{\partial\hat{\mathcal{L}}}{\partial(\partial_\mu\hat{\phi}_1)}\right)\right]\delta\hat{\phi}_1 + \left[\partial_\mu\left(\frac{\partial\hat{\mathcal{L}}}{\partial(\partial_\mu\hat{\phi}_2)}\right)\right]\delta\hat{\phi}_2.\end{aligned} \tag{7.7}$$

Since $\delta(\partial_\mu\hat{\phi}_i) = \partial_\mu(\delta\hat{\phi}_i)$, the right-hand side of (7.7) is just a total divergence, and (7.7) becomes

$$0 = \partial_\mu\left[\frac{\partial\hat{\mathcal{L}}}{\partial(\partial_\mu\hat{\phi}_1)}\delta\hat{\phi}_1 + \frac{\partial\hat{\mathcal{L}}}{\partial(\partial_\mu\hat{\phi}_2)}\delta\hat{\phi}_2\right]. \tag{7.8}$$

These formal steps are actually perfectly general, and will apply whenever a certain Lagrangian depending on two fields $\hat{\phi}_1$ and $\hat{\phi}_2$ is invariant under $\hat{\phi}_i \to \hat{\phi}_i + \delta\hat{\phi}_i$. In the present case, with $\delta\hat{\phi}_i$ given by (7.5), we have

$$\begin{aligned}0 &= \partial_\mu\left[-\frac{\partial\hat{\mathcal{L}}}{\partial(\partial_\mu\hat{\phi}_1)}\epsilon\hat{\phi}_2 + \frac{\partial\hat{\mathcal{L}}}{\partial(\partial_\mu\hat{\phi}_2)}\epsilon\hat{\phi}_1\right] \\ &= \epsilon\partial_\mu[(\partial^\mu\hat{\phi}_2)\hat{\phi}_1 - (\partial^\mu\hat{\phi}_1)\hat{\phi}_2]\end{aligned} \tag{7.9}$$

where the free-field Lagrangian (7.1) has been used in the second step. Since ϵ is arbitrary, we have proved that the 4-vector operator

$$\hat{N}^{\mu}_{\phi} = \hat{\phi}_1 \partial^{\mu} \hat{\phi}_2 - \hat{\phi}_2 \partial^{\mu} \hat{\phi}_1 \tag{7.10}$$

is conserved:

$$\partial_{\mu} \hat{N}^{\mu}_{\phi} = 0. \tag{7.11}$$

Such conserved 4-vector operators are called *symmetry currents*, often denoted generically by $\hat{J}^{\mu}$. There is a general theorem (due to Noether (1918) in the classical field case) to the effect that if a Lagrangian is invariant under a continuous transformation, then there will be an associated symmetry current. We shall consider Noether's theorem again in volume 2.

What does all this have to do with symmetry operators? Written out in full, (7.11) is

$$\partial \hat{N}^{0}_{\phi} / \partial t + \nabla \cdot \hat{\boldsymbol{N}}_{\phi} = 0. \tag{7.12}$$

Integrating this equation over all space, we obtain

$$\frac{\mathrm{d}}{\mathrm{d}t} \int_{V \to \infty} \hat{N}^{0}_{\phi} \, \mathrm{d}^3 \boldsymbol{x} + \int_{S \to \infty} \hat{\boldsymbol{N}}_{\phi} \cdot \mathrm{d}\boldsymbol{S} = 0 \tag{7.13}$$

where we have used the divergence theorem in the second term. Normally the fields may be assumed to die off sufficiently fast at infinity that the surface integral vanishes (by using wave packets, for example), and we can therefore deduce that the quantity $\hat{N}_{\phi}$ is constant in time, where

$$\hat{N}_{\phi} = \int \hat{N}^{0}_{\phi} \, \mathrm{d}^3 \boldsymbol{x} \tag{7.14}$$

that is, *the volume integral of the* $\mu = 0$ *component of a symmetry current is a symmetry operator.*

In order to see how $\hat{N}_{\phi}$ serves to distinguish 'particle' from 'anti-particle' in the simple example we are considering, it turns out to be convenient to regard $\hat{\phi}_1$ and $\hat{\phi}_2$ as components of a single *complex* field

$$\begin{aligned} \hat{\phi} &= \tfrac{1}{\sqrt{2}} (\hat{\phi}_1 - \mathrm{i} \hat{\phi}_2) \\ \hat{\phi}^{\dagger} &= \tfrac{1}{\sqrt{2}} (\hat{\phi}_1 + \mathrm{i} \hat{\phi}_2). \end{aligned} \tag{7.15}$$

The plane-wave expansions of the form (5.142) for $\hat{\phi}_1$ and $\hat{\phi}_2$ imply that $\hat{\phi}$ has the expansion

$$\hat{\phi} = \int \frac{\mathrm{d}^3 \boldsymbol{k}}{(2\pi)^3 \sqrt{2\omega}} [\hat{a}(k) \mathrm{e}^{-\mathrm{i}k \cdot x} + b^{\dagger}(k) \mathrm{e}^{\mathrm{i}k \cdot x}] \tag{7.16}$$

where

$$\begin{aligned} \hat{a}(k) &= \tfrac{1}{\sqrt{2}} (\hat{a}_1 - \mathrm{i} \hat{a}_2) \\ b^{\dagger}(k) &= \tfrac{1}{\sqrt{2}} (\hat{a}_1^{\dagger} - \mathrm{i} \hat{a}_2^{\dagger}) \end{aligned} \tag{7.17}$$

and $\omega = (M^2 + \boldsymbol{k}^2)^{1/2}$. The operators $\hat{a}$, $\hat{a}^\dagger$, $\hat{b}$, $\hat{b}^\dagger$ obey the commutation relations

$$[\hat{a}(k), \hat{a}^\dagger(k')] = (2\pi)^3\delta^3(\boldsymbol{k} - \boldsymbol{k}')$$
$$[\hat{b}(k), \hat{b}^\dagger(k)] = (2\pi)^3\delta^3(\boldsymbol{k} - \boldsymbol{k}') \tag{7.18}$$

with all others vanishing; this follows from the commutation relations

$$[\hat{a}_i(k), \hat{a}_j(k)] = \delta_{ij}(2\pi)^3\delta(\mathbf{k} - \mathbf{k}') \qquad \text{etc} \tag{7.19}$$

for the $\hat{a}_i$ operators. Note that two distinct mode operators, $\hat{a}$ and $\hat{b}$, are appearing in the expansion (7.16) of the complex field.

In terms of this complex $\hat{\phi}$ the Lagrangian of (7.1) becomes

$$\hat{\mathcal{L}} = \partial_\mu\hat{\phi}^\dagger\partial^\mu\hat{\phi} - M^2\hat{\phi}^\dagger\hat{\phi} \tag{7.20}$$

and the Hamiltonian is (dropping the zero-point energy, i.e. normally ordering)

$$\hat{H} = \int \frac{\mathrm{d}^3\boldsymbol{k}}{(2\pi)^3}[\hat{a}^\dagger(k)\hat{a}(k) + \hat{b}^\dagger(k)\hat{b}(k)]\omega. \tag{7.21}$$

The O(2) transformation (7.2) becomes a simple phase change

$$\hat{\phi}' = \mathrm{e}^{-\mathrm{i}\alpha}\hat{\phi} \tag{7.22}$$

which (see comment (iii) of section 3.6) is called a global U(1) phase transformation; plainly the Lagrangian (7.20) is invariant under (7.22). The associated symmetry current $\hat{N}^\mu_\phi$ becomes

$$\hat{N}^\mu_\phi = \mathrm{i}(\hat{\phi}^\dagger\partial^\mu\hat{\phi} - \hat{\phi}\partial^\mu\hat{\phi}^\dagger) \tag{7.23}$$

and the symmetry operator $\hat{N}_\phi$ is (see problem 7.2)

$$\hat{N}_\phi = \int \frac{\mathrm{d}^3\boldsymbol{k}}{(2\pi)^3}[\hat{a}^\dagger(k)\hat{a}(k) - \hat{b}^\dagger(k)\hat{b}(k)]. \tag{7.24}$$

Note that $\hat{N}_\phi$ has been normally ordered in anticipation of our later vacuum definition (7.30), so that $\hat{N}_\phi|0\rangle = 0$.

We now observe that the Hamiltonian (7.21) involves the *sum* of the number operators for 'a' quanta and 'b' quanta, whereas $\hat{N}_\phi$ involves the *difference* of these number operators. Put differently, $\hat{N}_\phi$ counts $+1$ for each particle of type 'a' and -1 for each of type 'b'. This strongly suggests the interpretation that the b's are the anti-particles of the a's: $\hat{N}_\phi$ is the conserved symmetry operator whose eigenvalues serve to distinguish them. For a general state, the eigenvalue of $\hat{N}_\phi$ is the number of a's minus the number of anti-a's and it is a constant of the motion, as is the total energy, which is the sum of the a energies and anti-a energies.

We have here the simplest form of the particle–anti-particle distinction: only one additive conserved quantity is involved. A more complicated example would be the (K^+, K^-) pair, which have opposite values of strangeness and of electric charge. Of course, in our simple Lagrangian (7.20) the electromagnetic interaction is absent, and so no electric charge can be defined (we shall remedy this later); the complex field $\hat{\phi}$ would be suitable (in respect of strangeness) for describing the (K^0, $\bar{K}^0$) pair.

The symmetry operator $\hat{N}_\phi$ has a number of further important properties. First of all, we have shown that $d\hat{N}_\phi/dt = 0$ from the general (Noether) argument, but we ought also to check that

$$[\hat{N}_\phi, \hat{H}] = 0 \tag{7.25}$$

as is required for consistency, and expected for a symmetry operator. This is indeed true (see problem 7.2(a)). We can also show

$$\begin{aligned} [\hat{N}_\phi, \hat{\phi}] &= -\hat{\phi} \\ [\hat{N}_\phi, \hat{\phi}^\dagger] &= \hat{\phi}^\dagger \end{aligned} \tag{7.26}$$

and, by expansion of the exponential (problem 7.2(b)), that

$$\hat{U}(\alpha)\hat{\phi}\hat{U}^{-1}(\alpha) = e^{-i\alpha}\hat{\phi} = \hat{\phi}' \tag{7.27}$$

with

$$\hat{U}(\alpha) = e^{i\alpha\hat{N}_\phi}. \tag{7.28}$$

This shows that the unitary operator $\hat{U}(\alpha)$ effects *finite* U(1) rotations.

Consider now a state $|N_\phi\rangle$ which is an eigenstate of $\hat{N}_\phi$ with eigenvalue N_ϕ. What is the eigenvalue of $\hat{N}_\phi$ for the state $\hat{\phi}|N_\phi\rangle$? It is easy to show, using (7.26), that

$$\hat{N}_\phi\hat{\phi}|N_\phi\rangle = (N_\phi - 1)\hat{\phi}|N_\phi\rangle \tag{7.29}$$

so the application of $\hat{\phi}$ to a state lowers its $\hat{N}_\phi$ eigenvalue by 1. This is consistent with our interpretation that the $\hat{\phi}$ field destroys particles 'a' via the $\hat{a}$ piece in (7.16). (This '$\hat{\phi}$ destroys particles' convention is the reason for choosing $\hat{\phi} = (\hat{\phi}_1 - i\hat{\phi}_2)/\sqrt{2}$ in (7.15), which in turn led to the minus sign in the relation (7.26) and to the earlier eigenvalue $N_\phi - 1$.) That $\hat{\phi}$ lowers the $\hat{N}_\phi$ eigenvalue by 1 is also consistent with the interpretation that the same field $\hat{\phi}$ creates an anti-particle via the $\hat{b}^\dagger$ piece in (7.16). In the same way, by considering $\hat{\phi}^\dagger|N_\phi\rangle$, one easily verifies that $\hat{\phi}^\dagger$ increases N_ϕ by 1, by creating a particle via $\hat{a}^\dagger$ or destroying an anti-particle via $\hat{b}$. The vacuum state (no particles and no anti-particles present) is defined by

$$\hat{a}(k)|0\rangle = \hat{b}(k)|0\rangle = 0 \qquad \text{for all } k. \tag{7.30}$$

As anticipated, therefore, the complex field $\hat{\phi}$ contains two distinct kinds of mode operator, one having to do with particles (with positive N_ϕ), the other with anti-particles (negative N_ϕ). Which we choose to call 'particle' and which 'anti-particle' is of course purely a matter of convention: after all, the negatively charged electron is always regarded as the 'particle', while in the case of the pions we call the positively charged π^+ the particle.

Feynman rules for theories involving complex scalar fields may be derived by a straightforward extension of the procedure explained in chapter 6. It is, however, worth pausing over the propagator . The only non-vanishing vev of the time-ordered product of two $\hat{\phi}$ fields is $\langle 0|T(\hat{\phi}(x_1)\hat{\phi}^\dagger(x_2))|0\rangle$ (the vev's of $T(\hat{\phi}\hat{\phi})$ and $T(\hat{\phi}^\dagger\hat{\phi}^\dagger)$ vanish with the vacuum defined as in (7.30)). In section 6.3.2 we gave a pictorial interpretation of the propagator for a real scalar field; let us now consider the analogous pictures for the complex field. For $t_1 > t_2$ the time-ordered product is $\hat{\phi}(x_1)\hat{\phi}^\dagger(x_2)$; using the expansion (7.16) and the vacuum conditions (7.30), the only surviving term in the vev is that in which an '$\hat{a}^\dagger$' creates a particle ($N_\phi = 1$) at $(\boldsymbol{x}_2, t_2)$ and an '$\hat{a}$' destroys it at $(\boldsymbol{x}_1, t_1)$; the '$\hat{b}$' operators in $\hat{\phi}(x_2)^\dagger$ give zero when acting on $|0\rangle$, as do the '$\hat{b}^\dagger$' operators in $\hat{\phi}^\dagger(x_1)$ when acting on $\langle 0|$. Thus for $t_1 > t_2$ we have the pictorial interpretation of figure 7.1(a). For $t_2 > t_1$, however, the time-ordered product is $\hat{\phi}^\dagger(x_2)\hat{\phi}(x_1)$. Here the surviving vev comes from the '$\hat{b}^\dagger$' in $\hat{\phi}(x_1)$ creating an anti-particle ($N_\phi = -1$) at x_1, which is then annihilated by the '$\hat{b}$' in $\hat{\phi}^\dagger(x_2)$. This $t_2 > t_1$ process is shown in figure 7.1(b). The inclusion of both processes shown in figure 7.1 makes sense physically, following considerations similar to those put forward 'intuitively' in section 4.5.4: the process of figure 7.1(a) creates (say) a positive unit of N_ϕ at x_2 and loses a positive unit at x_1, while another way of effecting the same 'N_ϕ transfer' is to create an anti-particle of unit negative N_ϕ at x_1, and propagate it to x_2 where it is destroyed, as in figure 7.1(b). It is important to be absolutely clear that the *Feynman propagator* $\langle 0|T(\hat{\phi}(x_1)\hat{\phi}^\dagger(x_2))|0\rangle$ includes *both* the processes in figures 7.1(a) and (b).

In practice, as we found in section 6.3.2, we want the momentum–space version of the propagator, i.e. its Fourier transform. As we also noted there (cf also appendix G), the propagator is a Green function for the KG operator $(\Box + m^2)$ with mass parameter m ; in momentum–space this is just the inverse, $(-k^2 + m^2)^{-1}$. In the present case, since both $\hat{\phi}$ and $\hat{\phi}^\dagger$ obey the same KG equation, with mass parameter M, we expect that the momentum–space version of $\langle 0|T(\hat{\phi}(x_1)\hat{\phi}^\dagger(x_2))|0\rangle$ is also

$$\frac{\mathrm{i}}{k^2 - M^2 + \mathrm{i}\epsilon}. \tag{7.31}$$

This can be verified by inserting the expansion (7.16) into the vev of the T-product, and following the steps used in section 6.3.2 for the scalar case.

In this (momentum–space) version, it is the 'iϵ' which keeps track of the 'particles going from 2 to 1 if $t_2 > t_1$' and 'anti-particles going from 1 to 2 if $t_2 > t_1$' (recall its appearance in the representation (6.93) of the all-important θ-function). As in the scalar case, momentum–space propagators in Feynman diagrams carry no implied order of emission/absorption process; *both*

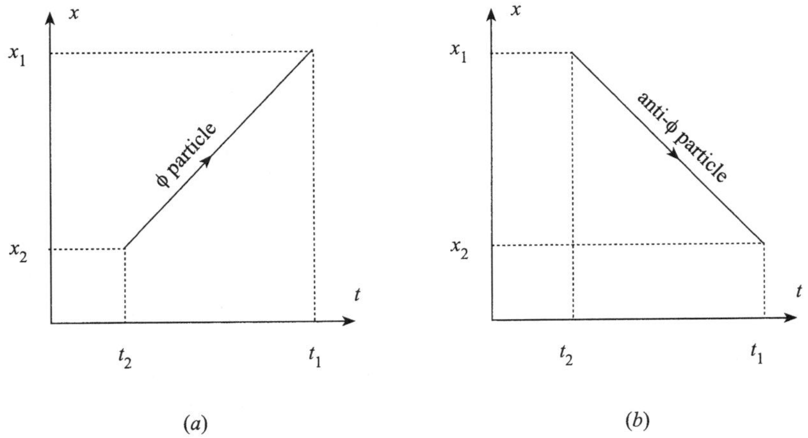

Figure 7.1. (a) For $t_1 > t_2$, a ϕ particle ($N_\phi = 1$) propagates from x_2 to x_1; (b) for $t_2 > t_1$ an anti-ϕ particle ($N_\phi = -1$) propagates from x_1 to x_2.

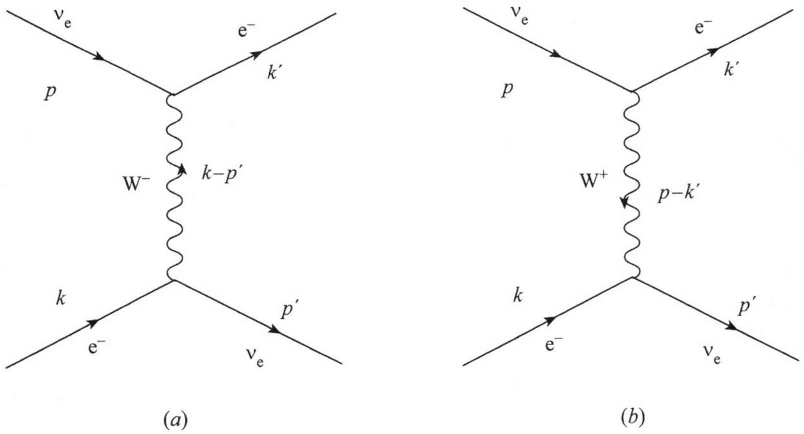

Figure 7.2. Equivalent Feynman graphs for single W-exchange in $\nu_e + e^- \to \nu_e + e^-$.

the processes in figure 7.1 are always included in all propagators. Arrows showing 'momentum flow' now also show the flow of all conserved quantum numbers. Thus the process shown in figure 7.2(a) can equally well be represented as in figure 7.2(b).

There is one more bit of physics to be gleaned from $\langle 0|T(\hat{\phi}(x_1)\hat{\phi}^\dagger(x_2))|0\rangle$. As in the real scalar field case, the vanishing of the commutator at space-like separations

$$[\hat{\phi}(x_1), \hat{\phi}^\dagger(x_2)] = 0 \qquad \text{for } (x_1 - x_2)^2 < 0 \tag{7.32}$$

guarantees the Lorentz invariance of the propagator for the complex scalar field and of the S-matrix. But in this (complex) case there is a further twist to the story. Evaluation of $[\hat{\phi}(x_1), \hat{\phi}^\dagger(x_2)]$ reveals (problem 7.3) that, in the region $(x_1 - x_2)^2 < 0$, the commutator is the difference of two functions (not field operators), one of which arises from the propagation of a particle from x_2 to x_1, the other of which comes from the propagation of an anti-particle from x_1 to x_2 (just as in figure 7.1). *Both* processes must exist for this difference to be zero, and furthermore for cancellations between them to occur in the space-like region the masses of the particle and anti-particle must be identical. In quantum field theory, therefore, 'causality' (in the previous sense) requires that every particle has to have a corresponding anti-particle, with the same mass and opposite quantum numbers.

7.2 The Dirac field and the spin-statistics connection

> I remember that when someone had tried to teach me about creation and annihilation operators, that this operator creates an electron, I said 'how do you create an electron? It disagrees with the conservation of charge,' and in that way I blocked my mind from learning a very practical scheme of calculation.

From the lecture delivered by Richard Feynman in Stockholm, Sweden, on 11 December 1965, when he received the Nobel Prize in physics, which he shared with Shinichiro Tomonaga and Julian Schwinger. (Feynman 1966).

We now turn to the problem of setting up a quantum field which, in its wave aspects, satisfies the Dirac equation (cf comment (5) in section 5.2.5), and in its 'particle' aspects creates or annihilates fermions and anti-fermions. Following the 'Heisenberg–Lagrange–Hamilton' approach of section 5.2.5, we begin by writing down the Lagrangian which, via the corresponding Euler–Lagrange equation, produces the Dirac equation as the 'field equation'. The answer (see problem 7.4) is

$$\mathcal{L}_{\rm D} = {\rm i}\psi^\dagger\dot{\psi} + {\rm i}\psi^\dagger\boldsymbol{\alpha}\cdot\boldsymbol{\nabla}\psi - m\psi^\dagger\beta\psi. \tag{7.33}$$

The relativistic invariance of this is more evident in γ-matrix notation (problem 4.11):

$$\mathcal{L}_{\rm D} = \bar{\psi}({\rm i}\gamma^\mu\partial_\mu - m)\psi. \tag{7.34}$$

We can now attempt to 'quantize' the field ψ by making a mode expansion in terms of plane-wave solutions of the Dirac equation, in a fashion similar to that for the complex scalar field in (7.16). We obtain (see problem 4.12 for the definition of the spinors u and v, and the attendant normalization choice)

$$\hat{\psi} = \int \frac{{\rm d}^3\boldsymbol{k}}{(2\pi)^3\sqrt{2\omega}} \sum_{s=1,2} [\hat{c}_s(k)u(k,s){\rm e}^{-{\rm i}k\cdot x} + \hat{d}_s^\dagger(k)v(k,s){\rm e}^{{\rm i}k\cdot x}], \tag{7.35}$$

where $\omega = (m^2 + \boldsymbol{k}^2)^{1/2}$. We wish to interpret $\hat{c}_s^\dagger(k)$ as the creation operator for a Dirac particle of spin s and momentum k. By analogy with (7.16), we expect that $\hat{d}_s^\dagger(k)$ creates the corresponding anti-particle. Presumably we must define the vacuum by (cf (7.30))

$$\hat{c}_s(k)|0\rangle = \hat{d}_s(k)|0\rangle = 0 \qquad \text{for all } k \text{ and } s = 1, 2. \tag{7.36}$$

A two-fermion state is then

$$|k_1, s_1; k_2, s_2\rangle \propto \hat{c}_{s_1}^\dagger(k_1)\hat{c}_{s_2}^\dagger(k_2)|0\rangle. \tag{7.37}$$

But it is here that there must be a difference from the boson case. We require a state containing two identical fermions to be *antisymmetric* under the exchange of state labels $k_1 \leftrightarrow k_2$, $s_1 \leftrightarrow s_2$, and thus to be forbidden if the two sets of quantum numbers are the same, in accordance with the *Pauli exclusion principle*, responsible for so many well-established features of the structure of matter.

The solution to this dilemma is simple but radical: for fermions, commutation relations are replaced by *anti-commutation* relations! The anti-commutator of two operators $\hat{A}$ and $\hat{B}$ is written:

$$\{\hat{A}, \hat{B}\} \equiv \hat{A}\hat{B} + \hat{B}\hat{A}. \tag{7.38}$$

If two different $\hat{c}$'s anti-commute, then

$$\hat{c}_{s_1}^\dagger(k_1)\hat{c}_{s_2}^\dagger(k_2) + \hat{c}_{s_2}^\dagger(k_2)\hat{c}_{s_1}^\dagger(k_1) = 0 \tag{7.39}$$

so that we have the desired antisymmetry

$$|k_1, s_1; k_2, s_2\rangle = -|k_2, s_2; k_1, s_1\rangle. \tag{7.40}$$

In general we postulate

$$\begin{aligned} \{\hat{c}_{s_1}(k_1), \hat{c}_{s_2}^\dagger(k_2)\} &= (2\pi)^3\delta^3(\boldsymbol{k}_1 - \boldsymbol{k}_2)\delta_{s_1 s_2} \\ \{\hat{c}_{s_1}(k_1), \hat{c}_{s_2}(k_2)\} &= \{\hat{c}_{s_1}^\dagger(k_1), \hat{c}_{s_2}^\dagger(k_2)\} = 0 \end{aligned} \tag{7.41}$$

and similarly for the $\hat{d}$'s and $\hat{d}^\dagger$'s. The factor in front of the δ-function depends on the convention for normalizing Dirac wavefunctions.

We must at once emphasize that in taking this 'replace commutators by anti-commutators' step we now depart decisively from the intuitive, quasi-mechanical, picture of a quantum field given in chapter 5, namely as a system of quantized harmonic oscillators. Of course, the field expansion (7.35) is a linear superposition of 'modes' (plane-wave solutions), as for the complex scalar field in (7.16) for example; but the 'mode operators' $\hat{c}_s$ and $\hat{d}_s^\dagger$ are fermionic (obeying anti-commutation relations) not bosonic (obeying commutation relations). As mentioned at the end of section 5.1, it does not seem possible to provide any mechanical model of a system (in three dimensions) whose normal vibrations are fermionic. Correspondingly, there is no concept of a 'classical electron field', analogous to the classical electromagnetic field (which doubtless explains why we tend to think of fermions as basically 'more particle-like'). However, we can certainly recover a quantum mechanical wavefunction from (7.35) by considering, as in comment (5) of section 5.2, the vacuum-to-one-particle matrix element $\langle 0|\hat{\psi}(\boldsymbol{x}, t)|k_1, s_1\rangle$.

In the bosonic case, we arrived at the commutation relations (5.117) for the mode operators by postulating the 'fundamental commutator of quantum field theory', equation (5.104), which was an extension to fields of the canonical commutation relations of quantum (particle) mechanics. For fermions, we have simply introduced the anti-commutation relations (7.41) 'by hand', so as to satisy the Pauli principle. We may ask: What then becomes of the analogous 'fundamental commutator' in the fermionic case? A plausible guess is that, as with the mode operators, the 'fundamental commutator' is to be replaced by a 'fundamental anti-commutator', between the fermionic field $\hat{\psi}$ and its 'canonically conjugate momentum field' $\hat{\pi}_{\mathrm{D}}$, of the form:

$$\{\hat{\psi}(\boldsymbol{x}, t), \hat{\pi}_{\mathrm{D}}(\boldsymbol{y}, t)\} = \mathrm{i}\delta(\boldsymbol{x} - \boldsymbol{y}). \tag{7.42}$$

As far as $\hat{\pi}_{\mathrm{D}}$ is concerned, we may suppose that its definition is formally analogous to (5.109), which would yield

$$\hat{\pi}_{\mathrm{D}} = \frac{\partial \hat{\mathcal{L}}_{\mathrm{D}}}{\partial \dot{\hat{\psi}}} = \mathrm{i}\hat{\psi}^{\dagger}. \tag{7.43}$$

We must also not forget that both $\hat{\psi}$ and $\hat{\pi}_{\mathrm{D}}$ are four-component objects, carrying spinor indices. Thus we are led to expect the result

$$\boxed{\{\hat{\psi}_{\alpha}(\boldsymbol{x}, t), \hat{\psi}_{\beta}^{\dagger}(\boldsymbol{y}, t)\} = \delta(\boldsymbol{x} - \boldsymbol{y})\delta_{\alpha\beta},} \tag{7.44}$$

where α and β are spinor indices. It is a good exercise to check that this is indeed the case (problem 7.5). In this (anti-commutator) sense, then, we have a 'canonical' formalism for fermions.

One may well wonder *why* things have to be this way—'bosons commute, fermions anti-commute'. To gain further insight, we turn again to a consideration of symmetries and the question of particle and anti-particle for the Dirac field.

The Dirac field $\hat{\psi}$ is a complex field, as is reflected in the two distinct mode operators in the expansion (7.35); as in the complex scalar field case, there is only one mass parameter and we expect the quanta to be interpretable as particle and anti-particle. The symmetry operator which distinguishes them is found by analogy with the complex scalar field case. We note that $\hat{\mathcal{L}}_{\mathrm{D}}$ (the quantized version of (7.34)) is invariant under the global U(1) transformation

$$\hat{\psi} \rightarrow \hat{\psi}' = \mathrm{e}^{-\mathrm{i}\alpha}\hat{\psi} \tag{7.45}$$

which is

$$\hat{\psi} \rightarrow \hat{\psi}' = \hat{\psi} - \mathrm{i}\epsilon\hat{\psi} \tag{7.46}$$

in infinitesimal form. The corresponding (Noether) symmetry current can be calculated as

$$\hat{N}^{\mu}_{\psi} = \bar{\hat{\psi}} \gamma^{\mu} \hat{\psi} \tag{7.47}$$

and the associated symmetry operator is

$$\hat{N}_{\psi} = \int \hat{\psi}^{\dagger} \hat{\psi} \, \mathrm{d}^3 \boldsymbol{x}. \tag{7.48}$$

$\hat{N}_{\psi}$ is clearly a number operator for the fermion case. As for the complex scalar field, invariance under a global U(1) phase transformation is associated with a number conservation law.

Inserting the plane-wave expansion (7.35), we obtain, after some effort (problem 7.6),

$$\hat{N}_{\psi} = \int \frac{\mathrm{d}^3 \boldsymbol{k}}{(2\pi)^3} \sum_{s=1,2} [\hat{c}_s^{\dagger}(k)\hat{c}_s(k) + \hat{d}_s(k)\hat{d}_s^{\dagger}(k)]. \tag{7.49}$$

Similarly the Dirac Hamiltonian may be shown to have the form (problem 7.5)

$$\hat{H}_{\mathrm{D}} = \int \frac{\mathrm{d}^3 \boldsymbol{k}}{(2\pi)^3} \sum_{s=1,2} [\hat{c}_s^{\dagger}(k)\hat{c}_s(k) - \hat{d}_s(k)\hat{d}_s^{\dagger}(k)]\omega. \tag{7.50}$$

It is important to state that in obtaining (7.49) and (7.50), we have *not* assumed either commutation or anti-commutation relations for the mode operators $\hat{c}$, $\hat{c}^{\dagger}$, $\hat{d}$ and $\hat{d}^{\dagger}$, only properties of the Dirac spinors; in particular, neither (7.49) nor (7.50) has been normally ordered. Suppose now that we assume commutation relations, so as to rewrite the last terms in (7.49) and (7.50) in normally ordered form as $\hat{d}_s^{\dagger}(k)\hat{d}_s(k)$. We see that $\hat{H}_{\mathrm{D}}$ will then contain the *difference* of two number operators for 'c' and 'd' particles, and is therefore not positive-definite as we require for a sensible theory. Moreover, we suspect that, as in the $\hat{\phi}$ case, the 'd's' ought to be the anti-particles of the 'c's', carrying opposite $\hat{N}_{\psi}$ value: but $\hat{N}_{\psi}$ is then (with the previous assumption about commutation relations) just proportional to the *sum* of 'c' and 'd' number operators, counting +1 for each type, which does not fit this interpretation. However, if *anti-commutation* relations are assumed, both these problems disappear: dropping the usual infinite terms, we obtain the normally ordered forms

$$\hat{N}_{\psi} = \int \frac{\mathrm{d}^3 \boldsymbol{k}}{(2\pi)^3 \sum_{s=1,2}} [\hat{c}_s^{\dagger}(k)\hat{c}_s(k) - \hat{d}_s^{\dagger}(k)\hat{d}_s(k)] \tag{7.51}$$

$$\hat{H}_{\mathrm{D}} = \int \frac{\mathrm{d}^3 \boldsymbol{k}}{(2\pi)^3} \sum_{s=1,2} [\hat{c}_s^{\dagger}(k)\hat{c}_s(k) + \hat{d}_s^{\dagger}(k)\hat{d}_s(k)] \tag{7.52}$$

which are satisfactory, and allow us to interpret the 'd' quanta as the anti-particles of the 'c' quanta. Similar difficulties would have occurred in the complex scalar field case if we had assumed anti-commutation relations for the boson operators, and the 'causality' discussion at the end of the preceding section would not have worked either (instead of a difference of terms we would have had a sum). It is in this way that quantum field theory enforces the connection between spin and statistics.

Our discussion here is only a part of a more general approach leading to the same conclusion, first given by Pauli (1940); see also Streater *et al* (1964).

As in the complex scalar case, the other crucial ingredient we need is the Dirac propagator $\langle 0|T(\hat{\psi}(x_1)\hat{\bar{\psi}}(x_2))|0\rangle$. We shall see in section 7.4 why it is $\hat{\bar{\psi}}$ here rather than $\hat{\psi}^\dagger$—the reason is essentially to do with Lorentz covariance (see problems 4.10, 4.11). Because the $\hat{\psi}$ fields are anti-commuting, the T-symbol now has to be understood as

$$T(\hat{\psi}(x_1)\hat{\bar{\psi}}(x_2)) = \hat{\psi}(x_1)\hat{\bar{\psi}}(x_2) \qquad \text{for } t_1 > t_2 \tag{7.53}$$

$$= -\hat{\bar{\psi}}(x_2)\hat{\psi}(x_1) \qquad \text{for } t_2 > t_1. \tag{7.54}$$

Once again, this propagator is proportional to a Green function, this time for the Dirac equation, of course. Using γ-matrix notation (problem 4.11) the Dirac equation is (cf (7.34))

$$(\mathrm{i}\gamma^\mu \partial_\mu - m)\hat{\psi} = 0. \tag{7.55}$$

The momentum–space version of the propagator is proportional to the inverse of the operator in (7.55), when written in k-space, namely to $(\not{k} - m)^{-1}$ where

$$\boxed{\not{k} = \gamma^\mu k_\mu} \tag{7.56}$$

is an important shorthand notation (pronounced 'k-slash'). In fact, the Feynman propagator for Dirac fields is

$$\frac{\mathrm{i}}{\not{k} - m + \mathrm{i}\epsilon}. \tag{7.57}$$

As in (7.31), the iϵ takes care of the particle/anti-particle, emission/absorption business. Formula (7.57) is the fermion analogue of 'rule (ii)' in (6.103).

The reader should note carefully one very important difference between (7.57) and (7.31), which is that (7.57) is a 4×4 *matrix*. What we are really saying (cf (6.98)) is that the Fourier transform of $\langle 0|T(\hat{\psi}_\alpha(x_1)\hat{\bar{\psi}}_\beta(x_2))|0\rangle$, where α and β run over the four components of the Dirac field, is equal to the (α, β) matrix element of the matrix $\mathrm{i}(\not{k} - m + \mathrm{i}\epsilon)^{-1}$:

$$\boxed{\int \mathrm{d}^4(x_1 - x_2)\, \mathrm{e}^{\mathrm{i}k\cdot(x_1-x_2)} \langle 0|T(\hat{\psi}_\alpha(x_1)\hat{\bar{\psi}}_\beta(x_2))|0\rangle = \mathrm{i}(\not{k} - m + \mathrm{i}\epsilon)^{-1}_{\alpha\beta}.} \tag{7.58}$$

The form (7.58) can be made to look more like (7.31) by making use of the result (problem 7.7)

$$(\not{k} - m)(\not{k} + m) = (k^2 - m^2) \tag{7.59}$$

(where the 4×4 unit matrix is understood on the right-hand side) so as to write (7.58) as

$$\frac{\mathrm{i}(\not{k}+m)}{k^2-m^2+\mathrm{i}\epsilon}. \tag{7.60}$$

As in the scalar case, (7.58) can be directly verified by inserting the field expansion (7.35) into the left-hand side, and following steps analogous to those in equations (6.92)–(6.98). In following this through one will meet the expressions $\sum_s u(k,s)\bar{u}(k,s)$ and $\sum_s v(k,s)\bar{v}(k,s)$, which are also 4×4 matrices. Problem 7.8 shows that these quantities are given by

$$\sum_s u_\alpha(k,s)\bar{u}_\beta(k,s) = (\not{k}+m)_{\alpha\beta} \qquad \sum_s v_\alpha(k,s)\bar{v}_\beta(k,s) = (\not{k}-m)_{\alpha\beta}. \tag{7.61}$$

With these results, and remembering the minus sign in (7.54), one can check (7.60) (problem 7.9).

One might now worry that the adoption of anti-commutation relations for Dirac fields might spoil 'causality', in the sense of the discussion after (7.32). One finds, indeed, that the fields $\hat{\psi}$ and $\hat{\bar{\psi}}$ anti-commute at space-like separation, but this is enough to preserve causality for physical observables, which will involve an even number of fermionic fields.

We now turn to the problem of quantizing the Maxwell (electromagnetic) field.

7.3 The Maxwell field $A^\mu(x)$

7.3.1 The classical field case

Following the now familiar procedure, our first task is to find the classical field Lagrangian which, via the corresponding Euler–Lagrangian equations, yields the Maxwell equation for the electromagnetic potential A^ν, namely (cf (3.20))

$$\Box A_\nu - \partial^\nu(\partial_\mu A^\mu) = j^\nu_{\mathrm{em}}. \tag{7.62}$$

The answer is (see problem 7.9)

$$\mathcal{L}_{\mathrm{em}} = -\tfrac{1}{4}F_{\mu\nu}\,F^{\mu\nu} - j^\nu_{\mathrm{em}}A_\nu \tag{7.63}$$

where $F_{\mu\nu} = \partial_\mu A_\nu - \partial_\nu A_\mu$. So the pure A-field part is the *Maxwell Lagrangian*

$$\mathcal{L}_A = -\tfrac{1}{4}F_{\mu\nu}\,F^{\mu\nu}. \tag{7.64}$$

Before proceeding to try to quantize (7.64), we need to understand some important aspects of the free *classical* field $A^\nu(x)$.

When j_{em} is set equal to zero, A^ν satisfies the equation

$$\partial_\mu F^{\mu\nu} = \Box A^\nu - \partial^\nu(\partial^\mu A_\mu) = 0. \tag{7.65}$$

As we have seen in section 3.3, these equations are left unchanged if we perform the gauge transformation

$$A^{\mu} \to A'^{\mu} = A^{\mu} - \partial^{\mu}\chi. \tag{7.66}$$

We can use this freedom to *choose* the A^{μ} with which we work to satisfy the condition

$$\boxed{\partial_{\mu}A^{\mu} = 0.} \tag{7.67}$$

This is called the *Lorentz condition.* The process of choosing a particular condition on A^{μ} so as to define it (ultimately) uniquely is called 'choosing a gauge'; actually the condition (7.67) does not yet define A^{μ} uniquely, as we shall see shortly. The Lorentz condition is a very convenient one, since it decouples the different components of A^{μ} in Maxwell's equations (7.65)—in a covariant way, moreover, leaving the very simple equation

$$\Box A^{\mu} = 0. \tag{7.68}$$

This has plane-wave solutions of the form

$$A^{\mu} = N\epsilon^{\mu}\mathrm{e}^{-\mathrm{i}k\cdot x} \tag{7.69}$$

with $k^2 = 0$ (i.e. $k_0^2 = \boldsymbol{k}^2$), where N is a normalization factor and ϵ^{μ} is a *polarization vector* for the wave. The gauge condition (7.67) now reduces to a condition on ϵ^{μ}:

$$k \cdot \epsilon = 0. \tag{7.70}$$

However, we have not yet exhausted all the gauge freedom. We are still free to make another shift in the potential

$$A^{\mu} \to A^{\mu} - \partial^{\mu}\tilde{\chi} \tag{7.71}$$

provided $\tilde{\chi}$ satisfies the massless KG equation

$$\Box\tilde{\chi} = 0. \tag{7.72}$$

This condition on $\tilde{\chi}$ ensures that, even after the further shift, the resulting potential still satisfies $\partial_{\mu}A^{\mu} = 0$. For our plane-wave solutions, this residual gauge freedom corresponds to changing ϵ^{μ} by a multiple of k^{μ}:

$$\epsilon^{\mu} \to \epsilon^{\mu} + \beta k^{\mu} \equiv \epsilon'^{\mu} \tag{7.73}$$

which still satisfies $\epsilon'^{\mu} \cdot k = 0$ *since* $k^2 = 0$ *for these free-field solutions.* The condition $k^2 = 0$ is, of course, the statement that a free photon is massless.

This freedom has important consequences. Consider a solution with

$$k^{\mu} = (k^0, \boldsymbol{k}) \qquad (k^0)^2 = \boldsymbol{k}^2 \tag{7.74}$$

and polarization vector

$$\epsilon^\mu = (\epsilon^0, \boldsymbol{\epsilon}) \tag{7.75}$$

satisfying the Lorentz condition

$$k \cdot \epsilon = 0. \tag{7.76}$$

Gauge invariance now implies that we can add multiples of k^μ to ϵ^μ and still have a satisfactory polarization vector.

It is therefore clear that we can arrange for the time component of ϵ^μ to vanish so that the Lorentz condition reduces to the 3-vector condition

$$\boldsymbol{k} \cdot \boldsymbol{\epsilon} = 0. \tag{7.77}$$

This means that there are only two independent polarization vectors, both transverse to $\boldsymbol{k}$, i.e. to the propagation direction. For a wave travelling in the z-direction ($k^\mu = (k^0, 0, 0, k^0)$) these may be chosen to be

$$\boldsymbol{\epsilon}_{(1)} = (1, 0, 0) \tag{7.78}$$

$$\boldsymbol{\epsilon}_{(2)} = (0, 1, 0). \tag{7.79}$$

Such a choice corresponds to *linear polarization* of the associated $\boldsymbol{E}$ and $\boldsymbol{B}$ fields—which can be easily calculated from (3.9 and 3.10), given

$$A^\mu_{(i)} = N(0, \boldsymbol{\epsilon}_{(i)})\mathrm{e}^{-\mathrm{i}k\cdot x} \qquad i = 1, 2. \tag{7.80}$$

A commonly used alternative choice is

$$\boldsymbol{\epsilon}(\lambda = +1) = -\frac{1}{\sqrt{2}}(1, \mathrm{i}, 0) \tag{7.81}$$

$$\boldsymbol{\epsilon}(\lambda = -1) = \frac{1}{\sqrt{2}}(1, -\mathrm{i}, 0) \tag{7.82}$$

(linear combinations of (7.78) and (7.79)), which correspond to circularly polarized radiation. The phase convention in (7.81) and (7.82) is the standard one in quantum mechanics for states of definite spin projection ('helicity') $\lambda = \pm 1$ along the direction of motion (the z-axis here). We may easily check that

$$\boldsymbol{\epsilon}^*(\lambda) \cdot \boldsymbol{\epsilon}(\lambda') = \delta_{\lambda\lambda'} \tag{7.83}$$

or, in terms of the corresponding 4-vectors $\epsilon^\mu = (0, \boldsymbol{\epsilon})$,

$$\epsilon^*(\lambda) \cdot \epsilon(\lambda') = -\delta_{\lambda\lambda'}. \tag{7.84}$$

We have therefore arrived at the result, familiar in classical electromagnetic theory, that the free electromagnetic fields are purely *transverse*. Though they are described in this formalism by a vector potential with apparently four independent components $(V, \boldsymbol{A})$, the condition (7.67) reduces this number by one, and the further gauge freedom exploited in (7.71)–(7.73) reduces it by one more.

A crucial point to note is that the reduction to only *two* independent field components (polarization states) can be traced back to the fact that the free photon is massless: see the remark after (7.73). By contrast, for massive spin-1 bosons, such as the $W^\pm$ and Z^0, all *three* expected polarization states are indeed present. However, weak interactions are described by a gauge theory, and the $W^\pm$ and Z^0 particles are gauge-field quanta, analogous to the photon. How gauge invariance can be reconciled with the existence of massive gauge quanta with three polarization states will be explained in volume 2.

We may therefore write the plane-wave mode expansion for the classical $A^\mu(x)$ field in the form

$$A^\mu(x) = \sum_\lambda \int \frac{\mathrm{d}^3\boldsymbol{k}}{(2\pi)^3\sqrt{2\omega}}[\epsilon^\mu(k,\lambda)\alpha(k,\lambda)\mathrm{e}^{-\mathrm{i}k\cdot x} + \epsilon^{\mu *}(k,\lambda)\alpha^*(k,\lambda)\mathrm{e}^{-\mathrm{i}k\cdot x}] \tag{7.85}$$

where the sum is over the two possible polarization states λ, for given k, as described by the suitable polarization vector $\epsilon^\mu(k,\lambda)$ and $\omega = |\boldsymbol{k}|$.

It would seem that all we have to do now, in order to 'quantize' (7.85), is to promote α and α^* to operators $\hat{\alpha}$ and $\hat{\alpha}^\dagger$, as usual. However, things are actually not nearly so simple.

7.3.2 Quantizing $A^\mu(x)$

Readers familiar with Lagrangian mechanics may already suspect that quantizing A^ν is not going to be straightforward. The problem is that, clearly, $A^\nu(x)$ has four (Lorentz) components—but, equally clearly in view of the previous section, they are not all *independent* field components or field degrees of freedom. In fact, there are only two independent degrees of freedom, both transverse. Thus there are *constraints* on the four fields, for instance the gauge condition (7.67). Constrained systems are often awkward to handle in classical mechanics (see for example Goldstein 1980) or classical field theory; and they present major problems when it comes to canonical quantization. It is actually at just this point that the 'path-integral' approach to quantization, alluded to briefly at the end of section 5.2.2, comes into its own. This is basically because it does not involve non-commuting (or anti-commuting) operators and it is therefore to that extent closer to the classical case. This means that the relatively straightforward procedures available for constrained classical mechanics systems can—when suitably generalized!—be efficiently brought to bear on the quantum problem. For an introduction to these ideas, we refer to Swanson (1992).

However, we do not wish at this stage to take what would be a very long detour, in setting up the path-integral quantization of QED. We shall continue along the 'canonical' route. To see the kind of problems we encounter, let us try and repeat for the A^ν field the 'canonical' procedure we introduced in section 5.2.5. This was based, crucially, on obtaining from the Lagrangian the momentum π conjugate to ϕ, and then imposing the commutation relation (5.104) on the corresponding operators $\hat{\pi}$ and $\hat{\phi}$. But inspection of our Maxwell Lagrangian (7.64) quickly reveals that

$$\frac{\partial \mathcal{L}_A}{\partial \dot{A}^0} = 0 \tag{7.86}$$

and hence there is *no* canonical momentum π^0 conjugate to A^0. We appear to be stymied before we can even start.

There is another problem as well. Following the procedure explained in chapter 6, we expect that the Feynman propagator for the $\hat{A}^\mu$ field, namely $\langle 0|T(\hat{A}^\mu(x_1)\hat{A}^\nu(x_2))|0\rangle$, will surely appear, describing the propagation of a photon between x_1 and x_2. In the case of real scalar fields, problem 6.3 showed that the analogous quantity was actually a Green function for the KG differential operator, $(\Box + m^2)$. It turned out, in that case, that what we really wanted was the Fourier transform of the Green function, which was essentially (apart from the tricky 'iϵ prescription' and a trivial $-$i factor) the inverse of the momentum–space operator corresponding to $(\Box + m^2)$, namely $(-k^2 + m^2)^{-1}$ (see equation (6.98) and appendix G, and also (7.55)–(7.57) for the Dirac case). Suppose, then, that we try to follow this route to obtaining the propagator for the $\hat{A}^\nu$ field. For this it is sufficient to consider the classical equations (7.65) with $j_{\text{em}} = 0$, written in k space (problem 7.11(a)):

$$(-k^2 g^{\nu\mu} + k^\nu k^\mu)\tilde{A}_\mu(k) \equiv M^{\nu\mu}\tilde{A}_\mu(k) = 0 \tag{7.87}$$

where $\tilde{A}_\mu(k)$ is the Fourier transform of $A_\mu(x)$. We therefore require the inverse

$$(-k^2 g^{\nu\mu} + k^\nu k^\mu)^{-1} \equiv (M^{-1})^{\nu\mu}. \tag{7.88}$$

Unfortunately it is easy to show that this inverse does not exist. From Lorentz covariance, it has to transform as a second-rank tensor, and the only ones available are $g^{\mu\nu}$ and $k^\mu k^\nu$. So the general form of $(M^{-1})^{\nu\mu}$ must be

$$(M^{-1})^{\nu\mu} = A(k^2)g^{\nu\mu} + B(k^2)k^\nu k^\mu. \tag{7.89}$$

Now the inverse is defined by

$$(M^{-1})^{\nu\mu} M_{\mu\sigma} = g^\nu_\sigma. \tag{7.90}$$

Putting (7.89) and (7.87) into (7.90) yields (problem 7.11(b))

$$-k^2 A(k^2) g^\nu_\sigma + A(k^2) k^\nu k_\sigma = g^\nu_\sigma \tag{7.91}$$

which cannot be satisfied. So we are thwarted again.

Nothing daunted, the attentive reader may have an answer ready for the propagator problem. Suppose that, instead of (7.65), we start from the much simpler equation

$$\Box A^{\nu} = 0 \tag{7.92}$$

which results from *imposing* the Lorentz condition (7.67). Then, in momentum–space, (7.92) becomes

$$-k^2 \tilde{A}^{\nu} = 0. \tag{7.93}$$

The '$-k^2$' on the left-hand side certainly has an inverse, implying that the Feynman propagator for the photon is (proportional to) $g_{\mu\nu}/k^2$. This form is indeed plausible, as it is very much what we would expect by taking the massless limit of the spin-0 propagator and tacking on $g_{\mu\nu}$ to account for the Lorentz indices in $\langle 0|T(\hat{A}_{\mu}(x_1)\hat{A}_{\nu}(x_2))|0\rangle$ (but then why no term in $k_{\mu}k_{\nu}$?—see the final two paragraphs of this section!).

Perhaps this approach helps with the 'no canonical momentum π^0' problem too. Let us ask: What Lagrangian leads to the field equation (7.92)? The answer is (problem 7.12)

$$\mathcal{L}_L = -\tfrac{1}{4}F_{\mu\nu}F^{\mu\nu} - \tfrac{1}{2}(\partial_{\mu}A^{\mu})^2. \tag{7.94}$$

This form does seem to offer better prospects for quantization, since at least all our π^{μ}'s are non-zero; in particular

$$\pi^0 = \frac{\partial \mathcal{L}}{\partial \dot{A}^0} = -\partial_{\mu}A^{\mu}. \tag{7.95}$$

The other π's are unchanged by the addition of the extra term in (7.94) and are given by

$$\pi^i = -\dot{A}^i + \partial^i A^0. \tag{7.96}$$

Interestingly, these are precisely the electric fields E^i (see (3.10)). Let us see, then, if all our problems are solved with $\mathcal{L}_L$.

Now that we have at least got four non-zero π^{μ}'s, we can write down a plausible set of commutation relations between the corresponding operator quantities $\hat{\pi}^{\mu}$ and $\hat{A}^{\nu}$:

$$[\hat{A}_{\mu}(\boldsymbol{x}, t), \hat{\pi}_{\nu}(\boldsymbol{y}, t)] = \mathrm{i}g_{\mu\nu}\delta^3(\boldsymbol{x} - \boldsymbol{y}). \tag{7.97}$$

Again, the $g_{\mu\nu}$ is there to give the same Lorentz transformation character on both sides of the equation. But we must now remember that, in the classical case, our development rested on imposing the condition $\partial_{\mu}A^{\mu} = 0$ (7.67). Can we, in the quantum version we are trying to construct, simply impose $\partial_{\mu}\hat{A}^{\mu} = 0$? We certainly cannot do so in $\hat{\mathcal{L}}_L$, or we are back to $\hat{\mathcal{L}}_A$ again (besides, constraints cannot be 'substituted back' into Lagrangians, in general). Furthermore, if we set $\mu = \nu = 0$ in (7.97), then the right-hand side is non-zero while the left-hand side is zero if $\partial_{\mu}\hat{A}^{\mu} = 0 = \hat{\pi}^0$. So it is inconsistent simply to set $\partial_{\mu}\hat{A}^{\mu} = 0$.

We will return to the treatment of '$\partial_\mu \hat{A}^\mu = 0$' eventually. First, let us press on with (7.94) and see if we can get as far as a (quantized) mode expansion, of the form (7.85), for $\hat{A}^\mu(x)$.

To set this up, we need to massage the commutator (7.97) into a form as close as possible to the canonical '$[\phi, \dot{\phi}] = \mathrm{i}\delta$' form. Assuming the other commutation relations (cf (5.105))

$$[\hat{A}_\mu(\boldsymbol{x}, t), \hat{A}_\nu(\boldsymbol{y}, t)] = [\hat{\pi}_\mu(\boldsymbol{x}, t), \hat{\pi}_\nu(\boldsymbol{y}, t)] = 0 \tag{7.98}$$

we see that the spatial derivatives of the $\hat{A}$'s commute with the $\hat{A}$'s, and with each other, at equal times. This implies that we can rewrite the (quantum) $\hat{\pi}$'s as

$$\hat{\pi}_\mu = -\dot{\hat{A}}_\mu + \text{pieces that commute.} \tag{7.99}$$

Hence (7.97) can be rewritten as

$$[\hat{A}_\mu(\boldsymbol{x}, t), \dot{\hat{A}}_\nu(\boldsymbol{y}, t)] = -\mathrm{i}g_{\mu\nu}\delta^3(\boldsymbol{x} - \boldsymbol{y}) \tag{7.100}$$

and (7.98) remains the same. Now (7.100) is indeed very much the same as '$[\phi, \dot{\phi}] = \mathrm{i}\delta$' for the *spatial* component $\hat{A}^i$—but the sign is wrong in the $\mu = \nu = 0$ case. We are not out of the maze yet.

Nevertheless, proceeding onwards on the basis of (7.100), we write the quantum mode expansion as (cf (7.85))

$$\hat{A}^\mu(x) = \sum_{\lambda=0}^{3} \int \frac{\mathrm{d}^3\boldsymbol{k}}{(2\pi)^3\sqrt{2\omega}} [\epsilon^\mu(k, \lambda)\hat{\alpha}_\lambda(k)\mathrm{e}^{-\mathrm{i}k\cdot x} + \epsilon^{*\mu}(k, \lambda)\hat{\alpha}_\lambda^\dagger(k)\mathrm{e}^{\mathrm{i}k\cdot x}] \tag{7.101}$$

where the sum is over *four* independent polarization states $\lambda = 0, 1, 2, 3$, since all four fields are still in play. Before continuing, we need to say more about these ϵ's (previously, we only had two of them, now we have four and they are 4-vectors). We take $\boldsymbol{k}$ to be along the z-direction, as in our discussion of the ϵ's in section 7.3.1, and choose two transverse polarization vectors as (cf (7.78), (7.79))

$$\left.\begin{aligned} \epsilon^\mu(\lambda = 1) &= (0, 1, 0, 0) \\ \epsilon^\mu(\lambda = 2) &= (0, 0, 1, 0) \end{aligned}\right\} \quad \text{'transverse polarizations'.} \tag{7.102}$$

The other two ϵ's are

$$\epsilon^\mu(\lambda = 0) = (1, 0, 0, 0) \qquad \text{'time-like polarization'} \tag{7.103}$$

and

$$\epsilon^\mu(\lambda = 3) = (0, 0, 0, 1) \qquad \text{'longitudinal polarization'.} \tag{7.104}$$

Making (7.101) consistent with (7.100) then requires

$$[\hat{\alpha}_\lambda(k), \hat{\alpha}_{\lambda'}^\dagger(k')] = -g_{\lambda\lambda'}(2\pi)^3\delta^3(\boldsymbol{k} - \boldsymbol{k}'). \tag{7.105}$$

This is where the wrong sign in (7.100) has come back to haunt us: we have the wrong sign in (7.105) for the case $\lambda = \lambda' = 0$ (time-like modes).

What is the consequence of this? It seems natural to assume that the vacuum is defined by

$$\hat{\alpha}_\lambda(k)|0\rangle = 0 \qquad \text{for all } \lambda = 0, 1, 2, 3. \tag{7.106}$$

But suppose we use (7.105) and (7.106) to calculate the normalization overlap of a 'one time-like photon' state; this is

$$\begin{aligned}\langle \boldsymbol{k}', \lambda = 0|\boldsymbol{k}, \lambda = 0\rangle &= \langle 0|\hat{\alpha}_0(\boldsymbol{k})\hat{\alpha}_0^\dagger(\boldsymbol{k}')|0\rangle \\ &= -(2\pi)^3\delta^3(\boldsymbol{k} - \boldsymbol{k}')\end{aligned} \tag{7.107}$$

and the state effectively has a negative norm (the $\boldsymbol{k} = \boldsymbol{k}'$ infinity is the standard plane-wave artefact). Such states would threaten fundamental properties such as the conservation of total probability if they contributed, uncancelled, in physical processes.

At this point we would do well to recall the condition '$\partial_\mu \hat{A}^\mu = 0$', which still needs to be taken into account, somehow, and it does, indeed, save us. Gupta (1950) and Bleuler (1950) proposed that, rather than trying (unsuccessfully) to impose it as an operator condition, one should replace it by the weaker condition

$$\partial_\mu \hat{A}^{\mu(+)}(x)|\Psi\rangle = 0 \tag{7.108}$$

where the (+) signifies the positive frequency part of $\hat{A}$, i.e. the part involving annihilation operators, and $|\Psi\rangle$ is any physical state (including $|0\rangle$). From (7.108) and its Hermitian conjugate

$$\langle\Psi|\partial_\mu \hat{A}^{\mu(-)}(x) = 0 \tag{7.109}$$

we can deduce that the Lorentz condition (7.67) does hold for all expectation values:

$$\langle\Psi|\partial_\mu \hat{A}^\mu|\Psi\rangle = \langle\Psi|\partial_\mu \hat{A}^{\mu(+)} + \partial_\mu \hat{A}^{\mu(-)}|\Psi\rangle = 0, \tag{7.110}$$

and so the classical limit of this quantization procedure will recover the classical Maxwell theory in Lorentz gauge.

Using (7.101), (7.103) and (7.104) with $k^\mu = (|\boldsymbol{k}|, 0, 0|\boldsymbol{k}|)$, condition (7.108) becomes

$$[\hat{\alpha}_0(k) - \hat{\alpha}_3(k)]|\Psi\rangle = 0. \tag{7.111}$$

To see the effect of this condition, consider the expression for the Hamiltonian of this theory. In normally ordered form, it turns out to be

$$\hat{H} = \int \frac{\mathrm{d}^3\boldsymbol{k}}{(2\pi)^3}(\hat{\alpha}_1^\dagger\hat{\alpha}_1 + \hat{\alpha}_2^\dagger\hat{\alpha}_2 + \hat{\alpha}_3^\dagger\hat{\alpha}_3 - \hat{\alpha}_0^\dagger\hat{\alpha}_0)\omega \tag{7.112}$$

so the contribution from the time-like modes looks dangerously negative. However, for any physical state $|\Psi\rangle$, we have

$$\begin{aligned}\langle\Psi|(\hat{\alpha}_3^\dagger\hat{\alpha}_3^\dagger - \hat{\alpha}_0^\dagger\hat{\alpha}_0)|\Psi\rangle &= \langle\Psi|(\hat{\alpha}_3^\dagger\hat{\alpha}_3 - \hat{\alpha}_3^\dagger\hat{\alpha}_0)|\Psi\rangle \\ &= \langle\Psi|\hat{\alpha}_3^\dagger(\hat{\alpha}_3 - \hat{\alpha}_0)|\Psi\rangle \\ &= 0, \end{aligned} \tag{7.113}$$

so that only the transverse modes survive.

We hope that by now the reader will have at least begun to develop a healthy respect for quantum gauge fields—and the non-Abelian versions in volume 2 are even worse! The fact is that the canonical approach has a difficult time coping with these constrained systems. Indeed, the complete Feynman rules in the non-Abelian case were found by an alternative quantization procedure ('path integral' quantization). This, however, is outside the scope of the present volume. The important points for our purposes are as follows. It is possible to carry out a consistent quantization in the Gupta–Bleuler formalism, which is the quantum version of the Maxwell theory constrained by the Lorentz condition. The propagator for the photon in this theory is

$$-\mathrm{i}g^{\mu\nu}/k^2 + \mathrm{i}\epsilon \tag{7.114}$$

which is the expected massless limit of the KG propagator as far as the spatial components are concerned (the time-like component has that negative sign).

As in all the other cases we have dealt with so far, the Feynman propagator $\langle 0|T(\hat{A}^\mu(x_1)\hat{A}^\nu(x_2))|0\rangle$ can be evaluated using the expansion (7.101) and the commutation relations (7.105). One finds that it is indeed equal to the Fourier transform of $-\mathrm{i}g^{\mu\nu}/k^2 + \mathrm{i}\epsilon$ just as asserted in (7.114). For this result, we need the 'pseudo completeness relation' (problem 7.12)

$$\begin{aligned}&-\epsilon^\mu(k,\lambda=0)\epsilon^\nu(k,\lambda=0) + \epsilon^\mu(k,\lambda=1)\epsilon^\nu(k,\lambda=1) \\ &\quad + \epsilon^\mu(k,\lambda=2)\epsilon^\nu(k,\lambda=2) + \epsilon^\mu(k,\lambda=3)\epsilon^\nu(k,\lambda=3) = -g^{\mu\nu}.\end{aligned} \tag{7.115}$$

We call this a pseudo completeness relation because of the minus sign appearing in the first term: its origin in the evaluation of this vev is precisely the 'wrong sign commutator' for the $\hat{\alpha}_0$ mode, (7.105).

Thus the gauge choice (7.67) can be made to work in quantum field theory via the condition (7.108). *But other choices are possible too.* In particular, a useful generalization of the Lagrangian (7.94) is

$$\mathcal{L}_\xi = -\frac{1}{4}F_{\mu\nu}F^{\mu\nu} - \frac{1}{2\xi}(\partial_\mu A^\mu)^2 \tag{7.116}$$

where ξ is a constant, the 'gauge parameter'. $\mathcal{L}_\xi$ leads to the equation of motion (problem 7.14)

$$\left(\Box g_{\mu\nu} - \partial_\mu\partial_\nu + \frac{1}{\xi}\partial_\mu\partial_\nu\right)A^\nu = 0. \tag{7.117}$$

In momentum–space this becomes (problem 7.14)

$$\left(-k^2 g_{\mu\nu} + k_\mu k_\nu - \frac{1}{\xi} k_\mu k_\nu\right) \tilde{A}^\nu = 0. \tag{7.118}$$

The inverse of the matrix acting on $\tilde{A}^\nu$ exists, and gives us the more general photon propagator (or Green function)

$$\boxed{\frac{\mathrm{i}[-g^{\mu\nu} + (1-\xi)k^\mu k^\nu/k^2]}{k^2 + \mathrm{i}\epsilon}} \tag{7.119}$$

as shown in problem 7.14. The previous case is recovered as $\xi \to 1$. Confusingly, the choice $\xi = 1$ is often called the 'Feynman gauge', though in classical terms it corresponds to the Lorentz gauge choice. For some purposes the 'Landau gauge' $\xi = 0$ (which is well defined in (7.119)) is convenient. In any event, it is important to be clear that *the photon propagator depends on the choice of gauge*. Formula (7.119) is the photon analogue of 'rule (ii)' in (6.103).

This may seem to imply that when we use the photon propagator (7.119) in Feynman amplitudes we will not get a definite answer, but rather one that depends on the arbitrary parameter ξ. This is a serious worry. But the propagator is not by itself a physical quantity—it is only one part of a physical amplitude. In the following chapter we shall derive the amplitudes for some simple processes in scalar and spinor electrodynamics, and one can verify that they are gauge invariant—either in the sense (for external photons) of being invariant under the replacement (7.73), or (in the case of internal photons) of being independent of ξ. It can be shown (Weinberg 1995, section 10.5) that at a given order in perturbation theory the sum of all diagrams contributing to the S-matrix is gauge invariant.

7.4 Introduction of electromagnetic interactions

After all these preliminaries, the job of introducing the first of our gauge field interactions, namely electromagnetism, into our non-interacting theory of complex scalar fields, and of Dirac fields, is very easy. From our discussion in chapter 3, we have a strong indication of how to introduce electromagnetic interactions into our theories. The 'gauge principle' in quantum mechanics consisted in elevating a global (space–time-independent) U(1) phase invariance into a local (space–time-dependent) U(1) invariance—the compensating fields being then identified with the electromagnetic ones. In quantum field theory, exactly the same principle exists and leads to the form of the electromagnetic interactions. Indeed, in the field theory formalism we have a true local U(1) phase (gauge) *in*variance of the Lagrangian (rather than a gauge *co*variance of a wave equation) and we shall be able to exhibit explicitly the symmetry current, and symmetry operator, associated with the U(1) invariance—and identify them precisely with the electromagnetic current and charge.

We have seen that for both the complex scalar and the Dirac fields the free Lagrangian is invariant under U(1) transformations (see (7.22) and (7.45)) which, we once again emphasize, are *global*. Let us therefore promote these global invariances into local ones in the way learned in chapter 3—namely by invoking the 'gauge principle' replacement

$$\partial^\mu \rightarrow \hat{D}^\mu = \partial^\mu + \mathrm{i}q\hat{A}^\mu \tag{7.120}$$

for a particle of charge q, this time written in terms of the quantum field $\hat{A}^\mu$. In the case of the Dirac Lagrangian

$$\hat{\mathcal{L}}_\mathrm{D} = \bar{\hat{\psi}}(\mathrm{i}\gamma^\mu\partial_\mu - m)\hat{\psi} \tag{7.121}$$

we expect to be able to 'promote' it to one which is invariant under the *local* U(1) phase transformation

$$\hat{\psi}(\boldsymbol{x}, t) \rightarrow \hat{\psi}'(\boldsymbol{x}, t) = \mathrm{e}^{-\mathrm{i}q\hat{\chi}(\boldsymbol{x},t)}\hat{\psi}(\boldsymbol{x}, t) \tag{7.122}$$

provided we make the replacement (7.120) and demand that the (quantized) 4-vector potential transforms as

$$\hat{A}^\mu \rightarrow \hat{A}'^\mu = \hat{A}^\mu + \partial^\mu\hat{\chi}. \tag{7.123}$$

Thus the locally U(1)-invariant Dirac Lagrangian is expected to be

$$\hat{\mathcal{L}}_{\mathrm{D\ local}} = \bar{\hat{\psi}}(\mathrm{i}\gamma^\mu\hat{D}_\mu - m)\hat{\psi}. \tag{7.124}$$

The invariance of (7.124) under (7.122) is easy to check, using the crucial property (3.42), which clearly carries over to the quantum field case:

$$\hat{D}'_\mu\hat{\psi}' = \mathrm{e}^{-\mathrm{i}q\hat{\chi}}(\hat{D}_\mu\hat{\psi}). \tag{7.125}$$

Equation (7.125) implies at once that

$$(\mathrm{i}\gamma^\mu\hat{D}'_\mu - m)\hat{\psi}' = \mathrm{e}^{-\mathrm{i}q\hat{\chi}}(\mathrm{i}\gamma^\mu\hat{D}_\mu - m)\hat{\psi}, \tag{7.126}$$

while taking the conjugate of (7.122) yields

$$\bar{\hat{\psi}}' = \bar{\hat{\psi}}\mathrm{e}^{\mathrm{i}q\hat{\chi}}. \tag{7.127}$$

Thus we have

$$\bar{\hat{\psi}}'(\mathrm{i}\gamma^\mu\hat{D}'_\mu - m)\hat{\psi}' = \bar{\hat{\psi}}\mathrm{e}^{\mathrm{i}q\hat{\chi}}\mathrm{e}^{-\mathrm{i}q\hat{\chi}}(\mathrm{i}\gamma^\mu\hat{D}_\mu - m)\hat{\psi} \tag{7.128}$$

$$= \bar{\hat{\psi}}(\mathrm{i}\gamma^\mu\hat{D}_\mu - m)\hat{\psi} \tag{7.129}$$

and the invariance is proved.

The Lagrangian has therefore gained an interaction term

$$\hat{\mathcal{L}}_{\text{D}} \to \hat{\mathcal{L}}_{\text{D local}} = \hat{\mathcal{L}}_{\text{D}} + \hat{\mathcal{L}}_{\text{int}} \tag{7.130}$$

where

$$\boxed{\hat{\mathcal{L}}_{\text{int}} = -q\bar{\hat{\psi}}\gamma^{\mu}\hat{\psi}\hat{A}_{\mu}.} \tag{7.131}$$

Since the addition of $\mathcal{L}_{\text{int}}$ has not changed the canonical momenta, the Hamiltonian then becomes $\hat{H} = \hat{H}_{\text{D}} + \hat{H}'_{\text{D}}$, where

$$\hat{H}'_{\text{D}} = -\hat{\mathcal{L}}_{\text{int}} = q\bar{\hat{\psi}}\gamma^{\mu}\hat{\psi}\hat{A}_{\mu} = q\hat{\psi}^{\dagger}\hat{\psi}\hat{A}_{0} - q\hat{\psi}^{\dagger}\boldsymbol{\alpha}\hat{\psi}\cdot\hat{\boldsymbol{A}} \tag{7.132}$$

which is the field theory analogue of the potential in (4.134). It has the expected form '$\rho A_0 - \boldsymbol{j}\cdot\boldsymbol{A}$' if we identify the electromagnetic charge density operator with $q\hat{\psi}^{\dagger}\hat{\psi}$ (the charge times the number density operator) and the electromagnetic current density operator with $q\hat{\psi}^{\dagger}\boldsymbol{\alpha}\hat{\psi}$. The electromagnetic 4-vector current operator $\hat{j}^{\mu}_{\text{em}}$ is thus identified as

$$\hat{j}^{\mu}_{\text{em}} = q\bar{\hat{\psi}}\gamma^{\mu}\hat{\psi}. \tag{7.133}$$

We now note that $\hat{j}^{\mu}_{\text{em}}$ is just q times the symmetry current $\hat{N}^{\mu}_{\psi}$ of section 7.2 (see equation (7.47)). Conservation of $\hat{j}^{\mu}_{\text{em}}$ would follow from *global* U(1) invariance alone (i.e. $\hat{\chi}$ a constant in equation (7.122)); but many Lagrangians, including interactions, could be constructed obeying this global U(1) invariance. The force of the *local* U(1) invariance requirement is that it has specified a unique form of the interaction (i.e. $\hat{\mathcal{L}}_{\text{int}}$ of equation (7.131)). Indeed, this is just $-\hat{j}^{\mu}_{\text{em}}\hat{A}_{\mu}$, so that in this type of theory the current $\hat{j}^{\mu}_{\text{em}}$ is not only a symmetry current, but also determines the precise way in which the vector potential $\hat{A}^{\mu}$ couples to the matter field $\hat{\psi}$. Adding the Lagrangian for the $\hat{A}^{\mu}$ field then completes the theory of a charged fermion field interacting with the Maxwell field. In a general gauge, the $\hat{A}^{\mu}$ field Lagrangian is the operator form of (7.116), $\hat{\mathcal{L}}_{\xi}$.

The interaction term $\hat{H}'_{\text{D}} = q\bar{\hat{\psi}}\gamma^{\mu}\hat{\psi}\hat{A}_{\mu}$ is a 'three-fields-at-a-point' kind of interaction just like our 3-scalar interaction $g\hat{\phi}_{\text{A}}\hat{\phi}_{\text{B}}\hat{\phi}_{\text{C}}$ in chapter 6. We know, by now, exactly what all the operators in $\hat{H}'_{\text{D}}$ are capable of: some of the possible emission and absorption processes are shown in figure 7.3. Unlike the 'ABC' model with $m_{\text{C}} > m_{\text{A}} + m_{\text{B}}$ however, none of these elementary 'vertex' processes can occur as a real physical process, because all are forbidden by the requirement of overall 4-momentum conservation. However, they will of course contribute as virtual transitions when 'paired up' to form Feynman diagrams, such as those in figure 7.4 (compare figures 6.4 and 6.5).

It is worth remarking on the fact that the 'coupling constant' q is dimensionless, in our units. Of course, we know this from its identification with the electromagnetic charge in this case (see appendix C). But it is instructive to check it as follows. A Lagrangian has mass dimension M^4, since the action

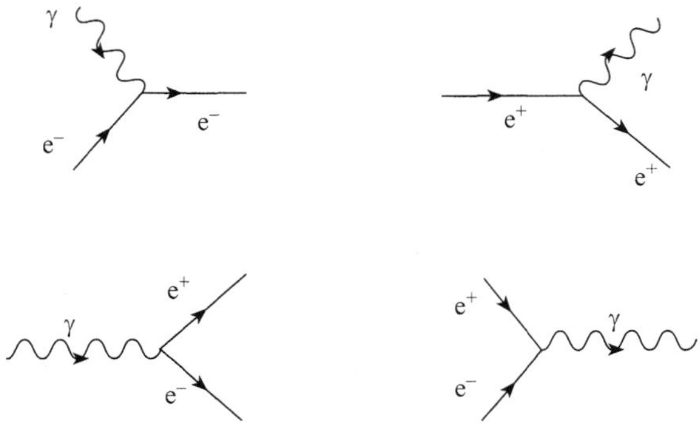

Figure 7.3. Possible basic '*vertices*' associated with the interaction density $e\bar{\hat{\psi}}\gamma^{\mu}\hat{\psi}\hat{A}_{\mu}$; these cannot occur as physical processes due to energy–momentum constraints.

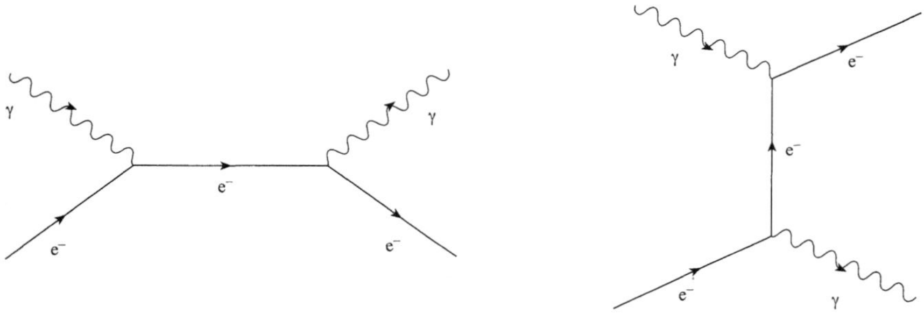

Figure 7.4. Lowest-order contributions to $\gamma e^{-} \to \gamma e^{-}$.

is dimensionless (with $\hbar = 1$). Referring then to (7.33) we see that the (mass) dimension of the $\hat{\psi}$ field is $M^{3/2}$, while (7.64) shows that that of $\hat{A}^{\mu}$ is M. It follows that $\bar{\hat{\psi}}\gamma^{\mu}\hat{\psi}\hat{A}_{\mu}$ has mass dimension M^4, and hence q must be dimensionless.

The application of the Dyson formalism of chapter 6 to fermions interacting via $\hat{H}'_{\rm D}$ leads directly to the Feynman rules for associating precise mathematical formulae with diagrams such as those in figure 7.4, as usual. This will be presented in the following chapter: see comment (3) in section 8.3.1 and appendix L. We may simply note here that a '$\hat{\psi}$' appears along with a '$\bar{\hat{\psi}}$' in $\hat{H}'_{\rm D}$, so that the process of 'contraction' (cf chapter 6) will lead to the form $\langle 0|T(\hat{\psi}(x_1)\bar{\hat{\psi}}(x_2))|0\rangle$ of the Dirac propagator, as stated in section 7.2.

In the same way, the global U(1) invariance (7.22) of the complex scalar field may be generalized to a local U(1) invariance incorporating electromagnetism. We have

$$\hat{\mathcal{L}}_{\rm KG} \to \hat{\mathcal{L}}_{\rm KG} + \hat{\mathcal{L}}_{\rm int} \tag{7.134}$$

where

$$\hat{\mathcal{L}}_{\text{KG}} = \partial_\mu \hat{\phi}^\dagger \partial^\mu \hat{\phi} - m^2 \hat{\phi}^\dagger \hat{\phi} \tag{7.135}$$

and (under $\partial_\mu \to \hat{D}_\mu$)

$$\hat{\mathcal{L}}_{\text{int}} = -\mathrm{i}q(\hat{\phi}^\dagger \partial^\mu \hat{\phi} - (\partial^\mu \hat{\phi}^\dagger)\hat{\phi})\hat{A}_\mu + q^2 \hat{A}^\mu \hat{A}_\mu \hat{\phi}^\dagger \hat{\phi} \tag{7.136}$$

which is the field theory analogue of the interaction in (4.132). The electromagnetic current is

$$\hat{j}^\mu_{\text{em}} = -\partial \hat{\mathcal{L}}_{\text{int}} / \partial \hat{A}_\mu \tag{7.137}$$

as before, which from (7.136) is

$$\hat{j}^\mu_{\text{em}} = \mathrm{i}q(\hat{\phi}^\dagger \partial^\mu \hat{\phi} - (\partial^\mu \hat{\phi}^\dagger)\hat{\phi}) - 2q^2 \hat{A}^\mu \hat{\phi}^\dagger \hat{\phi}. \tag{7.138}$$

We note that for the boson case the electromagnetic current is *not* just q times the (number) current $\hat{N}_\phi$ appropriate to the global phase invariance. This has its origin in the fact that the boson current involves a derivative, and so the gauge covariant boson current must develop a term involving $\hat{A}^\mu$ itself, as is evident in (7.138), and as we also saw in the wavefunction case (cf equation (3.39)). The full scalar QED Lagrangian is completed by the inclusion of $\hat{\mathcal{L}}_\xi$ as before.

The application of the formalism of chapter 6 is not completely straightforward in this scalar case. The problem is that $\hat{\mathcal{L}}_{\text{int}}$ of (7.136) involves derivatives of the fields and, in particular, their time derivatives. Hence the canonical momenta will be changed from their non-interacting forms. This, in turn, implies that the additional (interaction) term in the Hamiltonian is not just $-\hat{\mathcal{L}}_{\text{int}}$, as in the Dirac case, but is given by (problem 7.15)

$$\hat{H}'_S = -\hat{\mathcal{L}}_{\text{int}} - q^2 (\hat{A}^0)^2 \hat{\phi}^\dagger \hat{\phi}. \tag{7.139}$$

The problem here is that the Hamiltonian and $-\hat{\mathcal{L}}_{\text{int}}$ differ by a term which is non-covariant (only $\hat{A}^0$ appears).This seems to threaten the whole approach of chapter 6. Fortunately, another subtlety rescues the situation. There is a second source of non-covariance arising from the time-ordering of terms involving time derivatives, which will occur when (7.139) is used in the Dyson series (6.42). In particular, one can show (problem 7.16) that

$$\begin{aligned} &\langle 0|T(\partial_{1\mu}\hat{\phi}(x_1)\partial_{2\nu}\hat{\phi}^\dagger(x_2))|0\rangle \\ &\quad = \partial_{1\mu}\partial_{2\nu}\langle 0|T(\hat{\phi}(x_1)\hat{\phi}^\dagger(x_2))|0\rangle - \mathrm{i}g_{\mu 0}g_{\nu 0}\delta^4(x_1 - x_2) \end{aligned} \tag{7.140}$$

which also exhibits a non-covariant piece. A careful analysis (Itzykson and Zuber 1980, section 6.14) shows that the two covariant effects exactly compensate, so that in the Dyson series we may use $\hat{H}'_S = -\hat{\mathcal{L}}_{\text{int}}$ after all. The Feynman rules for charged scalar electrodynamics are given in appendix L.

We may now proceed to look at some simple processes in scalar and spinor electrodynamics, in the following two chapters.

Problems

7.1 Verify that the Lagrangian $\hat{\mathcal{L}}$ of (7.1) is invariant (i.e. $\hat{\mathcal{L}}(\hat{\phi}_1, \hat{\phi}_2) = \hat{\mathcal{L}}(\hat{\phi}_1', \hat{\phi}_2')$) under the transformation (7.2) of the fields $(\hat{\phi}_1, \hat{\phi}_2) \to (\hat{\phi}_1', \hat{\phi}_2')$.

7.2

(a) Verify that, for $\hat{N}_\phi^\mu$ given by (7.23), the corresponding $\hat{N}_\phi$ of (7.14) reduces to the form (7.24); and that, with $\hat{H}$ given by (7.21),

$$[\hat{N}_\phi, \hat{H}] = 0.$$

(b) Verify equation (7.27).

7.3 Show that

$$[\hat{\phi}(x_1), \hat{\phi}^\dagger(x_2)] = 0 \qquad \text{for } (x_1 - x_2)^2 < 0$$

[*Hint*: insert expression (7.16) for the $\hat{\phi}$'s and use the commutation relations (7.18) to express the commutator as the difference of two integrals; in the second integral, $x_1 - x_2$ can be transformed to $-(x_1 - x_2)$ by a Lorentz transformation—the time-ordering of space-like separated events is frame-dependent!].

7.4 Verify that varying $\psi^\dagger$ in the action principle with Lagrangian (7.34) gives the Dirac equation.

7.5 Verify (7.44).

7.6 Verify equations (7.49) and (7.50).

7.7 Verify (7.59).

7.8 Verify the expression given in (7.61) for $\sum_s u(k, s)\bar{u}(k, s)$. [*Hint*: first, note that u is a four-component Dirac spinor arranged as a column, while $\bar{u}$ is another four-component spinor but this time arranged as a row because of the transpose in the † symbol. So '$u\bar{u}$' has the form

$$\begin{pmatrix} u_1 \\ u_2 \\ u_3 \\ u_4 \end{pmatrix} (\bar{u}_1 \quad \bar{u}_2 \quad \bar{u}_3 \quad \bar{u}_4) = \begin{pmatrix} u_1\bar{u}_1 & u_1\bar{u}_2 & \cdots \\ u_2\bar{u}_1 & u_2\bar{u}_2 & \cdots \\ \vdots & \vdots & \end{pmatrix}$$

and is therefore a 4×4 matrix. Use the expression (4.105) for the u's, and take

$$\phi^1 = \begin{pmatrix} 1 \\ 0 \end{pmatrix} \qquad \phi^2 = \begin{pmatrix} 0 \\ 1 \end{pmatrix}.$$]

Verify that

$$\phi^1\phi^{1\dagger} + \phi^2\phi^{2\dagger} = \begin{pmatrix} 1 & 0 \\ 0 & 1 \end{pmatrix}.$$

Similarly, verify the expression for $\sum_s v(k, s)\bar{v}(k, s)$.

7.9 Verify the result quoted in (7.60) for the Feynman propagator for the Dirac field.

7.10 Verify that if $\mathcal{L} = -\frac{1}{4}F_{\mu\nu}F^{\mu\nu} - j_{\text{em}}A_\mu$, where $F_{\mu\nu} = \partial_\mu A_\nu - \partial_\nu A_\mu$, the Euler–Lagrange equations for A_μ yield the Maxwell form

$$\Box A^\mu - \partial^\mu(\partial_\nu A^\nu) = j^\mu_{\text{em}}.$$

[*Hint*: it is helpful to use antisymmetry of $F_{\mu\nu}$ to rewrite the '$F \cdot F$' term as $-\frac{1}{2}F_{\mu\nu}\partial^\mu A^\nu$.]

7.11

(a) Show that the Fourier transform of the free-field equation for A_μ (i.e. the one in the previous question with j^μ_{em} set to zero) is given by (7.87).
(b) Verify (7.91).

7.12 Show that the equation of motion for A_μ, following from the Lagrangian $\mathcal{L}_L$ of (7.94) is

$$\Box A^\mu = 0.$$

7.13 Verify equation (7.115).

7.14 Verify equations (7.117), (7.118) and (7.119).

7.15 Verify the form (7.139) of the interaction Hamiltonian, H'_S, in charged spin-0 electrodynamics.

7.16 Verify equation (7.140).

PART 3

TREE-LEVEL APPLICATIONS IN QED

8

ELEMENTARY PROCESSES IN SCALAR AND SPINOR ELECTRODYNAMICS

8.1 Coulomb scattering of charged spin-0 particles

We begin our study of electromagnetic interactions by considering the simplest case, that of the scattering of a (hypothetical) positively charged spin-0 particle 's^+' by a fixed Coulomb potential, treated as a classical field. This will lead us to the relativistic generalization of the Rutherford formula for the cross section. We shall use this example as an exercise to gain familiarity with the quantum field-theoretic approach of chapter 6, since it can also be done straightforwardly using the 'wavefunction' approach familiar from non-relativistic quantum mechanics, when supplemented by the work of chapter 4. We shall also look at 's^-' Coulomb scattering, to test the anti-particle prescriptions of chapter 4. Incidentally, we call these scalar particles $s^\pm$ to emphasize that they are not to be identified with, for instance, the physical pions $\pi^\pm$, since the latter are composite ($q\bar{q}$) systems, and hence their interactions are more complicated than those of our hypothetical 'point-like' $s^\pm$ (as we shall see in section 8.4). No point-like charged scalar particles have been discovered, as yet.

8.1.1 Coulomb scattering of s^+ (wavefunction approach)

Consider the scattering of a spin-0 particle of charge e and mass M, the 's^+', in an electromagnetic field described by the classical potential A^μ. The process we are considering is

$$s^+(p) \to s^+(p') \tag{8.1}$$

as shown in figure 8.1, where p and p' are the initial and final 4-momenta respectively. The appropriate potential for use in the KG equation has been given in section 4.6:

$$\hat{V}_{\mathrm{KG}} = \mathrm{i}e(\partial_\mu A^\mu + A^\mu \partial_\mu) - e^2 A^2. \tag{8.2}$$

As we shall see in more detail as we go along, the parameter characterizing each order of perturbation theory based on this potential is found to be $e^2/4\pi$. In natural units (see appendices B and C) this has the value

$$\alpha = e^2/4\pi \approx \frac{1}{137} \tag{8.3}$$

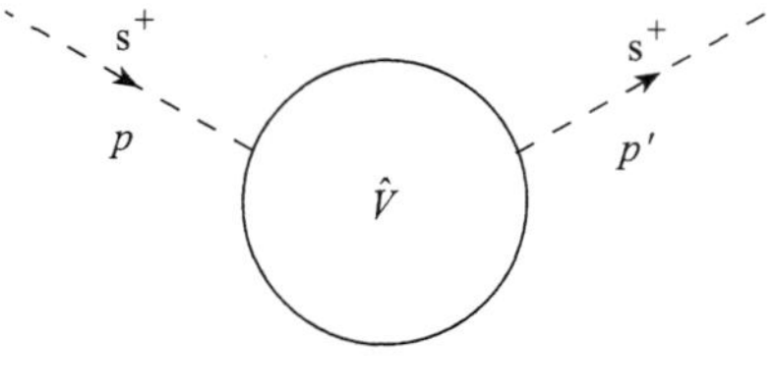

Figure 8.1. Coulomb scattering of s^+.

for the elementary charge e. α is called the fine structure constant. The smallness of α is the reason why a perturbation approach has been very successful for QED.

To lowest order in α we can neglect the e^2A^2 term and the perturbing potential is then

$$\hat{V} = \mathrm{i}e(\partial_\mu A^\mu + A^\mu \partial_\mu). \tag{8.4}$$

For a scattering process we shall assume[1] the same formula for the transition amplitude as in non-relativistic quantum mechanics (NRQM) time-dependent perturbation theory (see appendix A, equations (A.23) and (A.24)):

$$\mathcal{A}_{\mathrm{s}^+} = -\mathrm{i}\int \mathrm{d}^4x\, \phi'^* \hat{V} \phi \tag{8.5}$$

where ϕ and ϕ' are the initial and final state free-particle solutions. The latter are (recall equation (4.11))

$$\phi = N\mathrm{e}^{-\mathrm{i}p\cdot x} \tag{8.6}$$

$$\phi' = N'\mathrm{e}^{-\mathrm{i}p'\cdot x} \tag{8.7}$$

and we shall fix the normalization factors later. Inserting the expression for $\hat{V}$ into (8.5), and doing some integration by parts (problem 8.1), we obtain

$$\mathcal{A}_{\mathrm{s}^+} = -\mathrm{i}\int \mathrm{d}^4x\, \{\mathrm{i}e[\phi'^*(\partial_\mu\phi) - (\partial_\mu\phi'^*)\phi]\}A^\mu. \tag{8.8}$$

The expression inside the braces is very reminiscent of the probability current expression (4.20). Indeed we can write (8.8) as

$$\mathcal{A}_{\mathrm{s}^+} = -\mathrm{i}\int \mathrm{d}^4x\, j^\mu_{\mathrm{em,s}^+}(x) A_\mu(x) \tag{8.9}$$

where

$$j^\mu_{\mathrm{em,s}^+}(x) = \mathrm{i}e(\phi'^*\partial^\mu\phi - (\partial^\mu\phi'^*)\phi) \tag{8.10}$$

can be regarded as an electromagnetic 'transition current', analogous to the simple probability current for a single state. In the following section we shall see the

[1] Justification may be found in chapter 9 of Bjorken and Drell (1964).

exact meaning of this idea, using quantum field theory. Meanwhile, we insert the plane-wave free-particle solutions (8.6) and (8.7) for ϕ and ϕ' into 8.10 to obtain

$$j^{\mu}_{\mathrm{em,s}^+}(x) = NN'e(p+p')^{\mu}\mathrm{e}^{-\mathrm{i}(p-p')\cdot x} \tag{8.11}$$

so that (8.9) becomes

$$\mathcal{A}_{\mathrm{s}^+} = -\mathrm{i}NN' \int \mathrm{d}^4x\, e(p+p')_{\mu}\mathrm{e}^{-\mathrm{i}(p-p')\cdot x}A^{\mu}(x). \tag{8.12}$$

In the case of Coulomb scattering from a static point charge Ze ($e > 0$), the vector potential A^{μ} is given by

$$A^0 = \frac{Ze}{4\pi|\boldsymbol{x}|} \qquad \boldsymbol{A} = 0. \tag{8.13}$$

Inserting (8.13) into (8.12) we obtain

$$\mathcal{A}_{\mathrm{s}^+} = -\mathrm{i}NN'Ze^2(E+E') \int \mathrm{e}^{-\mathrm{i}(E-E')t}\,\mathrm{d}t \int \frac{\mathrm{e}^{\mathrm{i}(\boldsymbol{p}-\boldsymbol{p}')\cdot\boldsymbol{x}}}{4\pi|\boldsymbol{x}|}\,\mathrm{d}^3\boldsymbol{x}. \tag{8.14}$$

The initial and final 4-momenta are

$$p = (E, \boldsymbol{p}) \qquad p' = (E', \boldsymbol{p}')$$

with $E = \sqrt{M^2+\boldsymbol{p}^2}$, $E' = \sqrt{M^2+\boldsymbol{p}'^2}$. The first (time) integral in (8.14) gives an energy-conserving δ-function $2\pi\delta(E-E')$ (see appendix E), as is expected for a static (non-recoiling) scattering centre. The second (spatial) integral is the Fourier transform of $1/4\pi|\boldsymbol{x}|$, which can be obtained from (2.2), (2.15) and (2.16) by setting $m_{\mathrm{U}} = 0$; the result is $1/\boldsymbol{q}^2$ where $\boldsymbol{q} = \boldsymbol{p} - \boldsymbol{p}'$. Hence

$$\begin{aligned}\mathcal{A}_{\mathrm{s}^+} &= -\mathrm{i}NN'2\pi\delta(E-E')\frac{Ze^2}{\boldsymbol{q}^2}2E &\quad (8.15)\\ &\equiv -\mathrm{i}(2\pi)\delta(E-E')V_{\mathrm{s}^+} \qquad \text{(cf equation (A.25))} &\quad (8.16)\end{aligned}$$

where in (8.15) we have used $E = E'$ in the matrix element. This is in the standard form met in time-dependent perturbation theory (cf equations (A.25) and (A.26)).

The transition probability per unit time is then (appendix H, equation (H.18))

$$\dot{P}_{\mathrm{s}^+} = 2\pi|V_{\mathrm{s}^+}|^2\rho(E') \tag{8.17}$$

where $\rho(E')$ is the density of final states per energy interval $\mathrm{d}E'$. This will depend on the normalization adopted for ϕ, ϕ' via the factors N, N'. We choose these to be unity, which means that we are adopting the 'covariant' normalization of $2E$ particles per unit volume. Then (cf equation (H.22))

$$\rho(E')\,\mathrm{d}E' = \frac{|\boldsymbol{p}'|^2\,\mathrm{d}|\boldsymbol{p}'|}{(2\pi)^3\,2E'}\,\mathrm{d}\Omega. \tag{8.18}$$

Using $E' = (M^2 + \boldsymbol{p}'^2)^{1/2}$ one easily finds

$$\rho(E') = \frac{|\boldsymbol{p}'|\,\mathrm{d}\Omega}{16\pi^3}. \tag{8.19}$$

Note that this differs from equation (H.22) since here we are using relativistic kinematics.

To obtain the cross section, we need to divide $\dot{P}_{\mathrm{s}^+}$ by the incident flux, which is $2|\boldsymbol{p}|$ in our normalization. Hence

$$\mathrm{d}\sigma = (4Z^2e^4E^2/16\pi^2\boldsymbol{q}^4)\,\mathrm{d}\Omega. \tag{8.20}$$

Finally, since $\boldsymbol{q}^2 = (\boldsymbol{p} - \boldsymbol{p}')^2 = 4|\boldsymbol{p}|^2\sin^2\theta/2$ (cf section 2.4) where θ is the angle between $\boldsymbol{p}$ and $\boldsymbol{p}'$, we obtain

$$\boxed{\frac{\mathrm{d}\sigma}{\mathrm{d}\Omega} = (Z\alpha)^2\frac{E^2}{4|\boldsymbol{p}|^4}\frac{1}{\sin^4\theta/2}}. \tag{8.21}$$

This is the Rutherford formula for relativistic kinematics, showing the characteristic $\sin^{-4}\theta/2$ angular dependence (cf figure 1.1). This deservedly famous formula will serve as a 'reference point' for all the subsequent calculations in this chapter, as we proceed to add in various complications, such as spin, recoil and structure. The non-relativistic form may be retrieved by replacing E by M.

8.1.2 Coulomb scattering of s$^+$ (field-theoretic approach)

We follow steps closely similar to those in section 6.3.1, making use of the result quoted in section 7.4, that the appropriate interaction Hamiltonian for use in the Dyson series (6.42) is $\hat{\mathcal{H}}'_{\mathrm{s}} = -\hat{\mathcal{L}}_{\mathrm{int}}$ where $\hat{\mathcal{L}}_{\mathrm{int}}$ is given by (7.136), with $q = e$. As in the step from (8.2) to (8.4) we discard the e^2 term to first order and use

$$\hat{\mathcal{H}}'_{\mathrm{s}}(x) = \mathrm{i}e(\hat{\phi}^\dagger(x)\partial^\mu\hat{\phi}(x) - (\partial^\mu\hat{\phi}^\dagger(x))\hat{\phi}(x))A_\mu(x). \tag{8.22}$$

Equation (8.22) can be written as $\hat{j}^\mu_{\mathrm{em,s}}A_\mu$ where

$$\hat{j}^\mu_{\mathrm{em,s}} = \mathrm{i}e(\hat{\phi}^\dagger\partial^\mu\hat{\phi} - (\partial^\mu\hat{\phi}^\dagger)\hat{\phi}). \tag{8.23}$$

Note that the field A_μ is *not* quantized: it is being treated as an 'external' classical potential. The expansion for the field $\hat{\phi}$ is given in (7.16). As in (6.48), the lowest-order amplitude is

$$\mathcal{A}_{\mathrm{s}^+} = -\mathrm{i}\langle \mathrm{s}^+, p'|\int \mathrm{d}^4x\,\hat{H}'_{\mathrm{s}}(x)|\mathrm{s}^+, p\rangle \tag{8.24}$$

where (cf (6.49))

$$|\mathrm{s}^+, p\rangle = \sqrt{2E}\hat{a}^\dagger(p)|0\rangle. \tag{8.25}$$

We are, of course, anticipating in our notation that (8.24) will indeed be the same as (8.12). The required amplitude is then

$$\mathcal{A}_{s^+} = -\mathrm{i} \int \mathrm{d}^4 x \, \langle s^+, p' | \hat{j}^{\mu}_{\mathrm{em,s}}(x) | s^+, p \rangle A_{\mu}(x). \tag{8.26}$$

Using the expansion (7.16), the definition (8.25) and the vacuum conditions (7.30), and following the method of section 6.3.1, it is a good exercise to check that the value of the matrix element in (8.26) is (problem 8.2)

$$\langle s^+, p' | \hat{j}^{\mu}_{\mathrm{em,s}}(x) | s^+, p \rangle = e(p + p')^{\mu} \mathrm{e}^{-\mathrm{i}(p-p')\cdot x}. \tag{8.27}$$

This is exactly the same as the expression we obtained in (8.11) for the wave mechanical transition current in this case, using the normalization $N = N' = 1$, which is consistent with the field-theoretic normalization in (8.25). Thus our *wave mechanical transition current is indeed the matrix element of the field-theoretical electromagnetic current operator*:

$$j^{\mu}_{\mathrm{em,s^+}}(x) = \langle s^+, p' | \hat{j}^{\mu}_{\mathrm{em,s}}(x) | s^+, p \rangle. \tag{8.28}$$

Combining all these results, we have therefore connected the 'wavefunction' amplitude and the 'field-theory' amplitude via

$$\begin{aligned} \mathcal{A}_{s^+} &= -\mathrm{i} \int \mathrm{d}^4 x \, j^{\mu}_{\mathrm{em,s^+}}(x) A_{\mu}(x) \\ &= -\mathrm{i} \int \mathrm{d}^4 x \, \langle s^+, p' | \hat{j}^{\mu}_{\mathrm{em,s}}(x) | s^+, p \rangle A_{\mu}(x). \end{aligned} \tag{8.29}$$

We note that because of the static nature of the potential, and the non-covariant choice of A^{μ} (only $A^0 \neq 0$), our answer in either case cannot be expected to yield a Lorentz invariant amplitude.

8.1.3 Coulomb scattering of s$^-$

The physical process is (figure 8.2(*a*))

$$s^-(p) \to s^-(p') \tag{8.30}$$

where, of course, E and E' are both positive ($E = (M^2 + \boldsymbol{p}^2)^{1/2}$ and similarly for E'). Since the charge on the anti-particle s$^-$ is $-e$, the amplitude for this process can, in fact, be immediately obtained from (8.12) by merely changing the sign of e. Because of the way e and the 4-momenta p and p' enter (8.12), however, this in turn is the same as letting $p \to -p'$ and $p' \to -p$: this changes the sign of the '$e(p+p')_{\mu}$' part as required, and leaves the exponential unchanged. Hence we see in action here (admittedly in a very simple example) the Feynman interpretation of the negative 4-momentum solutions, described in section 4.5.4: the amplitude

Figure 8.2. Coulomb scattering of s^-: (*a*) the physical process with anti-particles of positive 4-momentum, and (*b*) the related unphysical process with particles of negative 4-momentum, using the Feynman prescription.

for $s^-(p) \to s^-(p')$ is the same as the amplitude for $s^+(-p') \to s^+(-p)$. The latter process is shown in figure 8.2(*b*).

The same conclusion can be *derived* from the field-theory formalism. In this case we need to evaluate the matrix element

$$\langle s^-, p' | \hat{j}^{\mu}_{\mathrm{em,s}}(x) | s^-, p \rangle, \tag{8.31}$$

where the *same* $\hat{j}_{\mathrm{em,s}}$ of equation (8.23) enters: $\hat{\phi}$ contains the anti-particle operator too! It is again a good exercise to check, using

$$|s^-, p\rangle = \sqrt{2E}\, \hat{b}^{\dagger}(p)|0\rangle \tag{8.32}$$

and remembering to *normally order* the operators in $\hat{j}^{\mu}_{\mathrm{em,s}}$, that (8.31) is given by the expected result, namely, (8.27) with $e \to -e$ (problem 8.3).

Since the matrix elements only differ by a sign, the cross sections for s^+ and s^- Coulomb scattering will be the same to this (lowest) order in α.

8.2 Coulomb scattering of charged spin-$\frac{1}{2}$ particles

8.2.1 Coulomb scattering of e^- (wavefunction approach)

We shall call the particle an electron, of charge $-e(e > 0)$ and mass m; note that by convention it is the negatively charged fermion that is the 'particle', but the positively charged boson. The process we are considering is (figure 8.3)

$$e^-(k, s) \to e^-(k', s') \tag{8.33}$$

where k, s are the 4-momentum and spin of the incident e^-, and similarly for k', s', with $k = (E, \boldsymbol{k})$ and $E = (m^2 + \boldsymbol{k}^2)^{1/2}$ and similarly for k'.

The appropriate potential to use in the Dirac equation has been given in section 4.6:

$$\hat{V}_{\mathrm{D}} = -eA^0\mathbf{1} + e\boldsymbol{\alpha} \cdot \boldsymbol{A} = -e\begin{pmatrix} A^0 & \boldsymbol{\sigma} \cdot \boldsymbol{A} \\ \boldsymbol{\sigma} \cdot \boldsymbol{A} & A^0 \end{pmatrix} \tag{8.34}$$

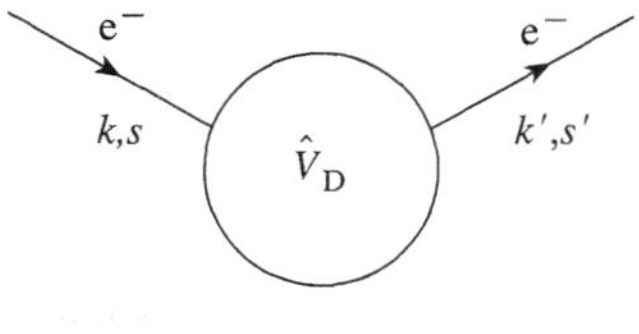

Figure 8.3. Coulomb scattering of e$^-$.

for a particle of charge $-e$. This potential is a 4×4 matrix and to obtain an amplitude in the form of a single complex number, we must use $\psi^\dagger$ instead of ψ^* in the matrix element. The first-order amplitude (figure 8.3) is therefore

$$\mathcal{A}_{e^-} = -i \int d^4x\, \psi^\dagger(k', s')\hat{V}_D\psi(k, s) \tag{8.35}$$

where s and s' label the spin components. The spin labels are necessary since the spin configuration may be changed by the interaction. In (8.35), ψ and ψ' are free-particle positive-energy solutions of the Dirac equation, as in (4.106), with u given by equation (4.105) and normalized to $u^\dagger u = 2E$, $E = (m^2 + \boldsymbol{k}^2)^{1/2}$.

The Lorentz properties of (8.35) become much clearer if we use the γ-matrix notation of problem 4.11. For convenience we re-state the definitions here:

$$\gamma^0 = \beta \qquad (\gamma^0)^2 = \mathbf{1} \tag{8.36}$$

$$\gamma^i = \beta\alpha_i \qquad (\gamma^i)^2 = -\mathbf{1} \qquad i = 1, 2, 3. \tag{8.37}$$

The Dirac equation may then be written (problem 4.12) as

$$(i\not{\partial} - m)\psi = 0 \tag{8.38}$$

where the 'slash' notation introduced in (7.56) has been used ($i\not{\partial} = i\gamma^\mu\partial_\mu$). Defining $\bar{\psi} = \psi^\dagger\gamma^0$, (8.35) becomes

$$\mathcal{A}_{e^-} = -i \int d^4x\, (-e\bar{\psi}'(x)\gamma^\mu\psi(x))A_\mu(x) \tag{8.39}$$

$$\equiv -i \int d^4x\, j^\mu_{em,e^-}(x)A_\mu(x) \tag{8.40}$$

where we have defined an electromagnetic transition current for a negatively charged fermion:

$$j^\mu_{em,e^-}(x) = -e\bar{\psi}'(x)\gamma^\mu\psi(x), \tag{8.41}$$

exactly analogous to the one for a positively charged boson introduced in section 8.1.1. The results of section 4.4 may be used to verify (problem 8.4) that $\bar{\psi}'\gamma^\mu\psi$ is a 4-vector, showing that $\mathcal{A}_{e^-}$ of (8.40) is Lorentz invariant.

Inserting free-particle solutions for ψ and $\psi'^\dagger$ in (8.41), we obtain

$$j^\mu_{em,e^-}(x) = -e\bar{u}(k', s')\gamma^\mu u(k, s)e^{-i(k-k')\cdot x} \tag{8.42}$$

so that (8.39) becomes

$$\mathcal{A}_{e^-} = -i \int d^4x \, (-e\bar{u}'\gamma^\mu u e^{-i(k-k')\cdot x}) A_\mu(x) \tag{8.43}$$

where $u = u(k, s)$ and similarly for u'. Note that the u's do not depend on x. For the case of the Coulomb potential in equation (8.13), $\mathcal{A}_{e^-}$ becomes

$$\mathcal{A}_{e^-} = ie2\pi\delta(E - E')\frac{Ze^2}{\boldsymbol{q}^2}u'^\dagger u \tag{8.44}$$

just as in (8.15), where $\boldsymbol{q} = \boldsymbol{k} - \boldsymbol{k}'$ and we have used $\bar{u}'\gamma^0 = u'^\dagger$. Comparing (8.44) with (8.15), we see that (using the covariant normalization $N = N' = 1$) the amplitude in the spinor case is obtained from that for the scalar case by the replacement '$2E \to u'^\dagger u$' and the sign of the amplitude is reversed as expected for e^- rather than s^+ scattering.

We now have to understand how to define the cross section for particles with spin and then how to calculate it. Clearly the cross section is proportional to $|\mathcal{A}_{e^-}|^2$, which involves $|u^\dagger(k', s')u(k, s)|^2$ here. Usually the incident beam is *unpolarized*, which means that it is a random mixture of both spin states s ('up' or 'down'). It is important to note that this is an *incoherent* average, in the sense that we average the *cross section* rather than the amplitude. Furthermore, most experiments usually measure only the direction and energy of the scattered electron and are not sensitive to the spin state s'. Thus what we wish to calculate, in this case, is the unpolarized cross section defined by

$$\begin{aligned} d\bar{\sigma} &\equiv \tfrac{1}{2}(d\sigma_{\uparrow\uparrow} + d\sigma_{\uparrow\downarrow} + d\sigma_{\downarrow\uparrow} + d\sigma_{\downarrow\downarrow}) \\ &= \tfrac{1}{2}\sum_{s',s} d\sigma_{s's} \end{aligned} \tag{8.45}$$

where $d\sigma_{s',s} \propto |u^\dagger(k', s')u(k, s)|^2$. In (8.45), we are averaging over the two possible initial spin polarizations and summing over the final spin states arising from each initial spin state.

It is possible to calculate the quantity

$$S = \tfrac{1}{2}\sum_{s',s} |u'^\dagger u|^2 \tag{8.46}$$

by brute force, using (4.105) and taking the two-component spinors to be, say,

$$\phi^1 = \begin{pmatrix} 1 \\ 0 \end{pmatrix} \qquad \phi^2 = \begin{pmatrix} 0 \\ 1 \end{pmatrix}. \tag{8.47}$$

One finds (problem 8.5)

$$S = (2E)^2(1 - v^2 \sin^2\theta/2) \tag{8.48}$$

where $v = |\boldsymbol{k}|/E$ is the particle's speed and θ is the scattering angle. If we now recall that (a) the matrix element (8.44) can be obtained from (8.15) by the replacement '$2E \rightarrow u'^{\dagger}u$' and (b) the normalization of our spinor states is the same ('$\rho = 2E$') as in the scalar case, so that the flux and density of states factors are unchanged, we may infer from (8.21) that

$$\boxed{\frac{\mathrm{d}\bar{\sigma}}{\mathrm{d}\Omega} = (Z\alpha)^2 \frac{E^2}{4|\boldsymbol{k}|^4} \frac{(1 - v^2 \sin^2\theta/2)}{\sin^4\theta/2}.} \tag{8.49}$$

This is the Mott cross section (Mott 1929). Comparing this with the basic Rutherford formula (8.21), we see that the factor $(1 - v^2 \sin^2\theta/2)$ (which comes from the spin summation) represents the effect of replacing spin-0 scattering particles by spin-$\frac{1}{2}$ ones.

Indeed, this factor has an important physical interpretation. Consider the extreme relativistic limit ($v \rightarrow 1, m \rightarrow 0$), when the factor becomes $\cos^2\theta/2$, which vanishes in the backward direction $\theta = \pi$. This may be understood as follows. In the $m \rightarrow 0$ limit, it is appropriate to use the representation (4.97) of the Dirac matrices and, in this case (as was discussed in problem 4.15), the Dirac spinor takes the form

$$u = \begin{pmatrix} u_{\mathrm{R}} \\ u_{\mathrm{L}} \end{pmatrix} \tag{8.50}$$

where u_{R} and u_{L} have positive and negative helicity respectively. The spinor part of the matrix element (8.44) then becomes $u_{\mathrm{R}}'^{\dagger}u_{\mathrm{R}} + u_{\mathrm{L}}'^{\dagger}u_{\mathrm{L}}$, from which it is clear that *helicity is conserved*: the helicity of the u' spinors equals that of the u spinors; in particular there are no helicity mixing terms of the form $u_{\mathrm{R}}'^{\dagger}u_{\mathrm{L}}$ or $u_{\mathrm{L}}'^{\dagger}u_{\mathrm{R}}$. Consider then an initial state electron with positive helicity, and take the z-axis to be along the incident momentum. The z-component of angular momentum is then $+\frac{1}{2}$. Suppose the electron is scattered through an angle of π. Since helicity is conserved, the scattered electron's helicity will still be positive, but since the direction of its momentum has been reversed, its angular momentum along the original axis will be $-\frac{1}{2}$. Hence this configuration is forbidden by angular momentum conservation—and similarly for an incoming negative helicity state. The spin labels s', s in (8.46) can be taken to be helicity labels and so it follows that the quantity S must vanish for $\theta = \pi$ in the $m \rightarrow 0$ limit. As was shown in problem 4.15, the 'R' and 'L' states are mixed by a mass term in the Dirac equation and hence we expect backward scattering to be increasingly allowed as m/E increases (recall that $v = (1 - m^2/E^2)^{1/2}$).

8.2.2 Coulomb scattering of e^- (field-theoretic approach)

Once again, the interaction Hamiltonian has been given in section 7.4, namely

$$\hat{H}_{\mathrm{D}}' = -e\bar{\hat{\psi}}\gamma^{\mu}\hat{\psi}A_{\mu} \equiv \hat{j}^{\mu}_{\mathrm{em,e}}A_{\mu} \tag{8.51}$$

where the current operator $\hat{j}^{\mu}_{\mathrm{em,e}}$ is just $-e\bar{\hat{\psi}}\gamma^{\mu}\hat{\psi}$ in this case. The lowest-order amplitude is then

$$\mathcal{A}_{\mathrm{e}^-} = -\mathrm{i}\langle \mathrm{e}^-, k', s'| \int \mathrm{d}^4x\ \hat{H}'_{\mathrm{D}}(x)|\mathrm{e}^-, k, s\rangle \tag{8.52}$$

$$= -\mathrm{i}\int \mathrm{d}^4x\ \langle \mathrm{e}^-, k', s'|\hat{j}^{\mu}_{\mathrm{em,e}}(x)|\mathrm{e}^- k, s\rangle A_{\mu}(x). \tag{8.53}$$

With our normalization, and referring to the fermionic expansion (7.35), the states are defined by

$$|\mathrm{e}^-, k, s\rangle = \sqrt{2E}\hat{c}^{\dagger}_{s}(k)|0\rangle \tag{8.54}$$

and similarly for the final state. We then find (problem 8.6) that the current matrix element in (8.53) takes the form

$$\langle \mathrm{e}^-, k', s'|\hat{j}^{\mu}_{\mathrm{em,e}}(x)|\mathrm{e}^-, k, s\rangle = -e\bar{u}'\gamma^{\mu}u\mathrm{e}^{-\mathrm{i}(k-k')\cdot x} = j^{\mu}_{\mathrm{em,e}^-}(x) \tag{8.55}$$

exactly as in (8.42). Thus once again, the 'wavefunction' and 'field-theoretic' approaches have been shown to be equivalent, in a simple case.

8.2.3 Trace techniques for spin summations

The calculation of cross sections involving fermions rapidly becomes laborious following the 'brute force' method of section 8.2.1, in which the explicit forms for u and $u'^{\dagger}$ were used. Fortunately we can avoid this by using a powerful labour-saving device due to Feynman, in which the γ's come into their own.

We need to calculate the quantity S given in (8.46). This will turn out to be just the first in a series of such objects. With later needs in mind, we shall here calculate a more general quantity than (8.46), namely the *lepton tensor*

$$L^{\mu\nu}(k', k) = \tfrac{1}{2}\sum_{s',s}\bar{u}(k', s')\gamma^{\mu}u(k, s)[\bar{u}(k', s')\gamma^{\nu}u(k, s)]^{*}. \tag{8.56}$$

Clearly this will be relevant to the more general case in which A^{μ} contains non-zero spatial components, for example. For our present application, we shall need only L^{00}.

We first note that $L^{\mu\nu}$ is correctly called a *tensor* (a contravariant second-rank one, in fact—see appendix D), because the two '$\bar{u}\gamma^{\mu}u$, $\bar{u}\gamma^{\nu}u$' factors are each 4-vectors, as we have seen. (We might worry a little over the complex conjugation of the second factor, but this will disappear after the next step.) Consider therefore the factor $[\bar{u}(k', s')\gamma^{\nu}u(k, s)]^{*}$. For each value of the index ν, this is just a number (the corresponding component of the 4-vector), and so it can make no difference if we take its transpose, in a matrix sense (the transpose of a 1×1 matrix is certainly equal to itself!). In that case the complex conjugate becomes the Hermitian conjugate, which is:

$$[\bar{u}(k', s')\gamma^{\nu}u(k, s)]^{\dagger} = u^{\dagger}(k, s)\gamma^{\nu\dagger}\gamma^{0\dagger}u(k', s') \tag{8.57}$$

$$= \bar{u}(k, s)\gamma^{\nu}u(k', s') \tag{8.58}$$

since (problem 8.7)

$$\gamma^0 \gamma^{\nu\dagger} \gamma^0 = \gamma^\nu \tag{8.59}$$

and $\gamma^0 = \gamma^{0\dagger}$. Thus $L^{\mu\nu}$ may be written in the more streamlined form

$$L^{\mu\nu} = \tfrac{1}{2} \sum_{s',s} \bar{u}(k', s')\gamma^\mu u(k, s)\bar{u}(k, s)\gamma^\nu u(k', s') \tag{8.60}$$

which is, moreover, evidently the (tensor) product of two 4-vectors. However, there is more to this than saving a few symbols. We have seen the expression

$$\sum_s u(k, s)\bar{u}(k, s) \tag{8.61}$$

before! (See (7.61) and problem 7.7.) Thus we can replace the sum (8.61) over spin states 's' by the corresponding matrix $(\not{k} + m)$:

$$L^{\mu\nu} = \tfrac{1}{2} \sum_{s'} \bar{u}_\alpha(k', s')(\gamma^\mu)_{\alpha\beta}(\not{k} + m)_{\beta\gamma}(\gamma^\nu)_{\gamma\delta} u_\delta(k', s') \tag{8.62}$$

where we have made the matrix indices explicit, and *summation on all repeated matrix indices is understood.* In particular, note that every matrix index is repeated, so that each one is in fact summed over: there are no 'spare' indices. Now, since we can reorder matrix elements as we wish, we can bring the u_δ to the front of the expression, and use the same trick to perform the second spin sum:

$$\sum_{s'} u_\delta(k', s')\bar{u}_\alpha(k', s') = (\not{k}' + m)_{\delta\alpha}. \tag{8.63}$$

Thus $L^{\mu\nu}$ takes the form of a matrix product, summed over the diagonal elements:

$$L^{\mu\nu} = \tfrac{1}{2}(\not{k}' + m)_{\delta\alpha}(\gamma^\mu)_{\alpha\beta}(\not{k} + m)_{\beta\gamma}(\gamma^\nu)_{\gamma\delta} \tag{8.64}$$

$$= \tfrac{1}{2} \sum_\delta [(\not{k}' + m)\gamma^\mu(\not{k} + m)\gamma^\nu]_{\delta\delta} \tag{8.65}$$

where we have explicitly reinstated the sum over δ. The right-hand side of (8.65) is the *trace* (i.e. the sum of the diagonal elements) of the matrix formed by the product of the four indicated matrices:

$$L^{\mu\nu} = \tfrac{1}{2}\,\mathrm{Tr}[(\not{k}' + m)\gamma^\mu(\not{k} + m)\gamma^\nu]. \tag{8.66}$$

Such matrix traces have some useful properties which we now list. Denote the trace of a matrix $\mathbf{A}$ by

$$\mathrm{Tr}\,\mathbf{A} = \sum_i A_{ii}. \tag{8.67}$$

Consider now the trace of a matrix product,

$$\mathrm{Tr}(\mathbf{AB}) = \sum_{i,j} A_{ij} B_{ji} \tag{8.68}$$

where we have written the summations in explicitly. We can (as before) freely exchange the order of the matrix elements A_{ij} and B_{ji}, to rewrite (8.68) as

$$\mathrm{Tr}(\mathbf{AB}) = \sum_{i,j} B_{ji} A_{ij}. \tag{8.69}$$

But the right-hand side is precisely Tr(**BA**); hence we have shown that

$$\mathrm{Tr}(\mathbf{AB}) = \mathrm{Tr}(\mathbf{BA}). \tag{8.70}$$

Similarly it is easy to show that

$$\mathrm{Tr}(\mathbf{ABC}) = \mathrm{Tr}(\mathbf{CAB}). \tag{8.71}$$

We may now return to (8.66). The advantage of the trace form is that we can invoke some powerful results about *traces of products of γ-matrices*. Here we shall just list the trace 'theorems' that we shall use to evaluate $L^{\mu\nu}$: more complete statements of trace theorems and γ-matrix algebra, together with proofs of these theorems, are given in appendix J .

We need the following results:

(a) $$\mathrm{Tr}\,\mathbf{1} = 4 \tag{8.72}$$

(b) $$\mathrm{Tr}\ (\text{odd number of } \gamma\text{'s}) = 0 \tag{8.73}$$

(c) $$\mathrm{Tr}(\not{a}\not{b}) = 4(a \cdot b) \tag{8.74}$$

(d) $$\mathrm{Tr}(\not{a}\not{b}\not{c}\not{d}) = 4[(a \cdot b)(c \cdot d) + (a \cdot d)(b \cdot c) - (a \cdot c)(b \cdot d)]. \tag{8.75}$$

Then

$$\begin{aligned} \mathrm{Tr}[(\not{k}' + m)\gamma_\mu(\not{k}' + m)] = {} & \mathrm{Tr}(\not{k}'\gamma^\mu\not{k}'\gamma^\nu) + m\,\mathrm{Tr}(\gamma^\mu\not{k}\gamma^\nu) \\ & + m\,\mathrm{Tr}(\not{k}'\gamma^\mu\gamma^\nu) + m^2\,\mathrm{Tr}(\gamma^\mu\gamma^\nu) \end{aligned} \tag{8.76}$$

The terms linear in m are zero by theorem (b), and using (c) in the form

$$\mathrm{Tr}(\gamma_\mu\gamma_\nu)a^\mu b^\nu = 4g_{\mu\nu}a^\mu b^\nu = 4a \cdot b \tag{8.77}$$

and (d) in a similar form, we obtain (problem 8.8)

$$L^{\mu\nu} = \tfrac{1}{2}\,\mathrm{Tr}[(\not{k}' + m)\gamma^\mu(\not{k} + m)\gamma^\nu] = 2[k'^\mu k^\nu + k'^\nu k^\mu - (k' \cdot k)g^{\mu\nu}] + 2m^2 g^{\mu\nu}. \tag{8.78}$$

In the present case we simply want L^{00}, which is found to be (problem 8.9)

$$L^{00} = 4E^2(1 - v^2 \sin^2\theta/2) \tag{8.79}$$

where $v = |\boldsymbol{k}|/E$, just as in (8.48).

8.2.4 Coulomb scattering of e^+

The physical process is

$$e^+(k, s) \to e^+(k', s') \tag{8.80}$$

where, as usual, we emphasize that E and E' are both positive. In the wavefunction approach, we saw in section 4.5.4. that, because $\rho \geq 0$ always for a Dirac particle, we had to introduce a minus sign 'by hand', according to the rule stated at the end of section 4.5.4. This rule gives us, in the present case,

$$\begin{aligned} &\text{amplitude } (e^+(k, s) \to e^+(k', s')) \\ &\quad = -\text{amplitude } (e^-(-k', -s') \to e^-(-k, -s)). \end{aligned} \tag{8.81}$$

Referring to (8.43), therefore, the required amplitude for the process (8.80) is

$$\mathcal{A}_{e^+} = -\mathrm{i} \int \mathrm{d}^4 x \, (e\bar{v}(k, s)\gamma^\mu v(k', s')\mathrm{e}^{-\mathrm{i}(k-k')\cdot x}) A_\mu(x) \tag{8.82}$$

since the 'v' solutions have been set up precisely to correspond to the '$-k, -s$' situation. In evaluating the cross section from (8.82), the only difference from the e^- case is the appearance of the spinors 'v' rather than 'u'; the lepton tensor in this case is

$$L^{\mu\nu} = \tfrac{1}{2} \operatorname{Tr}[(\not{k} - m)\gamma^\mu (\not{k}' - m)\gamma^\nu] \tag{8.83}$$

using the result (7.61) for $\sum_s v(k, s)\bar{v}(k, s)$. Expression (8.83) differs from (8.66) by the sign of m and by $k \leftrightarrow k'$, but the result (8.78) for the trace is insensitive to these changes. Thus the positron Coulomb scattering cross section is equal to the electron one to lowest order in α.

In the field-theoretic approach, the *same* interaction Hamiltonian $\hat{H}'_{\mathrm{D}}$ which we used for e^- scattering will again automatically yield the e^+ matrix element (recall the discussion at the end of section 7.4). In place of (8.53), the amplitude we wish to calculate is

$$\begin{aligned} \mathcal{A}_{e^+} &= -\mathrm{i} \int \mathrm{d}^4 x \, \langle e^+, k', s' | \hat{j}^\mu_{\mathrm{em,e}}(x) | e^+, k, s \rangle A_\mu(x) \\ &= -\mathrm{i} \int \mathrm{d}^4 x \, \langle e^+, k', s' | -e\hat{\bar{\psi}}(x)\gamma^\mu \hat{\psi}(x) | e^+, k, s \rangle A_\mu(x) \end{aligned} \tag{8.84}$$

where, referring to the fermionic expansion (4.167),

$$|e^+, k, s\rangle = \sqrt{2E} d_s^\dagger(k)|0\rangle, \tag{8.85}$$

and similarly for the final state. In evaluating the matrix element in (8.84) we must again remember to *normally order* the fields, according to the discussion in section 7.2. Bearing this in mind, and inserting the expansion (7.35), one finds (problem 8.10)

$$\langle e^+, k', s' | \hat{j}^\mu_{\mathrm{em,e}}(x) | e^+, k, s \rangle = +e\bar{v}(k, s)\gamma^\mu v(k', s')\mathrm{e}^{-\mathrm{i}(k-k')\cdot x} \tag{8.86}$$

$$\equiv j^\mu_{\mathrm{em,e^+}}(x) \tag{8.87}$$

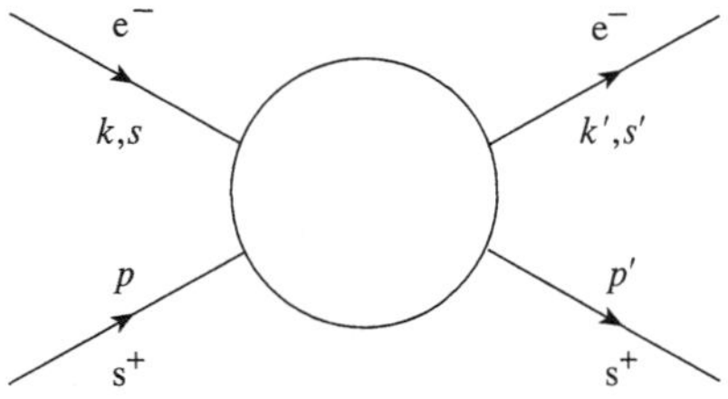

Figure 8.4. e^-s^+ scattering amplitude.

just as required in (8.82). Note especially that the correct sign has emerged naturally without having to be put in 'by hand', as was necessary in the wavefunction approach when applied to an anti-fermion.

We are now ready to look at some more realistic (and covariant) processes.

8.3 e^-s^+ scattering

8.3.1 The amplitude for $e^-s^+ \rightarrow e^-s^+$

We consider the two-body scattering process

$$e^-(k, s) + s^+(p) \rightarrow e^-(k', s') + s^+(p') \tag{8.88}$$

where the 4-momenta and spins are as indicated in figure 8.4. How will the e^- and s^+ interact? In this case, there is no 'external' classical electromagnetic potential in the problem. Instead, each of e^- and s^+, as charged particles, act as sources for the electromagnetic field, with which they in turn interact. We can picture the process as one in which each particle scatters off the 'virtual' field produced by the other (we shall make this more precise in comment (2) after equation (8.101)). The formalism of quantum field theory is perfectly adapted to account for such effects, as we shall see. It is very significant that no *new* interaction is needed to describe the process (8.88) beyond what we already have: the complete Lagrangian is now simply the free-field Lagrangians for the spin-$\frac{1}{2}$ e^-, the spin-0 s^+ and the Maxwell field, together with the sum of the scalar electromagnetic interaction Lagrangian of (7.136), with $q = e$, and the Dirac Lagrangian of (7.131) with $q = -e$. The full interaction Hamiltonian is then

$$\hat{H}'(x) = [\mathrm{i}e(\hat{\phi}^\dagger(x)\partial^\mu\hat{\phi}(x) - \partial^\mu\hat{\phi}^\dagger(x)\hat{\phi}(x)) - e\bar{\hat{\psi}}(x)\gamma^\mu\hat{\psi}(x)]\hat{A}_\mu(x) \tag{8.89}$$

$$\equiv (\hat{j}^\mu_{\mathrm{em,s}}(x) + \hat{j}^\mu_{\mathrm{em,e}}(x))\hat{A}_\mu(x) \tag{8.90}$$

where the 'total current' in (8.90) is just the indicated sum of the $\hat{\phi}$ (scalar) and $\hat{\psi}$ (spinor) currents. This $\hat{H}'$ must now be used in the Dyson expansion (6.42), in a perturbative calculation of the $e^-s^+ \rightarrow e^-s^+$ amplitude.

Note now that, in contrast to our Coulomb scattering 'warm-ups', the electromagnetic field *is* quantized in (8.89). We first observe that, since there are no free photons in either the initial or final states in our process $e^-s^+ \to e^-s^+$, the first-order matrix element of $\hat{H}'$ must vanish (as did the corresponding first-order amplitude in AB $\to$ AB scattering, in section 6.3.2). The first non-vanishing scattering processes arise at second order (cf (6.74)):

$$\mathcal{A}_{e^-s^+} = \frac{(-i)^2}{2} \iint d^4x_1\, d^4x_2\, \langle 0|\hat{c}_{s'}(p')\hat{a}(k')T\{\hat{H}'(x_1)\hat{H}'(x_2)\}\hat{a}^\dagger(k)\hat{c}_s^\dagger(p)|0\rangle \times (16E_kE_{k'}E_pE_{p'})^{1/2}. \tag{8.91}$$

Just as for AB $\to$ AB and the $\hat{C}$ field in the 'ABC' model (cf (6.81)), as far as the $\hat{A}_\mu$ operators in (8.91) are concerned the only surviving contraction is

$$\langle 0|T(\hat{A}_\mu(x_1)\hat{A}_\nu(x_2))|0\rangle \tag{8.92}$$

which is the Feynman propagator for the photon, in coordinate space. As regards the rest of the matrix element (8.91), since the $\hat{a}$'s and $\hat{c}$'s commute the 's^+' and 'e^-' parts are quite independent, and (8.91) reduces to

$$\frac{(-i)^2}{2} \iint d^4x_1\, d^4x_2\, \{\langle s^+, p'|\hat{j}^\mu_{\mathrm{em,s}}(x_1)|s^+, p\rangle\langle 0|T(\hat{A}_\mu(x_1)\hat{A}_\nu(x_2)|0\rangle \times \langle e^-, k', s'|\hat{j}^\nu_{\mathrm{em,e}}(x_2)|e^-, k, s\rangle + (x_1 \leftrightarrow x_2)\}. \tag{8.93}$$

But we know the explicit form of the current matrix elements in (8.93), from (8.27) and (8.55). Inserting these expressions into (8.93), and noting that the term with $x_1 \leftrightarrow x_2$ is identical to the first term, one finds (cf (6.102) and problem 8.11)

$$\mathcal{A}_{e^-s^+} = i(2\pi)^4\delta^4(p + k - p' - k')\mathcal{M}_{e^-s^+} \tag{8.94}$$

where (using the general form (7.119) of the photon propagator)

$$i\mathcal{M}_{e^-s^+} = (-i)^2(e(p+p')^\mu)\left(\frac{i[-g_{\mu\nu} + (1-\xi)q_\mu q_\nu/q^2]}{q^2}\right) \times (-e\bar{u}(k', s')\gamma^\nu u(k, s)) \tag{8.95}$$

$$\equiv (-i)^2 j^\mu_{s^+}(p, p')\left(\frac{i[-g_{\mu\nu} + (1-\xi)q_\mu q_\nu/q^2]}{q^2}\right) j^\nu_{e^-}(k, k') \tag{8.96}$$

and $q = (k - k') = (p' - p)$. We have introduced here the 'momentum–space' currents

$$j^\mu_{s^+}(p, p') = e(p + p')^\mu \tag{8.97}$$

and

$$j^\mu_{e^-}(k, k') = -e\bar{u}(k', s')\gamma^\mu u(k, s) \tag{8.98}$$

shortening the notation by dropping the 'em' suffix, which is understood.

Before proceeding to calculate the cross section, some comments on (8.96) are in order:

Comment (1)

The $j^{\mu}_{s^+}(p, p')$ and $j^{\nu}_{e^-}(k, k')$ in (8.97) and (8.98) are the momentum–space versions of the x-dependent current matrix elements in (8.27) and (8.55); they are, in fact, simply those matrix elements evaluated at $x = 0$. The x-dependent matrix elements (8.27) and (8.55) both satisfy the current conservation equations $\partial_\mu j^\mu(x) = 0$ as is easy to check (problem 8.12). Correspondingly, it follows from (8.97) and (8.98) that we have

$$q_\mu j^{\mu}_{s^+}(p, p') = q_\mu j^{\mu}_{e^-}(k, k') = 0 \tag{8.99}$$

where $q = p' - p = k - k'$, and we have used the mass-shell conditions $p^2 = p'^2 = M^2$, $\not{k}u = mu$, $\not{k}'u' = mu'$; the relations (8.99) are the momentum–space versions of current conservation. The ξ-dependent part of the photon propagator, which is proportional to $q^\mu q^\nu$, therefore vanishes in the matrix element (8.96). This shows that the amplitude is independent of the gauge parameter ξ—in other words, it is *gauge invariant* and proportional simply to

$$j^{\mu}_{s^+} \frac{g_{\mu\nu}}{q^2} j^{\nu}_{e^-}. \tag{8.100}$$

Comment (2)

The amplitude (8.96) has the appealing form of two currents 'hooked together' by the photon propagator. In the form (8.100), it has a simple 'semi-classical' interpretation. Suppose we regard the process $e^- s^+ \to e^- s^+$ as the scattering of the e^-, say, in the field produced by the s^+ (we can see from (8.100) that the answer is going to be symmetrical with respect to whichever of e^- and s^+ is singled out in this way). Then the amplitude will be, as in (8.43),

$$\mathcal{A}_{e^- s^+} = -i \int d^4x \, j^{\nu}_{e^-}(k, k') e^{-i(k-k')\cdot x} A_\nu(x) \tag{8.101}$$

where now the classical field $A_\nu(x)$ is not an 'external' Coulomb field but the field caused by the motion of the s^+. It seems very plausible that this $A_\nu(x)$ should be given by the solution of the Maxwell equations (3.21), with the $j_{\nu\text{em}}(x)$ on the right-hand side given by the transition current (8.11) (with $N = N' = 1$) appropriate to the motion $s^+(p) \to s^+(p')$:

$$\Box A^\nu - \partial^\nu(\partial^\mu A_\mu) = j^{\nu}_{s^+}(x) \tag{8.102}$$

where

$$j^{\nu}_{s^+}(x) = e(p + p')^\nu e^{-i(p-p')\cdot x} \tag{8.103}$$

Equation (8.102) will be much easier to solve if we can decouple the components of A^ν by using the Lorentz condition $\partial^\mu A_\mu = 0$. We are aware of the problems with this condition in the field-theory case (cf section 7.3.2) but we are here

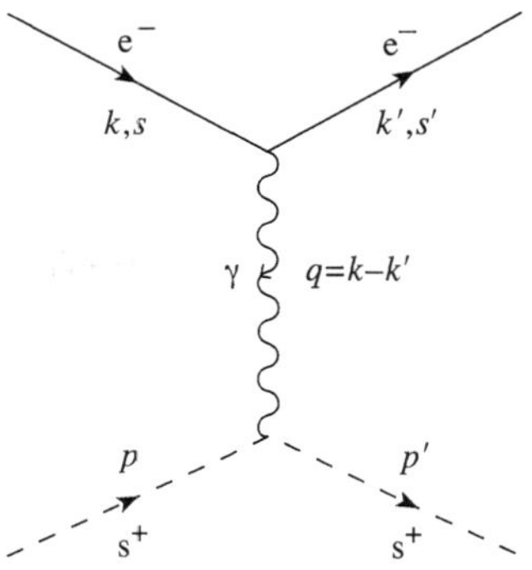

Figure 8.5. Feynman diagram for e$^-$s$^+$ scattering in the one-photon exchange approximation.

treating A^ν classically. Although A^ν is not a free field in (8.102), it is easy to see that we may consistently take $\partial^\mu A_\mu = 0$ provided that the current is conserved, $\partial_\nu j^\nu_{s^+}(x) = 0$, which we know to be the case. Thus we have to solve

$$\Box A^\nu(x) = e(p+p')^\nu \mathrm{e}^{-\mathrm{i}(p-p')\cdot x}. \tag{8.104}$$

Noting that

$$\Box \mathrm{e}^{-\mathrm{i}(p-p')\cdot x} = -(p-p')^2 \mathrm{e}^{-\mathrm{i}(p-p')\cdot x} \tag{8.105}$$

we obtain, by inspection,

$$A^\nu(x) = -\frac{1}{q^2} e(p+p')^\nu \mathrm{e}^{-\mathrm{i}(p-p')\cdot x} \tag{8.106}$$

where $q = p' - p$. Inserting this expression into the amplitude (8.101) we find

$$\mathcal{A}_{\mathrm{e^-s^+}} = \mathrm{i}(2\pi)^4 \delta^4(p+k-p'-k')\mathcal{M}_{\mathrm{e^-s^+}} \tag{8.107}$$

where

$$\mathrm{i}\mathcal{M}_{\mathrm{e^-s^+}} = j^\mu_{\mathrm{s^+}}(p,p') \frac{\mathrm{i}g_{\mu\nu}}{q^2} j^\nu_{\mathrm{e^-}}(k,k') \tag{8.108}$$

exactly as in (8.96) for $\xi = 1$ (the gauge appropriate to '$\partial_\mu A^\mu = 0$').

Comment (3)

From the work of chapter 6, it is clear that we can give a *Feynman graph interpretation* of the amplitude (8.108), as shown in figure 8.5, and set out the corresponding *Feynman rules*:

(i) At a vertex where a photon is emitted or absorbed by an s$^+$ particle, the factor is $-\mathrm{i}e(p+p')^\mu$ where p, p' are the incident and outgoing 4-momenta of the s$^+$; the vertex for s$^-$ has the opposite sign.

(ii) At a vertex where a photon is emitted or absorbed by an e^-, the factor is $ie\gamma^\mu (e > 0)$; for an e^+ it is $-ie\gamma^\mu$. (This and the previous rule arise from associating one '(−i)' factor in (8.93) or (8.96) with each current.)
(iii) For each initial state fermion line a factor $u(k, s)$ and for each final state fermion line a factor $\bar{u}(k', s')$; for each initial state anti-fermion a factor $\bar{v}(k, s)$ and for each final state anti-fermion line a factor $v(k', s')$ (these rules reconstruct the e^+ Coulomb amplitudes of section 8.2.4).
(iv) For an internal photon of 4-momentum q, there is a factor $-ig_{\mu\nu}/q^2$ in the gauge $\xi = 1$.
(v) Multiply these factors together and then multiply the result by an overall 4-momentum-conserving δ-function factor $(2\pi)^4\delta(p' + k' + \cdots - p - k - \cdots)$. This gives the quantity $i\mathcal{M}$.

Comment (4)

We know that our amplitude is proportional to

$$j^\mu_{s+} \frac{g_{\mu\nu}}{q^2} j^\nu_{e^-}. \tag{8.109}$$

Choosing the coordinate system such that $q = (q^0, 0, 0, |\boldsymbol{q}|)$, the current conservation equations $q \cdot j_{s^+} = q \cdot j_{e^-} = 0$ read:

$$j^3 = q^0 j^0 / |\boldsymbol{q}| \tag{8.110}$$

for both currents. Expression (8.100) can then be written as

$$\begin{aligned}(j^1_{s+} j^1_{e-} + j^2_{s+} j^2_{e-})/q^2 + (j^3_{s+} j^3_{e-} - j^0_{s+} j^0_{e-})/q^2 \\ = (j^1_{s+} j^1_{e-} + j^2_{s+} j^2_{e-})/q^2 + j^0_{s+} j^0_{e-}/\boldsymbol{q}^2\end{aligned} \tag{8.111}$$

using (8.110). The first term may be interpreted as being due to the exchange of a transversely polarized photon (only the 1, 2 components enter, perpendicular to $\boldsymbol{q}$). For *real* photons $q^2 \to 0$, so that this term will completely dominate the second. The latter, however, must obviously be included when $q^2 \neq 0$, as of course is the case for this *virtual* γ (cf section 6.3.3). We note that the second term depends on the 3-momentum squared, $\boldsymbol{q}^2$, rather than the 4-momentum squared q^2, and that it involves the charge densities j^0_{s+} and j^0_{e-}. Referring back to section 8.1, we can interpret it as the *instantaneous Coulomb interaction* between these charge densities, since

$$\int d^4x\, e^{iq\cdot x}\delta(t)/r = \int d^3\boldsymbol{x}\, e^{i\boldsymbol{q}\cdot\boldsymbol{x}}/r = 4\pi/\boldsymbol{q}^2. \tag{8.112}$$

Thus, in summary, the single covariant amplitude (8.108) includes contributions from the exchange of transversely polarized photons *and* from the familiar Coulomb potential. This is the true relativistic extension of the static Coulomb results of (8.15) and (8.44).

8.3.2 The cross section for $e^-s^+ \to e^-s^+$

The invariant amplitude $\mathcal{M}_{e^-s^+}(s, s')$ for our process is given by (8.108) as

$$\mathcal{M}_{e^-s^+}(s, s') = e\bar{u}(k', s')\gamma^\mu u(k, s)(-g_{\mu\nu}/q^2)e(p + p')^\nu \tag{8.113}$$

where we have now included the spin dependence of the amplitude $\mathcal{M}_{e^-s^+}$ in the notation. The steps to the cross sections are now exactly as for the spin-0 case (section 6.3.4), as modified by the spin summing and averaging already met in sections 8.2.1 and 8.2.3, particularly the latter. The cross section for the scattering of an electron in spin state s to one in spin state s' is (cf (6.110))

$$\begin{aligned} \mathrm{d}\sigma_{ss'} &= \frac{1}{4E\omega|\boldsymbol{v}|}|\mathcal{M}_{e^-s^+}(s, s')|^2(2\pi)^4\delta^4(k' + p' - k - p) \\ &\quad \times \frac{1}{(2\pi)^3}\frac{\mathrm{d}^3k'}{2\omega'}\frac{\mathrm{d}^3p'}{2E'} \end{aligned} \tag{8.114}$$

where we have defined

$$\begin{aligned} k^\mu &= (\omega, \boldsymbol{k}) \qquad & k'^\mu &= (\omega', \boldsymbol{k}') \\ p^\mu &= (E, \boldsymbol{p}) \qquad & p'^\mu &= (E', \boldsymbol{p}'). \end{aligned} \tag{8.115}$$

For the unpolarized cross section we are required, as in (8.46), to evaluate the quantity

$$\begin{aligned} \frac{1}{2}\sum_{s,s'}|\mathcal{M}_{e^-s^+}(s, s')|^2 &= \left(\frac{e^2}{q^2}\right)^2 \frac{1}{2}\sum_{s,s'}\bar{u}(k', s')\gamma^\mu u(k, s)\bar{u}(k, s)\gamma^\nu u(k', s') \\ &\quad \times (p + p')_\mu(p + p')_\nu \end{aligned} \tag{8.116}$$

$$\equiv \left(\frac{e^2}{q^2}\right)^2 L^{\mu\nu}(k, k')T_{\mu\nu}(p, p') \tag{8.117}$$

where the boson tensor $T_{\mu\nu}$ is just $(p + p')_\mu(p + p')_\nu$ and the lepton tensor $L^{\mu\nu}$ has been evaluated in (8.78). Using $q^2 = (k - k')^2 = 2m^2 - 2k \cdot k'$, the expression (8.78) can be rewritten as

$$\boxed{L^{\mu\nu}(k, k') = 2[k'^\mu k^\nu + k'^\nu k^\mu + (q^2/2)g^{\mu\nu}].} \tag{8.118}$$

We then find (problem 8.13)

$$L^{\mu\nu}T_{\mu\nu} = 8[2(p \cdot k)(p \cdot k') + (q^2/2)M^2] \tag{8.119}$$

since $k' \cdot p' = k \cdot p$ and $k \cdot p' = k' \cdot p$ from 4-momentum conservation, and $p^2 = p'^2 = M^2$ (we are using m for the e^- mass and M for the s^+ mass).

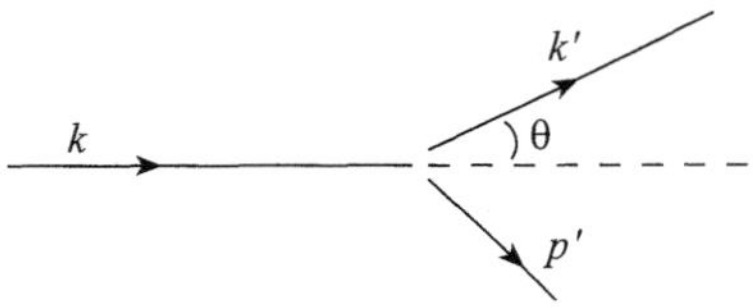

Figure 8.6. Two-body scattering in the 'laboratory' frame.

We can now give the differential cross section in the CM frame by taking over the formula (6.129) with

$$|\mathcal{M}|^2 \to \tfrac{1}{2}\sum_{s,s'} |\mathcal{M}_{\mathrm{e^-s^+}}(s,s')|^2$$

so as to obtain

$$\left(\frac{\mathrm{d}\bar{\sigma}}{\mathrm{d}\Omega}\right)_{\mathrm{CM}} = \frac{2\alpha^2}{W^2(q^2)^2}[2(p\cdot k)(p\cdot k') + (q^2/2)M^2] \tag{8.120}$$

where $\alpha = e^2/4\pi$ and $W^2 = (k+p)^2$. A somewhat more physically meaningful formula is found if we ask for the cross section in the 'laboratory' frame which we define by the condition $p^\mu = (M, \mathbf{0})$. The evaluation of the phase space integral requires some care and this is detailed in appendix K. The result is

$$\frac{\mathrm{d}\bar{\sigma}}{\mathrm{d}\Omega} = \frac{\alpha^2}{4k^2\sin^4(\theta/2)}\cos^2(\theta/2)\frac{k'}{k}. \tag{8.121}$$

In this formula we have neglected the electron mass in the kinematics so that

$$k \equiv |\boldsymbol{k}| = \omega \tag{8.122}$$

$$k' \equiv |\boldsymbol{k}'| = \omega' \tag{8.123}$$

and

$$q^2 = -4kk'\sin^2(\theta/2) \tag{8.124}$$

where θ is the electron scattering angle in this frame, as shown in figure 8.6, and

$$(k/k') = 1 + (2k/M)\sin^2(\theta/2) \tag{8.125}$$

from equation (K.20). Note that there is a slight abuse of notation here: in the context of results for such laboratory frame calculations, 'k' and 'k'' are not 4-vectors, but rather the moduli of 3-vectors, as defined in equations (8.122) and (8.123).

We shall denote the cross section (8.121) by

$$\left(\frac{\mathrm{d}\sigma}{\mathrm{d}\Omega}\right)_{\mathrm{ns}} \qquad \text{'no-structure' cross section.} \tag{8.126}$$

It describes essentially the 'kinematics' of a relativistic electron scattering from a target which recoils. Comparing the result (8.121) with equation (8.49), and remembering that here $Z = 1$ and we are taking $v \rightarrow 1$ for the electron, we see that the effect of recoil is contained in the factor (k'/k), in this limit. We recover the 'no-recoil' result (8.49) in the limit $M \rightarrow \infty$, as expected. In particular, referring to (8.124), we understand Rutherford's '$\sin^{-4}\theta/2$' factor in terms of the exchange of a massless quantum, via the propagator factor $(1/q^2)^2$.

This 'no-structure' cross section also occurs in the cross section for the scattering of electrons by protons or muons: the appellation 'no-structure' will be made clearer in the discussion of form factors which follows. As in the case of e^+ Coulomb scattering, the cross sections for e^-s^+ and for e^+s^+ scattering are identical at this (lowest) order of perturbation theory.

8.4 Scattering from a non-point-like object: the pion form factor in $e^-\pi^+ \rightarrow e^-\pi^+$

As remarked earlier, we have been careful not to call the 's^+' particle a π^+, because the latter is a composite system which cannot be expected to have point-like interactions with the electromagnetic field, as has been assumed for the s^+; rather, in the case of the π^+ it is the quark constituents which interact locally with the electromagnetic field. The quarks also, of course, interact *strongly* with each other via the interactions of QCD, and since these are strong they cannot (in this case) be treated perturbatively. Indeed, a full understanding of the electromagnetically-probed 'structure' of hadrons has not yet been achieved. Instead, we must describe the e^- scattering from physical π^+'s in terms of a phenomenological quantity—the *pion form-factor*—which encapsulates in a relativistically invariant manner the 'non-point-like' aspect of the hadronic state π^+.

The physical process is

$$e^-(k, s) + \pi^+(p) \rightarrow e^-(k', s') + \pi^+(p') \tag{8.127}$$

which we represent, in general, by figure 8.7. To lowest order in α, the amplitude is represented diagrammatically by a generalization of figure 8.5, shown in figure 8.8, in which the point-like ssγ vertex is replaced by the $\pi\pi\gamma$ 'blob', which signifies all the unknown strong interaction corrections.

8.4.1 Rutherford scattering from a charge distribution

It is helpful to begin the discussion by returning to e^- Coulomb scattering again, but this time let us consider the case in which the potential $A^0(\boldsymbol{x})$ corresponds, not to a point charge, but to a spread-out charge density $\rho(\boldsymbol{x})$. Then $A^0(\boldsymbol{x})$ satisfies Poisson's equation

$$\nabla^2 A^0(\boldsymbol{x}) = -Ze\rho(\boldsymbol{x}). \tag{8.128}$$

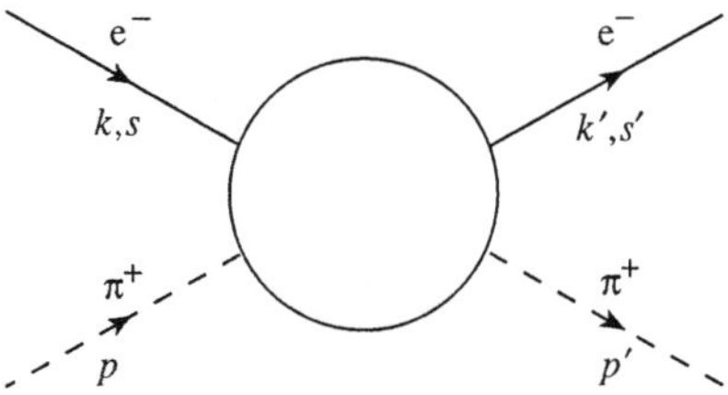

Figure 8.7. $e^-\pi^+$ scattering amplitude.

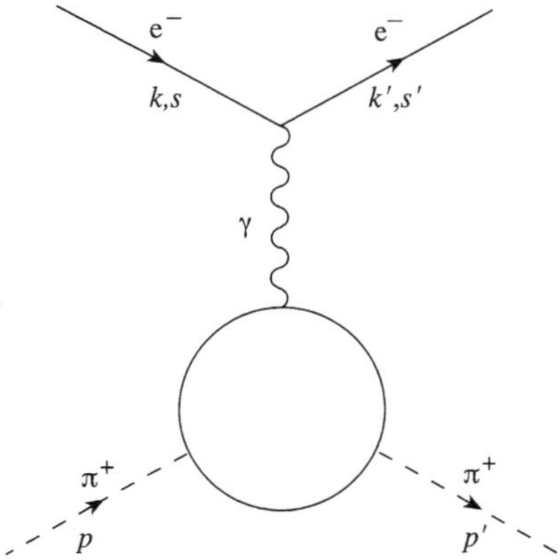

Figure 8.8. One-photon exchange amplitude in $e^-\pi^+$ scattering, including hadronic corrections at the $\pi\pi\gamma$ vertex.

Note that if $A^0(\boldsymbol{x}) = Ze/4\pi|\boldsymbol{x}|$ as in (8.13) then $\rho(\boldsymbol{x}) = \delta(\boldsymbol{x})$ (see appendix G) and we recover the point-like source. The calculation of the Coulomb matrix element will proceed as before, except that now we require, at equation (8.14), the Fourier transform

$$\tilde{A}^0(\boldsymbol{q}) = \int e^{-i\boldsymbol{q}\cdot\boldsymbol{x}} A^0(\boldsymbol{x}) d^3x \tag{8.129}$$

where $q = \boldsymbol{k} - \boldsymbol{k}'$. To evaluate (8.129), note first that from the definition of $A^0(\boldsymbol{x})$, we can write

$$\begin{aligned}\int e^{-i\boldsymbol{q}\cdot\boldsymbol{x}} \nabla^2 A^0(\boldsymbol{x})\, d^3\boldsymbol{x} &= -Ze\int e^{-i\boldsymbol{q}\cdot\boldsymbol{x}} \rho(\boldsymbol{x})\, d^3\boldsymbol{x} \\ &\equiv -ZeF(\boldsymbol{q})\end{aligned} \tag{8.130}$$

where the (static) form factor $F(\boldsymbol{q})$ has been introduced, the Fourier transform of $\rho(\boldsymbol{x})$, satisfying

$$F(0) = \int \rho(\boldsymbol{x})\, d^3\boldsymbol{x} = 1. \tag{8.131}$$

Condition (8.131) simply means that the total charge is Ze. The left-hand side of (8.130) can be transformed by two (three-dimensional) partial integrations to give

$$\int (\nabla^2 \mathrm{e}^{-\mathrm{i}\boldsymbol{q}\cdot\boldsymbol{x}}) A^0(\boldsymbol{x})\,\mathrm{d}^3\boldsymbol{x} = -\boldsymbol{q}^2 \int \mathrm{e}^{-\mathrm{i}\boldsymbol{q}\cdot\boldsymbol{x}} A^0(\boldsymbol{x})\,\mathrm{d}^3\boldsymbol{x}. \tag{8.132}$$

Using this result in (8.130), we find

$$\tilde{A}^0(\boldsymbol{q}) = \frac{F(\boldsymbol{q})}{\boldsymbol{q}^2} Ze. \tag{8.133}$$

Thus referring to equation (8.15) for example, the net result of the non-point-like charge distribution is to multiply the 'point-like' amplitude $Ze^2/\boldsymbol{q}^2$ by the form factor $F(\boldsymbol{q})$ which in this simple static case has the interpretation of the Fourier transform of the charge distribution. So, for this (infinitely heavy π^+ case), the 'blob' in figure 8.8 would be represented by $F(\boldsymbol{q})$.

To gain some idea of what $F(\boldsymbol{q}^2)$ might look like, consider a simple exponential shape for $\rho(\boldsymbol{x})$:

$$\rho(\boldsymbol{x}) = \frac{1}{(8\pi a^3)^{1/2}} \mathrm{e}^{-|\boldsymbol{x}|/a} \tag{8.134}$$

which has been normalized according to (8.131). Then $F(\boldsymbol{q}^2)$ is (problem 8.14)

$$F(\boldsymbol{q}^2) = \frac{1}{(\boldsymbol{q}^2 a^2 + 1)^2}. \tag{8.135}$$

We see that $F(\boldsymbol{q}^2)$ decreases smoothly away from unity at $\boldsymbol{q}^2 = 0$. The characteristic scale of the fall-off in $|\boldsymbol{q}|$ is $\sim a^{-1}$ from (8.135), which, as expected from Fourier transform theory, is the reciprocal of the spatial fall-off, which is approximately a from (8.134). Since $\boldsymbol{q}^2 = 4\boldsymbol{k}^2 \sin^2\theta/2$, a larger $\boldsymbol{q}^2$ means a larger θ: hence, in scattering from an extended charge distribution, the cross section at larger angles will drop below the point-like value. This is, of course, how Rutherford deduced that the nucleus had a spatial extension (see figure 1.4).

We now seek a Lorentz-invariant generalization of this static form factor. In the absence of a fundamental understanding of the π^+ structure coming from QCD, we shall rely on Lorentz invariance and electromagnetic current conservation (one aspect of gauge invariance) to restrict the general form of the $\pi\pi\gamma$ vertex shown in figure 8.8. The use of invariance arguments to place restrictions on the form of amplitudes is an extremely general and important tool, in the absence of a complete theory.

8.4.2 Lorentz invariance

First, consider Lorentz invariance. We seek to generalize the point-like ssγ vertex (cf (8.97) and comment (1) after (8.98))

$$j^{\mu}_{\mathrm{s}^+}(p, p') = \langle \mathrm{s}^+, p' | \hat{j}^{\mu}_{\mathrm{em,s}}(0) | \mathrm{s}^+, p\rangle = e(p + p')^{\mu} \tag{8.136}$$

to $j^{\mu}_{\pi^+}(p, p')$, which will include strong interaction effects. Whatever these effects are, they cannot destroy the 4-vector character of the current. To construct the general form of $j^{\mu}_{\pi^+}(p, p')$ therefore, we must first enumerate the independent momentum 4-vectors we have at our disposal to parametrize the 4-vector nature of the current. These are just

$$p \qquad p' \qquad \text{and} \qquad q \tag{8.137}$$

subject to the condition

$$p' = p + q. \tag{8.138}$$

There are two independent combinations; these we can choose to be the linear combinations

$$(p' + p)_{\mu} \tag{8.139}$$

and

$$(p' - p)_{\mu} = q_{\mu}. \tag{8.140}$$

Both of these 4-vectors can, in general, parametrize the 4-vector nature of the electromagnetic current of a real pion. Moreover, they can be multiplied by an unknown scalar function of the available Lorentz scalar products for this process. Since

$$p^2 = p'^2 = M^2 \tag{8.141}$$

and

$$q^2 = 2M^2 - 2p \cdot p' \tag{8.142}$$

there is only one independent scalar in the problem, which we may take to be q^2, the 4-momentum transfer to the vertex. Thus, from Lorentz invariance, we are led to write the electromagnetic vertex of a pion in the form

$$j^{\mu}_{\pi^+}(p, p') = \langle \pi^+, p' | \hat{j}^{\mu}_{\text{em},\pi}(0) | \pi^+, p \rangle = e[F(q^2)(p' + p)^{\mu} + G(q^2) q^{\mu}]. \tag{8.143}$$

The functions F and G are called 'form factors'.

This is as far as Lorentz invariance can take us. To identify the pion form factor, we must consider our second symmetry principle, gauge invariance—in the form of current conservation.

8.4.3 Current conservation

The Maxwell equations (7.62) reduce, in the Lorentz gauge

$$\partial_{\mu} A^{\mu} = 0 \tag{8.144}$$

to the simple form

$$\Box A^{\mu} = j^{\mu} \tag{8.145}$$

and the gauge condition is consistent with the familiar current conservation condition

$$\partial_\mu j^\mu = 0. \tag{8.146}$$

As we have seen in (8.99), the current conservation condition is equivalent to the condition

$$q_\mu \langle \pi^+(p')|\hat{j}^\mu_{\mathrm{em},\pi}(0)|\pi^+(p)\rangle = 0 \tag{8.147}$$

on the pion electromagnetic vertex.

In the case of the point-like s^+ this is clearly satisfied since

$$q \cdot (p' + p) = 0 \tag{8.148}$$

with the aid of (8.141). In the general case we obtain the condition

$$q_\mu[F(q^2)(p' + p)^\mu + G(q^2)q^\mu] = 0. \tag{8.149}$$

The first term vanishes as before, but $q^2 \neq 0$ in general, and we therefore conclude that current conservation implies that

$$G(q^2) = 0. \tag{8.150}$$

In other words, all the virtual strong interaction effects at the $\pi^+\pi^+\gamma$ vertex are described by one scalar function of the virtual photon's squared 4-momentum:

$e(p' + p)^\mu$	$\rightarrow$	$eF(q^2)(p' + p)^\mu.$
'point pion'		'real pion'

(8.151)

$F(q^2)$ is the *electromagnetic form factor of the pion*, which generalizes the static form factor $F(\boldsymbol{q}^2)$ of section 8.4.1. The pion electromagnetic vertex is then

$$j^\mu_{\pi^+}(p, p') = eF(q^2)(p + p')^\mu. \tag{8.152}$$

The electric charge is defined to be the coupling at zero momentum transfer, so the form factor is normalized by the condition (cf (8.131))

$$F(0) = 1. \tag{8.153}$$

To lowest order in α, the invariant amplitude for $e^-\pi^+ \rightarrow e^-\pi^+$ is therefore given by replacing $j^\mu_{s^+}(p, p')$ in (8.96) or (8.108) by $j^\mu_{\pi^+}(p, p')$:

$$\mathrm{i}\mathcal{M}_{e^-\pi^+} = -\mathrm{i}e(p + p')^\mu F((p' - p)^2)\left(\frac{-\mathrm{i}g_{\mu\nu}}{(p' - p)^2}\right)[+\mathrm{i}e\bar{u}(k', s')\gamma_\nu u(k, s)]. \tag{8.154}$$

It is clear that the effect of the pion structure is simply to multiply the 'no-structure' cross section (8.121) by the square of the form factor, $F(q^2 = (p' - p)^2)$.

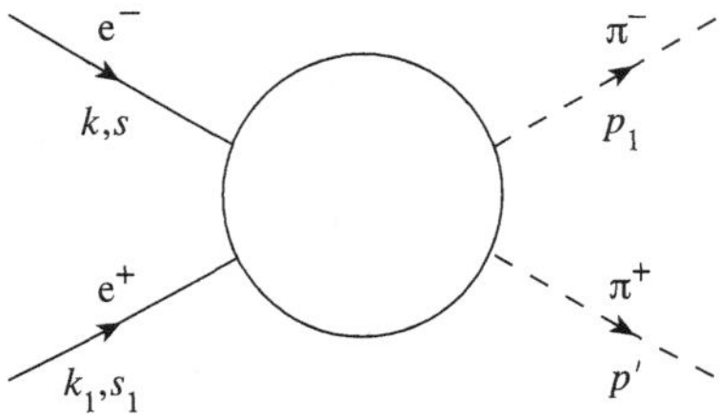

Figure 8.9. $e^+e^- \to \pi^+\pi^-$ scattering amplitude.

For $e^-\pi^+ \to e^-\pi^+$ in the CM frame we may take $p = (E, \boldsymbol{p})$ and $p' = (E, \boldsymbol{p}')$ with $|\boldsymbol{p}| = |\boldsymbol{p}'|$ and $E = (m_\pi^2 + \boldsymbol{p}^2)^{1/2}$. Then

$$q^2 = (p' - p)^2 = -4\boldsymbol{p}^2 \sin^2\theta/2 \tag{8.155}$$

as in section 8.1, where θ is now the CM scattering angle between $\boldsymbol{p}$ and $\boldsymbol{p}'$. Hence $F(q^2)$ can be probed for negative (space-like) values of q^2, in the process $e^-\pi^+ \to e^-\pi^+$. As in the static case, we expect the form factor to fall off as $-q^2$ increases since, roughly speaking, it represents the amplitude for the target to remain intact when probed by the electromagnetic current. As $-q^2$ increases, the amplitudes of inelastic processes which involve the creation of extra particles become greater, and the elastic amplitude is correspondingly reduced. We shall consider inelastic scattering in the following chapter.

Interestingly, $F(q^2)$ may also be measured at positive (time-like) q^2, in the related reaction $e^+e^- \to \pi^+\pi^-$ as we now discuss.

8.5 The form factor in the time-like region: $e^+e^- \to \pi^+\pi^-$ and crossing symmetry

The physical process is

$$e^+(k_1, s_1) + e^-(k, s) \to \pi^+(p') + \pi^-(p_1) \tag{8.156}$$

as shown in figure 8.9. We can use this as an instructive exercise in the Feynman interpretation of section 4.5.4. From that section, we know that the invariant amplitude for (8.156) is equal to minus the amplitude for a process in which the ingoing anti-particle e^+ with (k_1, s_1) becomes an outgoing particle e^- with $(-k_1, -s_1)$, and the outgoing anti-particle π^- with p_1 becomes an ingoing particle π^+ with $-p_1$. In this way the 'physical' (positive 4-momentum) anti-particle states (e^+ and π^-) are replaced by appropriate 'unphysical' (negative 4-momentum) particle states (e^- and π^+). These changes transform figure 8.9 to figure 8.10.

If we now look at figure 8.10 'from the top downwards' (instead of from left to right—remember that Feynman diagrams are *not* in coordinate space!), we see

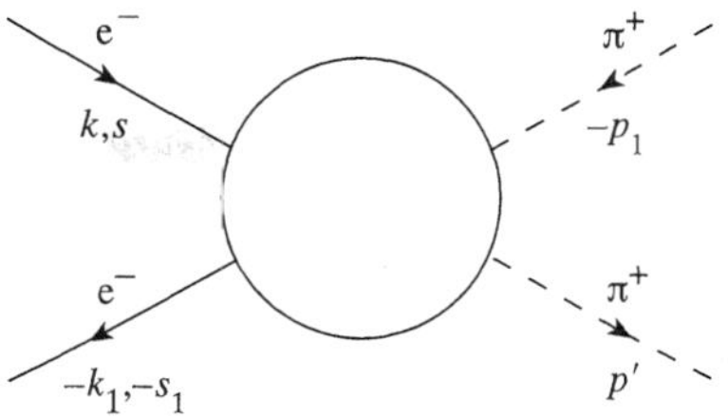

Figure 8.10. The amplitude of figure 8.9, with positive 4-momentum anti-particles replaced by negative 4-momentum particles.

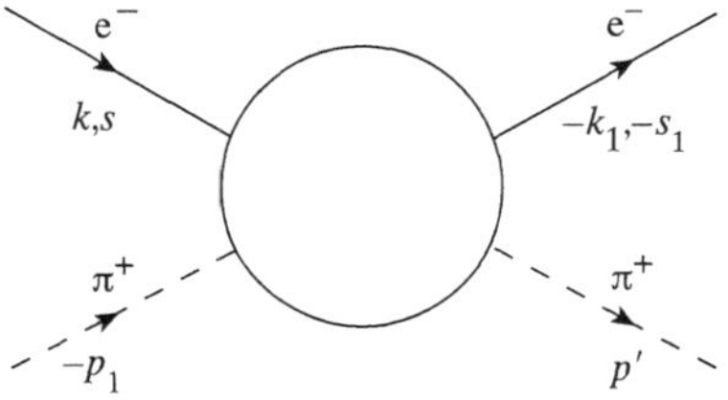

Figure 8.11. The amplitude of figure 8.10 redrawn so as to obtain a reaction in which the initial state has only 'ingoing' lines and the final state has only 'outgoing' lines.

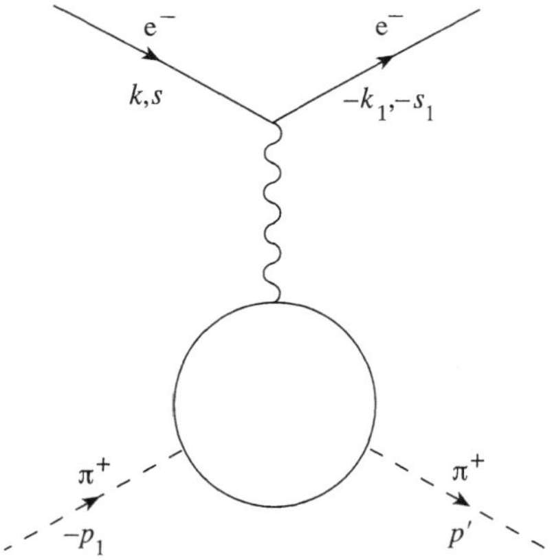

Figure 8.12. One-photon exchange amplitude for the process of figure 8.11.

a process of $e^-\pi^+$ scattering, namely

$$e^-(k, s) + \pi^+(-p_1) \to e^-(-k_1, -s_1) + \pi^+(p'). \tag{8.157}$$

But (8.157) is something we have already calculated! (Though we shall have to substitute a negative-energy spinor v for a positive energy one u.) In fact, let us redraw figure 8.10 as figure 8.11 to make it look more like figure 8.7.

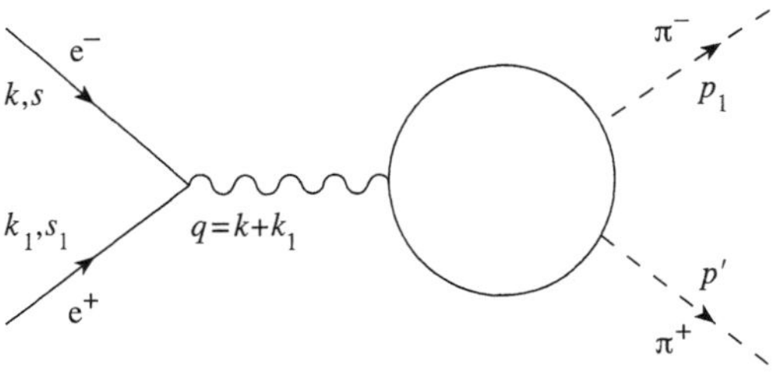

Figure 8.13. One-photon exchange amplitude for the process of figure 8.9.

Then, to lowest order in α, the amplitude for figure 8.11 is shown in figure 8.12 (compare figure 8.8). To obtain the corresponding mathematical expression for the amplitude $\mathrm{i}\mathcal{M}_{\mathrm{e}^+\mathrm{e}^-\to\pi^+\pi^-}$, we simply need to modify (8.154): (a) by inserting a minus sign; (b) by replacing p by $-p_1$ and k' by $-k_1$ as in figure 8.12; and (c) by replacing $\bar{u}(k', s')$ by $\bar{v}(k_1, s_1)$. This yields the invariant amplitude for figure 8.12 as

$$\mathrm{i}\mathcal{M}_{\mathrm{e}^+\mathrm{e}^-\to\pi^+\pi^-} = -\mathrm{i}e(-p_1 + p')^\mu F((p_1 + p')^2)\left(\frac{-\mathrm{i}g_{\mu\nu}}{(p_1 + p')^2}\right) \times [-\mathrm{i}e\bar{v}(k_1, s_1)\gamma_\nu u(k, s)] \tag{8.158}$$

which is represented by the Feynman diagram of figure 8.13 for the *original* process of (8.156) and figure 8.9.

In the language introduced in section 6.3.3, figure 8.13 is an 's-channel process' ($s = (k+k_1)^2 = (p_1+p')^2$) for $\mathrm{e}^+\mathrm{e}^- \to \pi^+\pi^-$, whereas figure 8.8 is a 't-channel process' ($t = (k-k')^2 = (p'-p)^2$) for $\mathrm{e}^-\pi^+ \to \mathrm{e}^-\pi^+$. However, we have seen that the amplitude for the $\mathrm{e}^+\mathrm{e}^- \to \pi^+\pi^-$ process can be obtained from the $\mathrm{e}^-\pi^+ \to \mathrm{e}^-\pi^+$ amplitude by making the replacement $k' \to -k_1$, $p \to -p_1$ (together with the sign, and $\bar{u} \to \bar{v}$). Under these replacements of the 4-momenta, the variable $t = (k - k')^2 = (p - p')^2$ of figure 8.8 becomes the variable $s = (k+k_1)^2 = (p_1+p')^2$ of figure 8.13. In particular, as is evident in the formula (8.158), the *same* form factor F is a function of the invariant $s = (p_1 + p')^2$ in process (8.156), and of $t = (p - p')^2$ in process (8.127). The interesting thing is that whereas (as we have seen) 't' is negative in process (8.127), 's' for process (8.156) is the square of the total CM energy, which is $\geq 4M^2$ where M is the pion mass ($2M$ is the threshold energy for the reaction to proceed in the CM system). Thus the form factor can be probed at negative values of its argument in the process $\mathrm{e}^-\pi^+ \to \mathrm{e}^-\pi^+$, and at positive values $\geq 4M^2$ in the process $\mathrm{e}^+\mathrm{e}^- \to \pi^+\pi^-$. In the next chapter (section 9.5) we shall see how, in the latter process, meson resonances dominate $F(s)$.

The procedure whereby an ingoing/outgoing anti-particle is switched to an outgoing/ingoing particle is called 'crossing' (the state is being 'crossed' from one side of the reaction to the other). By an extension of this language, $\mathrm{e}^+\mathrm{e}^- \to \pi^+\pi^-$ is called the crossed process relative to $\mathrm{e}^-\pi^+ \to \mathrm{e}^-\pi^+$ (or vice versa). The fact that the amplitude for a given process and its 'crossed' analogue are directly related via the Feynman interpretation (or by quantum field theory!) is called 'crossing symmetry'. In the example studied here, what is an s-channel process for one reaction becomes a t-channel process for the crossed reaction. Essentially, little more is involved than looking in the one case from left to right and, in the other, from top to bottom!

8.6 Electron Compton scattering

8.6.1 The lowest-order amplitudes

We proceed to explore some other elementary electromagnetic processes. So far we have not considered a reaction with external photons, so let us now discuss electron Compton scattering

$$\gamma(k,\lambda) + \mathrm{e}^-(p,s) \rightarrow \gamma(k',\lambda') + \mathrm{e}^-(p',s') \tag{8.159}$$

where the λ's stand for the polarizations of the photons. Since only the γ's and e^-'s are involved, the interaction Hamiltonian is simply $\hat{H}'_{\mathrm{D}}$, and it is clear that this must act at least twice in the reaction (8.159). By following the method of section 6.3.2 one can formally derive what we are here going to assume is by now obvious, which is that to order e^2 (i.e. α in the amplitude) there are two contributing Feynman graphs, as shown in figures 8.14(a) and (b). The first is an s-channel process, the second a u-channel process. We already know the factors for the vertices and for the external electron lines; we need to know the factors for the internal electron lines (propagators) and the external photon lines. The fermion propagator was given in section 7.2 and is $\mathrm{i}/(\not{q} - m + \mathrm{i}\epsilon)$ for a line carrying 4-momentum q. As regards the 'external-γ' factor, this will arise from contractions of the form (cf (6.90))

$$\sqrt{2E_{k'}}\langle 0|\alpha(k',\lambda')\hat{A}^\mu(x_1)|0\rangle = \epsilon^{\mu *}(k',\lambda')\mathrm{e}^{\mathrm{i}k'\cdot x_1} \tag{8.160}$$

where the evaluation of the vev has used the mode expansion (7.101) and the commutation relations (7.105), as usual; note, however, that only *transverse* polarization states ($\lambda, \lambda' = 1$ and 2) enter in the external (physical) photon lines in figures 8.14(a) and (b).

Thus we add two more rules to the (i)–(v) of section 8.3.1:

(vi) For an incoming photon of 4-momentum k and polarization λ, there is a factor $\epsilon^\mu(k,\lambda)$; for an outgoing one, $\epsilon^{\mu *}(k',\lambda')$.

(vii) For an internal spin-$\frac{1}{2}$ particle carrying 4-momentum q, there is a factor $\mathrm{i}/(\not{q} - m + \mathrm{i}\epsilon) = \mathrm{i}(\not{q} + m)/(q^2 - m^2 + \mathrm{i}\epsilon)$.

The invariant amplitude $\mathcal{M}_{\gamma\mathrm{e}^-}$ corresponding to figures 8.14(a) and (b) is therefore

$$\mathcal{M}_{\gamma\mathrm{e}^-} = -e^2\epsilon^*_\nu(k',\lambda')\epsilon_\mu(k,\lambda)\bar{u}(p',s')\gamma^\nu\frac{(\not{p} + \not{k} + m)}{(p+k)^2 - m^2}\gamma^\mu u(p,s)$$

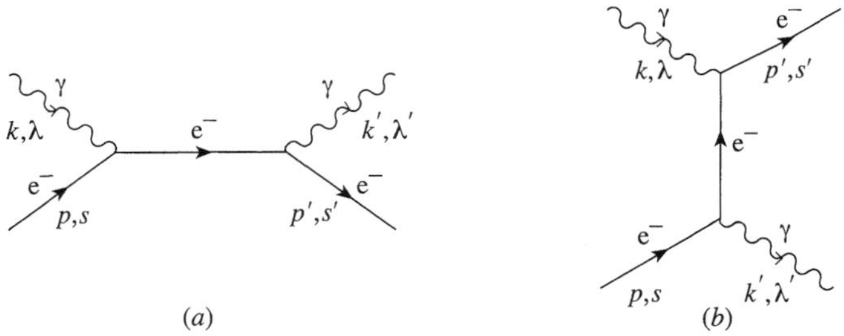

Figure 8.14. $O(e^2)$ contributions to electron Compton scattering.

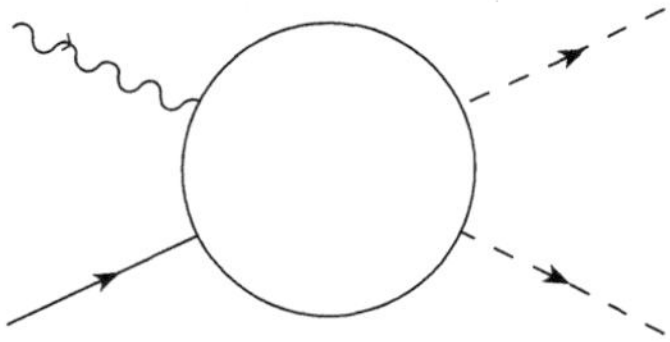

Figure 8.15. General one-photon process.

$$-e^2\epsilon_\nu^*(k',\lambda')\epsilon_\mu(k,\lambda)\bar{u}(p',s')\gamma^\mu\frac{(\not{p}+\not{k}+m)}{(p-k')^2-m^2}\gamma^\nu u(p',s'). \tag{8.161}$$

To get the spinor factors in expression such as these, the rule is to start at the ingoing fermion line ('$u(p,s)$') and follow the line through until the end, inserting vertices and propagators in the right order, until you reach the outgoing state ('$\bar{u}$'). Note that here $s=(p+k)^2$ and $u=(p-k')^2$.

8.6.2 Gauge invariance

We learned in section 7.3.1 that the gauge symmetry ($A^\mu \rightarrow A^\mu - \partial^\mu\chi$) of electromagnetism, as applied to real free photons, implied that any photon polarization vector $\epsilon^\mu(k,\lambda)$ could be replaced by

$$\epsilon'^\mu(k\lambda) = \epsilon^\mu(k,\lambda) + \beta k^\mu \tag{8.162}$$

where β is an arbitrary constant. Such a transformation amounted to a change of gauge, always remaining within the Lorentz gauge for which $\epsilon\cdot k = \epsilon'\cdot k = 0$. Thus our amplitude (8.161) must be unchanged if we make either or both the replacements $\epsilon \rightarrow \epsilon + \beta k$ and $\epsilon^* \rightarrow \epsilon^* + \beta k'$ indicated in (8.162). This means that if in (8.161) we replace either or both of $\epsilon_\mu(k,\lambda)$ and $\epsilon_\nu^*(k',\lambda')$ by k_μ and k'_ν, respectively, the result has to be zero. This can indeed be verified (problem 8.15).

A similar result is generally true and very important. Consider a process, shown in figure 8.15, involving a photon of momentum k^μ, whose polarization state is described by the vector ϵ^μ. The amplitude $\mathcal{A}_\gamma$ for this process must be linear in the photon polarization vector and thus we may write

$$\mathcal{A}_\gamma = \epsilon^\mu T_\mu \tag{8.163}$$

where T_μ depends on the particular process under consideration. With the Lorentz choice for ϵ^μ we have

$$k \cdot \epsilon = 0. \tag{8.164}$$

But gauge invariance implies that if we replace ϵ^μ in (8.163) by k^μ we must get zero:

$$\boxed{k^\mu T_\mu = 0.} \tag{8.165}$$

This important condition on T_μ is known as a *Ward identity* (Ward 1950).

8.6.3 The Compton cross section

The calculation of the cross section is of considerable interest, since it is required when considering lowest-order QCD corrections to the parton model for deep inelastic scattering of leptons from nucleons (see the following chapter and volume 2). We must average $|\mathcal{M}_{\gamma e^-}|^2$ over initial electron spins and photon polarizations and sum over final ones. Consider first the s-channel process of figure 8.14(a), with amplitude $\mathcal{M}^{(s)}_{\gamma e^-}$. For this contribution we must evaluate

$$\frac{e^4}{4(s-m^2)^2} \cdot \sum_{\lambda,\lambda',s,s'} \epsilon'^*_\nu \epsilon_\mu \epsilon^*_\rho \epsilon'_\sigma \bar{u}' \gamma^\nu (\not{p}+\not{k}+m)\gamma^\mu u \bar{u} \gamma^\rho (\not{p}+\not{k}+m)\gamma^\sigma u' \tag{8.166}$$

where we have shortened the notation in an obvious way and introduced the invariant Mandelstam variable (section 6.3.3) $s = (p+k)^2$. We know how to write the spin sums in a convenient form, as a trace. We need to find a similar trick for the polarization sum.

Consider the general 'one-photon' process shown in figure 8.15, with amplitude $\mathcal{A}_\gamma = \epsilon^\mu(k,\lambda)T_\mu$, where $\epsilon^\mu(k,1) = (0,1,0,0)$ and $\epsilon^\mu(k,2) = (0,0,1,0)$, and $k^\mu = (k,0,0,k)$. Then the required polarization sum would be

$$\sum_{\lambda=1,2} \epsilon^\mu(k,\lambda)T_\mu \epsilon^{\nu*}(k,\lambda)T^*_\nu = |T_1|^2 + |T_2|^2. \tag{8.167}$$

However, we also know that $k^\mu T_\mu = 0$ from the Ward identity (8.165). This tells us that

$$kT_0 - kT_3 = 0 \tag{8.168}$$

and hence $T_0 = T_3$. It follows that we may write (8.167) as

$$\sum_{\lambda=1,2} \epsilon^\mu(k,\lambda)\epsilon^{\nu*}(k,\lambda)T_\mu T^*_\nu = |T_1|^2 + |T_2|^2 + |T_3|^2 - |T_0|^2 \tag{8.169}$$

$$= -g^{\mu\nu}T_\mu T^*_\nu. \tag{8.170}$$

Thus we may replace the non-covariant expression '$\sum_{\lambda=1,2}\epsilon^{\mu}(k,\lambda)\epsilon^{\nu *}(k,\lambda)$' by the covariant one '$-g^{\mu\nu}$'. The reader may here recall equation (7.115), where the 'pseudo-completeness' relation involving all *four* ϵ's was given, a similarly covariant expression. This relation corresponds exactly to the right-hand side of (8.169), which (in these terms) shows that the $\lambda = 0$ state enters with negative norm.

Using this result, the term (8.166) becomes

$$\frac{e^4}{4(s-m^2)^2}\sum_{s,s'}\bar{u}'\gamma^{\nu}(\not{p}+\not{k}+m)\gamma^{\mu}u\bar{u}\gamma_{\mu}(\not{p}+\not{k}+m)\gamma_{\nu}u'$$
$$= \frac{e^4}{4(s-m^2)^2}\operatorname{Tr}[\gamma_{\nu}(\not{p}'+m)\gamma^{\nu}(\not{p}+\not{k}+m)\gamma^{\mu}(\not{p}+m)\gamma_{\mu}(\not{p}+\not{k}+m)] \tag{8.171}$$

where, in the second step, we have moved the γ_{ν} to the front of the trace, using (8.70). Expression (8.171) involves the trace of eight γ matrices, which is beyond the power of the machinery given so far. However, it simplifies greatly if we neglect the electron mass—that is, if we are interested in the high-energy limit, as we shall be in parton model applications. In that case, (8.171) becomes

$$\frac{e^4}{4s^2}\operatorname{Tr}[\gamma_{\nu}\not{p}'\gamma^{\nu}(\not{p}+\not{k})\gamma^{\mu}\not{p}\gamma_{\mu}(\not{p}+\not{k})] \tag{8.172}$$

which we can simplify using the result (J.3) to

$$\frac{e^4}{s^2}\operatorname{Tr}[\not{p}'(\not{p}+\not{k})\not{p}(\not{p}+\not{k})] \tag{8.173}$$
$$= \frac{e^4}{s^2}\operatorname{Tr}[\not{p}'\not{k}\not{p}\not{k}] \qquad \text{using } \not{p}^2 = p^2 = 0 \tag{8.174}$$
$$= \frac{4e^4}{s^2}\cdot 2(p'\cdot k)(p\cdot k) \qquad \text{using (8.75) and } k^2 = 0 \tag{8.175}$$
$$= -2e^4 u/s \tag{8.176}$$

where $u = (p-k')^2$. Problem 8.16 finishes the calculation, with the result that the spin-averaged squared amplitude is

$$\frac{1}{4}\sum_{s,s',\lambda,\lambda'}|\mathcal{M}_{\gamma e^-}|^2 = -2e^4\left(\frac{u}{s}+\frac{s}{u}\right). \tag{8.177}$$

The cross section in the CMS is then (cf (6.129))

$$\frac{d\sigma}{d(\cos\theta)} = \frac{2\pi 2e^4}{64\pi^2 s}\left(\frac{-u}{s}-\frac{s}{u}\right) = \frac{\pi\alpha^2}{s}\left(\frac{-u}{s}-\frac{s}{u}\right). \tag{8.178}$$

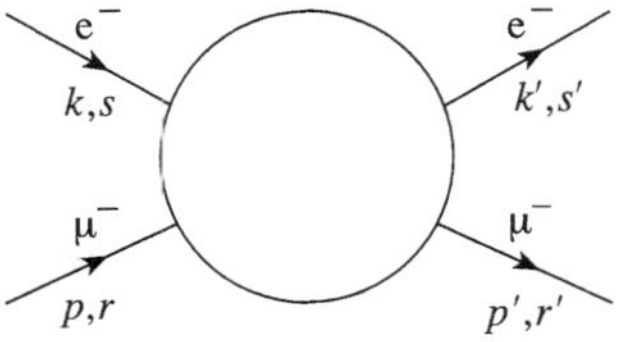

Figure 8.16. $e^-\mu^-$ scattering amplitude.

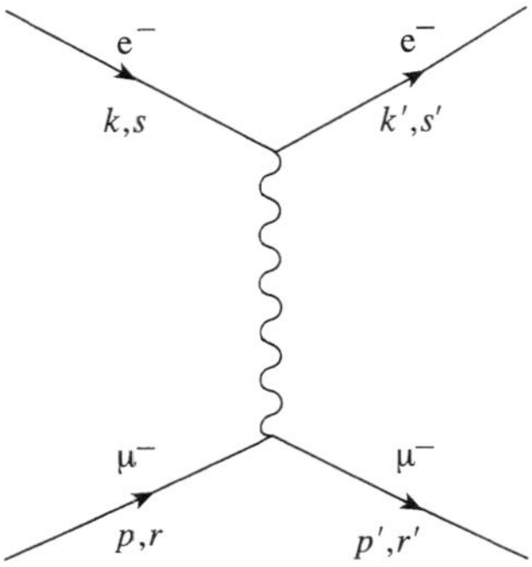

Figure 8.17. One-photon exchange amplitude in $e^-\mu^-$ scattering.

For parton model calculations, what is actually required is the analogous quantity calculated for the case in which the initial photon is virtual (see section 9.2). However, the discussion of section 7.3.2 shows that we may still use the polarization sum (8.169). A difference will arise in passing from (8.174) to (8.175) where we must remember that $k^2 \neq 0$. Since k^2 will be space-like, we put $k^2 = -Q^2$ and find (problem 8.17) that the spin-averaged squared amplitude for the virtual Compton process

$$\gamma^*(k^2 = -Q^2) + e^- \to \gamma + e^- \tag{8.179}$$

is given by

$$-2e^4 \left(\frac{u}{s} + \frac{s}{u} - \frac{2Q^2 t}{su} \right). \tag{8.180}$$

8.7 Electron muon elastic scattering

Our final group of electrodynamic processes are ones in which two fermions interact electromagnetically. In this section we discuss the scattering of two point-like fermions (i.e. leptons); in the following one we look at the change (analogous to those for the π^+ as compared to the s^+) necessitated when one fermion is a hadron, for example the proton.

We shall consider $e^-\mu^-$ elastic scattering: our notation is indicated in figure 8.16. In the lowest order of perturbation theory—the one-photon exchange approximation—we can draw the relevant Feynman graph for this process. This is shown in figure 8.17. All the elements for the graph have been met before and so we can immediately write down the invariant amplitude which now depends on four spin labels:

$$\mathcal{M}_{e^-\mu^-}(r,s;r',s') = e\bar{u}(k',s')\gamma_\mu u(k,s)(g^{\mu\nu}/q^2)e\bar{u}(p',r')\gamma_\nu u(p,r). \quad (8.181)$$

Although experiments with polarized leptons are not uncommon, we shall only be concerned with the unpolarized cross section

$$d\bar{\sigma} \sim \tfrac{1}{4} \sum_{r,r';s,s'} |\mathcal{M}_{e^-\mu^-}(r,s;r',s')|^2. \quad (8.182)$$

We perform the same manipulations as in our e^-s^+ example and the cross section reduces to a factorized form involving two traces:

$$\frac{1}{4}\sum_{r,r';s,s'} |\mathcal{M}_{e^-\mu^-}(r,s;r',s')|^2 = \left(\frac{e^2}{q^2}\right)^2 \left\{\frac{1}{2}\operatorname{Tr}[(\not{k}'+m)\gamma_\mu(\not{k}+m)\gamma_\nu]\right\}$$
$$\times \{\tfrac{1}{2}\operatorname{Tr}[(\not{p}'+M)\gamma^\mu(\not{p}+M)\gamma^\nu]\} \quad (8.183)$$
$$= (e^2/q^2)^{-2} L_{\mu\nu} M^{\mu\nu} \quad (8.184)$$

where $L_{\mu\nu}$ is the 'electron tensor' calculated before:

$$L_{\mu\nu} = 2[k'_\mu k_\nu + k'_\nu k_\mu + (q^2/2)g_{\mu\nu}] \quad (8.185)$$

but now $M^{\mu\nu}$ is the appropriate tensor for the muon coupling, with the same structure as $L_{\mu\nu}$:

$$M^{\mu\nu} = 2[p'^\mu p^\nu + p'^\nu p^\mu + (q^2/2)g^{\mu\nu}]. \quad (8.186)$$

To evaluate the cross section we must perform the 'contraction' $L_{\mu\nu}M^{\mu\nu}$. A useful trick to simplify this calculation is to use current conservation for the electron tensor $L_{\mu\nu}$. For the electron transition current, the electromagnetic current conservation condition is (cf equation (8.99))

$$q^\mu[\bar{u}(k',s')\gamma_\mu u(k,s)] = 0 \quad (8.187)$$

i.e. independent of the particular spin projections s and s'. Since $L_{\mu\nu}$ is the product of two such currents, summed and averaged over polarizations, current conservation implies the conditions

$$q^\mu L_{\mu\nu} = q^\nu L_{\mu\nu} = 0 \quad (8.188)$$

which can be explicitly checked using our result for $L_{\mu\nu}$. The usefulness of this result is that in the contraction $L_{\mu\nu}M^{\mu\nu}$ we can replace p' in $M^{\mu\nu}$ by $(p+q)$ and then drop all the terms involving q's, i.e.

$$L_{\mu\nu}M^{\mu\nu} = L_{\mu\nu}M^{\mu\nu}_{\text{eff}} \tag{8.189}$$

where

$$M^{\mu\nu}_{\text{eff}} = 2[2p^{\mu}p^{\nu} + (q^2/2)g^{\mu\nu}]. \tag{8.190}$$

The calculation of the cross section is now straightforward. In the 'laboratory' system, defined (unrealistically) by the target muon at rest

$$p^{\mu} = (M, 0, 0, 0) \tag{8.191}$$

with M now the muon mass, the result is (problem 8.18(a))

$$\frac{\mathrm{d}\sigma}{\mathrm{d}\Omega} = \left(\frac{\mathrm{d}\sigma}{\mathrm{d}\Omega}\right)_{\text{ns}} \left(1 - \frac{q^2 \tan^2\theta/2)}{2M^2}\right). \tag{8.192}$$

Note the following points:

Comment (a)

The 'no-structure' cross section (8.121) for $\mathrm{e}^-\mathrm{s}^+$ scattering now appears modified by an additional term proportional to $\tan^2(\theta/2)$. This is due to the spin-$\frac{1}{2}$ nature of the muon which gives rise to scattering from both the charge *and* the magnetic moment.

Comment (b)

In the kinematics the electron mass has been neglected, which is usually a good approximation at high energies. We should add a word of explanation for the 'laboratory' cross sections we have calculated, with the target muon unrealistically at rest. The form of the cross section, $(\mathrm{d}\sigma/\mathrm{d}\Omega)_{\text{ns}}$, and of the cross section for the scattering of two Dirac point particles, will be of great value in our discussion of the quark parton model in the next chapter.

Comment (c)

The crossed version of this process, namely $\mathrm{e}^+\mathrm{e}^- \to \mu^+\mu^-$, is a very important monitoring reaction for electron–positron colliding beam machines. It is also basic to a discussion of the predictions of the quark parton model for $\mathrm{e}^+\mathrm{e}^- \to$ hadrons. An instructive calculation similar to this one leads to the result (see problem 8.19)

$$\frac{\mathrm{d}\sigma}{\mathrm{d}\Omega} = \frac{\alpha^2}{4q^2}(1 + \cos^2\theta) \tag{8.193}$$

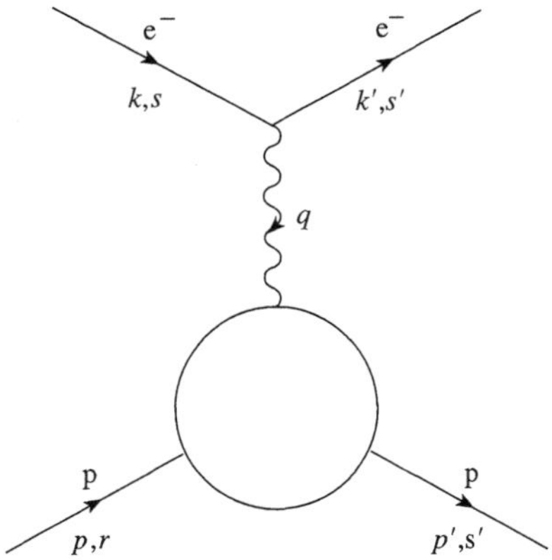

Figure 8.18. One-photon exchange amplitude in e^-p scattering, including hadronic corrections at the ppγ vertex.

where all variables are defined in the e^+e^- CM frame, and all masses have been neglected.

8.8 Electron–proton elastic scattering and nucleon form factors

In the one-photon exchange approximation, the Feynman diagram for elastic electron–proton scattering may be drawn as in figure 8.18, where the 'blob' at the ppγ vertex signifies the expected modification of the point coupling due to strong interactions. The structure of the proton vertex can be analysed using symmetry principles in the same way as for the pion vertex. The presence of Dirac spinors and γ-matrices makes this a somewhat involved procedure: problem 8.20 is an example of the type of complication that arises. Full details of such an analysis can be found in Bernstein (1968), for example. Here, however, we shall proceed in a different way, in order to generalize more easily to inelastic scattering in the following chapter. We focus directly on the 'proton tensor' $B^{\mu\nu}$, which is the product of two proton current matrix elements, summed and averaged over polarizations, as is required in the calculation of the unpolarized cross section:

$$B^{\mu\nu} = \frac{1}{2}\sum_{s,s'} \langle \mathrm{p};\, p', s'|\hat{j}^{\mu}_{\mathrm{em,p}}(0)|\mathrm{p};\, p, s\rangle(\langle \mathrm{p};\, p', s'|\hat{j}^{\nu}_{\mathrm{em,p}}(0)|\mathrm{p};\, p, s\rangle)^{*}. \quad (8.194)$$

We remarked in comment (a) after equation (8.192) that for e^- scattering from a point-like charged fermion an additional term in the cross section was present, corresponding to scattering from the target's magnetic moment. Since a real proton is not a point particle, the virtual strong interaction effects will modify both the charge and the magnetic moment distribution. Hence we may expect that *two* form factors will be needed to describe the deviation from point-like behaviour. This is in fact the case, as we now show using symmetry arguments similar to those of section 8.4.

8.8.1 Lorentz invariance

$T^{\mu\nu}$ must retain its tensor character: this must be made up using the available 4-vectors and tensors at our disposal. For the spin-averaged case we have only

$$p,\ q \text{ and } g_{\mu\nu} \tag{8.195}$$

since $p' = p + q$. The antisymmetric tensor $\epsilon_{\mu\nu\alpha\beta}$ (see appendix J) must actually be ruled out using parity invariance: the tensor $B^{\mu\nu}$ is not a pseudo tensor since $\hat{j}^{\mu}_{\text{em,p}}$ is a vector. It is helpful to remember that $\epsilon_{\mu\nu\alpha\beta}$ is the generalization of ϵ_{ijk} in three dimensions, and that the vector product of two 3-vectors—a pseudo vector—may be written

$$(\boldsymbol{a} \times \boldsymbol{b})_i = \epsilon_{ijk} a_j b_k. \tag{8.196}$$

8.8.2 Current conservation

For a real proton, current conservation gives the condition (cf (8.147))

$$q_\mu \langle \text{p};\ p', s' | \hat{j}^{\mu}_{\text{em,p}}(0) | \text{p};\ p, s \rangle = 0 \tag{8.197}$$

which translates to the conditions (cf (8.188))

$$q_\mu B^{\mu\nu} = q_\nu B^{\mu\nu} = 0 \tag{8.198}$$

on the tensor $B^{\mu\nu}$.

There are only two possible tensors we can make that satisfy both these requirements. One involves p and is constructed to be orthogonal to q. We introduce a vector

$$\tilde{p}_\mu = p_\mu + \alpha q_\mu \tag{8.199}$$

and require

$$q \cdot \tilde{p} = 0. \tag{8.200}$$

Hence we find

$$\tilde{p}_\mu = p_\mu - (p \cdot q / q^2) q_\mu \tag{8.201}$$

and thus the tensor

$$\tilde{p}^\mu \tilde{p}^\nu = [p^\mu - (p \cdot q/q^2) q^\mu][p^\nu - (p \cdot q/q^2) q^\nu] \tag{8.202}$$

satisfies all our requirements. The second tensor must involve $g^{\mu\nu}$ and may be chosen to be

$$-g^{\mu\nu} + q^\mu q^\nu / q^2 \tag{8.203}$$

which again satisfies our conditions. Thus from invariance arguments alone, the tensor $B^{\mu\nu}$ for the proton vertex may be parametrized by these two tensors, each multiplied by an unknown function of q^2. If we define

$$\begin{aligned} B^{\mu\nu} = {} & 4A(q^2)[p^\mu - (p \cdot q/q^2) q^\mu][p^\nu - (p \cdot q/q^2) q^\nu] \\ & + 2M^2 B(q^2)(-g^{\mu\nu} + q^\mu q^\nu / q^2) \end{aligned} \tag{8.204}$$

the cross section in the laboratory frame is

$$\frac{\mathrm{d}\sigma}{\mathrm{d}\Omega} = \left(\frac{\mathrm{d}\sigma}{\mathrm{d}\Omega}\right)_{\mathrm{ns}} [A + B \tan^2(\theta/2)]. \tag{8.205}$$

The functions A and B may be related to the 'charge' and 'magnetic' form factors of the proton. The Dirac 'charge' and Pauli 'anomalous magnetic moment' form factors are defined by

$$\begin{aligned}
&\langle \mathrm{p};\, p', s' | \hat{j}^{\mu}_{\mathrm{em,p}}(0) | \mathrm{p};\, p, s \rangle \\
&\quad = (+e)\bar{u}(p', s') \left[\gamma^{\mu} \mathcal{F}_1(q^2) + \frac{\mathrm{i}\kappa \mathcal{F}_2(q^2)}{2M} \sigma^{\mu\nu} q_{\nu} \right] u(p, s)
\end{aligned} \tag{8.206}$$

with the normalization

$$\mathcal{F}_1(0) = 1 \tag{8.207}$$

$$\mathcal{F}_2(0) = 1 \tag{8.208}$$

and the magnetic moment of the muon is not one (nuclear) magneton, as for an electron or muon (neglecting higher-order corrections), but rather $\mu_{\mathrm{p}} = 1 + \kappa$ with $\kappa = 1.79$. Problem 8.20 shows that the $\bar{u}\gamma^{\mu}u$ piece in (8.206) can be rewritten in terms of $\bar{u}(p + p')^{\mu}u/2M$ and $\bar{u}\mathrm{i}\sigma^{\mu\nu}q_{\nu}u/2M$. The first of these is analogous to the interaction of a charged spin-0 particle. As regards the second, we note that $\sigma^{\mu\nu}$ is just

$$\sigma^{\mu\nu} = \tfrac{1}{2}\mathrm{i}[\gamma^{\mu}, \gamma^{\nu}] \tag{8.209}$$

which reduces to the Pauli spin matrices for the space-like components

$$\sigma^{ij} = \begin{pmatrix} \sigma^k & 0 \\ 0 & \sigma^k \end{pmatrix} \tag{8.210}$$

with our representation of γ-matrices (σ^{ij} is a 4×4 matrix, σ^k is 2×2, and i, j and k are in cyclic order). The second term in this 'Gordon decomposition' of $\bar{u}\gamma^{\mu}u$ thus corresponds to an interaction via the spin magnetic moment—with, in fact, $g = 2$. Thus the addition of the κ term in (8.206) corresponds to an 'anomalous' magnetic moment piece. In terms of $\mathcal{F}_1$ and $\mathcal{F}_2$ one can show that

$$A = \mathcal{F}_1^2 + \tau\kappa^2\mathcal{F}_2^2 \tag{8.211}$$

$$B = 2\tau(\mathcal{F}_1 + \kappa\mathcal{F}_2)^2 \tag{8.212}$$

where

$$\tau = -q^2/4M^2. \tag{8.213}$$

The cross section formula for electron–proton elastic scattering is called the 'Rosenbluth' cross section. The point-like cross section (8.192) is recovered from (8.205) by setting $\mathcal{F}_1 = 1$ and $\kappa = 0$ in (8.211) and (8.212).

Before we leave elastic scattering it is helpful to look in some more detail at the kinematics. It will be sufficient to consider the 'point-like' case, which we shall call $e^-\mu^+$, for definiteness. Energy and momentum conservation at the μ^+ vertex gives the condition

$$p + q = p' \tag{8.214}$$

with the mass-shell conditions (M is the μ^+ mass)

$$p^2 = p'^2 = M^2. \tag{8.215}$$

Hence for elastic scattering we have the relation

$$2p \cdot q = -q^2. \tag{8.216}$$

It is conventional to relate these invariants to the corresponding laboratory frame ($p^\mu = (M, \mathbf{0})$) expressions. Neglecting the electron mass so that[2]

$$k \equiv |\boldsymbol{k}| = \omega \tag{8.217}$$

$$k' \equiv |\boldsymbol{k}'| = \omega' \tag{8.218}$$

we have

$$q^2 = -2kk'(1 - \cos\theta) = -4kk' \sin^2(\theta/2) \tag{8.219}$$

and

$$p \cdot q = M(k - k') = M\nu \tag{8.220}$$

where ν is the energy transfer q^0 in this frame. To avoid unnecessary minus signs, it is convenient to define

$$Q^2 = -q^2 = 4kk' \sin^2(\theta/2) \tag{8.221}$$

and the elastic scattering relation between $p \cdot q$ and q^2 reads

$$\nu = Q^2/2M \tag{8.222}$$

or

$$\frac{k'}{k} = \frac{1}{1 + (2k/M)\sin^2(\theta/2)}. \tag{8.223}$$

Remembering, therefore, that for elastic scattering k' and θ are not independent variables, we can perform a change of variables (see appendix K) in the laboratory frame

$$\mathrm{d}\Omega = 2\pi\, \mathrm{d}(\cos\theta) = (\pi/k'^2)\, \mathrm{d}Q^2 \tag{8.224}$$

and write the differential cross section for $e^-\mu^+$ scattering as

$$\frac{\mathrm{d}\sigma}{\mathrm{d}Q^2} = \frac{\pi\alpha^2}{4k^2 \sin^4(\theta/2)} \frac{1}{kk'} [\cos^2(\theta/2) + 2\tau \sin^2(\theta/2)]. \tag{8.225}$$

[2] As after equation (8.125), note again that in the present context 'k' and 'k'' are not 4-vectors but the moduli of 3-vectors.

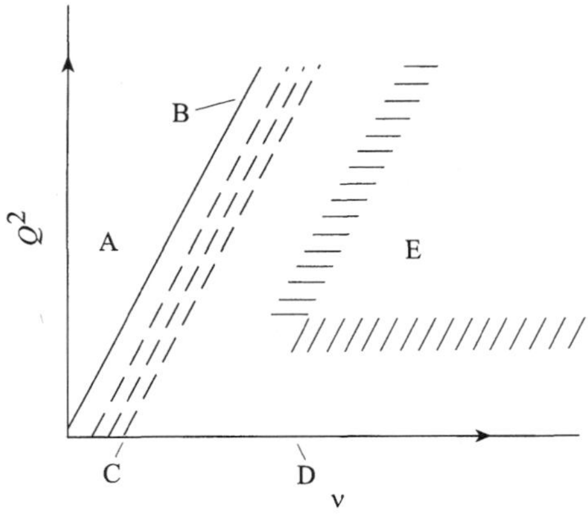

Figure 8.19. Physical regions in the Q^2, ν variables: A, kinematically forbidden region; B, line of elastic scattering ($Q^2 = 2M\nu$); C, lines of resonance electroproduction; D, photoproduction; E, deep inelastic region (Q^2 and ν large).

For elastic scattering ν is not independent of Q^2 but we may formally write this as a double-differential cross section by inserting the δ-function to ensure this condition is satisfied:

$$\boxed{\frac{\mathrm{d}^2\sigma}{\mathrm{d}Q^2\mathrm{d}\nu} = \frac{\pi\alpha^2}{4k^2\sin^4(\theta/2)}\frac{1}{kk'}\left[\cos^2(\theta/2) + \left(\frac{Q^2}{2M^2}\right)\sin^2(\theta/2)\right]\delta\left(\nu - \frac{Q^2}{2M}\right).} \tag{8.226}$$

This is the cross section for the scattering of an electron from a point-like fermion target of charge e and mass M.

It is illuminating to plot out the physically allowed regions of Q^2 and ν (figure 8.19). Elastic $\mathrm{e}^-\mathrm{p}$ scattering corresponds to the line $Q^2 = 2M\nu$. Resonance production $\mathrm{e}^-\mathrm{p} \to \mathrm{e}^-\mathrm{N}^*$ with $p'^2 = M'^2$ corresponds to lines parallel to the elastic line, shifted to the right by $M'^2 - M^2$ since

$$2M\nu = Q^2 + M'^2 - M^2. \tag{8.227}$$

Experiments with real photons, $Q^2 = 0$, correspond to exploring along the ν-axis. In the next chapter we switch our attention to so-called deep inelastic electron scattering—the region of large Q^2 and large ν.

Problems

8.1 Consider a matrix element of the form

$$M = \int \mathrm{d}^3\boldsymbol{x} \int \mathrm{d}t\, \mathrm{e}^{+\mathrm{i}p_\mathrm{f}\cdot x}\partial_\mu A^\nu \mathrm{e}^{-\mathrm{i}p_\mathrm{i}\cdot x}.$$

Assuming the integration is over all space–time and that

$$A^0 \to 0 \qquad \text{as } t \to \pm\infty$$

and

$$|\boldsymbol{A}| \to 0 \qquad \text{as } |\boldsymbol{x}| \to \infty$$

use integration by parts to show

(a) $\displaystyle\int \mathrm{d}t\, \mathrm{e}^{+\mathrm{i}p_\mathrm{f}\cdot x}\partial_0 A^0 \mathrm{e}^{-\mathrm{i}p_\mathrm{i}\cdot x} = (-\mathrm{i}p_{\mathrm{f}0})\int \mathrm{d}t\, \mathrm{e}^{+\mathrm{i}p_\mathrm{f}\cdot x} A^0 \mathrm{e}^{-\mathrm{i}p_\mathrm{i}\cdot x}$

(b) $\displaystyle\int \mathrm{d}^3\boldsymbol{x}\, \mathrm{e}^{+\mathrm{i}p_\mathrm{f}\cdot x}\nabla\cdot \boldsymbol{A}\mathrm{e}^{-\mathrm{i}p_\mathrm{i}\cdot x} = +\mathrm{i}\boldsymbol{p}_\mathrm{f}\cdot\left(\int \mathrm{d}^3\boldsymbol{x}\, \mathrm{e}^{+\mathrm{i}p_\mathrm{f}\cdot x}\boldsymbol{A}\mathrm{e}^{-\mathrm{i}p_\mathrm{i}\cdot x}\right).$

Hence show that

$$\int \mathrm{d}^3\boldsymbol{x}\int \mathrm{d}t\, \mathrm{e}^{+\mathrm{i}p_\mathrm{f}\cdot x}(\partial_\mu A^\mu + A^\mu\partial_\mu)\mathrm{e}^{-\mathrm{i}p_\mathrm{i}\cdot x}$$
$$= -\mathrm{i}(p_\mathrm{f} + p_\mathrm{i})_\mu \int \mathrm{d}^3\boldsymbol{x}\int \mathrm{d}t\, \mathrm{e}^{+\mathrm{i}p_\mathrm{f}\cdot x}A^\mu \mathrm{e}^{-\mathrm{i}p_\mathrm{i}\cdot x}.$$

8.2 Verify equation (8.27).

8.3 Evaluate (8.31) and interpret the result physically (i.e. compare it with (8.27)).

8.4 Verify for a simple choice of Lorentz transformation (e.g. one along the x'-axis) that $\bar{\psi}'\gamma^\mu\psi$ is a 4-vector, where ψ and ψ' are four-component Dirac spinors transforming as in section 4.4.

8.5

(a) Using the u-spinors normalized as in (4.105), the $\phi^{1,2}$ of (8.47), and the result for $\boldsymbol{\sigma}\cdot\boldsymbol{A}\boldsymbol{\sigma}\cdot\boldsymbol{B}$ from problem 4.13, show that

$$u^\dagger(k', s' = 1)u(k, s = 1) = (E+m)\left\{1 + \frac{\boldsymbol{k}'\cdot\boldsymbol{k}}{(E+m)^2} + \frac{\mathrm{i}\phi^{1\dagger}\boldsymbol{\sigma}\cdot\boldsymbol{k}'\times\boldsymbol{k}\phi^1}{(E+m)^2}\right\}.$$

(b) For any vector $\boldsymbol{A} = (A^1, A^2, A^3)$, show that $\phi^{1\dagger}\boldsymbol{\sigma}\cdot\boldsymbol{A}\phi^1 = A^3$. Find similar expressions for $\phi^{1\dagger}\boldsymbol{\sigma}\cdot\boldsymbol{A}\phi^2$, $\phi^{2\dagger}\boldsymbol{\sigma}\cdot\boldsymbol{A}\phi^1$, $\phi^{2\dagger}\boldsymbol{\sigma}\cdot\boldsymbol{A}\phi^2$.

(c) Show that the S of (8.46) is equal to

$$S = (E+m)^2\left\{\left[1 + \frac{\boldsymbol{k}'\cdot\boldsymbol{k}}{(E+m)^2}\right]^2 + \frac{(\boldsymbol{k}'\times\boldsymbol{k})^2}{(E+m)^4}\right\}.$$

(d) Using $\cos\theta = \boldsymbol{k}\cdot\boldsymbol{k}'/(|\boldsymbol{k}||\boldsymbol{k}'|)$, $|\boldsymbol{k}| = |\boldsymbol{k}'|$ and $v = |\boldsymbol{k}|/E$, show that

$$S = (2E)^2(1 - v^2\sin^2\theta/2).$$

8.6 Verify equation (8.55).

8.7 Check that $\gamma^0\gamma^{\mu\dagger}\gamma^0 = \gamma^\mu$.

8.8 Verify equation (8.78) for the lepton tensor $L^{\mu\nu}$.

8.9 Evaluate L^{00} as in equation (8.79).

8.10 Verify equation (8.86).

8.11 Verify equation (8.95) for the $\mathrm{e^-s^+ \to e^-s^+}$ amplitude to $O(e^2)$.

8.12 Check that both the scalar and the spinor current matrix elements (8.27) and (8.55), satisfy $\partial_\mu j^\mu(x) = 0$.

8.13 Verify equation (8.119).

8.14 Verify equation (8.135) for the Fourier transform of $\rho(\boldsymbol{x})$ given by (8.134).

8.15 Check the gauge invariance of $\mathcal{M}_{\gamma\mathrm{e}^-}$ given by (8.161), by showing that if ϵ_μ is replaced by k_μ, or ϵ_ν^* by k'_ν, the result is zero.

8.16

(a) The spin-averaged squared amplitude for lowest-order electron Compton scattering contains the interference term

$$\sum_{\lambda,\lambda',s,s'} \mathcal{M}^{(s)}_{\gamma\mathrm{e}^-}\mathcal{M}^{(u)*}_{\gamma\mathrm{e}^-}$$

where (s) and (u) refer to the s- and u-channel processes of figures 8.14(*a*) and (*b*) respectively. Obtain an expression analogous to (8.171) for this term, and prove that it is, in fact, zero. [*Hint*: use relations (J.4) and (J.5).]

(b) Explain why the term

$$\sum_{\lambda,\lambda',s,s'} \mathcal{M}^{(u)}_{\gamma\mathrm{e}^-}\mathcal{M}^{(u)*}_{\gamma\mathrm{e}^-}$$

is given by (8.176) with s and u interchanged.

8.17 Recalculate the interference term of problem 8.16(a) for the case $k^2 = -Q^2$ and hence verify (8.180).

8.18

(a) Derive an expression for the spin-averaged differential cross section for lowest-order $\mathrm{e}^-\mu^-$ scattering in the laboratory frame, defined by $p^\mu = (M, \mathbf{0})$ where M is now the muon mass, and show that it may be written in the form

$$\frac{\mathrm{d}\sigma}{\mathrm{d}\Omega} = \left(\frac{\mathrm{d}\sigma}{\mathrm{d}\Omega}\right)_{\mathrm{ns}} [1 - (q^2/2M^2)\tan^2(\theta/2)]$$

where the 'no-structure' cross section is that of $\mathrm{e^-s^+}$ scattering (appendix K) and the electron mass has, as usual, been neglected.

(b) Neglecting *all* masses, evaluate the spin-averaged expression (8.183) in terms of s, t and u and use the result

$$\frac{d\sigma}{dt} = \frac{1}{16\pi s^2}\frac{1}{4}\sum_{r,r';s,s'} |\mathcal{M}_{e^-\mu^-}(r,s;r',s')|^2$$

to show that the $e^-\mu^-$ cross section may be written in the form

$$\frac{d\sigma}{dt} = \frac{4\pi\alpha^2}{t^2}\frac{1}{2}\left(1+\frac{u^2}{s^2}\right).$$

Show also that by introducing the variable y, defined in terms of laboratory variables by $y = (k - k')/k$, this reduces to the result

$$\frac{d\sigma}{dy} = \frac{4\pi\alpha^2}{t^2} s \frac{1}{2}[1 + (1-y)^2].$$

8.19 Consider the process $e^+e^- \to \mu^+\mu^-$ in the CM frame.

(a) Draw the lowest-order Feynman diagram and write down the corresponding amplitude.

(b) Show that the spin-averaged squared matrix element has the form

$$\overline{|\mathcal{M}|^2} = \frac{(4\pi\alpha)^2}{Q^4} L(e)_{\mu\nu} L(\mu)^{\mu\nu}$$

where Q^2 is the square of the total CM energy, and $L(e)$ depends on the e^- and e^+ momenta and $L(\mu)$ on those of the μ^+, μ^-.

(c) Evaluate the traces and the tensor contraction (neglecting lepton masses): (i) directly, using the trace theorems; and (ii) by using crossing symmetry and the results of section 8.7 for $e^-\mu^-$ scattering. Hence show that

$$\overline{|\mathcal{M}|^2} = (4\pi\alpha)^2(1 + \cos^2\theta)$$

where θ is the CM scattering angle.

(d) Hence show that the total elastic cross section is (see equation (B.18) of appendix B)

$$\sigma = 4\pi\alpha^2/3Q^2.$$

Figure 8.20 shows data (a) for σ in $e^+e^- \to \mu^+\mu^-$ and $e^+e^- \to \tau^+\tau^-$ and (b) for the angular distribution in $e^+e^- \to \mu^+\mu^-$. The data in (a) agree well with the prediction of part (d). The broken curve in figure 8.20(*b*) shows the pure QED prediction

$$\frac{d\sigma}{d\Omega} = \frac{\alpha^2}{4Q^2}(1 + \cos^2\theta).$$

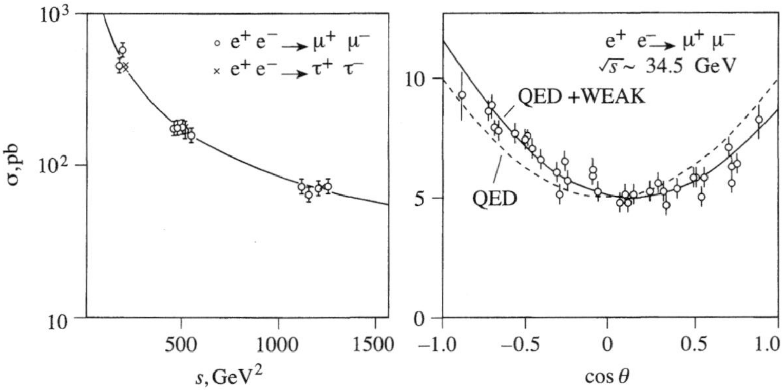

Figure 8.20. (*a*) Total cross sections for $e^+e^- \to \mu^+\mu^-$ and $e^+e^- \to \tau^+\tau^-$; (*b*) differential cross section for $e^+e^- \to \mu^+\mu^-$. (From Perkins D 2000 *Introduction to High Energy Physics* 4th edn, courtesy Cambridge University Press.)

It is clear that, while the distribution has the general $1 + \cos^2\theta$ form as predicted, there is a small but definite forward–backward asymmetry. This arises because, in addition to the γ-exchange amplitude there is also a Z^0-exchange amplitude (see volume 2) which we have neglected. Such asymmetries are an important test of the electroweak theory. They are too small to be visible in the total cross sections in (a).

8.20 Starting from the expression

$$\bar{u}(p')\mathrm{i}\frac{\sigma^{\mu\nu}}{2M}q_\nu u(p)$$

where $q = p' - p$ and $\sigma^{\mu\nu} = \frac{1}{2}\mathrm{i}[\gamma^\mu, \gamma^\nu]$, use the Dirac equation and properties of γ-matrices to prove the 'Gordon decomposition' of the current

$$\bar{u}(p')\gamma^\mu u(p) = \bar{u}(p')\left(\frac{(p+p')^\mu}{2M} + \mathrm{i}\frac{\sigma^{\mu\nu}q_\nu}{2M}\right)u(p).$$

9

DEEP INELASTIC ELECTRON–NUCLEON SCATTERING AND THE QUARK PARTON MODEL

We have obtained the rules for doing calculations of simple processes in quantum electrodynamics for particles of spin-0 and spin-$\frac{1}{2}$, and many explicit examples have been considered. In this chapter we build on these results to give an (admittedly brief) introduction to a topic of central importance in particle physics, the structure of hadrons as revealed by deep inelastic scattering experiments (the equally important neutrino scattering experiments will be discussed in volume 2). We do this partly because the necessary calculations involve straightforward, illustrative and eminently practical applications of the rules already obtained, but, more particularly, because it is from a comparison of these calculations with experiment that one obtains compelling evidence for the existence of the point-like constituents of hadrons—'quarks'—the interactions of which are described by gauge theories.

9.1 Inelastic electron–proton scattering: kinematics and structure functions

At large momentum transfers there is very little elastic scattering: inelastic scattering, in which there is more than just the electron and proton in the final state, is much more probable. The simplest inelastic cross section to measure is the so-called 'inclusive' cross section, for which only the final electron is observed. This is therefore a sum over the cross sections for all the possible hadronic final states: no attempt is made to select any particular state from the hadronic debris created at the proton vertex. This process may be represented by the diagram of figure 9.1, assuming that the one-photon exchange amplitude dominates. The 'blob' at the proton vertex indicates our ignorance of the detailed structure: X indicates a sum over all possible hadronic final states. However, the assumption of one-photon exchange, which is known experimentally to be a very good approximation, means that, as in our previous examples (cf (8.117) and (8.184)), the cross section must factorize into a leptonic tensor contracted with a tensor describing the hadron vertex:

$$\mathrm{d}\sigma \sim L_{\mu\nu} W^{\mu\nu}(q, p). \tag{9.1}$$

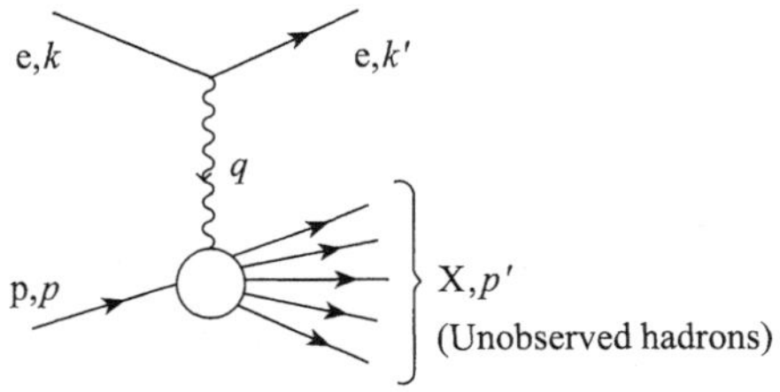

Figure 9.1. Inelastic electron–proton scattering, in one-photon exchange approximation.

The lepton vertex is well described by QED and takes the same form as before:

$$L_{\mu\nu} = 2[k'_\mu k_\nu + k'_\nu k_\mu + (q^2/2)g_{\mu\nu}]. \tag{9.2}$$

For the hadron tensor, however, we expect strong interactions to play an important role and we must deduce its general structure by our powerful invariance arguments. We will only consider unpolarized scattering and therefore perform an average over the initial proton spins. The sum over final states, X, includes all possible quantum numbers for each hadronic state with total momentum p'. For an inclusive cross section, the final phase space involves only the scattered electron. Moreover, since we are not restricting the scattering process by picking out any specific state of X, the energy k' and the scattering angle θ of the final electron are now independent variables. In $W^{\mu\nu}(q, p)$ the sum over X includes the phase space for each hadronic state restricted by the usual 4-momentum-conserving δ-function to ensure that each state in X has momentum p'. Including some conventional factors, we define $W^{\mu\nu}(q, p)$ by (see problem 9.1)

$$e^2 W^{\mu\nu}(q, p) = \frac{1}{4\pi M}\frac{1}{2}\sum_s \sum_{\mathrm{X}} \langle \mathrm{p};\, p, s|\hat{j}^{\mu}_{\mathrm{em,p}}(0)|\mathrm{X};\, p'\rangle\langle \mathrm{X};\, p'|\hat{j}^{\nu}_{\mathrm{em,p}}(0)|\mathrm{p};\, p, s\rangle$$
$$\times (2\pi)^4\delta^4(p + q - p'). \tag{9.3}$$

How do we parametrize the tensor structure of $W^{\mu\nu}$? As usual, Lorentz invariance and current conservation come to our aid. There is one important difference compared with the elastic form factor case of section 8.8. For inclusive inelastic scattering there are now two independent scalar variables. The relation

$$p' = p + q \tag{9.4}$$

leads to

$$p'^2 = M^2 + 2p \cdot q + q^2 \tag{9.5}$$

where M is the proton mass. In this case, the invariant mass of the hadronic final state is a *variable*

$$p'^2 = W^2 \tag{9.6}$$

and is related to the other two scalar variables

$$p \cdot q = M\nu \tag{9.7}$$

and (cf (8.221))

$$q^2 = -Q^2 \tag{9.8}$$

by the condition (cf (8.227))

$$2M\nu = Q^2 + W^2 - M^2. \tag{9.9}$$

Our invariance arguments lead us to the same *tensor* structure as for *elastic* electron–proton scattering, but now the functions $A(q^2)$, $B(q^2)$ are replaced by 'structure functions' which are functions of two variables, usually taken to be ν and Q^2. The conventional definition of the proton structure functions W_1 and W_2 is

$$\boxed{\begin{aligned} W^{\mu\nu}(q, p) = {} & (-g^{\mu\nu} + q^\mu q^\nu/q^2)W_1(Q^2, \nu) \\ & + [p^\mu - (p \cdot q/q^2)q^\mu][p^\nu - (p \cdot q/q^2)q^\nu]M^{-2}W_2(Q^2, \nu). \end{aligned}} \tag{9.10}$$

Inserting the usual flux factor together with the final electron phase space leads to the following expression for the inclusive differential cross section for inelastic electron–proton scattering (see problem 9.1):

$$\mathrm{d}\sigma = \left(\frac{4\pi\alpha}{q^2}\right)^2 \frac{1}{4[(k \cdot p)^2 - m^2M^2]^{1/2}} 4\pi M L_{\mu\nu} W^{\mu\nu} \frac{\mathrm{d}^3k'}{2\omega'(2\pi)^3}. \tag{9.11}$$

In terms of 'laboratory' variables, neglecting electron mass effects, this yields

$$\boxed{\frac{\mathrm{d}^2\sigma}{\mathrm{d}\Omega\mathrm{d}k'} = \frac{\alpha^2}{4k^2\sin^4(\theta/2)}[W_2\cos^2(\theta/2) + 2W_1\sin^2(\theta/2)].} \tag{9.12}$$

Remembering now that $\cos\theta$ and k' are independent variables for inelastic scattering, we can change variables from $\cos\theta$ and k' to Q^2 and ν, assuming azimuthal symmetry for the unpolarized cross section. We have

$$Q^2 = 2kk'(1 - \cos\theta) \tag{9.13}$$

$$\nu = k - k' \tag{9.14}$$

so that

$$\mathrm{d}(\cos\theta)\,\mathrm{d}k' = \frac{1}{2kk'}\mathrm{d}Q^2\,\mathrm{d}\nu \tag{9.15}$$

and

$$\boxed{\frac{\mathrm{d}^2\sigma}{\mathrm{d}Q^2\mathrm{d}\nu} = \frac{\pi\alpha^2}{4k^2\sin^4(\theta/2)}\frac{1}{kk'}[W_2\cos^2(\theta/2) + 2W_1\sin^2(\theta/2)].} \tag{9.16}$$

Yet another choice of variables is sometimes used instead of these, namely the variables

$$x = Q^2/2M\nu \tag{9.17}$$

whose significance we shall see in the next section, and

$$y = \nu/k \tag{9.18}$$

which is the fractional energy transfer in the 'laboratory' frame. The Jacobian for the transformation from Q^2 and ν to x and y is (see problem 9.2)

$$\mathrm{d}Q^2\,\mathrm{d}\nu = 2Mk^2 y\,\mathrm{d}x\,\mathrm{d}y. \tag{9.19}$$

We emphasize that the foregoing—in particular (9.3), (9.12) and (9.16)—is all completely general, given the initial one-photon approximation. The physics is all contained in the ν and Q^2 dependence of the two structure functions W_1 and W_2.

A priori, one might expect W_1 and W_2 to be complicated functions of ν and Q^2, reflecting the complexity of the inelastic scattering process. However, in 1969 Bjorken predicted that in the 'deep inelastic region'—large ν and Q^2, but Q^2/ν finite—there should be a very simple behaviour. He predicted that the structure functions should *scale*, i.e. become functions not of Q^2 and ν independently but only of their ratio Q^2/ν. It was the verification of approximate 'Bjorken scaling' that led to the development of the modern quark parton model. We therefore specialize our discussion of inelastic scattering to the deep inelastic region.

9.2 Bjorken scaling and the parton model

From considerations based on the quark model current algebra of Gell-Mann (1962), Bjorken (1969) was led to propose the following 'scaling hypothesis': in the limit

$$\left.\begin{array}{r} Q^2 \to \infty \\ \nu \to \infty \end{array}\right\} \quad \text{with } x = Q^2/2M\nu \text{ fixed} \tag{9.20}$$

the structure functions scale as

$$MW_1(Q^2, \nu) \to F_1(x) \tag{9.21}$$

$$\nu W_2(Q^2, \nu) \to F_2(x). \tag{9.22}$$

We must emphasize that the *physical* content of Bjorken's hypothesis is that the functions $F_1(x)$ and $F_2(x)$ are *finite*[1].

[1] It is always possible to write $W(Q^2, \nu) = f(x, Q^2)$, say, where $f(x, Q^2)$ will tend to some function $F(x)$ as $Q^2 \to \infty$ with x fixed. $F(x)$ may, however, be zero, finite or infinite. The physics lies in the hypothesis that, in this limit, a finite part remains.

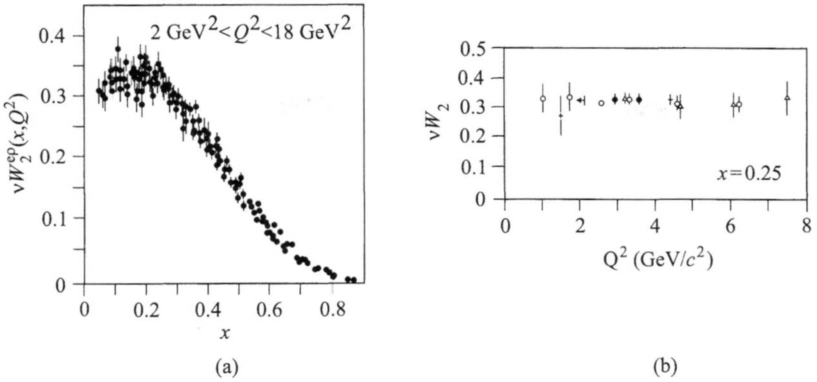

Figure 9.2. Bjorken scaling: the structure function νW_2 (*a*) plotted against x for different Q^2 values (Attwood 1980) and (*b*) plotted against Q^2 for the single x value, $x = 0.25$ (Friedman and Kendall 1972).

Early experimental support for these predictions (figure 9.2) led initially to an examination of the theoretical basis of Bjorken's arguments and to the formulation of the simple intuitive picture provided by the quark parton model. Closer scrutiny of figure 9.2(*a*) will encourage the (correct) suspicion that, in fact, there is a small but significant spread in the data for any given x value. In volume 2 we shall give an introduction to the way in which the *violations* of simple scaling behaviour can be predicted by QCD. Nevertheless, scaling is certainly the most immediate gross feature of the data and an understanding of it is of fundamental importance.

How can the scaling be understood? Feynman, when asked to explain Bjorken's arguments, gave an intuitive explanation in terms of elastic scattering from free point-like constituents of the nucleon, which he dubbed 'partons'. The essence of the argument lies in the *kinematics of elastic scattering of electrons by point-like partons*: we can therefore use the results of the previous chapters to derive the parton model results. At high Q^2 and ν it is intuitively reasonable (and in fact the basis for the light-cone and short-distance operator approach (Wilson 1969) to scaling) that the virtual photon is probing very short distances and time scales within the proton. In this situation, Feynman supposed that the photon interacts with small (point-like) constituents within the proton, which carry only a certain fraction f of the proton's energy and momentum (figure 9.3). Over the short time scales involved in the transfer of a large amount of energy ν, the struck constituents can perhaps be treated as effectively free. We then have the idealized elastic scattering process shown in figure 9.4. It is the kinematics of the elastic scattering condition for the partons that leads directly to a relation between Q^2 and ν and hence to the observed scaling behaviour. The original discussion of the parton model took place in the infinite-momentum frame of the proton. While

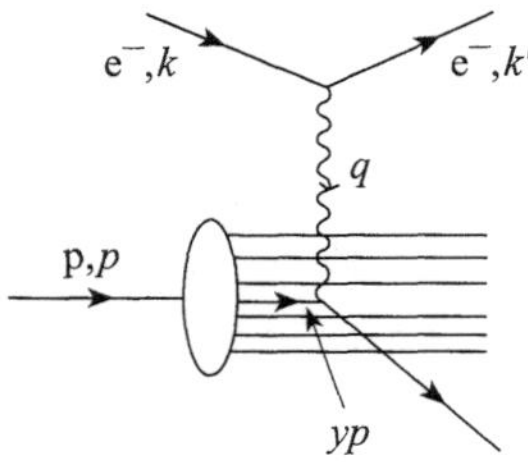

Figure 9.3. Photon–parton interaction.

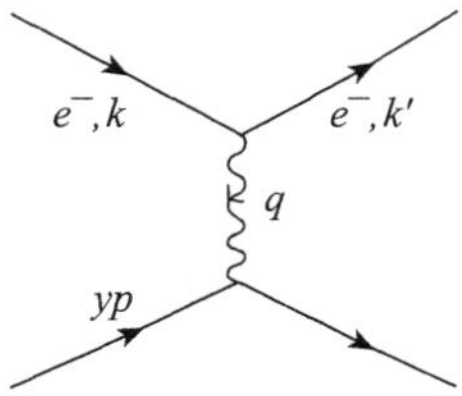

Figure 9.4. Elastic electron–parton scattering.

this has the merit that it eliminates the need for explicit statements about parton masses and so on, it also obscures the simple kinematic origin of the scaling. For this reason, at the expense of some theoretical niceties, we prefer to perform a direct calculation of electron–parton scattering in close analogy with our previous examples.

We first show that the fraction f is none other than Bjorken's variable x. For a parton of type i we write

$$p_i^\mu \approx f p^\mu \tag{9.23}$$

and, roughly speaking[2], we can imagine that the partons have mass

$$m_i \approx fM. \tag{9.24}$$

Then, exactly as in (8.214) and (8.215), energy and momentum conservation at the parton vertex, together with the assumption that the struck parton remains on-shell (as indicated by the fact that in figure 9.4 the partons are free), imply that

$$(q + fp)^2 = m_i^2 \tag{9.25}$$

which, using (9.8), (8.220) and (9.24), gives

$$f = Q^2/2M\nu \equiv x. \tag{9.26}$$

[2] Explicit statements about parton transverse momenta and masses, such as those made in equations (9.23) and (9.24), are unnecessary in a rigorous treatment, where such quantities can be shown to give rise to non-leading scaling behaviour (Sachrajda 1983).

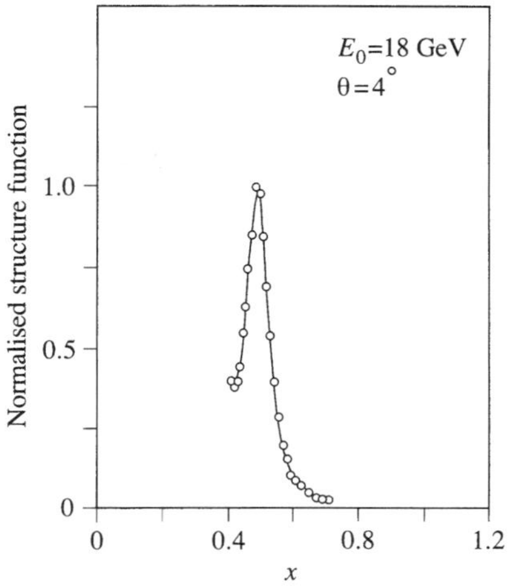

Figure 9.5. Structure function for quasi-elastic ed scattering, plotted against x (Attwood 1980).

Thus the fact that the nucleon structure functions do seem to depend (to a good approximation) only on the variable x is interpreted physically as showing that the scattering is dominated by the 'quasi-free' electron–parton process shown in figure 9.4. In section 11.5.3 we shall see how the 'asymptotic freedom' property of QCD suggests a dynamical understanding of this picture, as will be discussed further in volume 2.

What sort of values for x do we expect? Consider an analogous situation—electron scattering from deuterium. Here the target (the deuteron) is undoubtedly composite, and its 'partons' are, to a first approximation, just the two nucleons. Since $m_{\rm N} \simeq \frac{1}{2}m_{\rm D}$, we expect to see the value $x \simeq \frac{1}{2}$ (cf (9.24)) favoured; $x = 1$ here would correspond to elastic scattering from the deuteron. A peak at $x \approx \frac{1}{2}$ is indeed observed (figure 9.5) in quasi-elastic $\rm e^-d$ scattering (the broadening of the peak is due to the fact that the constituent nucleons have some motion within the deuteron.) By 'quasi-elastic' here we mean that the incident electron scatters off 'quasi-free' nucleons, an approximation we expect to be good for incident energies significantly greater than the binding energy of the n and p in the deuteron ($\sim$2 MeV). What about the nucleon itself, then? A simple three-quark model would, on this analogy, lead us to expect a peak at $x \simeq \frac{1}{3}$, but the data already shown (figure 9.2(a)) do not look much like that. Perhaps there is something else present too—which we shall uncover as our story proceeds.

Certainly it seems sensible to suppose that a nucleon contains *at least* some quarks (and also anti-quarks) of the type introduced in the simple composite models of the nucleon (section 1.2.3). If quarks are supposed to have spin-$\frac{1}{2}$, then the scattering of an electron from a quark or anti-quark—generically a *parton*—

of type i, charge e_i (in units of e) is just given by the eμ scattering cross section (8.226), with obvious modifications:

$$\frac{d^2\sigma^i}{dQ^2 d\nu} = \frac{\pi\alpha^2}{4k^2\sin^4(\theta/2)}\frac{1}{kk'}\left(e_i^2\cos^2(\theta/2) + e_i^2\frac{Q^2}{4m_i^2}2\sin^2(\theta/2)\right)$$
$$\times\,\delta(\nu - Q^2/2m_i). \tag{9.27}$$

This is to be compared with the general inclusive inelastic cross section formula written in terms of W_1 and W_2:

$$\frac{d^2\sigma}{dQ^2 d\nu} = \frac{\pi\alpha^2}{4k^2\sin^4(\theta/2)}\frac{1}{kk'}[W_2\cos^2(\theta/2) + W_1 2\sin^2(\theta/2)]. \tag{9.28}$$

Thus the contribution to W_1 and W_2 from one parton of type i is immediately seen to be

$$W_1^i = e_i^2\frac{Q^2}{4M^2x^2}\delta(\nu - Q^2/2Mx) \tag{9.29}$$
$$W_2^i = e_i^2\delta(\nu - Q^2/2Mx) \tag{9.30}$$

where we have set $m_i = xM$. At large ν and Q^2 it is assumed that the contributions from different partons add incoherently in cross section. Thus, to obtain the total contribution from all quark partons, we must sum over the contributions from all types of partons, i, and integrate over all values of x, the momentum fraction carried by the parton. The integral over x must be weighted by the probability $f_i(x)$ for the parton of type i to have a fraction x of momentum. These probability distributions are not predicted by the model and are, in this quark parton picture, fundamental parameters of the proton. The structure function W_2 becomes

$$W_2(\nu, Q^2) = \sum_i\int_0^1 dx\, f_i(x)e_i^2\delta(\nu - Q^2/2Mx). \tag{9.31}$$

Using the result for the Dirac δ-function (see appendix E, equation (E.34))

$$\delta(g(x)) = \frac{\delta(x - x_0)}{|dg/dx|_{x=x_0}} \tag{9.32}$$

where x_0 is defined by $g(x_0) = 0$, we can rewrite

$$\delta(\nu - Q^2/2Mx) = (x/\nu)\delta(x - Q^2/2M\nu) \tag{9.33}$$

under the x integral. Hence we obtain

$$\nu W_2(\nu, Q^2) = \sum_i e_i^2 x f_i(x) \equiv F_2(x) \tag{9.34}$$

which is the desired scaling behaviour. Similar manipulations lead to

$$MW_1(\nu, Q^2) = F_1(x) \tag{9.35}$$

where

$$2xF_1(x) = F_2(x). \tag{9.36}$$

This relation between F_1 and F_2 is called the Callan–Gross relation (see Callan and Gross 1969): it is a direct consequence of our assumption of spin-$\frac{1}{2}$ quarks. The physical origin of this relation is best discussed in terms of virtual photon total cross sections for transverse ($\lambda = \pm 1$) virtual photons and for a longitudinal/scalar ($\lambda = 0$) virtual photon contribution. The longitudinal/scalar photon is present because $q^2 \neq 0$ for a virtual photon (see comment (4) in section 8.3.1). However, in the discussion of polarization vectors a slight difference occurs for space-like q^2. In a frame in which

$$q^\mu = (q^0, 0, 0, q^3) \tag{9.37}$$

the transverse polarization vectors are as before

$$\epsilon^\mu(\lambda = \pm 1) = \mp 2^{-1/2}(0, 1, \pm \mathrm{i}, 0) \tag{9.38}$$

with normalization

$$\epsilon^2(\lambda = \pm 1) = -1. \tag{9.39}$$

To construct the longitudinal/scalar polarization vector, we must satisfy

$$q \cdot \epsilon = 0 \tag{9.40}$$

and so are led to the result

$$\epsilon^\mu(\lambda = 0) = (1/\sqrt{Q^2})(q^3, 0, 0, q^0) \tag{9.41}$$

with

$$\epsilon^2(\lambda = 0) = +1. \tag{9.42}$$

The precise definition of a virtual photon cross section is obviously just a convention. It is usually taken to be

$$\sigma_\lambda(\gamma \mathrm{p} \to \mathrm{X}) = (4\pi^2\alpha/K)\epsilon^*_\mu(\lambda)\epsilon_\nu(\lambda)W^{\mu\nu} \tag{9.43}$$

by analogy with the total cross section for real photons of polarization λ incident on an unpolarized proton target. Note the presence of the factor $W^{\mu\nu}$ defined in (9.3). The factor K is the flux factor: for real photons, producing a final state of mass W, this is just

$$K = (W^2 - M^2)/2M. \tag{9.44}$$

In the so-called 'Hand convention', this same factor is used for virtual photons which produce a final state of mass W. With these definitions we find (see problem 9.3) that the transverse ($\lambda = \pm 1$) photon cross section

$$\sigma_{\rm T} = \left(\frac{4\pi^2\alpha}{K}\right)\frac{1}{2}\sum_{\lambda=\pm 1}\epsilon_\mu^*(\lambda)\epsilon_\nu(\lambda)W^{\mu\nu} \tag{9.45}$$

is given by

$$\sigma_{\rm T} = (4\pi^2\alpha/K)W_1 \tag{9.46}$$

and the longitudinal/scalar cross section

$$\sigma_{\rm S} = (4\pi^2\alpha/K)\epsilon_\mu^*(\lambda = 0)\epsilon_\nu(\lambda = 0)W^{\mu\nu} \tag{9.47}$$

by

$$\sigma_{\rm S} = (4\pi^2\alpha/K)[(1 + \nu^2/Q^2)W_2 - W_1]. \tag{9.48}$$

In fact these expressions give an intuitive explanation of the positivity properties of W_1 and W_2, namely

$$W_1 \geq 0 \tag{9.49}$$
$$(1 + \nu^2/Q^2)W_2 - W_1 \geq 0. \tag{9.50}$$

The combination in the $\lambda = 0$ cross section is sometimes denoted by $W_{\rm L}$:

$$W_{\rm L} = (1 + \nu^2/Q^2)W_2 - W_1. \tag{9.51}$$

The scaling limit of these expressions can be taken using

$$\nu W_2 \to F_2 \tag{9.52}$$
$$M W_1 \to F_1 \tag{9.53}$$

and $x = Q^2/2M\nu$ finite, as Q^2 and ν grow large. We find

$$\sigma_{\rm T} \to \frac{4\pi^2\alpha}{MK}F_1(x) \tag{9.54}$$

and

$$\sigma_{\rm S} \to (4\pi^2\alpha/MK)(1/2x)(F_2 - 2xF_1) \tag{9.55}$$

where we have neglected a term of order F_2/ν in the last expression. Thus the Callan–Gross relation corresponds to the result

$$\sigma_{\rm S}/\sigma_{\rm T} \to 0 \tag{9.56}$$

in terms of photon cross sections.

A parton calculation using point-like spin-0 partons shows the opposite result, namely

$$\sigma_{\rm T}/\sigma_{\rm S} \to 0. \tag{9.57}$$

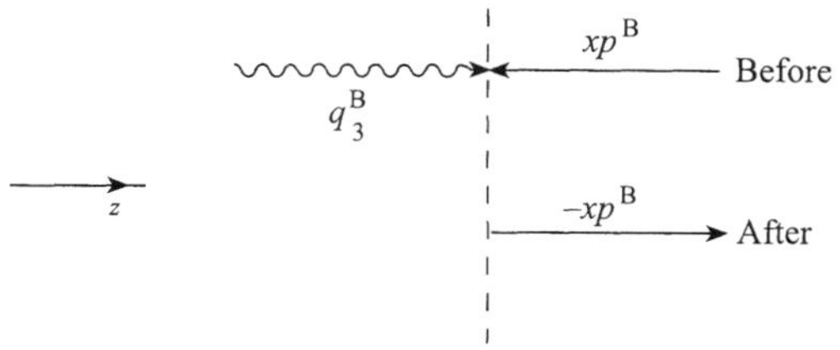

Figure 9.6. Photon–parton interaction in the Breit frame.

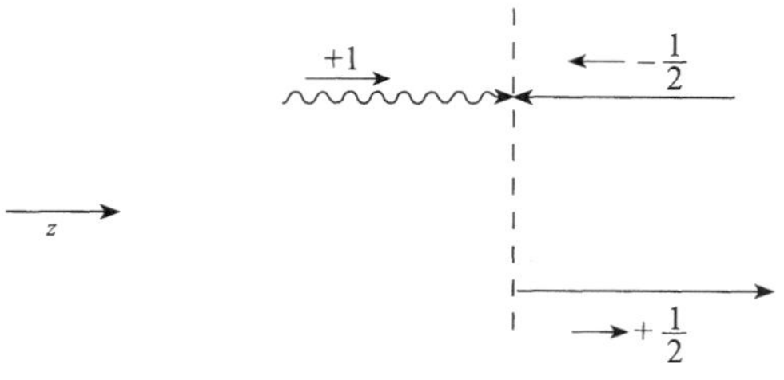

Figure 9.7. Angular momentum balance for absorption of photon by helicity-conserving spin-$\frac{1}{2}$ parton.

Both these results may be understood by considering the helicities of partons and photons in the so-called parton Breit or 'brick-wall' frame. The particular frame is the one in which the photon and parton are collinear and the 3-momentum of the parton is exactly reversed by the collision (see figure 9.6). In this frame, the photon transfers no energy, only 3-momentum. The vanishing of transverse photon cross sections for scalar partons is now obvious. The transverse photons bring in ± 1 units of the z-component of angular momentum: spin-0 partons cannot absorb this. Thus only the scalar $\lambda = 0$ cross section is non-zero. For spin-$\frac{1}{2}$ partons the argument is slightly more complicated in that it depends on the helicity properties of the γ_μ coupling of the parton to the photon. As is shown in problem 9.4, for massless spin-$\frac{1}{2}$ particles the γ_μ coupling conserves helicity—i.e. the projection of spin along the direction of motion of the particle. Thus in the Breit frame, and neglecting parton masses, conservation of helicity necessitates a change in the z-component of the parton's angular momentum by ± 1 unit, thereby requiring the absorption of a transverse photon (figure 9.7). The Lorentz transformation from the parton Breit frame to the 'laboratory' frame does not affect the ratio of transverse to longitudinal photons, if we neglect the parton transverse momenta. These arguments therefore make clear the origin of the Callan–Gross relation. Experimentally, the Callan–Gross relation is reasonably well satisfied in that $R = \sigma_S/\sigma_T$ is small for most, if not all, of the deep inelastic regime (figure 9.8). This leads us to suppose that the *electrically charged partons coupling to photons have spin-$\frac{1}{2}$*.

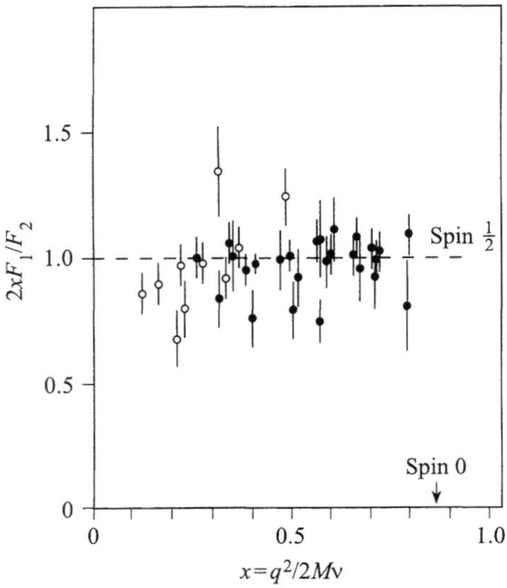

Figure 9.8. The ratio $2xF_1/F_2$: ∘, $1.5 < q^2 < 4$ GeV^2; •, $0.5 < q^2 < 11$ GeV^2; ×, $12 < q^2 < 16$ GeV^2. (From Perkins D 1987 *Introduction to High Energy Physics* 3rd edn, courtesy Addison-Wesley Publishing Company.)

9.3 The quark parton model

We now proceed a stage further with the idea that the 'partons' *are* quarks (and anti-quarks). If we assume that the photon only couples to these objects, we can make more specific scaling predictions. The quantum numbers of the quarks have been given in chapter 1. For a proton we have the result (cf (9.34))

$$F_2^{\mathrm{ep}}(x) = x\{\tfrac{4}{9}[u(x) + \bar{u}(x)] + \tfrac{1}{9}[d(x) + \bar{d}(x) + s(x) + \bar{s}(x)] + \cdots\} \quad (9.58)$$

where $u(x)$ is the probability distribution for u quarks in the proton, $\bar{u}(x)$ for u anti-quarks and so on in an obvious notation, and the dots indicate further possible flavours. So far we do not seem to have gained much, replacing one unknown function by six or more unknown functions. The full power of the quark parton model lies in the fact that the same distribution functions appear, in different combinations, for neutron targets, and in the analogous scaling functions for deep inelastic scattering with neutrino and antineutrino beams (see volume 2). For electron scattering from neutron targets we can use I-spin invariance (see for example Close 1979, or Leader and Predazzi 1996) to relate the distribution of u and d quarks in a neutron to the distributions in a proton. The result is

$$u^{\mathrm{p}}(x) = d^{\mathrm{n}}(x) \equiv u(x) \quad (9.59)$$

$$d^{\mathrm{p}}(x) = u^{\mathrm{n}}(x) \equiv d(x). \quad (9.60)$$

Hence the scaling function for en scattering may be written

$$F_2^{\text{en}}(x) = x\{\tfrac{4}{9}[d(x) + \bar{d}(x)] + \tfrac{1}{9}[u(x) + \bar{u}(x) + s(x) + \bar{s}(x)] + \cdots\}. \quad (9.61)$$

The quark distributions inside the proton and neutron must satisfy some constraints. Since both proton and neutron have strangeness zero, we have a *sum rule* (treating only u, d and s flavours from now on)

$$\int_0^1 \mathrm{d}x\,[s(x) - \bar{s}(x)] = 0. \quad (9.62)$$

Similarly, from the proton and neutron charges we obtain two other sum rules:

$$\int_0^1 \mathrm{d}x\,\{\tfrac{2}{3}[u(x) - \bar{u}(x)] - \tfrac{1}{3}[d(x) - \bar{d}(x)]\} = 1 \quad (9.63)$$

$$\int_0^1 \mathrm{d}x\,\{\tfrac{2}{3}[d(x) - \bar{d}(x)] - \tfrac{1}{3}[u(x) - \bar{u}(x)]\} = 0. \quad (9.64)$$

These are equivalent to the sum rules

$$2 = \int_0^1 \mathrm{d}x\,[u(x) - \bar{u}(x)] \quad (9.65)$$

$$1 = \int_0^1 \mathrm{d}x\,[d(x) - \bar{d}(x)] \quad (9.66)$$

which are, of course, just the excess of u and d quarks over anti-quarks inside the proton. Testing these sum rules requires neutrino data to separate the various structure functions, as we shall explain in volume 2.

One can gain some further insight if one is prepared to make a model. For example, one can introduce the idea of 'valence' quarks (those of the elementary constituent quark model) and 'sea' quarks ($q\bar{q}$ pairs created virtually). Then, in a proton, the u and d quark distributions would be parametrized by the sum of valence and sea contributions

$$u = u_{\mathrm{V}} + q_{\mathrm{S}} \quad (9.67)$$

$$d = d_{\mathrm{V}} + q_{\mathrm{S}} \quad (9.68)$$

while the anti-quark and strange quark distributions are taken to be pure sea

$$\bar{u} = \bar{d} = s = \bar{s} = q_{\mathrm{S}} \quad (9.69)$$

where we have assumed that the 'sea' is flavour-independent. Such a model replaces the six unknown functions now in play by three, and is consequently more predictive. The strangeness sum rule (9.62) is now satisfied automatically, while (9.65) and (9.66) are satisfied by the valence distributions alone:

$$\int_0^1 \mathrm{d}x\, u_V(x) = 2 \tag{9.70}$$

$$\int_0^1 \mathrm{d}x\, d_V(x) = 1. \tag{9.71}$$

It is interesting to consider the ratio $F_2^{\mathrm{en}}/F_2^{\mathrm{ep}}$. From (9.58), (9.61), (9.67) and (9.68) we have

$$F_2^{\mathrm{en}}/F_2^{\mathrm{ep}} = \frac{4d_V + u_V + \Sigma}{4u_V + d_V + \Sigma} \tag{9.72}$$

where Σ is the total sea contribution. The data are shown in figure 9.9. Independent of any model, since all the probability functions must be positive-definite, we obtain immediately the bounds

$$\tfrac{1}{4} \leq F_2^{\mathrm{en}}/F_2^{\mathrm{ep}} \leq 4 \tag{9.73}$$

which are consistent with the data. However, we note that, at small x, the ratio appears to tend to 1, indicating—from (9.72)—that the sea contribution dominates. This could then explain why the distribution shown in figure 9.2 differed so considerably from our simplest expectation: perhaps it is the sum of two contributions, one peaked roughly at $x = \frac{1}{3}$ (valence) and the other coming in strongly at small x (sea). Such a picture may be at least qualitatively tested by considering the *difference* between the proton and neutron structure functions, from which the sea contribution may be assumed to cancel out. An example of such a distribution is shown in figure 9.10 which does roughly indicate the behaviour expected for the valence quarks. More quantitative separations between quark and anti-quark contributions can be made with the help of the neutrino data. An important sum rule emerges from the picture of $xf_i(x)$ as the fractional momentum carried by quark i. This is the *momentum sum rule*

$$\int_0^1 \mathrm{d}x\, x[u(x) + \bar{u}(x) + d(x) + \bar{d}(x) + s(x) + \bar{s}(x)] = 1 - \epsilon \tag{9.74}$$

where ϵ is interpreted as the fraction of the proton momentum that is not carried by quarks. The integral in (9.74) is directly related to ν and $\bar{\nu}$ cross sections, and its evaluation implies $\epsilon \simeq \frac{1}{2}$ (the CHARM (1981) result was $1-\epsilon = 0.44\pm0.02$). This suggests that about half the total momentum is carried by uncharged objects. These remaining 'partons' will be identified with the gluons of QCD.

9.4 The Drell–Yan process

Much of the importance of the parton model lies outside its original domain of deep inelastic scattering. In deep inelastic scattering it is possible to provide

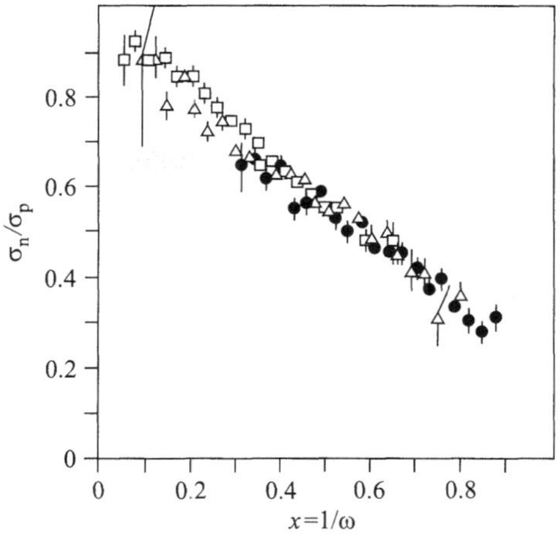

Figure 9.9. The differential cross section $\sigma_{\rm n}/\sigma_{\rm p}$ versus x (Bodek *et al* 1974). •, 15°, 19°, 26°, 34°; △, 18°, 26°, 34°; □, 6°, 10°. If, as the data indicate, $\sigma_{\rm S}/\sigma_{\rm T}$ is the same for n and p, $\sigma_{\rm n}/\sigma_{\rm p}$ may be interpreted as $F_2^{\rm en}/F_2^{\rm ep}$.

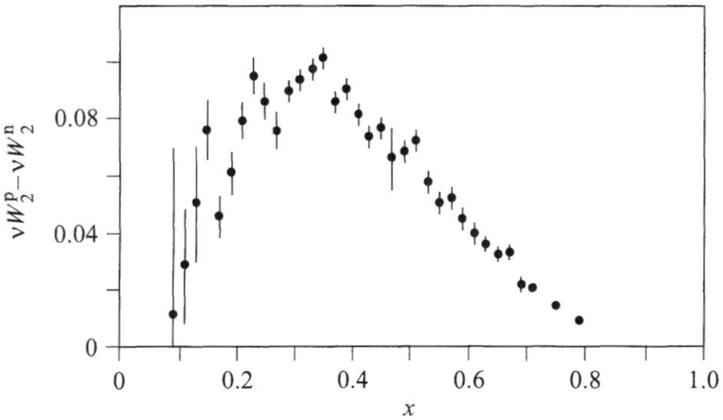

Figure 9.10. $\nu W_2^{\rm p} - \nu W_2^{\rm n}$ versus x (Bodek *et al* 1973).

a more formal basis for the parton model in terms of light-cone and short-distance operator expansions (see chapter 18 of Peskin and Schroeder 1995). The advantage of the parton formulation lies in the fact that it suggests other processes for which a parton description may be relevant but for which formal operator arguments are not possible. The most important such process is the Drell–Yan process (Drell and Yan 1970)

$$\mathrm{p} + \mathrm{p} \to \mu^+\mu^- + \mathrm{X} \tag{9.75}$$

in which a $\mu^+\mu^-$ pair is produced in proton–proton collisions along with unobserved hadrons X, as shown in figure 9.11. The assumption of the parton

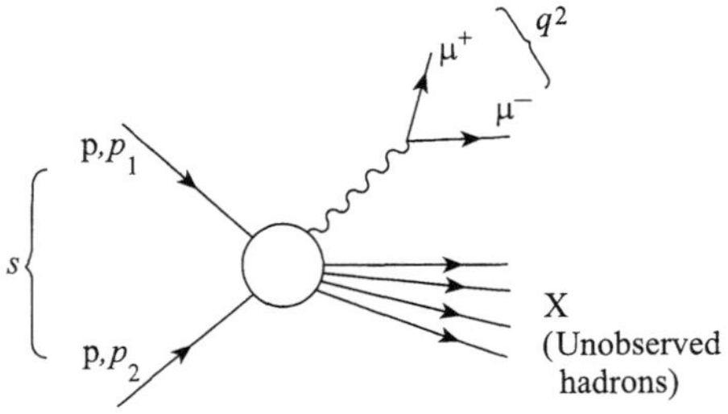

Figure 9.11. Drell–Yan process.

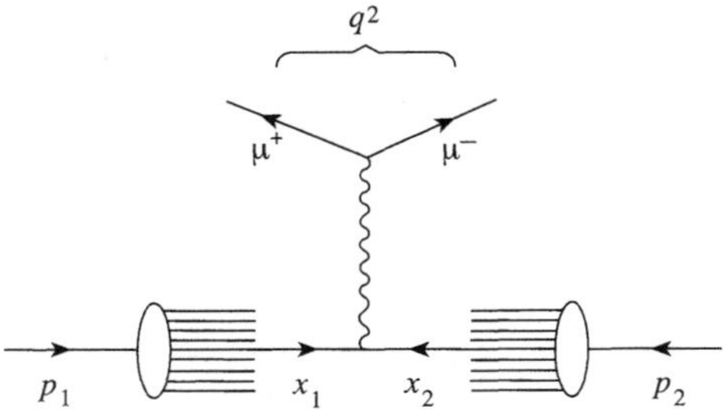

Figure 9.12. Parton model amplitude for the Drell–Yan process.

model is that in the limit

$$s \to \infty \qquad \text{with } \tau = q^2/s \text{ finite} \tag{9.76}$$

the dominant process is that shown in figure 9.12: a quark and anti-quark from different hadrons are assumed to annihilate to a virtual photon which then decays to a $\mu^+\mu^-$ pair (compare figures 9.3 and 9.4), the remaining quarks and anti-quarks subsequently emerging as hadrons.

Let us work in the CM system and neglect all masses. In this case we have

$$p_1^\mu = (P, 0, 0, P) \qquad p_2^\mu = (P, 0, 0, -P) \tag{9.77}$$

and

$$s = 4P^2. \tag{9.78}$$

Neglecting quark masses and transverse momenta, we have quark momenta

$$p_{q_1}^\mu = x_1(P, 0, 0, P) \tag{9.79}$$
$$p_{q_2}^\mu = x_2(P, 0, 0, -P) \tag{9.80}$$

and the photon momentum

$$q = p_{q_1} + p_{q_2} \tag{9.81}$$

has non-zero components

$$q^0 = (x_1 + x_2)P \tag{9.82}$$
$$q^3 = (x_1 - x_2)P. \tag{9.83}$$

Thus we find

$$q^2 = 4x_1x_2P^2 \tag{9.84}$$

and hence

$$\boxed{\tau = q^2/s = x_1x_2.} \tag{9.85}$$

The cross section for the basic process

$$\mathrm{q\bar{q}} \to \mu^+\mu^- \tag{9.86}$$

is calculated using the result of problem 8.19. Since the QED process

$$\mathrm{e^+e^-} \to \mu^+\mu^- \tag{9.87}$$

has the cross section (neglecting all masses)

$$\sigma(\mathrm{e^+e^-} \to \mu^+\mu^-) = 4\pi\alpha^2/3q^2 \tag{9.88}$$

we expect the result for a quark of type a with charge e_a (in units of e) to be

$$\sigma(q_a\bar{q}_a \to \mu^+\mu^-) = (4\pi\alpha^2/3q^2)e_a^2. \tag{9.89}$$

To obtain the parton model prediction for proton–proton collisions, one merely multiplies this cross section by the probabilities for finding a quark of type a with momentum fraction x_1, and an anti-quark of the same type with fraction x_2, namely

$$q_a(x_1)\,\mathrm{d}x_1\,\bar{q}_a(x_2)\,\mathrm{d}x_2. \tag{9.90}$$

There is, of course, another contribution for which the anti-quark has fraction x_1 and the quark x_2:

$$\bar{q}_a(x_1)\,\mathrm{d}x_1\,q_a(x_2)\,\mathrm{d}x_2. \tag{9.91}$$

Thus the Drell–Yan prediction is

$$\boxed{\begin{aligned}&\mathrm{d}^2\sigma(\mathrm{pp} \to \mu^+\mu^- + \mathrm{X})\\ &\qquad = \frac{4\pi\alpha^2}{9q^2}\sum_a e_a^2[q_a(x_1)\bar{q}_a(x_2) + \bar{q}_a(x_1)q_a(x_2)]\,\mathrm{d}x_1\,\mathrm{d}x_2\end{aligned}} \tag{9.92}$$

where we have included a factor $\frac{1}{3}$ to account for the *colour* of the quarks: in order to make a colour singlet photon, one needs to match the colours of quark and anti-quark. Equation (9.92) is the master formula. Its importance lies in the fact

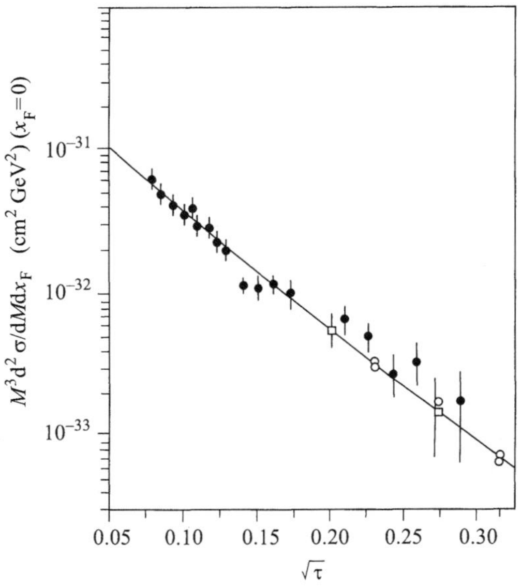

Figure 9.13. The dimensionless cross section $M^3 d^2\sigma/dM dx_F$ ($M = \sqrt{q^2}$) at $x_F = 0$ for pN scattering, plotted against $\sqrt{\tau} = M/\sqrt{s}$ (Scott 1985): •, $\sqrt{s} = 62$ GeV; □, 44; □, 27.4; ○, 23.8.

that the *same* quark distribution functions are measured in deep inelastic lepton scattering so one can make absolute predictions. For example, if the photon in figure 9.12 is replaced by a W(Z), one can predict W(Z) production cross sections, as we shall see in volume 2.

We would expect some 'scaling' property to hold for this cross section, following from the point-like constituent cross section (9.89). One way to exhibit this is to use the variables q^2 and $x_F = x_1 - x_2$ as discussed in problem 9.6. There it is shown that the dimensionless quantity

$$q^4 \frac{d^2\sigma}{dq^2\, dx_F} \tag{9.93}$$

should be a function of x_F and the *ratio* $\tau = q^2/s$. The data bear out this prediction well—see figure 9.13.

Furthermore, the assumption that the lepton pair is produced via quark–anti-quark annihilation to a virtual photon can be checked by observing the angular distribution of either lepton in the dilepton rest frame, relative to the incident proton beam direction. This distribution is expected to be the same as in $e^+e^- \rightarrow \mu^+\mu^-$, namely (cf (8.193))

$$d\sigma/d\Omega \propto (1 + \cos^2\theta) \tag{9.94}$$

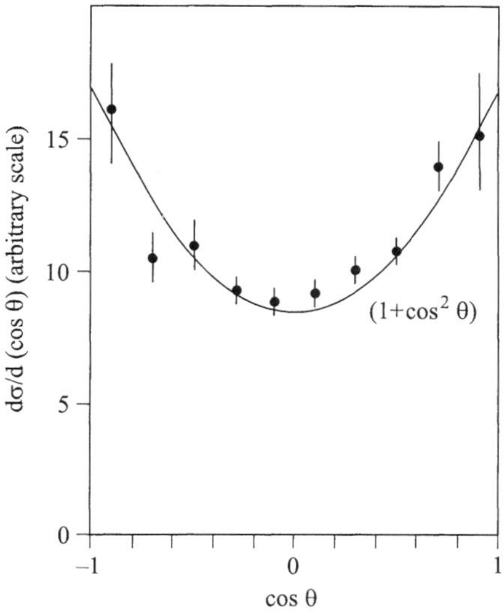

Figure 9.14. Angular distribution of muons, measured in the $\mu^+\mu^-$ rest frame, relative to the incident beam direction, in the Drell–Yan process. (From Perkins D 1987 *Introduction to High Energy Physics* 3rd edn, courtesy Addison-Wesley Publishing Company.)

as is indeed observed (figure 9.14). Note that figure 9.14 provides evidence that the quarks have spin-$\frac{1}{2}$: if they are assumed to have spin-0, the angular distribution would be (see problem 9.7) proportional to $(1 - \cos^2\theta)$, and this is clearly ruled out.

9.5 e^+e^- annihilation into hadrons

The last electromagnetic process we wish to consider is electron–positron annihilation into hadrons (figure 9.15):

$$e^+e^- \rightarrow X. \tag{9.95}$$

As usual, the dominance of the one-photon intermediate state is assumed. Figure 9.15 is clearly a generalization of figure 8.9, the latter describing the particular case in which the final hadronic state is $\pi^+\pi^-$. As a preliminary to discussing (9.95), let us therefore revisit $e^+e^- \rightarrow \pi^+\pi^-$ first.

The $O(e^2)$ amplitude is given in equation (8.158). Performing the spin-average $\frac{1}{2}\sum |\mathcal{M}|^2$ as in the case of $e^-s^+ \rightarrow e^-s^+$ ((8.116)–(8.119)), or applying crossing symmetry to the end result (8.119), we find (problem 9.7) that the '$L \cdot T$' product for $e^+e^- \rightarrow \pi^+\pi^-$ (or s^+s^-) is given by $8[-2(k\cdot p_1)(k_1\cdot p_1)+q^2M^2/2]$, where $q^2 = (k + k_1)^2 = (p_1 + p')^2$. Borrowing the general CM cross section formula (6.129) from chapter 6, as in (8.120), and neglecting both the electron

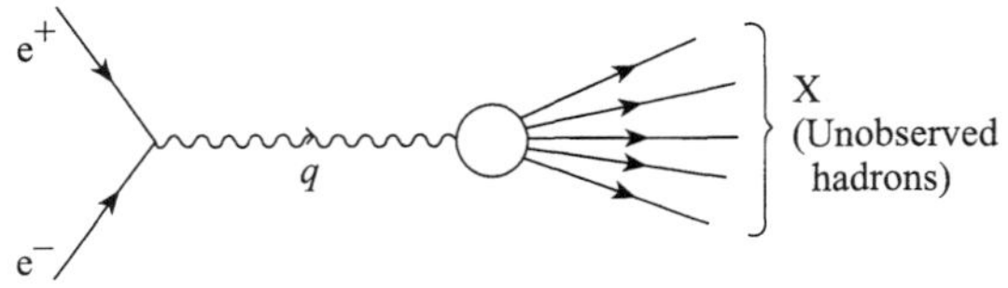

Figure 9.15. e^+e^- annihilation to hadrons in one-photon approximation.

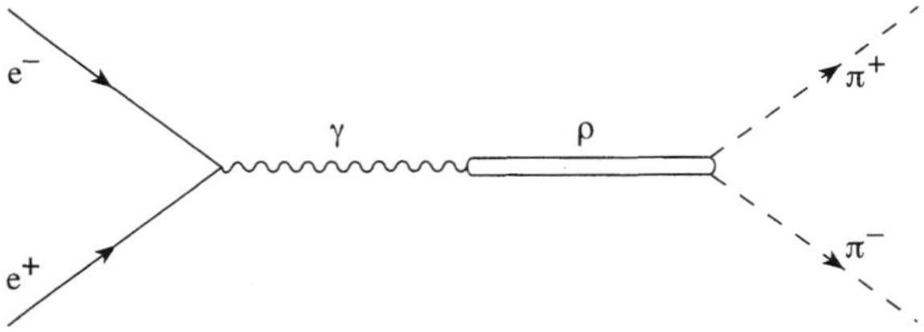

Figure 9.16. ρ-dominance of the pion electromagnetic form factor in the time-like ($q^2 > 0$) region.

and the pion mass, we obtain for the unpolarized CM differential cross section (problem 9.7)

$$\left(\frac{d\bar{\sigma}}{d\Omega}\right)_{CM} = \frac{F^2(q^2)\alpha^2}{4q^2}(1-\cos^2\theta) \tag{9.96}$$

and the total unpolarized cross section is

$$\bar{\sigma} = F^2(q^2)\frac{2\pi\alpha^2}{3q^2}. \tag{9.97}$$

The cross section $\bar{\sigma}$ therefore decreases with q^2 as $1/q^2$ in general, just like that for $e^+e^- \to \mu^+\mu^-$ as in (9.88), but its behaviour is modified by the square of the form factor, evaluated at time-like q^2. When the measured $\bar{\sigma}$ is plotted against q^2 for $q^2 \leq 1$ (GeV)^2, a pronounced resonance is seen at $q^2 \approx m_\rho^2$, superimposed on the smooth $1/q^2$ background, where m_ρ is the mass of the rho resonance ($J^P = 1^-$ $q\bar{q}$ state). The interpretation of this is shown in figure 9.16. $F(q^2)$ should therefore be parametrized as a resonance, as in (6.107)—or a more sophisticated version to take account of the fact that the π's are emitted in an $\ell = 1$ state. Just as $F^2(q^2)$ modified the point-like cross section in the space-like region for $e^-\pi^+ \to e^-\pi^+$, so here it modifies the point-like ($\sim 1/q^2$) behaviour in the time-like region.

Returning now to the process (9.95), the cross section for it is shown as a function of CM energy $(q^2)^{1/2}$ in figure 9.17. The general point-like fall-off as $1/q^2$ is seen, with peaks due to a succession of boson resonances superimposed (ρ, J/ψ, Υ, Z^0, ...). The $1/q^2$ fall-off is suggestive of a (point-like) parton

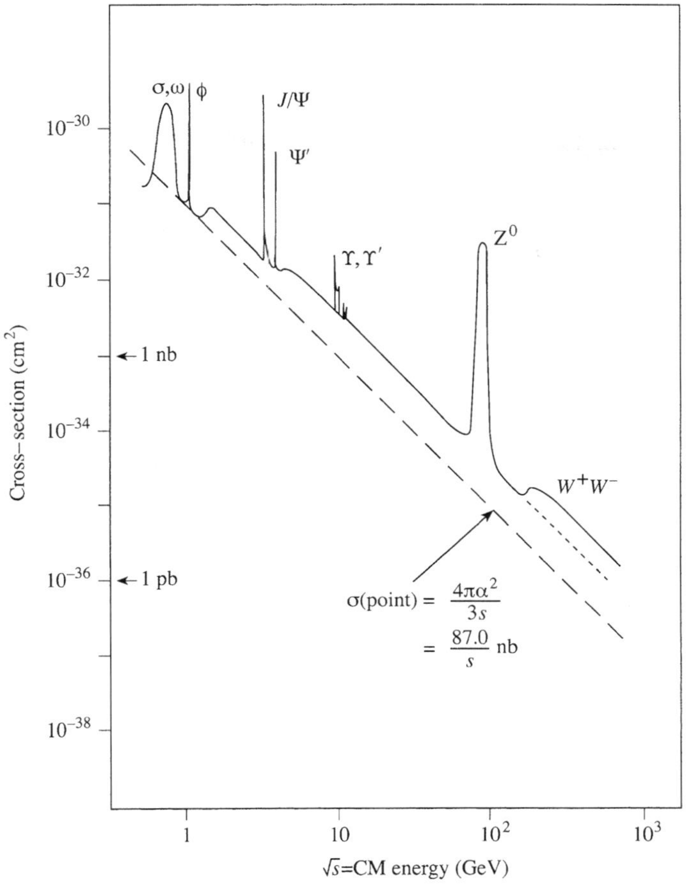

Figure 9.17. The cross section σ for the annihilation process $e^+e^- \to$ anything, as a function of cm energy. The prominent peaks are due to various boson resonances, but the general $1/q^2$ dependence of the cross section, typical of a point-like process, is clear. (From Perkins D 2000 *Introduction to High Energy Physics* 4th edn, courtesy Cambridge University Press.)

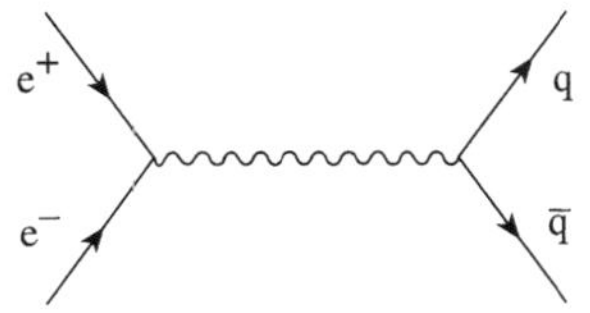

Figure 9.18. Parton model subprocess in $e^+e^- \to$ hadrons.

picture and indeed the process (9.95) is similar to the Drell–Yan one:

$$\mathrm{pp} \to \mu^+\mu^- + \mathrm{X}. \tag{9.98}$$

It is natural to imagine that at large q^2 the basic subprocess is quark–anti-quark pair creation (figure 9.18). The total cross section for $\mathrm{q\bar{q}}$ pair production is then (cf (9.89))

$$\sigma(\mathrm{e^+e^-} \to q_a\bar{q}_a) = (4\pi\alpha^2/3q^2)e_a^2. \tag{9.99}$$

In the vicinity of mesonic resonances such as the ρ, we can infer that the dominant component in the final state is that in which the $\mathrm{q\bar{q}}$ pair is strongly bound into a mesonic state, which then decays into hadrons. Away from resonances, and increasingly at larger values of q^2, the produced $\mathrm{q\bar{q}}$ pair seek to separate from the interaction region. As they draw apart, however, the interaction between them increases (recall section 2.6), producing more $\mathrm{q\bar{q}}$ pairs, together with radiated gluons. In this process, the coloured quarks and gluons eventually must form colourless hadrons, since we know that no coloured particles have been observed ('confinement of colour'). If one assumes that the presumed colour confinement mechanism does not affect the prediction (9.99), then we arrive at the result that, at large q^2,

$$\sigma(\mathrm{e^+e^-} \to \text{hadrons}) = (4\pi\alpha^2/3q^2)\sum e_a^2. \tag{9.100}$$

This model is best tested by taking out the dominant $1/q^2$ behaviour and plotting the ratio

$$R = \frac{\sigma(\mathrm{e^+e^-} \to \text{hadrons})}{\sigma(\mathrm{e^+e^-} \to \mu^+\mu^-)} = \sum_a e_a^2. \tag{9.101}$$

For the light quarks u, d and s occurring in three colours, we therefore predict

$$R = 3[(\tfrac{2}{3})^2 + (-\tfrac{1}{3})^2 + (-\tfrac{1}{3})^2] = 2. \tag{9.102}$$

Above the c threshold but below the b threshold we expect $R = \frac{10}{3}$, and above the b threshold $R = \frac{11}{3}$. These expectations are in reasonable accord with experiment, especially at energies well beyond the resonance region and the b threshold, as figure 9.19 shows. The important subject of QCD corrections to the parton model result (9.101) will be discussed in volume 2.

The success of this prediction leads one to consider more detailed consequences of the picture. For example, the angular distribution of massless spin-$\frac{1}{2}$ quarks is expected to be (cf (8.193) again)

$$\mathrm{d}\sigma/\mathrm{d}\Omega = (\alpha^2/4q^2)e_a^2(1 + \cos^2\theta) \tag{9.103}$$

just as for the $\mu^+\mu^-$ process. However, in this case there is an important difference: the quarks are not observed! Nevertheless a remarkable 'memory' of (9.103) is retained by the observed final-state hadrons. Experimentally one observes events in which hadrons emerge from the interaction region in two

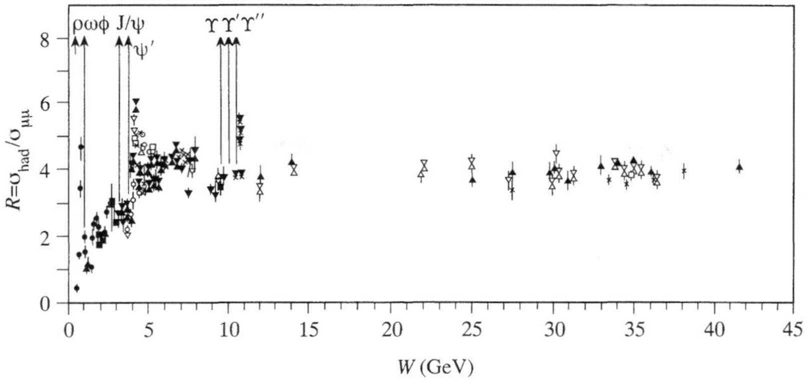

Figure 9.19. The ratio R (see equation (9.101)) (courtesy R Cashmore).

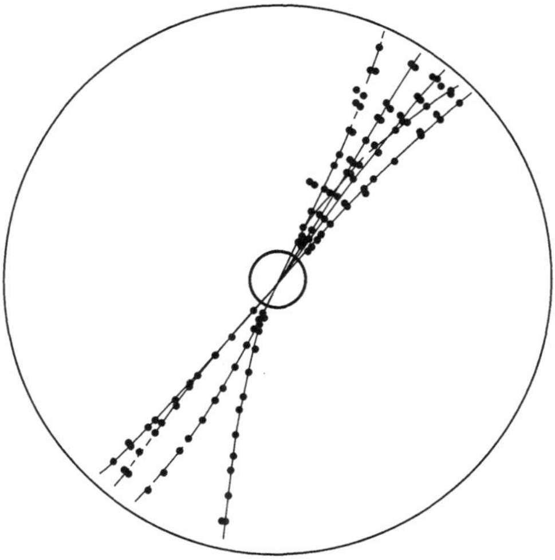

Figure 9.20. Two-jet event in e^+e^- annihilation from the TASSO detector at the e^+e^- storage ring PETRA.

relatively well-collimated cones or 'jets'—see figure 9.20. The distribution of events as a function of the (inferred) angle of the jet axis is shown in figure 9.21 and is in good agreement with (9.103). The interpretation is that the primary process is $e^+e^- \rightarrow q\bar{q}$, the quark and the anti-quark then turning into hadrons as they separate and experience the very strong colour forces, but without losing the memory of the original quark angular distribution.

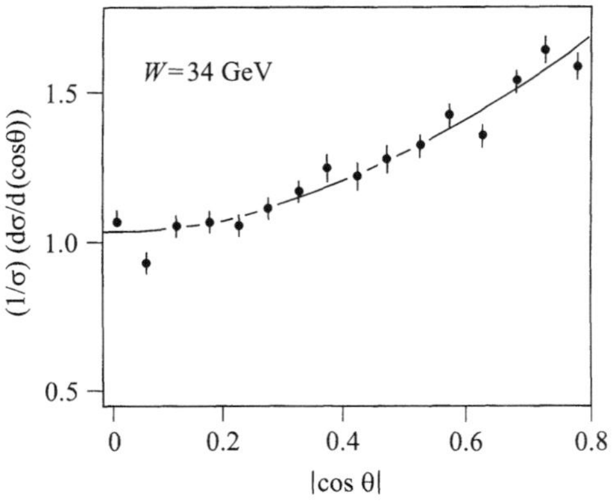

Figure 9.21. Angular distribution of jets in two-jet events, measured in the two-jet rest frame, relative to the incident beam direction, in the process $e^+e^- \rightarrow$ two jets (Althoff *et al* 1984). The full curve is the $(1+\cos^2\theta)$ distribution. Since it is not possible to say which jet corresponded to the quark and which to the anti-quark, only half the angular distribution can be plotted. The asymmetry visible in figure 8.20(*b*) is therefore not apparent.

Problems

9.1 The various normalization factors in equations (9.3) and (9.11) may be checked in the following way. The cross section for inclusive electron–proton scattering may be written (equation (9.11)):

$$\mathrm{d}\sigma = \left(\frac{4\pi\alpha}{q^2}\right)^2 \frac{1}{4[(k\cdot p)^2 - m^2M^2]^{1/2}} 4\pi M L_{\mu\nu} W^{\mu\nu} \frac{\mathrm{d}^3k'}{2\omega'(2\pi)^3} \tag{9.104}$$

in the usual one-photon exchange approximation, and the tensor $W^{\mu\nu}$ is related to hadronic matrix elements of the electromagnetic current operator by equation (9.3):

$$e^2 W^{\mu\nu}(q,p) = \frac{1}{4\pi M}\frac{1}{2}\sum_s\sum_{\mathrm{X}} \langle \mathrm{p};\, p, s|\hat{j}^\mu_{\mathrm{em}}(0)|\mathrm{X};\, p'\rangle \times \langle \mathrm{X};\, p'|\hat{j}^\nu_{\mathrm{em}}(0)|\mathrm{p};\, p, s\rangle (2\pi)^4\delta^4(p+q-p')$$

where the sum X is over all possible hadronic final states. If we consider the special case of elastic scattering, the sum over X is only over the final proton's degrees of freedom:

$$e^2 W^{\mu\nu}_{\mathrm{el}} = \frac{1}{4\pi M}\frac{1}{2}\sum_s\sum_{s'} \langle \mathrm{p};\, p, s|\hat{j}^\mu_{\mathrm{em}}(0)|\mathrm{p};\, p', s'\rangle\langle \mathrm{p};\, p', s'|\hat{j}^\nu_{\mathrm{em}}(0)|\mathrm{p};\, p, s\rangle \times (2\pi)^4\delta^4(p+q-p')\frac{1}{(2\pi)^3}\frac{\mathrm{d}^3p'}{2E'}.$$

Now use equation (8.206) with $\mathcal{F}_1 = 1$ and $\kappa = 0$ (i.e. the electromagnetic current matrix element for a 'point' proton) to show that the resulting cross section is identical to that for elastic eμ scattering.

9.2

(a) Perform the contraction $L_{\mu\nu}W^{\mu\nu}$ for inclusive inelastic electron–proton scattering (remember $q^\mu L_{\mu\nu} = q^\nu L_{\mu\nu} = 0$). Hence verify that the inclusive differential cross section in terms of 'laboratory' variables, and neglecting the electron mass, has the form

$$\frac{\mathrm{d}^2\sigma}{\mathrm{d}\Omega\mathrm{d}k'} = \frac{\alpha^2}{4k^2\sin^4(\theta/2)}[W_2\cos^2(\theta/2) + W_1 2\sin^2(\theta/2)].$$

(b) By calculating the Jacobian

$$J = \begin{vmatrix} \partial u/\partial x & \partial u/\partial y \\ \partial v/\partial x & \partial v/\partial y \end{vmatrix}$$

for a change of variables $(x, y) \to (u, v)$

$$\mathrm{d}u\,\mathrm{d}v = |J|\mathrm{d}x\,\mathrm{d}y$$

find expressions for $\mathrm{d}^2\sigma/\mathrm{d}Q^2\,\mathrm{d}\nu$ and $\mathrm{d}^2\sigma/\mathrm{d}x\,\mathrm{d}y$, where Q^2 and ν have their usual significance, and x is the scaling variable $Q^2/2M\nu$ and $y = \nu/k$.

9.3 Consider the description of inelastic electron–proton scattering in terms of virtual photon cross sections:

(a) In the 'laboratory' frame with

$$p^\mu = (M, 0, 0, 0) \qquad \text{and} \qquad q^\mu = (q^0, 0, 0, q^3)$$

evaluate the transverse spin sum

$$\tfrac{1}{2}\sum_{\lambda=\pm 1} \epsilon_\mu(\lambda)\epsilon_\nu^*(\lambda)W^{\mu\nu}.$$

Hence show that the 'Hand' cross section for transverse virtual photons is

$$\sigma_{\mathrm{T}} = (4\pi^2\alpha/K)W_1.$$

(b) Using the definition

$$\epsilon_{\mathrm{S}}^\mu = (1/\sqrt{Q^2})(q^3, 0, 0, q^0)$$

and rewriting this in terms of the 'laboratory' 4-vectors p^μ and q^μ, evaluate the longitudinal/scalar virtual photon cross section. Hence show that

$$W_2 = \frac{K}{4\pi^2\alpha}\frac{Q^2}{Q^2+\nu^2}(\sigma_{\mathrm{S}} + \sigma_{\mathrm{T}}).$$

9.4 In this question, we consider a representation of the 4 × 4 Dirac matrices in which (see section 4.4)

$$\boldsymbol{\alpha} = \begin{pmatrix} \boldsymbol{\sigma} & 0 \\ 0 & -\boldsymbol{\sigma} \end{pmatrix} \qquad \beta = \begin{pmatrix} 0 & 1 \\ 1 & 0 \end{pmatrix}.$$

Define also the 4 × 4 matrix $\gamma_5 = \begin{pmatrix} 1 & 0 \\ 0 & -1 \end{pmatrix}$ and the Dirac four-component spinor $u = \begin{pmatrix} \phi \\ \chi \end{pmatrix}$. Then the two-component spinors ϕ, χ satisfy

$$\begin{aligned} \boldsymbol{\sigma} \cdot \boldsymbol{p}\phi &= E\phi - m\chi \\ \boldsymbol{\sigma} \cdot \boldsymbol{p}\chi &= -E\chi + m\psi. \end{aligned}$$

(a) Show that for a massless Dirac particle, ϕ and χ become helicity eigenstates (see section 4.3) with positive and negative helicity respectively.

(b) Defining

$$P_{\mathrm{R}} = \frac{1+\gamma_5}{2} \qquad P_{\mathrm{L}} = \frac{1-\gamma_5}{2}$$

show that $P_{\mathrm{R}}^2 = P_{\mathrm{L}}^2 = 1$, $P_{\mathrm{R}}P_{\mathrm{L}} = 0 = P_{\mathrm{L}}P_{\mathrm{R}}$, and that $P_{\mathrm{R}} + P_{\mathrm{L}} = 1$. Show also that

$$P_{\mathrm{R}}\begin{pmatrix} \phi \\ \chi \end{pmatrix} = \begin{pmatrix} \phi \\ 0 \end{pmatrix} \qquad P_{\mathrm{L}}\begin{pmatrix} \phi \\ \chi \end{pmatrix} = \begin{pmatrix} 0 \\ \chi \end{pmatrix}$$

and hence that P_{R} and P_{L} are projection operators for massless Dirac particles, onto states of definite helicity. Discuss what happens when $m \neq 0$.

(c) The general massless spinor u can be written

$$u = (P_{\mathrm{L}} + P_{\mathrm{R}})u \equiv u_{\mathrm{L}} + u_{\mathrm{R}}$$

where u_{L}, u_{R} have the indicated helicities. Show that

$$\bar{u}\gamma^{\mu}u = \bar{u}_{\mathrm{L}}\gamma^{\mu}u_{\mathrm{L}} + \bar{u}_{\mathrm{R}}\gamma^{\mu}u_{\mathrm{R}}$$

where $\bar{u}_{\mathrm{L}} = u_{\mathrm{L}}^{\dagger}\gamma^0$, $\bar{u}_{\mathrm{R}} = u_{\mathrm{R}}^{\dagger}\gamma^0$; and deduce that in electromagnetic interactions of massless fermions helicity is conserved.

(d) In weak interactions an *axial vector current* $\bar{u}\gamma^{\mu}\gamma_5 u$ also enters. Is helicity still conserved?

(e) Show that the 'Dirac' mass term $m\bar{\hat{\psi}}\hat{\psi}$ may be written as $m(\bar{\hat{\psi}}_{\mathrm{L}}^{\dagger}\hat{\psi}_{\mathrm{R}} + \bar{\hat{\psi}}_{\mathrm{R}}^{\dagger}\hat{\psi}_{\mathrm{L}})$.

9.5 In the HERA colliding beam machine, positrons of total energy 27.5 GeV collide head on with protons of total energy 820 GeV. Neglecting both the positron and the proton rest masses, calculate the centre-of-mass energy in such a collision process.

Some theories have predicted the existence of 'leptoquarks', which could be produced at HERA as a resonance state formed from the incident positron and the struck quark. How would a distribution of such events look, if plotted versus the variable x?

9.6

(a) By the expedient of inserting a δ-function, the differential cross section for Drell–Yan production of a lepton pair of mass $\sqrt{q^2}$ may be written as

$$\frac{\mathrm{d}\sigma}{\mathrm{d}q^2} = \int \mathrm{d}x_1\,\mathrm{d}x_2\,\frac{\mathrm{d}^2\sigma}{\mathrm{d}x_1\,\mathrm{d}x_2}\delta(q^2 - sx_1x_2).$$

Show that this is equivalent to the form

$$\frac{\mathrm{d}\sigma}{\mathrm{d}q^2} = \frac{4\pi\alpha^2}{9q^4}\int \mathrm{d}x_1\,\mathrm{d}x_2\,x_1x_2\delta(x_1x_2 - \tau) \times \sum_a e_a^2[q_a(x_1)\bar{q}_a(x_2) + \bar{q}_a(x_1)q_a(x_2)]$$

which, since $q^2 = s\tau$, exhibits a scaling law of the form

$$s^2\mathrm{d}\sigma/\mathrm{d}q^2 = F(\tau).$$

(b) Introduce the Feynman scaling variable

$$x_{\mathrm{F}} = x_1 - x_2$$

with

$$q^2 = sx_1x_2$$

and show that

$$\mathrm{d}q^2\,\mathrm{d}x_{\mathrm{F}} = (x_1 + x_2)s\mathrm{d}x_1\,\mathrm{d}x_2.$$

Hence show that the Drell–Yan formula can be rewritten as

$$\frac{\mathrm{d}^2\sigma}{\mathrm{d}q^2\,\mathrm{d}x_{\mathrm{F}}} = \frac{4\pi\alpha^2}{9q^4}\frac{\tau}{(x_{\mathrm{F}}^2 + 4\tau)^{1/2}}\sum_a e_a^2[q_a(x_1)\bar{q}_a(x_2) + \bar{q}_a(x_1)q_a(x_2)].$$

9.7 Verify that if the quarks participating in the Drell–Yan subprocess $q\bar{q} \to \gamma \to \mu\bar{\mu}$ had spin-0, the CM angular distribution of the final $\mu^+\mu^-$pair would be proportional to $(1 - \cos^2\theta)$.

PART 4

LOOPS AND RENORMALIZATION

10

LOOPS AND RENORMALIZATION I: THE ABC THEORY

We have seen how Feynman diagrams represent terms in a perturbation theory expansion of physical amplitudes, namely the Dyson expansion of section 6.2. Terms of a given order all involve the same power of a 'coupling constant', which is the multiplicative constant appearing in the interaction Hamiltonian—for example, 'g' in the ABC theory, or the charge 'e' in electrodynamics. In practice, it often turns out that the relevant parameter is actually the square of the coupling constant, and factors of 4π have a habit of appearing on a regular basis; so, for QED, the perturbation series is conveniently ordered according to powers of the *fine structure constant* $\alpha = e^2/4\pi \approx 1/137$.

Equivalently, this is an expansion in terms of the number of vertices appearing in the diagrams, since one power of the coupling constant is associated with each vertex. For a given physical process, the lowest-order diagrams (the ones with the fewest vertices) are those in which each vertex is connected to every other vertex by just one internal line; these are called *tree diagrams*. The Yukawa (u-channel) exchange process of figure 6.4, and the s-channel process of figure 6.5, are both examples of tree diagrams, and indeed all of our calculations so far have not gone further than this lowest-order ('tree') level. Admittedly, since α is after all pretty small, tree diagrams in QED are likely to give us a good approximation to compare with experiment. Nevertheless, a long history of beautiful and ingenious experiments has resulted in observables in QED being determined to an accuracy far better than the $O(1\%)$ represented by the leading (tree) terms. More generally, precision experiments at LEP and other laboratories have an accuracy sensitive to higher-order corrections in the Standard Model. Hence, some understanding of the physics beyond the tree approximation is now essential for phenomenology.

All higher-order processes beyond the tree approximation involve *loops*, a concept easier to recognize visually than to define in words. In section 6.3.5 we already met (figure 6.8) one example of an $O(g^4)$ correction to the $O(g^2)$ C-exchange tree diagram of figure 6.4, which contains one loop. The crucial point is that whereas a tree diagram can be cut into two separate pieces by severing just one internal line, to cut a loop diagram into two separate pieces requires the severing of at least two internal lines.

In these last two chapters of volume 1, we aim to provide an introduction

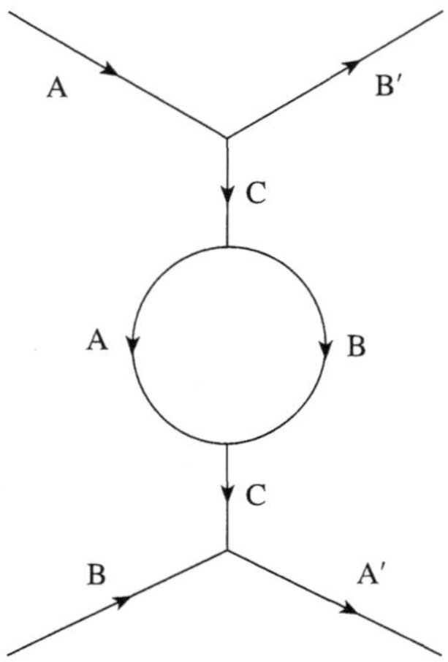

Figure 10.1. $O(g^4)$ contribution to the process A+B → A+B, involving the modification of the C propagator by the insertion of a loop.

to higher-order processes, confining ourselves to 'one-loop' order. In the present chapter we shall concentrate mainly on the particular loop appearing in figure 6.8. This will lead us into the physics of *renormalization* for the ABC theory, which—as a Yukawa-like theory—is a good theoretical laboratory for studying 'one-loop physics', without the complications of spinor and gauge fields. In the following chapter, we shall discuss one-loop diagrams in QED, emphasizing some important physical consequences, such as corrections to Coulomb's law, anomalous magnetic moments and the running coupling constant.

10.1 The propagator correction in ABC theory

10.1.1 The $O(g^2)$ self-energy $\Pi_{\mathrm{C}}^{[2]}(q^2)$

We consider figure 6.8, reproduced here again as figure 10.1. In section 6.3.5, we gave the extra rule ('(iii)') needed to write down the invariant amplitude for this process. We first show how this rule arises in the special case of figure 10.1.

Clearly, figure 10.1 is a fourth-order process, so it must emerge from the term

$$\begin{aligned}\frac{(-\mathrm{i}g)^4}{4!}\iiiint \mathrm{d}^4x_1\,\mathrm{d}^4x_2\,\mathrm{d}^4x_3\,\mathrm{d}^4x_4\,\langle 0|\hat{a}_{\mathrm{A}}(p_{\mathrm{A}}')\hat{a}_{\mathrm{B}}(p_{\mathrm{B}}')\\ \times T\{\hat{\phi}_{\mathrm{A}}(x_1)\hat{\phi}_{\mathrm{B}}(x_1)\hat{\phi}_{\mathrm{C}}(x_1)\ldots\hat{\phi}_{\mathrm{A}}(x_4)\hat{\phi}_{\mathrm{B}}(x_4)\hat{\phi}_{\mathrm{C}}(x_4)\}\\ \times\hat{a}_{\mathrm{A}}^{\dagger}(p_{\mathrm{A}})\hat{a}_{\mathrm{B}}^{\dagger}(p_{\mathrm{B}})|0\rangle(16E_{\mathrm{A}}E_{\mathrm{B}}E_{\mathrm{A}}'E_{\mathrm{B}}')^{1/2}\end{aligned} \tag{10.1}$$

of the Dyson expansion. Since it is basically a u-channel exchange process ($u=(p_{\mathrm{A}}-p_{\mathrm{B}}')^2=(p_{\mathrm{A}}'-p_{\mathrm{B}})^2$), the vev's involving the external creation

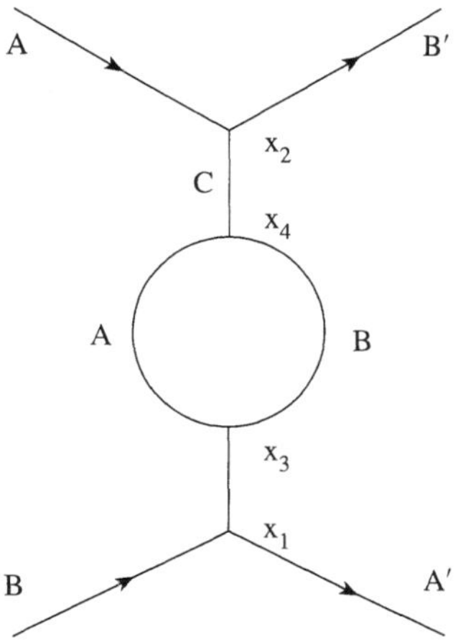

Figure 10.2. The space–time structure of the integrand in (10.3).

and annihilation operators must appear as they do in equation (6.89) ('ingoing A, outgoing B′ at one point x_2; ingoing B, outgoing A′ at another point x_1') rather than as in equation (6.88) ('ingoing A and B at x_2; outgoing A′ and B′ at x_1'). In (10.1), however, we unfortunately have *four* space–time points to choose from, rather than merely the two in (6.74). Figuring out exactly which choices are in fact equivalent and which are not is best left to private struggle, especially since we are not seriously interested in the numerical value of our fourth-order corrections in this case. Let us simply consider one choice, analogous to (6.89). This yields the amplitude (cf (6.91))

$$(-\mathrm{i}g)^4 \iiiint \mathrm{d}^4x_1\,\mathrm{d}^4x_2\,\mathrm{d}^4x_3\,\mathrm{d}^4x_4\,\mathrm{e}^{\mathrm{i}(p'_\mathrm{A}-p_\mathrm{B})\cdot x_1}\mathrm{e}^{\mathrm{i}(p'_\mathrm{B}-p_\mathrm{A})\cdot x_2}$$
$$\times\langle 0|T\{\hat{\phi}_\mathrm{C}(x_1)\hat{\phi}_\mathrm{C}(x_2)\hat{\phi}_\mathrm{A}(x_3)\hat{\phi}_\mathrm{B}(x_3)\hat{\phi}_\mathrm{C}(x_3)\hat{\phi}_\mathrm{A}(x_4)\hat{\phi}_\mathrm{B}(x_4)\hat{\phi}_\mathrm{C}(x_4)\}|0\rangle \tag{10.2}$$

and we have discarded the numerical factor 1/4! Once again, there are many terms in the expansion of the vev of the eight operators in (10.2). But, with an eye on the structure of the Feynman amplitude at which we are aiming (figure 10.1), let us consider again just a single contribution

$$(-\mathrm{i}g)^4 \iiiint \mathrm{d}^4x_1\,\mathrm{d}^4x_2\,\mathrm{d}^4x_3\,\mathrm{d}^4x_4\,\mathrm{e}^{\mathrm{i}(p'_\mathrm{A}-p_\mathrm{B})\cdot x_1}\mathrm{e}^{\mathrm{i}(p'_\mathrm{B}-p_\mathrm{A})\cdot x_2}$$
$$\times\langle 0|T(\hat{\phi}_\mathrm{C}(x_1)\hat{\phi}_\mathrm{C}(x_3))|0\rangle\langle 0|T(\hat{\phi}_\mathrm{C}(x_2)\hat{\phi}_\mathrm{C}(x_4))|0\rangle$$
$$\times\langle 0|T(\hat{\phi}_\mathrm{A}(x_3)\hat{\phi}_\mathrm{A}(x_4))|0\rangle\langle 0|T(\hat{\phi}_\mathrm{B}(x_3)\hat{\phi}_\mathrm{B}(x_4))|0\rangle \tag{10.3}$$

which contains four propagators connected as in figure 10.2.

As we saw in section 6.3.2, each of these propagators is a function only of the difference of the two space–time points involved. Introducing relative coordinates $x = x_1 - x_3$, $y = x_2 - x_4$, $z = x_3 - x_4$ and the CM coordinate $X = \frac{1}{4}(x_1 + x_2 + x_3 + x_4)$, we find (problem 10.1) that (10.3) becomes

$$(-\mathrm{i}g)^4 \iiiint \mathrm{d}^4 X\, \mathrm{d}^4 x\, \mathrm{d}^4 y\, \mathrm{d}^4 z\, \mathrm{e}^{\mathrm{i}(p_A' + p_B' - p_A - p_B)\cdot X} \mathrm{e}^{\mathrm{i}(p_A' - p_B)\cdot(3x - y + 2z)/4}$$
$$\times\, \mathrm{e}^{\mathrm{i}(p_B' - p_A)\cdot(-x+3y-2z)/4} D_C(x) D_C(y) D_A(z) D_B(z) \tag{10.4}$$

where D_i is the position–space propagator for type-i particles (i = A, B, C), defined as in (6.98). The integral over X gives the expected overall 4-momentum conservation factor, $(2\pi)^4 \delta^4(p_A' + p_B' - p_A - p_B)$. Setting $q = p_A - p_B' = p_A' - p_B$ (where 4-momentum conservation has been used), (10.4) becomes

$$(-\mathrm{i}g)^4 (2\pi)^4 \delta^4(p_A' + p_B' - p_A - p_B) \iiint \mathrm{d}^4 x\, \mathrm{d}^4 y\, \mathrm{d}^4 z\, \mathrm{e}^{\mathrm{i}q\cdot x} D_C(x)$$
$$\times\, \mathrm{e}^{-\mathrm{i}q\cdot y} D_C(y) \mathrm{e}^{\mathrm{i}q\cdot z} D_A(z) D_B(z). \tag{10.5}$$

The integrals over x and y separate out completely, each being just the Fourier transform of a C propagator—that is, the momentum–space propagator $\tilde{D}_C(q)$. Since the latter is a function of q^2 only, we end up with two factors of $\mathrm{i}/(q^2 - m_C^2 + \mathrm{i}\epsilon)$, corresponding to the two C propagators in the momentum–space Feynman diagram of figure 10.1. Note that the Mandelstam u-variable is defined by $u = (p_A - p_B')^2$ and is thus equal to q^2; we shall, however, continue to use q^2 rather than u in what follows.

The remaining factor represents the loop. Including $(-\mathrm{i}g)^2$ for the two vertices in the loop, it is given by

$$(-\mathrm{i}g)^2 \int \mathrm{d}^4 z\, \mathrm{e}^{\mathrm{i}q\cdot z} D_A(z) D_B(z) \tag{10.6}$$

which is the main result of our calculation so far. Since we want to end up finally with a momentum–space amplitude, let us introduce the A and B propagators in momentum space, and write (10.6) as (cf (6.99))

$$(-\mathrm{i}g)^2 \int \mathrm{d}^4 z\, \mathrm{e}^{\mathrm{i}q\cdot z} \int \frac{\mathrm{d}^4 k_1}{(2\pi)^4} \mathrm{e}^{-\mathrm{i}k_1\cdot z} \frac{\mathrm{i}}{k_1^2 - m_A^2 + \mathrm{i}\epsilon} \int \frac{\mathrm{d}^4 k_2}{(2\pi)^4} \mathrm{e}^{-\mathrm{i}k_2\cdot z} \frac{\mathrm{i}}{k_2^2 - m_B^2 + \mathrm{i}\epsilon}$$
$$= (-\mathrm{i}g)^2 \iint \frac{\mathrm{d}^4 k_1}{(2\pi)^4} \frac{\mathrm{d}^4 k_2}{(2\pi)^4} \frac{\mathrm{i}}{k_1^2 - m_A^2 + \mathrm{i}\epsilon} \frac{\mathrm{i}}{k_2^2 - m_B^2 + \mathrm{i}\epsilon}$$
$$\times\, (2\pi)^4 \delta^4(k_1 + k_2 - q)$$
$$= (-\mathrm{i}g)^2 \int \frac{\mathrm{d}^4 k}{(2\pi)^4} \frac{\mathrm{i}}{k^2 - m_A^2 + \mathrm{i}\epsilon} \frac{\mathrm{i}}{(q-k)^2 - m_B^2 + \mathrm{i}\epsilon} \tag{10.7}$$
$$\equiv -\mathrm{i}\Pi_C^{[2]}(q^2), \tag{10.8}$$

where we have defined the function $-\mathrm{i}\Pi_{\mathrm{C}}^{[2]}(q^2)$ as the loop (or 'bubble') amplitude appearing in figure 10.1. It is a function of q^2, as follows from Lorentz invariance. The $^{[2]}$ refers to the two powers of g, as will be explained shortly, after (10.15).

Careful consideration of the equivalences among the various contractions shows that the amplitude corresponding to figure 10.1 is, in fact, just the simple expression

$$(-\mathrm{i}g)^2(2\pi)^4\delta^4(p_{\mathrm{A}}' + p_{\mathrm{B}}' - p_{\mathrm{A}} - p_{\mathrm{B}})\frac{\mathrm{i}}{q^2 - m_{\mathrm{C}}^2 + \mathrm{i}\epsilon}(-\mathrm{i}\Pi_{\mathrm{C}}^{[2]}(q^2))\frac{\mathrm{i}}{q^2 - m_{\mathrm{C}}^2 + \mathrm{i}\epsilon} \tag{10.9}$$

where $\Pi_{\mathrm{C}}^{[2]}(q^2)$ is given in (10.8). We see that whereas the 'single-particle' pieces, involving one C-exchange, do not involve any integral in momentum–space, the loop (which involves both A and B particles) does involve a momentum integral. This can be simply understood in terms of 4-momentum conservation, which holds at every vertex of a Feynman graph. At the top (or bottom) vertex of figure 10.1, the 4-momentum q of the C-particle is fully determined by that of the incoming and outgoing particles ($q = p_{\mathrm{A}} - p_{\mathrm{B}}' = p_{\mathrm{A}}' - p_{\mathrm{B}}$). This same 4-momentum q flows in (and out) of the loop in figure 10.1, but nothing determines how it is to be shared between the A- and B-particles; all that can be said is that if the 4-momentum of A is k (as in (10.7)) then that of B is $q - k$, so that their sum is q. The 'free' variable k then has to be integrated over, and this is the physical origin of rule (iii) of section 6.3.5.

We have devoted some time to the steps leading to expression (10.7), not only in order to follow the emergence of rule (iii) mathematically, but so as to lend some plausibility to a very important statement: the Feynman rules for associating factors with vertices and propagators, which we learned for tree graphs in chapters 6 and 8, *also* work, with the addition of rule (iii), for all more complicated graphs as well! Having seen most of just one fairly short calculation of a higher-order amplitude, the reader may perhaps now begin to appreciate just how powerful is the precise correspondence between 'diagrams and amplitudes', given by the Feynman rules.

Having arrived at the expression for our first one-loop graph, we must at once draw the reader's attention to the *bad news*: the integral in (10.7) is divergent at large values of k. We shall postpone a more detailed mathematical analysis until section 10.3.1, but the divergence can be plausibly inferred just from a simple counting of powers: there are four powers of k in the numerator and four in the denominator, and the likelihood is that the integral diverges as $\int_0^\Lambda k^3\mathrm{d}k/k^4 \sim \ln\Lambda$, as $\Lambda \to \infty$. This is plainly a disaster: a quantity which was supposed to be a small correction in perturbation theory is actually infinite! Such divergences, occurring as loop momenta go to infinity, are called 'ultraviolet divergences', and they are ubiquitous in quantum field theory. Only after a long struggle with these infinities was it understood how to obtain physically sensible results from such perturbation expansions. Depending on the type of field theory involved, the infinities can often be 'tamed' through a procedure known as *renormalization*, to which we shall provide an introduction in this and the following chapter.

The physical ideas behind renormalization are, however, just as relevant in cases—such as condensed matter physics—where the analogous higher-order (loop) corrections are not infinite, though possibly large. In quantum mechanics, infinite momentum corresponds to zero distance, and our fields are certainly 'point-like'. But in condensed matter physics there is generally a natural non-zero smallest distance—the lattice size, or an atomic diameter, for example. In quantum field theory, such a 'shortest distance' would correspond to a 'highest momentum', meaning that the magnitudes of loop momenta would run from zero up to some finite limit Λ, say, rather than infinity. Such a Λ is called a (momentum) 'cut-off'. With such a cut-off in place, our loop integrals are of course finite—but it would seem that we have then maltreated our field theory in some way. However, we might well ask whether we seriously believe that any of our quantum field theories is literally valid for arbitrarily high energies (or arbitrarily small distances). The answer is surely no: we are virtually certain that 'new physics' will come into play at some stage, which is not contained in—say—the QED, or even the Standard Model, Lagrangian. At what scale this new physics will enter (the Planck energy? 1 TeV?) we do not know but surely the current models will break down at some point. We should not be too alarmed, therefore, by formal divergences as $\Lambda \to \infty$. Rather, it may be sensible to regard a cut-off Λ as standing for some 'new physics' scale, accepting some such manoeuvre as physically realistic as well as mathematically prudent.

At the same time, however, we would not want our physical predictions, made using quantum field theories, to depend sensitively on Λ—i.e. on the unknown short-distance physics, in this interpretation. Indeed, theories exist (for example, those in the Standard Model and the ABC theory) which can be reformulated in such a way that all dependence on Λ disappears, as $\Lambda \to \infty$; these are, precisely, *renormalizable* quantum field theories. Roughly speaking, a renormalizable quantum field theory is one such that, when formulae are expressed in terms of certain 'physical' parameters taken from experiment, rather than in terms of the original parameters appearing in the Lagrangian, calculated quantities will be finite and independent of Λ as $\Lambda \to \infty$.

Solid state physics provides a close analogy. There, the usefulness of a description of, say, electrons in a metal in terms of their 'effective charge' and 'effective mass', rather than their free-space values, is well established. In this analogy, the free-space quantities correspond to our Lagrangian values, while the effective parameters correspond to our 'physical' ones. In both cases, the interactions are causing changes to the parameters.

It is clear that we need to understand more precisely just what our 'physical parameters' might be and how they might be defined. This is what we aim to do in the remainder of the present section, and in the next one, before returning in section 10.3 to the mathematical details associated with evaluating (10.7), and

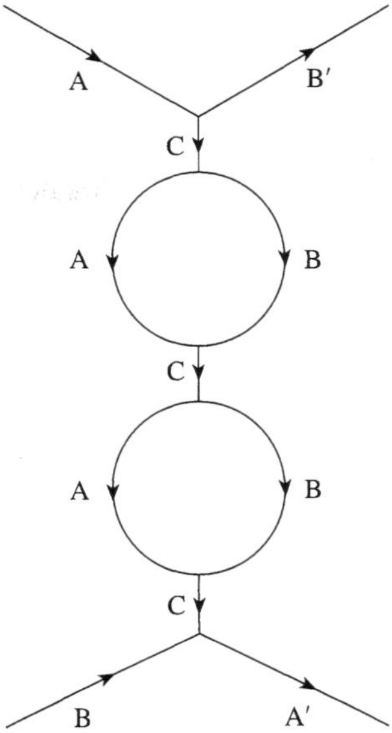

Figure 10.3. $O(g^6)$ term in A + B → A + B, involving the insertion of two loops in the C propagator.

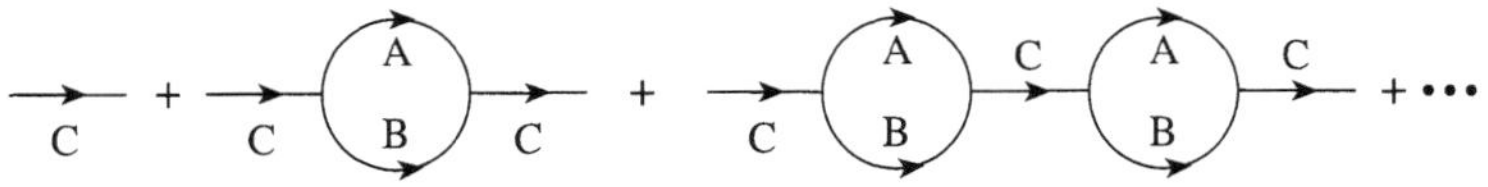

Figure 10.4. Series of one-loop (or 'bubble') insertions in the C propagator.

indicating how renormalization works for the self-energy. Having thus prepared the ground, we shall introduce a more powerful approach in section 10.4, and offer a few preliminary remarks about 'renormalizability' in section 10.5, returning to that topic at the end of the following chapter. Although usually not explicitly indicated, loop corrections considered in this and the following section will be understood to be defined with a cut-off Λ, so that they are finite.

To begin the discussion of the physical significance of our $O(g^4)$ correction, (10.9), it is convenient to consider both the $O(g^2)$ term (6.100) and the $O(g^4)$ correction together, obtaining

$$(-\mathrm{i}g)^2(2\pi)^4\delta^4(p'_{\mathrm{A}} + p'_{\mathrm{B}} - p_{\mathrm{A}} - p_{\mathrm{B}}) \times \left\{\frac{\mathrm{i}}{q^2 - m_{\mathrm{C}}^2} + \frac{\mathrm{i}}{q^2 - m_{\mathrm{C}}^2}(-\mathrm{i}\Pi_{\mathrm{C}}^{[2]}(q^2))\frac{\mathrm{i}}{q^2 - m_{\mathrm{C}}^2}\right\} \tag{10.10}$$

where the iϵ in the C propagators does not need to be retained. Both the form of (10.10), and inspection of figure 10.1, suggest that the $O(g^4)$ term we have calculated can be regarded as an $O(g^2)$ correction to the propagator for the C-particle. Indeed, we can easily imagine adding in the $O(g^6)$ term shown in figure 10.3, and in fact the whole infinite series of such 'bubbles' connected by simple C propagators. The infinite geometric series for the corrected propagator shown in figure 10.4 has the form

$$\frac{\mathrm{i}}{q^2 - m_{\mathrm{C}}^2} + \frac{\mathrm{i}}{q^2 - m_{\mathrm{C}}^2}(-\mathrm{i}\Pi_{\mathrm{C}}^{[2]}(q^2))\frac{\mathrm{i}}{q^2 - m_{\mathrm{C}}^2} + \frac{\mathrm{i}}{q^2 - m_{\mathrm{C}}^2}(-\mathrm{i}\Pi_{\mathrm{C}}^{[2]}(q^2))\frac{\mathrm{i}}{q^2 - m_{\mathrm{C}}^2}(-\mathrm{i}\Pi_{\mathrm{C}}^{[2]}(q^2))\frac{\mathrm{i}}{q^2 - m_{\mathrm{C}}^2} + \cdots \tag{10.11}$$

$$= \frac{\mathrm{i}}{q^2 - m_{\mathrm{C}}^2}(1 + r + r^2 + \cdots) \tag{10.12}$$

where

$$r = \Pi_{\mathrm{C}}^{[2]}(q^2)/(q^2 - m_{\mathrm{C}}^2). \tag{10.13}$$

The geometric series in (10.12) may be summed, at least formally[1], to give $(1 - r)^{-1}$ so that (10.12) becomes

$$\frac{\mathrm{i}}{q^2 - m_{\mathrm{C}}^2}\frac{1}{1 - \Pi_{\mathrm{C}}^{[2]}(q^2)/(q^2 - m_{\mathrm{C}}^2)} = \frac{\mathrm{i}}{q^2 - m_{\mathrm{C}}^2 - \Pi_{\mathrm{C}}^{[2]}(q^2)}. \tag{10.14}$$

In this form it is particularly clear that we are dealing with corrections to the simple C propagator $\mathrm{i}/(q^2 - m_{\mathrm{C}}^2)$. $\Pi_{\mathrm{C}}^{[2]}$ is called the $O(g^2)$ *self-energy*.

Before proceeding with the analysis of (10.14), we note that it is a special case of the more general expression

$$\tilde{D}_{\mathrm{C}}'(q^2) = \frac{\mathrm{i}}{q^2 - m_{\mathrm{C}}^2 - \Pi_{\mathrm{C}}(q^2)} \tag{10.15}$$

where $\tilde{D}_{\mathrm{C}}'(q^2)$ is the *complete* (including all corrections) C propagator, and $\Pi_{\mathrm{C}}(q^2)$ is the sum of all 'insertions' in the C line, *excluding* those which can be cut into two separate bits by severing a single line: $\Pi_{\mathrm{C}}(q^2)$ is the *one-particle irreducible self-energy* and we must exclude all one-particle bits from it as they are already included in the geometric series summation (cf (10.11)). The amplitude $\Pi_{\mathrm{C}}^{[2]}$ which we have calculated is simply the lowest-order ($O(g^2)$) contribution to $\Pi_{\mathrm{C}}(q^2)$; an $O(g^4)$ contribution to $\Pi_{\mathrm{C}}(q^2)$ is shown in figure 10.5.

[1] Properly speaking this is valid only for $|r| < 1$, yet we know that $\Pi_{\mathrm{C}}^{[2]}(q^2)$ actually diverges! As we shall see, however, renormalization will be carried out after making such quantities finite by 'regularization' (section 10.3.2), and then working systematically at a given order in g (section 10.4).

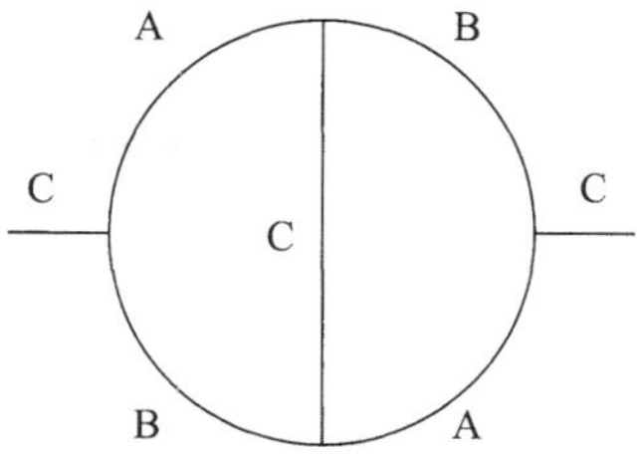

Figure 10.5. $O(g^4)$ contribution to $\Pi_{\rm C}(q^2)$.

10.1.2 Mass shift

We return to the expression (10.14) which includes the effect of all the iterated $O(g^2)$ bubbles in the C propagator, where $\Pi_{\rm C}^{[2]}(q^2)$ is given by

$$-{\rm i}\Pi_{\rm C}^{[2]}(q^2) = (-{\rm i}g)^2 \int \frac{{\rm d}^4 k}{(2\pi)^4} \frac{\rm i}{k^2 - m_{\rm A}^2 + {\rm i}\epsilon} \frac{\rm i}{(q-k)^2 - m_{\rm B}^2 + {\rm i}\epsilon}. \quad (10.16)$$

Postponing the evaluation of (10.16) (and in particular the treatment of its divergence) until section 10.3, we proceed to discuss the further implications of (10.14).

First, suppose $\Pi_{\rm C}^{[2]}$ were simply a constant, $\delta m_{\rm C}^2$ say. In the absence of this correction, we know (cf section 6.3.3) that the vanishing of the denominator of the C propagator would correspond to the 'mass-shell condition' $q^2 = m_{\rm C}^2$ appropriate to a free particle of momentum q and energy $q_0 = (\boldsymbol{q}^2 + m_{\rm C}^2)^{1/2}$, where $m_{\rm C}$ is the mass of a C particle. It seems very plausible, therefore, to interpret the constant $\delta m_{\rm C}^2$ as a shift in the (mass)^2 of the C particle, the denominator of (10.14) now vanishing at $q_0 = (\boldsymbol{q}^2 + m_{\rm C}^2 + \delta m_{\rm C}^2)^{1/2}$, if $\Pi_{\rm C}^{[2]} \simeq \delta m_{\rm C}^2$. The idea that the mass of a particle can be changed from its 'free space' value by the presence of interactions with its 'environment' is a familiar one in condensed matter physics, as noted above. In the case of electrons in a metal, for example, it is not surprising that the presence of the lattice ions, and the attendant band structure, affect the response of conduction electrons to external fields, so that their apparent inertia changes. In the present case, the 'environment' is, in fact, the *vacuum*. The process described by the bubble $\Pi_{\rm C}^{[2]}(q^2)$ is one in which a C particle dissociates virtually into an A–B pair, which then recombine into the C particle, no other 'external' source being present. As in earlier uses of the word, by 'virtual' here is meant a process in which the participating particles leave their mass-shells. Thus, in particular, in the expression (10.16) for $\Pi_{\rm C}^{[2]}$, it will in general be the case that $k^2 \neq m_{\rm A}^2$, and $(q-k)^2 \neq m_{\rm B}^2$.

In the case of the electron in a metal, both the 'free' and the 'effective' masses are measurable quantities. But we cannot get outside the vacuum! This strongly suggests that what we must mean by 'the physical (mass)^2' of a particle in our ABC theory is *not* the 'free' (Lagrangian) value m_i^2, which is unmeasurable, but the effective (mass)^2 which includes all vacuum interactions. This 'physical (mass)^2' may be *defined* to be that value of q^2 for which

$$q^2 - m_i^2 - \Pi_i(q^2) = 0 \tag{10.17}$$

where $\Pi_i(q^2)$ is the complete one-particle irreducible self-energy for particle type 'i'. If we call the physical mass $m_{\mathrm{ph},i}$, then, we will have $q^2 - m_i^2 - \Pi_i(q^2) = 0$ when $q^2 = m_{\mathrm{ph},i}^2$.

What we are dealing with in (10.14) is just the lowest-order contribution to $\Pi_{\mathrm{C}}(q^2)$, namely $\Pi_{\mathrm{C}}^{[2]}(q^2)$, so that in our case $m_{\mathrm{ph,C}}^2$ is determined by the condition

$$q^2 - m_{\mathrm{C}}^2 - \Pi_{\mathrm{C}}^{[2]}(q^2) = 0 \qquad \text{when } q^2 = m_{\mathrm{ph,C}}^2, \tag{10.18}$$

which (to this order) is

$$m_{\mathrm{ph,C}}^2 = m_{\mathrm{C}}^2 + \Pi_{\mathrm{C}}^{[2]}(m_{\mathrm{ph,C}}^2). \tag{10.19}$$

Once we have calculated $\Pi_{\mathrm{C}}^{[2]}$ (see section 10.3), equation (10.19) could be regarded as an equation to determine $m_{\mathrm{ph,C}}^2$ in terms of the parameter m_{C}^2, which appeared in the original ABC Lagrangian. This might, indeed, be the way such an equation would be viewed in condensed matter physics, where we should know the values of the parameters in the Lagrangian. But in the field-theory case m_{C}^2 is unobservable, so that such an equation has no predictive value. Instead, we may regard it as an equation determining (up to $O(g^2)$) m_{C}^2 in terms of $m_{\mathrm{ph,C}}^2$, thus enabling us to eliminate—to this order in g—all occurrences of the unobservable parameter m_{C}^2 from our amplitudes in favour of the physical parameter $m_{\mathrm{ph,C}}^2$. Note that $\Pi_{\mathrm{C}}^{[2]}$ contains two powers of g, so that in the spirit of systematic perturbation theory, the mass shift represented by (10.19) is a second-order correction.

The crucial point here is that $\Pi_{\mathrm{C}}^{[2]}$ depends on the cut-off Λ, whereas the physical mass $m_{\mathrm{ph,C}}^2$ clearly does not. But there is nothing to stop us supposing that the unknown and unobservable Lagrangian parameter m_{C}^2 depends on Λ in just such a way as to cancel the Λ-dependence of $\Pi_{\mathrm{C}}^{[2]}$, leaving $m_{\mathrm{ph,C}}^2$ independent of Λ. This is the beginning of the 'renormalization procedure' in quantum field theory.

10.1.3 Field strength renormalization

We now need to consider the more realistic case in which $\Pi_{\mathrm{C}}^{[2]}(q^2)$ is not a constant. Let us expand it about the point $q^2 = m_{\mathrm{ph,C}}^2$, writing

$$\Pi_{\mathrm{C}}^{[2]}(q^2) \approx \Pi_{\mathrm{C}}^{[2]}(m_{\mathrm{ph,C}}^2) + (q^2 - m_{\mathrm{ph,C}}^2)\left.\frac{\mathrm{d}\Pi_{\mathrm{C}}^{[2]}}{\mathrm{d}q^2}\right|_{q^2=m_{\mathrm{ph,C}}^2} + \cdots. \tag{10.20}$$

The corrected propagator (10.14) then becomes

$$\frac{\mathrm{i}}{q^2 - m_{\mathrm{C}}^2 - \Pi_{\mathrm{C}}^{[2]}(m_{\mathrm{ph,C}}^2) - (q^2 - m_{\mathrm{ph,C}}^2)\left.\frac{\mathrm{d}\Pi_{\mathrm{C}}^{[2]}}{\mathrm{d}q^2}\right|_{q^2=m_{\mathrm{ph,C}}^2} + \cdots} \tag{10.21}$$

$$= \frac{\mathrm{i}}{(q^2 - m_{\mathrm{ph,C}}^2)\left[1 - \left.\frac{\mathrm{d}\Pi_{\mathrm{C}}^{[2]}}{\mathrm{d}q^2}\right|_{q^2=m_{\mathrm{ph,C}}^2}\right] + O(q^2 - m_{\mathrm{ph,C}}^2)^2}. \tag{10.22}$$

The expression (10.22) has indeed the expected form for a 'physical C' propagator, having the simple behaviour $\sim 1/(q^2 - m_{\mathrm{ph,C}}^2)$ for $q^2 \approx m_{\mathrm{ph,C}}^2$. However, the *normalization* of this (corrected) propagator is different from that of the 'free' one, $\mathrm{i}/(q^2 - m_{\mathrm{C}}^2)$, because of the extra factor

$$\left[1 - \left.\frac{\mathrm{d}\Pi_{\mathrm{C}}^{[2]}}{\mathrm{d}q^2}\right|_{q^2=m_{\mathrm{ph,C}}^2}\right]^{-1}.$$

To the order at which we are working ($O(g^2)$), it is consistent to replace this expression by

$$1 + \left.\frac{\mathrm{d}\Pi_{\mathrm{C}}^{[2]}}{\mathrm{d}q^2}\right|_{q^2=m_{\mathrm{ph,C}}^2}.$$

Let us see how this factor may be understood.

Our $O(g^2)$ corrected propagator is an approximation to the exact propagator which we may write as $\langle\Omega|T(\hat{\phi}_{\mathrm{C}}(x_1)\hat{\phi}_{\mathrm{C}}(x_2))|\Omega\rangle$, in coordinate space, where $|\Omega\rangle$ is the exact vacuum. The free propagator, however, is $\langle 0|T(\hat{\phi}_{\mathrm{C}}(x_1)\hat{\phi}_{\mathrm{C}}(x_2))|0\rangle$ as calculated in section 6.3.2. Consider one term in the latter, $\theta(t_1 - t_2)\langle 0|\hat{\phi}_{\mathrm{C}}(x_1)\hat{\phi}_{\mathrm{C}}(x_2)|0\rangle$, and insert a complete set of free-particle states '$1 = \sum_n |n\rangle\langle n|$' between the two free fields, obtaining

$$\theta(t_1 - t_2)\sum_n \langle 0|\hat{\phi}_{\mathrm{C}}(x_1)|n\rangle\langle n|\hat{\phi}_{\mathrm{C}}(x_2)|0\rangle. \tag{10.23}$$

The only free particle state $|n\rangle$ having a non-zero matrix element of the free field $\hat{\phi}_{\mathrm{C}}$ to the vacuum is the $1 - \mathrm{C}$ state, for which $\langle 0|\hat{\phi}_{\mathrm{C}}(x)|\mathrm{C}, k\rangle = \mathrm{e}^{-\mathrm{i}k\cdot x}$ as we learned in chapters 5 and 6. Thus (10.23) becomes (cf equation (6.92))

$$\theta(t_1 - t_2)\int \frac{\mathrm{d}^3\boldsymbol{k}}{(2\pi)^3 2\omega_k}\mathrm{e}^{-\mathrm{i}\omega_k(t_1-t_2)+\mathrm{i}\boldsymbol{k}\cdot(\boldsymbol{x}_1-\boldsymbol{x}_2)} \tag{10.24}$$

which is exactly the first term of equation (6.92). Consider now carrying out a similar manipulation for the corresponding term of the interacting propagator, obtaining

$$\theta(t_1 - t_2)\sum_n \langle\Omega|\hat{\phi}_{\mathrm{C}}(x_1)\overline{|n\rangle}\,\overline{\langle n|}\hat{\phi}_{\mathrm{C}}(x_2)|\Omega\rangle \tag{10.25}$$

where the states $\overline{|n\rangle}$ are now the exact eigenstates of the full Hamiltonian. The crucial difference between (10.23) and (10.25) is that in (10.25), multi-particle states can appear in the states $\overline{|n\rangle}$. For example, the state $|\mathrm{A}, \mathrm{B}\rangle$ consisting of an A particle and a B particle will enter, because the interaction couples this state to the 1-C states created and destroyed in $\hat{\phi}_{\mathrm{C}}$: indeed, just such an A+B state is present in $\Pi_{\mathrm{C}}^{[2]}$! This means that, whereas in the free case the 'content' of the state $\langle 0|\hat{\phi}_{\mathrm{C}}(x)$ was fully exhausted by the $1-\mathrm{C}$ state $|\mathrm{C}, k\rangle$ (in the sense that all overlaps with other states $|n\rangle$ were zero), this is not so in the interacting case. The 'content' of $\langle\Omega|\hat{\phi}_{\mathrm{C}}(x)$ is not fully exhausted by the state $\overline{|\mathrm{C}, k\rangle}$: rather, it has overlaps with many other states. Now the sum total of all these overlaps (in the sense of '$\sum_n \overline{|n\rangle}\,\overline{\langle n|}$') must be unity. Thus it seems clear that the 'strength' of the single matrix element $\langle\Omega|\hat{\phi}_{\mathrm{C}}(x)\overline{|\mathrm{C}, k\rangle}$ in the interacting case cannot be the same as the free case (where the single state exhausted the completeness sum). However, we expect it to be true that $\langle\Omega|\hat{\phi}_{\mathrm{C}}(x)\overline{|\mathrm{C}, k\rangle}$ is still basically the wavefunction for the C-particle. Hence we shall write

$$\langle\Omega|\hat{\phi}_{\mathrm{C}}(x)\overline{|\mathrm{C}, k\rangle} = \sqrt{Z_{\mathrm{C}}}\mathrm{e}^{-\mathrm{i}k\cdot x} \tag{10.26}$$

where $\sqrt{Z_{\mathrm{C}}}$ is a constant to take account of the change in normalization—the *renormalization*, in fact—required by the altered 'strength' of the matrix element.

If (10.26) is accepted, we can now imagine repeating the steps leading from equation (6.92) to equation (6.98) but this time for $\langle\Omega|T(\hat{\phi}_{\mathrm{C}}(x_1)\hat{\phi}_{\mathrm{C}}(x_2))|\Omega\rangle$, retaining explicitly only the single-particle state $\overline{|\mathrm{C}, k\rangle}$ in (10.25), and using the physical (mass)^2, $m^2_{\mathrm{ph,C}}$. We should then arrive at a propagator in the interacting case which has the form

$$\langle\Omega|T(\hat{\phi}_{\mathrm{C}}(x_1)\hat{\phi}_{\mathrm{C}}(x_2))|\Omega\rangle = \int \frac{\mathrm{d}^4k}{(2\pi)^4}\mathrm{e}^{-\mathrm{i}k\cdot(x_1-x_2)}\left\{\frac{\mathrm{i}Z_{\mathrm{C}}}{k^2 - m^2_{\mathrm{ph,C}} + \mathrm{i}\epsilon} + \text{multiparticle contributions}\right\}. \tag{10.27}$$

The single-particle contribution in (10.27)—after undoing the Fourier transform—has exactly the same form as the one we found in (10.22), if we identify the *field strength renormalization constant* Z_{C} with the proportionality factor in (10.22), to this order:

$$Z_{\mathrm{C}} \approx Z_{\mathrm{C}}^{[2]} = 1 + \left.\frac{\mathrm{d}\Pi_{\mathrm{C}}^{[2]}}{\mathrm{d}q^2}\right|_{q^2=m^2_{\mathrm{ph,C}}}. \tag{10.28}$$

This is how the change in normalization in (10.22) is to be interpreted.

It may be helpful to sketch briefly an analogy between this 'renormalization' and a very similar one in ordinary quantum mechanical perturbation theory. Suppose we have a Hamiltonian $H = H_0 + V$ and that the $|n\rangle$ are a complete set of orthonormal states such that $H_0|n\rangle = E_n^{(0)}|n\rangle$. The exact eigenstates $\overline{|n\rangle}$ satisfy

$$(H_0 + V)\overline{|n\rangle} = E_n\overline{|n\rangle}. \tag{10.29}$$

To obtain $\overline{|n\rangle}$ and E_n in perturbation theory, we write

$$\overline{|n\rangle} = \sqrt{N_n}|n\rangle + \sum_{i\neq n} c_{i,n}|i\rangle \tag{10.30}$$

where, if $|n\rangle$ is also normalized, we have

$$1 = N_n + \sum_{i\neq n} |c_{i,n}|^2. \tag{10.31}$$

N_n cannot be unity, since non-zero amounts of the states $|i\rangle$ $(i \neq n)$ have been 'mixed in' by the perturbation- just as the A + B state was introduced into the summation '$\sum_n \overline{|n\rangle}\,\overline{\langle n|}$', in addition to the 1 − C state. Inserting (10.30) into (10.29) and taking the bracket with $\langle j|$ yields

$$c_{j,n} = -\frac{\langle j|V\overline{|n\rangle}}{E_j^{(0)} - E_n} \tag{10.32}$$

which is still an exact expression. The lowest non-trivial approximation to $c_{j,n}$ is to take $\overline{|n\rangle} \approx \sqrt{N_n}|n\rangle$ and $E_n \approx E_n^{(0)}$ in (10.32), giving

$$c_{j,n} \approx -\sqrt{N_n}\frac{\langle j|V|n\rangle}{E_j^{(0)} - E_n^{(0)}} \equiv -\sqrt{N_n}\frac{V_{jn}}{E_j^{(0)} - E_n^{(0)}}. \tag{10.33}$$

Equation (10.31) then gives N_n as

$$N_n \approx 1\Big/\left(1 + \sum_j |V_{jn}|^2/(E_j^{(0)} - E_n^{(0)})^2\right) \approx 1 - \sum_j |V_{jn}|^2/(E_j^{(0)} - E_n^{(0)})^2 \tag{10.34}$$

to second order in V_{jn}. The reader may ponder on the analogy between (10.34) and (10.28).

10.2 The vertex correction

At the same order (g^4) of perturbation theory, we should also include, for consistency, the processes shown in figures 10.6(a) and (b). Figure 10.6(a), for example, has the general form

$$-\mathrm{i}g\frac{\mathrm{i}}{q^2 - m_\mathrm{C}^2}(-\mathrm{i}gG^{[2]}(p'_\mathrm{A}, p_\mathrm{B})) \tag{10.35}$$

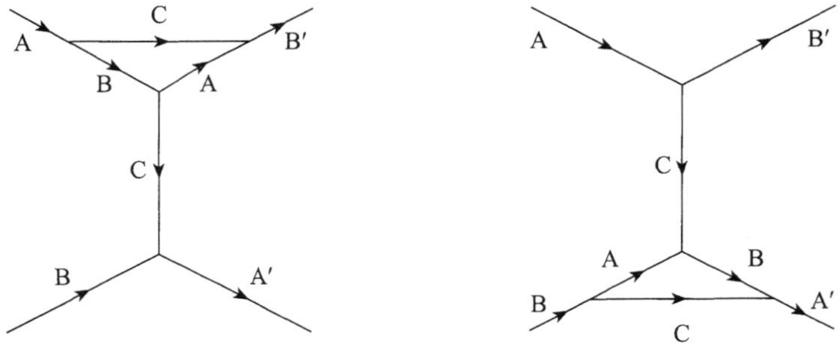

Figure 10.6. $O(g^4)$ contributions to A + B → A + B, involving corrections to the ABC vertices in figure 6.4.

where $-\mathrm{i}gG^{[2]}$ is the 'triangle' loop, given by an expression similar to (10.16) but with a factor $(-\mathrm{i}g)^3$ and three propagators. The 'vertex correction' $G^{[2]}$ depends on just two of its external 4-momenta because the third is determined by 4-momentum conservation, as usual. Thus, the addition of figure 10.6(a) and the $O(g^2)$ C-exchange tree diagram gives

$$-\mathrm{i}g\frac{\mathrm{i}}{q^2-m_{\mathrm{C}}^2}\{-\mathrm{i}g+(-\mathrm{i}gG^{[2]}(p'_{\mathrm{A}},p_{\mathrm{B}}))\} \tag{10.36}$$

from which it seems plausible that $G^{[2]}$ will contribute—among other effects—to a change in g. This change will be of order g^2, since we may write the $\{\ldots\}$ bracket in (10.36) as

$$-\mathrm{i}g\{1+G^{[2]}(p'_{\mathrm{A}},p_{\mathrm{B}})\} \tag{10.37}$$

where $G^{[2]}$ is dimensionless and contains a g^2 factor—hence the superscript [2].

Once again, the effect of interactions with the environment (i.e. vacuum fluctuations) has been to alter the value of a Lagrangian parameter away from the 'free' value. In the case of g the change is analogous to that in which an electron in a metal acquires an 'effective charge'. How we define the 'physical g' is less clear than in the case of the physical mass and we shall not pursue this point here, since we shall discuss it again in the more interesting case of the charge 'e' in QED, in the following chapter. At all events, some suitable definition of 'g_{ph}' can be given, so that it can be related to g after the relevant amplitudes have been computed.

Let us briefly recapitulate progress. We are studying higher-order (one-loop) corrections to tree graph amplitudes in the ABC model, which has the Lagrangian density:

$$\hat{\mathcal{L}}=\sum_i\{\tfrac{1}{2}\partial_\mu\hat{\phi}_i\partial^\mu\hat{\phi}_i-\tfrac{1}{2}m_i^2\hat{\phi}_i^2\}-g\hat{\phi}_{\mathrm{A}}\hat{\phi}_{\mathrm{B}}\hat{\phi}_{\mathrm{C}}. \tag{10.38}$$

We have found that the loops considered so far, namely those in figures 10.1 and 10.5, have the following qualitative effects:

(a) (b)

Figure 10.7. Elementary one-loop amplitudes: (*a*) self-energy; (*b*) vertex correction.

(a) the position of the single-particle mass-shell condition becomes shifted away from the 'Lagrangian' value m_i^2 to a 'physical' value $m_{\mathrm{ph},i}^2$ given by the vanishing of an expression such as (10.17);
(b) the vacuum-to-one-particle matrix elements of the fields $\hat{\phi}_i$ have to be 'renormalized' by a factor $\sqrt{Z_i}$, given by (10.28) to $O(g^2)$ for i=C, and these factors have to be included in S-matrix elements;
(c) the propagators contain some contribution from two-particle states (e.g. ' A + B ' for the C propagator);
(d) the Lagrangian coupling g is shifted by the interactions to a 'physical' value g_{ph}.

Responsible for these effects were two 'elementary' loops, that for $-\mathrm{i}\Pi^{[2]}$ shown in figure 10.7(*a*) and that for $-\mathrm{i}gG^{[2]}$ shown in figure 10.7(*b*). It is noteworthy that the effects (a), (b) and (d) all relate to changes (renormalizations, shifts) in the fields and parameters of the original Lagrangian. We say, collectively, that the 'fields, masses and coupling have been renormalized'—i.e. generically altered from their 'free' values, by the virtual interactions represented generically by figures 10.7(*a*) and (*b*). However, whereas in condensed matter physics one might well have the ambition to calculate such effects from first principles, in the field-theory case that makes no sense. Rather, by rewriting all calculated expressions (at a given order of perturbation theory) in terms of 'renormalized' quantities, we aim to eliminate the 'unknown physics scale', Λ, from the theory. Let us now see how this works in more mathematical detail.

10.3 Dealing with the bad news: a simple example

10.3.1 Exposure of the divergence in $\Pi_{\mathrm{C}}^{[2]}(q^2)$

We turn our attention to the actual evaluation of a one-loop amplitude, beginning with the simplest, which is $-\mathrm{i}\Pi_{\mathrm{C}}^{[2]}(q^2)$:

$$-\mathrm{i}\Pi_{\mathrm{C}}^{[2]}(q^2) = (-\mathrm{i}g)^2 \int \frac{\mathrm{d}^4 k}{(2\pi)^4} \frac{\mathrm{i}}{k^2 - m_{\mathrm{A}}^2 + \mathrm{i}\epsilon} \frac{\mathrm{i}}{(q-k)^2 - m_{\mathrm{B}}^2 + \mathrm{i}\epsilon}; \tag{10.39}$$

in particular, we want to know the precise mathematical form of the divergence which arises when the momentum integral in (10.39) is not cut off at an upper limit Λ. This will necessitate the introduction of a few modest tricks from a large armoury (mostly due to Feynman) for dealing with such integrals.

The first move in evaluating (10.39) is to 'combine the denominators' using the identity (problem 10.2)

$$\frac{1}{AB} = \int_0^1 \frac{\mathrm{d}x}{[(1-x)A + xB]^2} \tag{10.40}$$

(similar 'Feynman identities' exist for combining three or more denominator factors). Applying (10.40) to (10.39) we obtain

$$-\mathrm{i}\Pi_{\mathrm{C}}^{[2]}(q^2) = g^2 \int_0^1 \mathrm{d}x \int \frac{\mathrm{d}^4 k}{(2\pi)^4} \times \frac{1}{[(1-x)(k^2 - m_{\mathrm{A}}^2 + \mathrm{i}\epsilon) + x((q-k)^2 - m_{\mathrm{B}}^2 + \mathrm{i}\epsilon)]^2} \tag{10.41}$$

Collecting up terms inside the [...] bracket and changing the integration variable to $k' = k - \alpha q$ leads to (problem 10.3)

$$-\mathrm{i}\Pi_{\mathrm{C}}^{[2]}(q^2) = g^2 \int_0^1 \mathrm{d}x \int \frac{\mathrm{d}^4 k'}{(2\pi)^4} \frac{1}{(k'^2 - \Delta + \mathrm{i}\epsilon)^2} \tag{10.42}$$

where

$$\Delta = -x(1-x)q^2 + x m_{\mathrm{B}}^2 + (1-x) m_{\mathrm{A}}^2. \tag{10.43}$$

The d^4k' integral means $\mathrm{d}k'^0\,\mathrm{d}^3\boldsymbol{k}'$, and $k'^2 = (k'^0)^2 - \boldsymbol{k}'^2$.

We now perform the k'^0 integration in (10.42) for which we will need the contour integration techniques explained in appendix F. The integral we want to calculate is

$$\int_{-\infty}^{\infty} \frac{\mathrm{d}k'^0}{[(k'^0)^2 - A]^2} = \frac{\partial}{\partial A} \int_{-\infty}^{\infty} \frac{\mathrm{d}k'^0}{[(k'^0)^2 - A]} \equiv \frac{\partial}{\partial A} I(A) \tag{10.44}$$

where $A = \boldsymbol{k}'^2 + \Delta^2 - \mathrm{i}\epsilon$. We rewrite $I(A)$ as

$$I(A) = \lim_{R\to\infty} \int_{C_R} \frac{\mathrm{d}z}{[z^2 - A]} \tag{10.45}$$

where the contour C_R is the real axis from $-R$ to R. Next, we identify the points where the integrand $[z^2 - A]^{-1}$ ceases to be analytic (called 'poles'), which are at $z = \pm\sqrt{A} = \pm(\boldsymbol{k}'^2 + \Delta - \mathrm{i}\epsilon)^{1/2}$. Figure 10.8 shows the location of these points in the complex $z(k'^0)$-plane: note that the '$\mathrm{i}\epsilon$' determines in which half-plane each point lies (compare the similar role of the '$\mathrm{i}\epsilon$' in $(z + \mathrm{i}\epsilon)^{-1}$, in the proof in appendix F of the representation (6.93) for the θ-function). We must now 'close

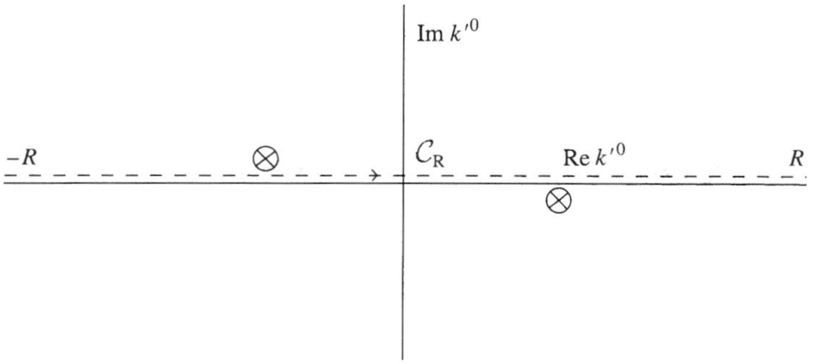

Figure 10.8. Location of the poles of (10.42) in the complex k'^0-plane.

the contour' in order to be able to use Cauchy's integral formula of (F.19). We may do this by means of a large semicircle in either the upper ($\mathcal{C}_+$) or lower ($\mathcal{C}_-$) half-plane (again compare the discussion in appendix F). The contribution from either such semicircle vanishes as $R \to \infty$, since on either we have $z = R\mathrm{e}^{\mathrm{i}\theta}$, and

$$\int_{\mathcal{C}_+ \text{ or } \mathcal{C}_-} \frac{\mathrm{d}z}{z^2 - A} = \int \frac{R\mathrm{e}^{\mathrm{i}\theta}\mathrm{i}\,\mathrm{d}\theta}{R^2\mathrm{e}^{2\mathrm{i}\theta} - A} \to 0 \qquad \text{as } R \to \infty. \tag{10.46}$$

For definiteness, let us choose to close the contour in the upper half-plane. Then we are evaluating

$$I(A) = \lim_{R\to\infty} \oint_{\mathcal{C}=\mathcal{C}_R+\mathcal{C}_+} \frac{\mathrm{d}z}{(z - \sqrt{A})(z + \sqrt{A})} \tag{10.47}$$

around the closed contour $\mathcal{C}$ shown in figure 10.9, which encloses the single non-analytic point at $z = -\sqrt{A}$. Applying Cauchy's integral formula (F.19) with $a = -\sqrt{A}$ and $f(z) = (z - \sqrt{A})^{-1}$, we find

$$I(A) = 2\pi\mathrm{i}\frac{1}{-2\sqrt{A}} \tag{10.48}$$

and thus

$$\int_{-\infty}^{\infty} \frac{\mathrm{d}k'^0}{[(k'^0)^2 - A]^2} = \frac{\pi\mathrm{i}}{2A^{3/2}}. \tag{10.49}$$

The reader may like to try taking the other choice ($\mathcal{C}_-$) of closing contour, and check that the answer is the same. Reinstating the remaining integrals in (10.42) we have finally (as $\epsilon \to 0$)

$$-\mathrm{i}\Pi_{\mathrm{C}}^{[2]}(q^2) = \frac{\mathrm{i}}{8\pi^2}g^2 \int_0^1 \mathrm{d}x \int_0^{\infty} \frac{u^2\,\mathrm{d}u}{(u^2 + \Delta)^{3/2}} \tag{10.50}$$

where $u = |\boldsymbol{k}'|$ and the integration over the angles of $\boldsymbol{k}'$ has yielded a factor of 4π. We see that the u-integral behaves as $\int \mathrm{d}u/u$ for large u, which is logarithmically divergent, as expected from the start.

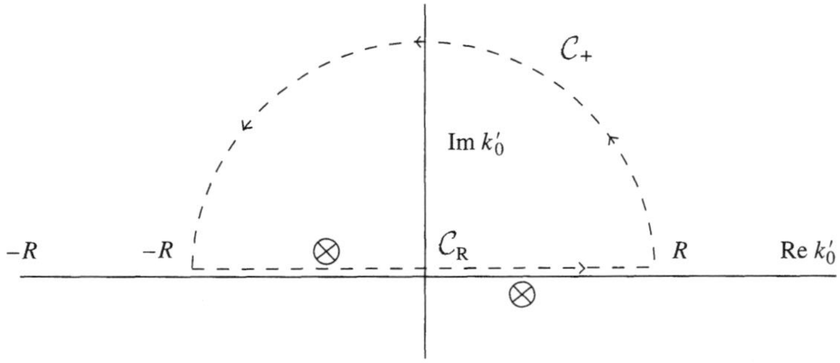

Figure 10.9. The closed contour $\mathcal{C}$ used in the integral (10.47).

10.3.2 Regularization and renormalization

Faced with results which are infinite, one can either try to go back to the very beginnings of the theory and see if a totally new start can avoid the infinities or one can see if they can somehow be 'lived with'. The first approach may yet, ultimately, turn out to be correct: perhaps a future theory will be altogether free of divergences (such theories do in fact exist, but none as yet successfully describes the pattern of particles and forces we actually seem to have in Nature). For the moment, it is the second approach which has been pursued—indeed with great success as we shall see in the next chapter and in volume 2.

Accepting the general framework of quantum field theory, then, the first thing we must obviously do is to modify the theory in some way so that integrals such as (10.50) do not actually diverge, so that we can at least discuss finite rather than infinite quantities. This step is called 'regularization' of the theory. There are many ways to do this but for our present purposes a simple one will do well enough, which is to cut off the u-integration in (10.50) at some finite value Λ (remember u is $|\boldsymbol{k}'|$, so Λ here will have dimensions of energy, or mass); such a step was given some physical motivation in section 10.1.1. Then we can evaluate the integral straightforwardly and move on to the next stage.

With the upper limit in (10.50) replaced by Λ, we can evaluate the u-integral, obtaining (problem 10.4)

$$\Pi_{\mathrm{C}}^{[2]}(q^2, \Lambda^2) = \frac{-g^2}{8\pi^2}\int_0^1 \mathrm{d}x \left\{\ln\left(\frac{\Lambda + (\Lambda^2 + \Delta)^{1/2}}{\Delta^{1/2}}\right) - 1\right\} \tag{10.51}$$

where from (10.43)

$$\Delta = -x(1-x)q^2 + xm_{\mathrm{B}}^2 + (1-x)m_{\mathrm{A}}^2. \tag{10.52}$$

Note that $\Delta > 0$ for $q^2 < 0$. Relation (10.19) then becomes

$$m_{\mathrm{C}}^2(\Lambda^2) = m_{\mathrm{ph,C}}^2 - \Pi_{\mathrm{C}}^{[2]}(q^2 = m_{\mathrm{ph,C}}^2, \Lambda^2) \tag{10.53}$$

and there will be similar relations for the A and B masses. As noted previously, after (10.19), the shift represented by (10.53) is in an $O(g^2)$ perturbative correction (because $\Pi_{\mathrm{C}}^{[2]}$ contains a factor g^2), so that—again in the spirit of systematic perturbation theory—it will be adequate to this order in g^2 to replace the Lagrangian masses m_{A}^2, m_{B}^2, and m_{C}^2 *inside* the expressions for $\Pi_{\mathrm{A}}^{[2]}$, $\Pi_{\mathrm{B}}^{[2]}$ and $\Pi_{\mathrm{C}}^{[2]}$ by their physical counterparts. In this way the relations (10.53) and the two similar ones give us the prescription for rewriting the m_i^2 in terms of the $m_{\mathrm{ph},i}^2$ and Λ^2. Of course, when this is done in the propagators, the result is just to produce the desired form $\sim(q^2 - m_{\mathrm{ph},i}^2)^{-1}$, to this order.

So, for the propagator at this one-loop order, the effect of such mass shifts is essentially trivial: the large Λ behaviour is simply absorbed into m_i^2. What about Z_{C}? This was defined via (10.28) in terms of the quantity

$$\left.\frac{\mathrm{d}\Pi_{\mathrm{C}}^{[2]}}{\mathrm{d}q^2}\right|_{q^2=m_{\mathrm{ph,C}}^2}.$$

Problem 10.5 shows us that this quantity is *not* divergent as $\Lambda \to \infty$, so in fact we have here an example of a *finite* field strength renormalization. Nevertheless, even if

$$\left.\frac{\mathrm{d}\Pi_{\mathrm{C}}^{[2]}}{\mathrm{d}q^2}\right|_{q^2=m_{\mathrm{ph,C}}^2}$$

were divergent (as analogous quantities are in QED), the divergence could still be formally absorbed in the rescaling represented by Z_{C}. We may also analyse the vertex correction $G^{[2]}$ of figure 10.6, and conclude that it too is finite, because there are now three propagators giving six powers of k in the denominator, with still only a four-dimensional d^4k integration. Once again, the analogous vertex correction in QED is divergent, as we shall see in chapter 11; there too this divergence can be absorbed into a redefinition of the physical charge. The ABC theory is, in fact, a 'super-renormalizable' one, meaning (loosely) that it has fewer divergences than might be expected. We shall come back to the classification of theories (renormalizable, non-renormalizable and super-renormalizable) at the end of the following chapter.

While it is not our purpose to present a full discussion of one-loop renormalization in the ABC theory (because it is not of any direct physical interest) we will use it to introduce one more important idea before turning, in the next chapter, to one-loop QED.

10.4 Bare and renormalized perturbation theory

10.4.1 Reorganizing perturbation theory

We have seen that, of the one-loop effects listed at the end of section 10.2, the mass shifts given by equations such as (10.14) do involve formal divergences as

$\Lambda \to \infty$, but the vertex correction and field strength renormalization are finite in the ABC theory. We shall find that in QED the corresponding quantities are all divergent, so that the perturbative replacement of all Lagrangian parameters by their 'physical' counterparts is mandatory in QED, in order to get rid of $\ln \Lambda$ terms. However, this process—of evaluating the connections between the two sets of parameters, and then inserting them into all the calculated amplitudes—is likely to be very cumbersome. In this section, we shall introduce an alternative formulation, which has both calculational and conceptual advantages.

By way of motivation, consider the QED analogue of the divergent part of equation (10.7), which contributes a correction to the bare electron mass of the form $\alpha m \ln(\Lambda/m)$ where m is the electron mass. At $\Lambda = 100$ GeV the magnitude of this is about 0.04 MeV (if we take m to have the physical value), which is a shift of some 10%. The application of perturbation theory would seem more plausible if this kind of correction were to be included from the start, so that the 'free' part of the Hamiltonian (or Lagrangian) involved the physical fields and parameters, rather than the (unobserved) ones appearing in the original theory. Then the main effects, in some sense, would already be included by the use of these (empirical) physical quantities, and corrections would be 'more plausibly' small. This is indeed the main reason for the usefulness of such 'effective' parameters in the analogous case of condensed matter physics. Actually, of course, in quantum field theory the corrections will be just as infinite (if we send Λ to infinity) in this approach also, since whichever way we set the calculation up, we shall get loops, which are divergent. All the same, this kind of 'reorganization' does offer a more systematic approach to renormalization.

To illustrate the idea, consider again our ABC Lagrangian

$$\hat{\mathcal{L}} = \hat{\mathcal{L}}_{0,\mathrm{A}} + \hat{\mathcal{L}}_{0,\mathrm{B}} + \hat{\mathcal{L}}_{0,\mathrm{C}} + \hat{\mathcal{L}}_{\mathrm{int}} \tag{10.54}$$

where

$$\hat{\mathcal{L}}_{0,\mathrm{C}} = \tfrac{1}{2}\partial_\mu \hat{\phi}_\mathrm{C} \partial^\mu \hat{\phi}_\mathrm{C} - \tfrac{1}{2} m_\mathrm{C}^2 \hat{\phi}_\mathrm{C}^2 \tag{10.55}$$

and similarly for $\hat{\mathcal{L}}_{0,\mathrm{A}}$, $\hat{\mathcal{L}}_{0,\mathrm{B}}$; and where

$$\hat{\mathcal{L}}_{\mathrm{int}} = -g \hat{\phi}_\mathrm{A} \hat{\phi}_\mathrm{B} \hat{\phi}_\mathrm{C}. \tag{10.56}$$

There are two obvious moves to make: (i) introduce the rescaled (renormalized) fields by

$$\hat{\phi}_{\mathrm{ph},i}(x) = Z_i^{-1/2} \hat{\phi}_i(x) \tag{10.57}$$

in order to get rid of the $\sqrt{Z_i}$ factors in the S-matrix elements; and (ii) introduce the physical masses $m_{\mathrm{ph},i}^2$. Consider first the non-interacting parts of $\hat{\mathcal{L}}$, namely

$$\hat{\mathcal{L}}_0 = \hat{\mathcal{L}}_{0,\mathrm{A}} + \hat{\mathcal{L}}_{0,\mathrm{B}} + \hat{\mathcal{L}}_{0,\mathrm{C}}. \tag{10.58}$$

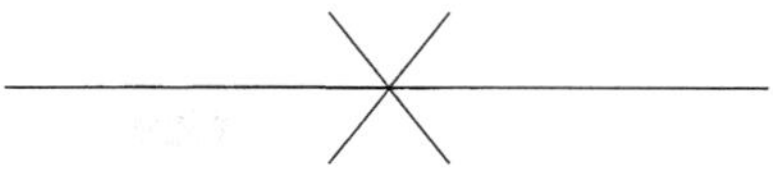

Figure 10.10. Counter term corresponding to the terms in braces in (10.60).

Singling out the C-parameters for definiteness, $\hat{\mathcal{L}}_0$ can then be written as

$$\begin{aligned}\hat{\mathcal{L}}_0 &= \tfrac{1}{2} Z_{\mathrm{C}} \partial_\mu \hat{\phi}_{\mathrm{ph,C}} \partial^\mu \hat{\phi}_{\mathrm{ph,C}} - \tfrac{1}{2} m_{\mathrm{C}}^2 Z_{\mathrm{C}} \hat{\phi}_{\mathrm{ph,C}}^2 + \cdots \\ &= \tfrac{1}{2} \partial_\mu \hat{\phi}_{\mathrm{ph,C}} \partial^\mu \hat{\phi}_{\mathrm{ph,C}} - \tfrac{1}{2} m_{\mathrm{ph,C}}^2 \hat{\phi}_{\mathrm{ph,C}}^2 \\ &\quad + \tfrac{1}{2}(Z_{\mathrm{C}} - 1) \partial_\mu \hat{\phi}_{\mathrm{ph,C}} \partial^\mu \hat{\phi}_{\mathrm{ph,C}} - \tfrac{1}{2}(m_{\mathrm{C}}^2 Z_{\mathrm{C}} - m_{\mathrm{ph,C}}^2) \hat{\phi}_{\mathrm{ph,C}}^2 + \cdots \end{aligned} \tag{10.59}$$

$$\equiv \hat{\mathcal{L}}_{0\mathrm{ph,C}} + \{\tfrac{1}{2} \delta Z_{\mathrm{C}} \partial_\mu \hat{\phi}_{\mathrm{ph,C}} \partial^\mu \hat{\phi}_{\mathrm{ph,C}} - \tfrac{1}{2}(\delta Z_{\mathrm{C}} m_{\mathrm{ph,C}}^2 + \delta m_{\mathrm{C}}^2 Z_{\mathrm{C}}) \hat{\phi}_{\mathrm{ph,C}}^2\} + \cdots \tag{10.60}$$

where $\hat{\mathcal{L}}_{0\mathrm{ph,C}}$ is the standard free-C Lagrangian in terms of the physical field and mass, which leads to a Feynman propagator $\mathrm{i}/(k^2 - m_{\mathrm{ph,C}}^2 + \mathrm{i}\epsilon)$ in the usual way; also, $\delta Z_{\mathrm{C}} = Z_{\mathrm{C}} - 1$ and $\delta m_{\mathrm{C}}^2 = m_{\mathrm{C}}^2 - m_{\mathrm{ph,C}}^2$. In (10.60) the dots signify similar rearrangements of $\hat{\mathcal{L}}_{0,\mathrm{A}}$ and $\hat{\mathcal{L}}_{0,\mathrm{B}}$. Note that Z_{C} and m_{C}^2 are understood to depend on Λ, as usual, although this has not been indicated explicitly.

We now regard '$\hat{\mathcal{L}}_{0\mathrm{ph,A}} + \hat{\mathcal{L}}_{0\mathrm{ph,B}} + \hat{\mathcal{L}}_{0\mathrm{ph,C}}$' as the 'unperturbed' part of $\hat{\mathcal{L}}$, and all the remainder of (10.60) as perturbations additional to the original $\hat{\mathcal{L}}_{\mathrm{int}}$ (much of theoretical physics consists of exploiting the identity '$a + b = (a + c) + (b - c)$'). The effect of this rearrangement is to introduce new perturbations, namely $\frac{1}{2} \delta Z_{\mathrm{C}} \partial_\mu \hat{\phi}_{\mathrm{ph,C}} \partial^\mu \hat{\phi}_{\mathrm{ph,C}}$ and the $\hat{\phi}_{\mathrm{ph,C}}^2$ term in (10.60), together with similar terms for the A and B fields. Such additional perturbations are called '*counter terms*' and they must be included in our new perturbation theory based on the $\hat{\mathcal{L}}_{0\mathrm{ph},i}$ pieces. As usual, this is conveniently implemented in terms of associated Feynman diagrams. Since both of these counter terms involve just the square of the field, it should be clear that they only have non-zero matrix elements between one-particle states, so that the associated diagram has the form shown in figure 10.10, which includes both these C-contributions. Problem 10.6 shows that the Feynman rule for figure 10.10 is that it contributes $\mathrm{i}[\delta Z_{\mathrm{C}} k^2 - (\delta Z_{\mathrm{C}} m_{\mathrm{ph,C}}^2 + \delta m_{\mathrm{C}}^2 Z_{\mathrm{C}})]$ to the $1\,\mathrm{C} \to 1\,\mathrm{C}$ amplitude.

The original interaction term $\hat{\mathcal{L}}_{\mathrm{int}}$ may also be rewritten in terms of the physical fields and a physical (renormalized) coupling constant g_{ph}:

$$\begin{aligned}-g\hat{\phi}_{\mathrm{A}}\hat{\phi}_{\mathrm{B}}\hat{\phi}_{\mathrm{C}} &= -g(Z_{\mathrm{A}} Z_{\mathrm{B}} Z_{\mathrm{C}})^{1/2} \hat{\phi}_{\mathrm{ph,A}} \hat{\phi}_{\mathrm{ph,B}} \hat{\phi}_{\mathrm{ph,C}} \\ &= -g_{\mathrm{ph}} \hat{\phi}_{\mathrm{ph,A}} \hat{\phi}_{\mathrm{ph,B}} \hat{\phi}_{\mathrm{ph,C}} - (Z_{\mathrm{V}} - 1) g_{\mathrm{ph}} \hat{\phi}_{\mathrm{ph,A}} \hat{\phi}_{\mathrm{ph,B}} \hat{\phi}_{\mathrm{ph,C}}\end{aligned} \tag{10.61}$$

where

$$Z_{\mathrm{V}} g_{\mathrm{ph}} = g(Z_{\mathrm{A}} Z_{\mathrm{B}} Z_{\mathrm{C}})^{1/2}. \tag{10.62}$$

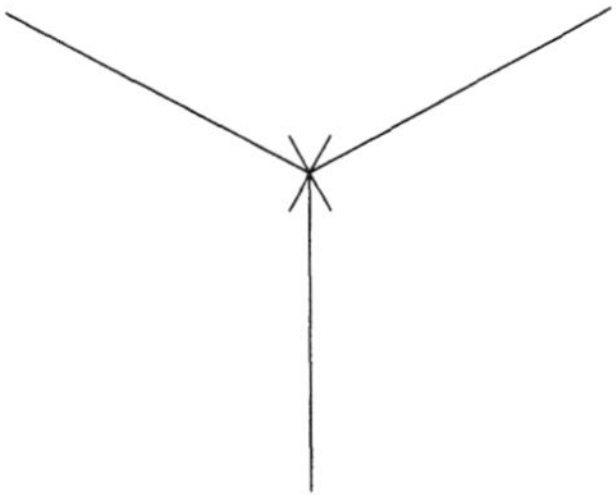

Figure 10.11. Counter term corresponding to the '$(Z_V - 1)$' term in (10.62).

The interpretation of (10.62) is clearly that 'g_{ph}' is the coupling constant describing the interactions among the $\hat{\phi}_{ph,i}$ fields, while the '$(Z_V - 1)$' term is another counter term, having the structure shown in figure 10.11.

In summary, we have reorganized $\hat{\mathcal{L}}$ so as to base perturbation theory on a part describing the free renormalized fields (rather than the fields in the original Lagrangian); in this formulation we find that, in addition to the (renormalized) ABC-interaction term, further terms have appeared which are interpreted as additional perturbations, called counter terms. These counter terms are determined, at each order in this (renormalized) perturbation theory, by what are basically self-consistency conditions—such as, for example, the requirement that the propagators really do reduce to the physical ones at the 'mass-shell' points. We shall now illustrate this procedure for the C propagator.

10.4.2 The $O(g_{ph}^2)$ renormalized self-energy revisited: how counter terms are determined by renormalization conditions

Let us return to the calculation of the C propagator, following the same procedure as in section 10.1, but this time 'perturbing' away from $\hat{\mathcal{L}}_{0ph,i}$ and including the contribution from the counter term of figure 10.10, in addition to the $O(g_{ph}^2)$ self energy. The expression (10.14) will now be replaced by

$$\frac{\mathrm{i}}{q^2 - m_{ph,C}^2 + q^2\delta Z_C - \delta Z_C m_{ph,C}^2 - \delta m_C^2 Z_C - \Pi_{ph,C}^{[2]}(q^2)} \tag{10.63}$$

where

$$-\mathrm{i}\Pi_{ph,C}^{[2]}(q^2, \Lambda^2) = (-\mathrm{i}g_{ph})^2 \int \frac{\mathrm{d}^4k}{(2\pi)^4} \frac{\mathrm{i}}{k^2 - m_{ph,A}^2 + \mathrm{i}\epsilon} \cdot \frac{\mathrm{i}}{(q-k)^2 - m_{ph,B}^2 + \mathrm{i}\epsilon} \tag{10.64}$$

and where we have indicated the cut-off dependence on the left-hand side, leaving it understood on the right. Comparing (10.64) with (10.39) we see that they are exactly the same, except that $\Pi_{ph,C}^{[2]}$ involves the 'physical' coupling constant g_{ph} and the physical masses, as expected in this renormalized perturbation theory. In particular, $\Pi^{[2]}_{\text{ph,C}}$ will be divergent in exactly the same way as $\Pi^{[2]}_{\text{C}}$, as the cut-off Λ goes to infinity.

The essence of this 'reorganized' perturbation theory is that we now determine δZ_{C} and δm_{C}^2 from the condition that as $q^2 \to m^2_{\text{ph,C}}$, the propagator (10.63) reduces to $\text{i}/(q^2 - m^2_{\text{ph,C}})$, i.e. it correctly represents the physical C propagator at the mass-shell point, with standard normalization. Expanding $\Pi^{[2]}_{\text{ph,C}}(q^2)$ about $q^2 = m^2_{\text{ph,C}}$ then, we reach the approximate form of (10.63), valid for $q^2 \approx m^2_{\text{ph,C}}$:

$$\frac{\text{i}}{(q^2 - m^2_{\text{ph,C}})Z_{\text{C}} - \delta m_{\text{C}}^2 Z_{\text{C}} - \Pi^{[2]}_{\text{ph,C}}(m^2_{\text{ph,C}}, \Lambda^2) - (q^2 - m^2_{\text{ph,C}}) \left.\dfrac{\text{d}\Pi^{[2]}_{\text{ph,C}}}{\text{d}q^2}\right|_{q^2=m^2_{\text{ph,C}}}}. \tag{10.65}$$

Requiring that this has the form $\text{i}/(q^2 - m^2_{\text{ph,C}})$ gives

$$\begin{aligned} &\text{condition (a)} \quad \delta m_{\text{C}}^2 = -Z_{\text{C}}^{-1}\Pi^{[2]}_{\text{ph,C}}(m^2_{\text{ph,C}}, \Lambda^2) \\ &\text{condition (b)} \quad Z_{\text{C}} = 1 + \left.\frac{\text{d}\Pi^{[2]}_{\text{ph,C}}}{\text{d}q^2}\right|_{q^2=m^2_{\text{ph,C}}}. \end{aligned} \tag{10.66}$$

Looking first at condition (b), we see that our renormalization constant Z_{C} has, in this approach, been determined up to $O(g^2_{\text{ph}})$ by an equation that is, in fact, very similar to (10.28), but it is expressed in terms of physical parameters. As regards (a), since $Z_{\text{C}} = 1 + O(g^2_{\text{ph}})$, it is sufficient to replace it by 1 on the right-hand side of (a), so that, to this order, $\delta m_{\text{C}}^2 \approx -\Pi^{[2]}_{\text{ph,C}}(m^2_{\text{ph,C}}, \Lambda^2)$. Once again, this is similar to (10.53), but written in terms of the physical quantities from the outset. We indicate that these evaluations of Z_{C} and δm_{C}^2 are correct to second order by adding a superscript, as in $Z_{\text{C}}^{[2]}$.

Of course, we have *not* avoided the infinities (in the limit $\Lambda \to \infty$) in this approach! It is still true that the loop integral in $\Pi^{[2]}_{\text{ph,C}}$ diverges logarithmically and so the mass shift $(\delta m_{\text{C}}^{[2]})^2$ is infinite as $\Lambda \to \infty$. Nevertheless, this is a conceptually cleaner way to do the business. It is called 'renormalized perturbation theory', as opposed to our first approach which is called 'bare perturbation theory'. What we there called the 'Lagrangian fields and parameters' are usually called the 'bare' ones; the 'renormalized' quantities are 'clothed' by the interactions.

We may now return to our propagator (10.63), and insert the results (10.66) to obtain the final important expression for the C propagator containing the one-loop $O(g^2_{\text{ph}})$ renormalized self-energy:

$$\boxed{\frac{\text{i}}{q^2 - m^2_{\text{ph,C}} - \overline{\Pi}^{[2]}_{\text{ph,C}}(q^2)}} \tag{10.67}$$

where

$$\overline{\Pi}^{[2]}_{\text{ph,C}}(q^2) = \Pi^{[2]}_{\text{ph,C}}(q^2) - \Pi^{[2]}_{\text{ph,C}}(m^2_{\text{ph,C}}) - (q^2 - m^2_{\text{ph,C}}) \left.\frac{\mathrm{d}\Pi^{[2]}_{\text{ph,C}}}{\mathrm{d}q^2}\right|_{q^2=m^2_{\text{ph,C}}}. \tag{10.68}$$

We stress that, in this final form, all infinities (even had Z_{C} been logarithmically divergent) are removed by the *subtractions* in (10.68) and the behaviour near the mass-shell point is indeed $\mathrm{i}/(q^2 - m^2_{\text{ph,C}})$.

A succinct way of summarizing our final renormalized result (10.67), with the definition (10.68), is to say that the C propagator may be defined by (10.67) where the $O(g^2_{\text{ph}})$ renormalized self-energy $\overline{\Pi}^{[2]}_{\text{ph,C}}$ satisfies the *renormalization conditions*

$$\overline{\Pi}^{[2]}_{\text{ph,C}}(q^2 = m^2_{\text{ph,C}}) = 0 \qquad \left.\frac{\mathrm{d}}{\mathrm{d}q^2}\overline{\Pi}^{[2]}_{\text{ph,C}}(q^2)\right|_{q^2=m^2_{\text{ph,C}}} = 0. \tag{10.69}$$

Relations analogous to (10.69) clearly hold for the A and B self-energies also. In this definition, the explicit introduction and cancellation of large-Λ terms has disappeared from sight, and *all that remains is the importation of one constant from experiment, $m^2_{\text{ph,C}}$, and a (hidden) rescaling of the fields.* It is useful to bear this viewpoint in mind when considering more general theories, including ones that are 'non-renormalizable' (see section 11.8 of the following chapter).

There is a lot of good physics in the expression (10.67), which we shall elucidate in the realistic case of QED in the next chapter. For the moment, we just whet the reader's appetite by pointing out that (10.67) must amount to the prediction of a finite, calculated correction to the Yukawa 1 − C exchange potential, which after all is given by the Fourier transform of the (static form of) the propagator, as we learned long ago. In the case of QED, this will amount to a calculated correction to Coulomb's law, due to radiative corrections, as we shall discuss in section 11.5.1.

There is an important technical implication we may draw from (10.69). Consider the Feynman diagram of figure 10.12 in which a propagator correction has been inserted in an *external* line. This diagram is of order g^4_{ph}, and should presumably be included along with the others at this order. However, the conditions (10.69)—in this case written for $\overline{\Pi}^{[2]}_{\text{ph,A}}$—imply that it vanishes. Omitting irrelevant factors, the amplitude for figure 10.11 is

$$\overline{\Pi}^{[2]}_{\text{ph,A}}(p_{\text{A}}) \frac{1}{p^2_{\text{A}} - m^2_{\text{ph,A}}} \frac{1}{q^2 - m^2_{\text{ph,C}}} \tag{10.70}$$

and we need to take the limit $p^2_{\text{A}} \to m^2_{\text{ph,A}}$ since the external A particle is on-shell. Expanding $\overline{\Pi}^{[2]}_{\text{ph,A}}$ about the point $p^2_{\text{A}} = m^2_{\text{ph,A}}$ and using conditions (10.69) for C → A we see that (10.70) vanishes. Thus with this definition, propagator corrections do not need to be applied to external lines.

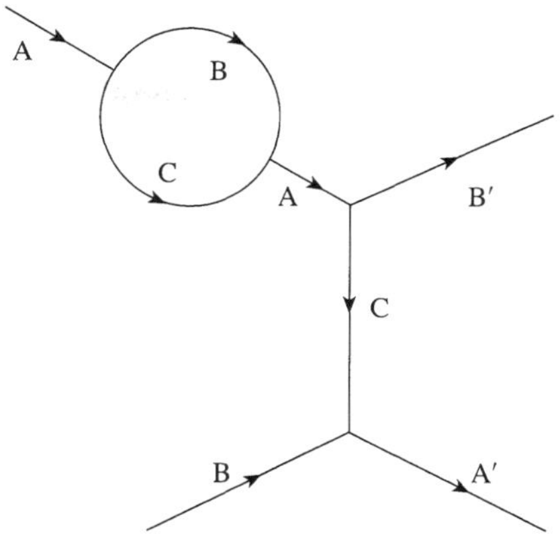

Figure 10.12. $O(g^4)$ contribution to A + B → A + B, involving a propagator correction inserted in an external line.

10.5 Renormalizability

We have seen how divergences present in self-energy loops like figure 10.6(a) can be eliminated by supposing that the 'bare' masses in the original Lagrangian depend on the cut-off in just such a way as to cancel the divergences, leaving a finite value for the physical masses. The latter are, however, parameters to be taken from experiment: they are not calculable. Alternatively, we may rephrase perturbation theory in terms of renormalized quantities from the outset, in which case the loop divergence is cancelled by appropriate counter terms; but again the physical masses have to be taken from experiment. We pointed out that, in the ABC theory, neither the field strength renormalizations Z_i nor the vertex diagrams of figure 10.5 were divergent, but we shall see in the next chapter that the analogous quantities in QED are divergent. These divergences too can be absorbed into redefinitions of the 'physical' fields and a 'physical' coupling constant (the latter again to be taken from experiment). Or, again, such divergences can be cancelled by appropriate counter terms in the renormalized perturbation theory approach.

In general, a theory will have various divergences at the one-loop level, and new divergences will enter as we go up in order of perturbation theory (or number of loops). Typically, therefore, quantum field theories betray sensitivity to unknown short-distance physics by the presence of formal divergences in loops, as a cut-off $\Lambda \to \infty$. In a *renormalizable* theory, this sensitivity can be sytematically removed by accepting that a finite number of parameters are uncalculable, and must be taken from experiment. These are the suitably defined 'physical' values of the masses and coupling constants appearing in the Lagrangian. Once these parameters are given, all other quantities are finite and calculable, to any desired order in perturbation theory—assuming, of course, that terms in successive orders

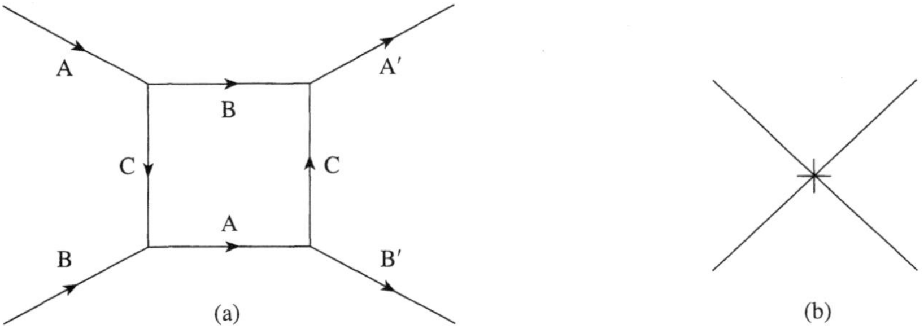

Figure 10.13. (*a*) $O(g^4)$ one-loop contribution to A + B → A + B; (*b*) counter term that would be required if (*a*) were divergent.

diminish sensibly in size.

Alternatively, we may say that a renormalizable theory is one in which a finite number of counter terms can be so chosen as to cancel all divergences order by order in renormalized perturbation theory. Note, now, that the only available counter terms are the ones which arise in the process of 'reorganizing' the original theory in terms of renormalized quantities plus extra bits (the counter terms). All the counter terms must correspond to masses, interactions, etc which are present in the original (or 'bare') Lagrangian—which is, in fact, *the* theory we are trying to make sense of! We are not allowed to add in any old kind of counter term—if we did, we would be redefining the theory.

We can illustrate this point by considering, for example, a one-loop ($O(g^4)$) contribution to AB → AB scattering, as shown in figure 10.13(*a*). If this graph is divergent, we will need a counter term with the structure shown in figure 10.13(*b*) to cancel the divergence—but there is no such 'contact' AB → AB interaction in the original theory (it would have the form $\lambda\hat{\phi}_A^2(x)\hat{\phi}_B^2(x)$). In fact, the graph is convergent, as indicated by the usual power-counting (four powers of k in the numerator, eight in the denominator from the four propagators). And indeed, the ABC theory is renormalizable—or rather, as noted earlier, 'super-renormalizable'.

We shall have something more to say about renormalizability and non-renormalizability (is it fatal?), at the end of the following chapter. The first and main business, however, will be to apply what we have learned here to QED.

Problems

10.1 Carry out the indicated change of variables so as to obtain (10.4) from (10.3).

10.2 Verify the Feynman identity (10.40).

10.3 Obtain (10.42) from (10.41).

10.4 Obtain (10.51) from (10.50), having replaced the upper limit of the u-integral by Λ.

10.5 Show that $\mathrm{d}\Pi^{[2]}_{\mathrm{ph,C}}/\mathrm{d}q^2$ is ultraviolet finite.

10.6 Obtain the Feynman rule quoted in the text for the sum of the counter terms appearing in (10.60).

11

LOOPS AND RENORMALIZATION II: QED

The present electrodynamics is certainly incomplete, but is no longer certainly incorrect.

F J Dyson (1949b)

We now turn to the analysis of loop corrections in QED. As we might expect, a theory with fermionic and gauge fields proves to be a tougher opponent than one with only spinless particles, even though we restrict ourselves to one-loop diagrams only.

At the outset we must make one important disclaimer. In QED many loop diagrams diverge not only as the loop momentum goes to infinity ('ultraviolet divergence') but also as it goes to zero ('infrared divergence'). This phenomenon can only arise when there are massless particles in the theory—for otherwise the propagator factors $\approx (k^2 - M^2)^{-1}$ will always prevent any infinity at low k. Of course, in a gauge theory we do have just such massless quanta. Our main purpose here is to demonstrate how the *ultraviolet* divergences can be tamed and we must refer the reader to Weinberg (1995, chapter 13), or to Peskin and Schroeder (1995, section 6.5), for instruction in dealing with the infrared problem. The remedy lies, essentially, in a careful consideration of the contribution, to physical cross sections, of amplitudes involving the *real* emission of very low frequency photons, along with infrared divergent virtual photon processes. It is a 'technical' problem, having to do with massless particles (of which there are not that many), whereas ultraviolet divergences are generic.

11.1 Counter terms

We shall consider the simplest case of a single fermion of bare mass m_0 and bare charge e_0 ($e_0 > 0$) interacting with the Maxwell field, for which the bare (i.e. actual!) Lagrangian is

$$\hat{\mathcal{L}} = \bar{\hat{\psi}}_0(\mathrm{i}\partial\!\!\!/ - m_0)\hat{\psi}_0 - e_0\bar{\hat{\psi}}_0\gamma^\mu\hat{\psi}_0\hat{A}_{0\mu} - \frac{1}{4}\hat{F}_{0\mu\nu}\hat{F}_0^{\mu\nu} - \frac{1}{2\xi_0}(\partial\cdot\hat{A}_0)^2 \quad (11.1)$$

according to chapter 7. We shall adopt the 'renormalized perturbation theory' approach and begin by introducing field strength renormalizations via

$$\hat{\psi} = Z_2^{-1/2}\hat{\psi}_0 \quad (11.2)$$

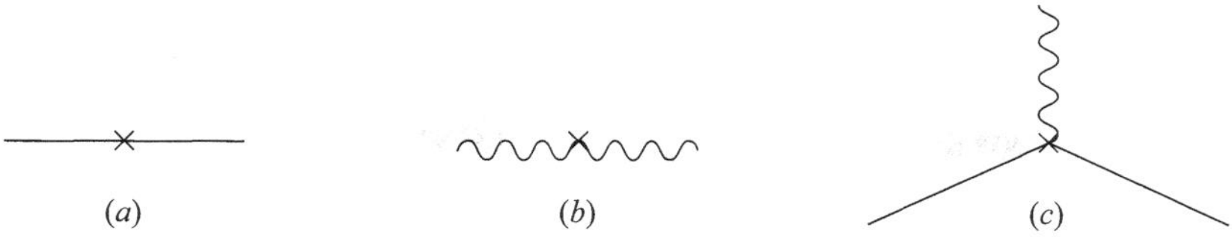

Figure 11.1. Counter terms in QED: (*a*) electron mass and wavefunction; (*b*) photon wavefunction; (*c*) vertex part.

$$\hat{A}^{\mu} = Z_3^{-1/2}\hat{A}_0^{\mu} \tag{11.3}$$

where the 'physical' fields and parameters will now simply have no '0' subscript. This will lead to a rewriting of the free and gauge-fixing part of (11.1):

$$\begin{aligned}
&\bar{\hat{\psi}}_0(\mathrm{i}\not{\partial} - m_0)\hat{\psi}_0 - \frac{1}{4}\hat{F}_{0\mu\nu}\hat{F}_0^{\mu\nu} - \frac{1}{2\xi_0}(\partial\cdot\hat{A}_0)^2 \\
&\quad = \bar{\hat{\psi}}(\mathrm{i}\not{\partial} - m)\hat{\psi} - \frac{1}{4}\hat{F}_{\mu\nu}\hat{F}^{\mu\nu} - \frac{1}{2\xi}(\partial\cdot\hat{A})^2 \\
&\qquad + [(Z_2 - 1)\bar{\hat{\psi}}\mathrm{i}\not{\partial}\hat{\psi} - \delta m\bar{\hat{\psi}}\hat{\psi}] - \tfrac{1}{4}(Z_3 - 1)\hat{F}_{\mu\nu}\hat{F}^{\mu\nu}
\end{aligned} \tag{11.4}$$

where $\xi = \xi_0/Z_3$ and $\delta m = m_0 Z_2 - m$ (compare (10.60)). We see the emergence of the expected '$\bar{\hat{\psi}}\ldots\hat{\psi}$' and '$\hat{F}\cdot\hat{F}$' counter terms in (11.4), affecting both the fermion and the gauge-field propagators. Next, we write the interaction in terms of a physical e, and the physical fields, together with a compensating third counter term:

$$-e_0\bar{\hat{\psi}}_0\gamma^{\mu}\hat{\psi}_0\hat{A}_{0\mu} = -e\bar{\hat{\psi}}\gamma^{\mu}\hat{\psi}\hat{A}_{\mu} - (Z_1 - 1)e\bar{\hat{\psi}}\gamma^{\mu}\hat{\psi}\hat{A}_{\mu} \tag{11.5}$$

where, with the aid of (11.2) and (11.3),

$$Z_1 e = e_0 Z_2 Z_3^{1/2}. \tag{11.6}$$

The three counter terms are represented diagrammatically as shown in figures 11.1(*a*), (*b*) and (*c*), for which the Feynman rules are, respectively,

$$\begin{aligned}
&\text{(a):}\quad \mathrm{i}[\not{k}(Z_2 - 1) - \delta m] \\
&\text{(b):}\quad -\mathrm{i}(g^{\mu\nu}k^2 - k^{\mu}k^{\nu})(Z_3 - 1) \\
&\text{(c):}\quad -\mathrm{i}e\gamma^{\mu}(Z_1 - 1).
\end{aligned} \tag{11.7}$$

These counter terms will compensate for the ultraviolet divergences of the three elementary loop diagrams of figure 11.2, and in fact they are sufficient to eliminate all such divergences in all QED loops.

Before proceeding further we remark that we already have a first indication that renormalizing a gauge theory presents some new features. Consider the two

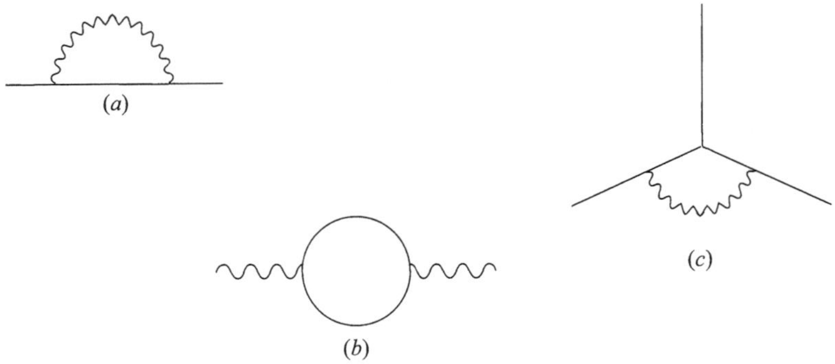

Figure 11.2. Elementary one-loop divergent diagrams in QED.

counter terms involving $Z_2 - 1$ and $Z_1 - 1$; their sum gives

$$\bar{\hat{\psi}}[\mathrm{i}(Z_2 - 1)\not{\partial} - e(Z_1 - 1)\hat{\not{A}}]\hat{\psi} \tag{11.8}$$

which is *not* of the 'gauge principle' form '$\mathrm{i}\not{\partial} - e\hat{\not{A}}$'! Unless, of course, $Z_1 = Z_2$. This relation between the two quite different renormalization constants is, in fact, true to all orders in perturbation theory, as a consequence of a *Ward identity* (Ward 1950), which is itself a consequence of gauge invariance. We shall discuss the Ward identity and $Z_1 = Z_2$ at the one loop level in section 11.6.

11.2 The $O(e^2)$ fermion self-energy

In analogy with $-\mathrm{i}\Pi_{\mathrm{C}}^{[2]}$, the amplitude corresponding to figure 11.2(a) is the fermion self-energy $-\mathrm{i}\Sigma^{[2]}$ where

$$-\mathrm{i}\Sigma^{[2]}(p) = (-\mathrm{i}e)^2 \int \gamma^{\nu} \frac{-\mathrm{i}g_{\mu\nu}}{k^2} \frac{\mathrm{i}}{\not{p} - \not{k} - m} \gamma^{\mu} \frac{\mathrm{d}^4 k}{(2\pi)^4} \tag{11.9}$$

and we have now chosen the gauge $\xi = 1$. As expected, the $\mathrm{d}^4 k$ integral in (11.9) diverges for large k—this time more seriously than the integral in $\Pi_{\mathrm{C}}^{[2]}$, because there are only three powers of k in the denominator of (11.9) as opposed to four in (10.7)). Once again, we need to choose some form of regularization to make (11.9) ultraviolet finite. We shall not be specific (as yet) about what choice we are making, since whatever it may be the outcome will be qualitatively similar to the $\Pi_{\mathrm{C}}^{[2]}$ case.

There is, however, one interesting new feature in this (fermion) case. As previously indicated, power-counting in the integral of (11.9) might lead us to expect that—if we adopt a simple cut-off—the leading ultraviolet divergence of $\Sigma^{[2]}$ would be proportional to Λ rather than $\ln \Lambda$. This is because we have that one extra power of k in the numerator and $\Sigma^{[2]}$ has dimensions of mass. However, this is not so. The leading p-independent divergence is, in fact, proportional to

$m \ln(\Lambda/m)$. The reason for this is important and it has interesting generalizations. Suppose that m in (11.4) were set equal to zero. Then, as we saw in problem 9.4, the two helicity components $\hat{\psi}_{\mathrm{L}}$ and $\hat{\psi}_{\mathrm{R}}$ of the electron field will not be coupled by the QED interaction. It follows that no terms of the form $\bar{\hat{\psi}}_{\mathrm{L}}\hat{\psi}_{\mathrm{R}}$ or $\bar{\hat{\psi}}_{\mathrm{R}}\hat{\psi}_{\mathrm{L}}$ can be generated, and hence no perturbatively-induced mass term, if $m = 0$. The perturbative mass shift must be proportional to m and therefore, on dimensional grounds, only logarithmically divergent.

There is also a p-dependent divergence of the self-energy, of which warning was given in section 10.3.2. As in the scalar case, this will be associated with the field strength renormalization factor Z_2. It is proportional to $\not{p} \ln(\Lambda/m)$ (Z_2 is the coefficient of $\not{\partial}$ in (11.8), which leads to $\not{p}$ in momentum space). The upshot is that the fermion propagator, including the one-loop renormalized self-energy, is given by

$$\frac{\mathrm{i}}{\not{p} - m - \bar{\Sigma}^{[2]}(p)} \tag{11.10}$$

where (cf (10.68))

$$\bar{\Sigma}^{[2]}(p) = \Sigma^{[2]}(p) - \Sigma^{[2]}(\not{p} = m) - (\not{p} - m)\left.\frac{\mathrm{d}\Sigma^{[2]}}{\mathrm{d}\not{p}}\right|_{\not{p}=m}. \tag{11.11}$$

Whatever form of regularization is used, the twice-subtracted $\bar{\Sigma}^{[2]}$ will be finite and independent of the regulator when it is removed. In terms of the 'compensating' quantities Z_2 and $m_0 - m$, we find (problem 11.1, cf (10.66))

$$Z_2 = 1 + \left.\frac{\mathrm{d}\Sigma^{[2]}}{\mathrm{d}\not{p}}\right|_{\not{p}=m} \qquad m_0 - m = -Z_2^{-1}\Sigma^{[2]}(\not{p} = m). \tag{11.12}$$

Note that, as in the case of $\bar{\Pi}_C^{[2]}$, the definition (11.11) of $\bar{\Sigma}^{[2]}$ implies that propagator corrections vanish for external (on-shell) fermions. The quantities Z_2 and m_0 determined by (11.12) now carry a superscript '[2]' to indicate that they are correct at $O(e^2)$.

We must now remind the reader that, although we have indeed eliminated the ultraviolet divergences in $\bar{\Sigma}^{[2]}$ by the subtractions of (11.11), there remains an untreated infrared divergence in $\mathrm{d}\Sigma^{[2]}/\mathrm{d}\not{p}$. To show how this is dealt with would take us beyond our intended scope, as explained at the start of the chapter. Suffice it to say that by the introduction of a 'regulating' photon mass μ^2, and consideration of relevant real photon processes along with virtual ones, these infrared problems can be controlled (Weinberg 1995, Peskin and Schroeder 1995).

11.3 The $O(e^2)$ photon self-energy

The amplitude corresponding to figure 11.2(b) is $\mathrm{i}\Pi_{\mu\nu}^{[2]}(q)$ where

$$\mathrm{i}\Pi_{\mu\nu}^{[2]}(q) = (-1)(-\mathrm{i}e)^2 \,\mathrm{Tr} \int \frac{\mathrm{d}^4 k}{(2\pi)^4} \frac{\mathrm{i}}{\not{q} + \not{k} - m}\gamma_\mu \frac{\mathrm{i}}{\not{k} - m}\gamma_\nu \tag{11.13}$$

$$= -e^2 \int \frac{d^4k}{(2\pi)^4} \frac{\text{Tr}[(\not{q} + \not{k} + m)\gamma_\mu(\not{k} + m)\gamma_\nu]}{[(q+k)^2 - m^2][k^2 - m^2]}. \quad (11.14)$$

Once again, this *photon self-energy* is analogous to the scalar particle self-energy of chapter 10. There are two new features to be commented on in (11.14). The first is the overall '−1' factor, which occurs whenever there is a *closed fermion loop*. The keen reader may like to pursue this via problem 11.2. The second feature is the appearance of the trace symbol 'Tr': this is plausible as the amplitude is basically a $1\gamma \to 1\gamma$ one with no spinor indices, but again the reader can follow that through in problem 11.3.

We now want to go some way into the calculation of $\Pi_{\mu\nu}^{[2]}$ because it will, in the end, contain important physics—for example , corrections to Coulomb's law. The first step is to evaluate the numerator trace factor using the theorems of section 8.2.3. We find (problem 11.4)

$$\begin{aligned}\text{Tr}[(\not{q} + \not{k} + m)\gamma_\mu(\not{k} + m)\gamma_\nu] = 4\{&(q_\mu + k_\mu)k_\nu + (q_\nu + k_\nu)k_\mu \\ &- g_{\mu\nu}((q\cdot k) + k^2 - m^2)\}. \quad (11.15)\end{aligned}$$

We then use the Feynman identity (10.40) to combine the denominators, yielding

$$\frac{1}{[(q+k)^2 - m^2][k^2 - m^2]} = \int_0^1 dx \frac{1}{[k'^2 - \Delta_\gamma + i\epsilon]^2} \quad (11.16)$$

where $k' = k + xq$, $\Delta_\gamma = -x(1-x)q^2 + m^2$ (note that Δ_γ is precisely the same as Δ of (10.43) with $m_A = m_B = m$) and we have reinstated the implied '$i\epsilon$'. Making the shift to the variable k' in the numerator factor (11.15) produces a revised numerator which is

$$4\{2k'_\mu k'_\nu - g_{\mu\nu}(k'^2 - \Delta_\gamma) - 2x(1-x)(q_\mu q_\nu - g_{\mu\nu}q^2) + \text{terms linear in } k'\} \quad (11.17)$$

where the terms linear in k' will vanish by symmetry when integrated over k' in (11.14). Our result so far is therefore

$$\begin{aligned}i\Pi_{\mu\nu}^{[2]}(q^2) = &-4e^2 \int_0^1 dx \left\{ \int \frac{d^4k'}{(2\pi)^4} \left[\frac{2k'_\mu k'_\nu}{(k'^2 - \Delta_\gamma + i\epsilon)^2} - \frac{g_{\mu\nu}}{(k'^2 - \Delta_\gamma + i\epsilon)} \right] \right\} \\ &+ 8e^2(q_\mu q_\nu - g_{\mu\nu}q^2) \int_0^1 dx \int \frac{d^4k'}{(2\pi)^4} \frac{x(1-x)}{(k'^2 - \Delta_\gamma + i\epsilon)^2}. \quad (11.18)\end{aligned}$$

Consider now the ultraviolet divergences of (11.18), adopting a simple cut-off as a regularization. The terms in the first line are both apparently quadratically divergent, while the integral in the second line is logarithmically divergent. What counter terms do we have to cancel these divergences? The answer is that the '$(Z_3 - 1)$' counter term of figure 11.1(b) is of exactly the right form to cancel the logarithmic divergence in the second line of (11.18), but we have no counter term proportional to the $g_{\mu\nu}$ term in the first line. Note, incidentally, that we can argue from Lorentz covariance (see appendix D) that

$$\int \frac{\mathrm{d}^4k'}{(2\pi)^4} \frac{k'_\mu k'_\nu}{(k'^2 - \Delta_\gamma + \mathrm{i}\epsilon)^2} = f(\Delta_\gamma) g_{\mu\nu} \tag{11.19}$$

so that taking the dot product of both sides with $g^{\mu\nu}$ we deduce that

$$\int \frac{\mathrm{d}^4k'}{(2\pi)^4} \frac{2k'_\mu k'_\nu}{(k'^2 - \Delta_\gamma + \mathrm{i}\epsilon)^2} = \frac{1}{2} \int \frac{\mathrm{d}^4k'}{(2\pi)^4} \frac{k'^2 g_{\mu\nu}}{(k'^2 - \Delta_\gamma + \mathrm{i}\epsilon)^2}. \tag{11.20}$$

It follows that both the terms in the first line of (11.18) produce a divergence of the form $\sim\Lambda^2 g_{\mu\nu}$, and they do not cancel, at least in our simple cut-off regularization.

A term proportional to $g_{\mu\nu}$ is, in fact, a photon mass term. If the Lagrangian included a mass term for the photon it would have the form $\frac{1}{2}m_{\gamma 0}^2 g_{\mu\nu} \hat{A}_0^\mu \hat{A}_0^\nu$, which after introducing the rescaled $\hat{A}_\mu$ will generate a counter term proportional to $g_{\mu\nu}\hat{A}^\mu \hat{A}^\nu$, and an associated Feynman amplitude proportional to $g_{\mu\nu}$. But such a term $m_{\gamma 0}^2$ violates gauge invariance! (It is plainly not invariant under (7.66).) Evidently the simple momentum cut-off that we have adopted as a regularization procedure does not respect gauge invariance. We saw in section 8.6.2 that this implied the condition

$$q^\mu T_\mu = 0 \tag{11.21}$$

where q is the 4-momentum of a photon entering a one-photon amplitude T_μ. Our discussion of (11.21) was limited in section 8.6.2 to the case of a real external photon, whereas the photon lines in $\mathrm{i}\Pi_{\mu\nu}^{[2]}$ are internal and virtual; nevertheless it is still true that gauge invariance implies (Peskin and Schroeder 1995, section 7.4)

$$q^\mu \Pi_{\mu\nu}^{[2]} = q^\nu \Pi_{\mu\nu}^{[2]} = 0. \tag{11.22}$$

Condition (11.22) is guaranteed by the tensor structure $(q_\mu q_\nu - g_{\mu\nu}q^2)$ of the *second* line in (11.18), provided the divergence is regularized. As previously implied, a simple cut-off Λ suffices for this term, since it does not alter the tensor structure, and the Λ-dependence can be compensated by the '$Z_3 - 1$' counter term which has the same tensor structure (cf figure 11.2(*b*)). But what about the first line of (11.18)? Various gauge-invariant regularizations have been used, the effect of all of which is to cause the first line of (11.18) to vanish. The most widely used, since the 1970s, is the *dimensional regularization* technique introduced by 't Hooft and Veltman (1972), which involves the 'continuation' of the number of space–time dimensions from four to d (< 4). As d is reduced, the integrals tend to diverge less, and the divergences can be isolated via the terms which diverge as $d \to 4$. Using gauge-invariant dimensional regularization, the two terms in the first line of (11.18) are found to cancel each other exactly, leaving just the manifestly gauge invariant second line.

We shall not enter into any further details of these technicalities here (see, for example, Peskin and Schroeder 1995). We proceed to the next step, renormalizing the gauge-invariant part of $\mathrm{i}\Pi_{\mu\nu}^{[2]}(q^2)$.

11.4 The $O(e^2)$ renormalized photon self-energy

The surviving (gauge-invariant) term of $\Pi^{[2]}_{\mu\nu}$ is

$$\mathrm{i}\Pi^{[2]}_{\mu\nu}(q^2) = 8e^2(q_\mu q_\nu - q^2 g_{\mu\nu}) \int_0^1 \mathrm{d}x \int \frac{\mathrm{d}^4k'}{(2\pi)^4} \frac{x(1-x)}{(k'^2 - \Delta_\gamma + \mathrm{i}\epsilon)^2} \tag{11.23}$$

$$\equiv \mathrm{i}(q^2 g_{\mu\nu} - q_\mu q_\nu)\Pi^{[2]}_\gamma(q^2) \tag{11.24}$$

and we regularize the gauge-invariant quantity $\Pi^{[2]}_\gamma(q^2)$ by a simple cut-off Λ. The analogue of (10.11) is then (in the gauge $\xi = 1$)

$$\begin{aligned}
&\frac{-\mathrm{i}g_{\mu\nu}}{q^2} + \frac{-\mathrm{i}g_{\mu\rho}}{q^2} \cdot \mathrm{i}(q^2 g^{\rho\sigma} - q^\rho q^\sigma)\Pi^{[2]}_\gamma(q^2) \cdot \frac{-\mathrm{i}g_{\sigma\nu}}{q^2} \\
&\quad + \frac{-\mathrm{i}g_{\mu\rho}}{q^2} \cdot \mathrm{i}(q^2 g^{\rho\sigma} - q^\rho q^\sigma)\Pi^{[2]}_\gamma(q^2) \cdot \frac{-\mathrm{i}g_{\sigma\tau}}{q^2} \\
&\quad \cdot \mathrm{i}(q^2 g^{\tau\eta} - q^\tau q^\eta) \cdot \Pi^{[2]}_\gamma(q^2) \cdot \frac{-\mathrm{i}g_{\eta\nu}}{q^2} + \cdots \\
&\quad = \frac{-\mathrm{i}g_{\mu\nu}}{q^2} + \frac{-\mathrm{i}g_{\mu\rho}}{q^2} P^\rho_\nu \Pi^{[2]}_\gamma(q^2) + \frac{-\mathrm{i}g_{\mu\rho}}{q^2} P^\rho_\tau P^\tau_\nu (\Pi^{[2]}_\gamma(q^2))^2 + \cdots
\end{aligned} \tag{11.25}$$

where

$$P^\rho_\nu = g^\rho_\nu - \frac{q^\rho q_\nu}{q^2}$$

and

$$g^\rho_\nu = \delta^\rho_\nu$$

(i.e. the 4×4 unit matrix). It is easy to check (problem 11.5) that $P^\rho_\tau P^\tau_\nu = P^\rho_\nu$. Hence the series (11.25) becomes

$$\begin{aligned}
&\frac{-\mathrm{i}g_{\mu\nu}}{q^2} + \frac{-\mathrm{i}g_{\mu\rho}}{q^2} P^\rho_\nu [\Pi^{[2]}_\gamma(q^2) + (\Pi^{[2]}_\gamma(q^2))^2 + \cdots] \\
&\quad = \frac{-\mathrm{i}g_{\mu\nu}}{q^2} + \frac{-\mathrm{i}g_{\mu\rho}}{q^2} P^\rho_\nu [1 + \Pi^{[2]}_\gamma(q^2) + (\Pi^{[2]}_\gamma(q^2))^2 + \cdots] + \frac{\mathrm{i}g_{\mu\rho}}{q^2} P^\rho_\nu \\
&\quad = \frac{-\mathrm{i}(g_{\mu\nu} - q_\mu q_\nu/q^2)}{q^2(1 - \Pi^{[2]}_\gamma(q^2))} - \frac{\mathrm{i}}{q^2}\left(\frac{q_\mu q_\nu}{q^2}\right)
\end{aligned} \tag{11.26}$$

after summing the geometric series, exactly as in (10.11)–(10.14).

But we have forgotten the counter term of figure 11.1(*b*), which contributes an amplitude $-\mathrm{i}(g^{\mu\nu}q^2 - q^\mu q^\nu)(Z_3 - 1)$. This has the effect of replacing $\Pi^{[2]}_\gamma$ in (11.26) by $\Pi^{[2]}_\gamma - (Z_3 - 1)$ and we arrive at the form

$$\frac{-\mathrm{i}(g_{\mu\nu} - q_\mu q_\nu/q^2)}{q^2(Z_3 - \Pi^{[2]}_\gamma(q^2))} - \frac{\mathrm{i}}{q^2}\frac{q_\mu q_\nu}{q^2}. \tag{11.27}$$

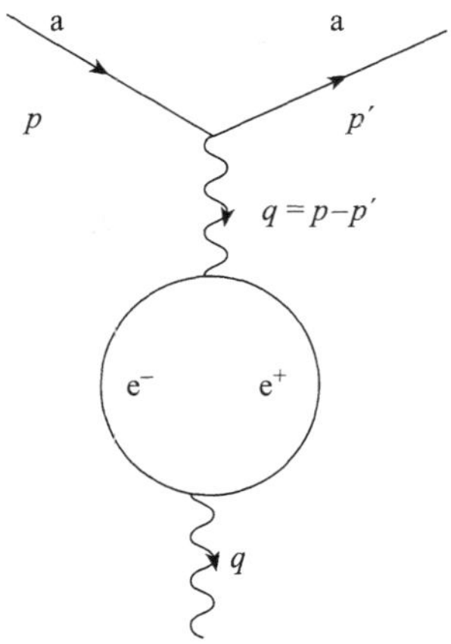

Figure 11.3. One-loop corrected photon propagator connected to a charged particle vertex.

Now in any S-matrix element, at least one end of this corrected propagator will connect to an external charged particle line via a vertex of the form $j_{\mathrm{a}}^{\mu}(p, p')$ (cf (8.97) and (8.98) for example), as in figure 11.3. But, as we have seen in (8.99), current conservation implies

$$q_{\mu} j_{\mathrm{a}}^{\mu}(p, p') = 0. \tag{11.28}$$

Hence the parts of (11.27) with $q_{\mu}q_{\nu}$ factors will not contribute to physical scattering amplitudes, and our $O(e^2)$ corrected photon propagator effectively takes the simple form

$$\frac{-\mathrm{i}g_{\mu\nu}}{q^2(Z_3 - \Pi_{\gamma}^{[2]}(q^2))}. \tag{11.29}$$

We must now determine Z_3 from the condition (just as for the C propagator) that (11.29) has the form $-\mathrm{i}g_{\mu\nu}/q^2$ as $q^2 \to 0$ (the mass-shell condition). This gives

$$Z_3^{[2]} = 1 + \Pi_{\gamma}^{[2]}(0) \tag{11.30}$$

the superscript on Z_3 indicating as usual that it is an $O(e^2)$ calculation as evidenced by the e^2 factor in (11.18). Inserting (11.30) into (11.29) we obtain the final important expression for the γ-propagator including the one-loop renormalized self-energy (cf (10.67)):

$$\boxed{\frac{-\mathrm{i}g_{\mu\nu}}{q^2(1 - \bar{\Pi}_{\gamma}^{[2]}(q^2))}} \tag{11.31}$$

where

$$\bar{\Pi}_{\gamma}^{[2]}(q^2) = \Pi_{\gamma}^{[2]}(q^2) - \Pi_{\gamma}^{[2]}(0) \tag{11.32}$$

Figure 11.4. The contribution of a massless particle to the photon self-energy.

and this 'once-subtracted' $\bar{\Pi}_\gamma^{[2]}$ is *finite* as $\Lambda \to \infty$, and tends to zero as $q^2 \to 0$.

The generalization of (11.31) to all orders will be given by

$$\frac{-\mathrm{i}g_{\mu\nu}}{q^2(1 - \bar{\Pi}_\gamma(q^2))} \tag{11.33}$$

where $\bar{\Pi}_\gamma(q^2)$ is the all-orders analogue of $\bar{\Pi}_\gamma^{[2]}$ in (11.31), and is similarly related to the 1-γ irreducible photon self-energy $\bar{\Pi}_{\mu\nu}$ via the analogue of (11.24):

$$\mathrm{i}\bar{\Pi}_{\mu\nu}(q^2) = \mathrm{i}(q^2 g_{\mu\nu} - q_\mu q_\nu)\bar{\Pi}_\gamma(q^2). \tag{11.34}$$

Because $\bar{\Pi}_{\mu\nu}$, and hence $\bar{\Pi}_\gamma$, has no 1–γ intermediate states, it is expected to have no contribution of the form A^2/q^2. If such a contribution were present, (11.33) shows that it would result in a photon propagator having the form

$$\frac{-\mathrm{i}g_{\mu\nu}}{q^2 - A^2} \tag{11.35}$$

which is, of course, that of a *massive* particle. Thus, provided no such contribution is present, the photon mass will remain zero through all radiative corrections. It is important to note, though, that gauge invariance is fully satisfied by the general form (11.34) relating $\bar{\Pi}_{\mu\nu}$ to $\bar{\Pi}_\gamma$; it does not prevent the occurrence of such an 'A^2/q^2' piece in $\bar{\Pi}_\gamma$. Remarkably, therefore, it seems possible, after all, to have a massive photon while respecting gauge invariance! This loophole in the argument 'gauge invariance implies $m_\gamma = 0$' was first pointed out by Schwinger (1962).

Such a $1/q^2$ contribution in $\bar{\Pi}_\gamma$ must, of course, correspond to a massless single particle intermediate state, via a diagram of the form shown in figure 11.4. Thus if the theory contains a massless particle, not the photon (since 1–γ states are omitted from $\bar{\Pi}_{\mu\nu}$) but coupling to it, the photon can acquire mass. This is one way of understanding the famous 'Higgs mechanism' for generating a mass for a gauge-field quantum while still respecting the gauge symmetry. The massless particle involved is called a 'Goldstone quantum'. As we shall see in volume 2, just such a photon mass is generated in a superconductor, and a similar mechanism is at work in giving masses to the $W^\pm$ and Z^0 particles which mediate the weak interactions.

11.5 The physics of $\bar{\Pi}_\gamma^{[2]}(q^2)$

The d^4k' integral in (11.23)—with a cut-off Λ—is exactly the same as the one evaluated in sections 10.3.1 and 10.3.2, setting $m_A = m_B = m$, and a straightforward application of this result yields (problem 11.6) the explicit expression for $\bar{\Pi}_{\gamma}^{[2]}$ (as $\Lambda \to \infty$):

$$\bar{\Pi}_{\gamma}^{[2]}(q^2) = -\frac{2\alpha}{\pi}\int_0^1 \mathrm{d}x\, x(1-x)\ln\left[\frac{m^2}{m^2 - q^2 x(1-x)}\right] \tag{11.36}$$

which was first given by Schwinger (1949a). We now consider some immediate physical consequences of this formula.

11.5.1 Modified Coulomb's law

In section 2.3 we saw how, in the static limit, a propagator of the form $-g_{\mathrm{s}}^2(\boldsymbol{q}^2 + m_{\mathrm{U}}^2)^{-1}$ could be interpreted (via a Fourier transform) in terms of a Yukawa potential

$$\frac{-g_{\mathrm{s}}^2}{4\pi}\frac{\mathrm{e}^{-r/a}}{r}$$

where $a = m_{\mathrm{U}}^{-1}$ (in units $\hbar = c = 1$). As $m_{\mathrm{U}} \to 0$ we arrive at the Coulomb potential, associated with the propagator $\sim 1/\boldsymbol{q}^2$ in the static ($q_0 = 0$) limit. It follows that the corrected propagator (11.31) must represent a correction to the $1/r$ Coulomb potential.

To see what it is, we expand the denominator of (11.31) so as to write (11.31) as

$$\frac{-\mathrm{i}g_{\mu\nu}}{q^2}(1 + \bar{\Pi}_{\gamma}^{[2]}(q^2)) \tag{11.37}$$

which is in fact the perturbative $O(\alpha)$ correction to the propagator (we shall return to (11.31) in a moment). At low energies, and in the static limit, $q^2 = -\boldsymbol{q}^2$ will be small compared to the fermion (mass)^2 in (11.36), and we may expand the logarithm in powers of $\boldsymbol{q}^2/m^2$, with the result that the static propagator becomes (problem 11.7)

$$\frac{\mathrm{i}g_{\mu\nu}}{\boldsymbol{q}^2}\left(1 + \frac{\alpha}{15\pi}\boldsymbol{q}^2/m^2\right) \tag{11.38}$$

$$= \frac{\mathrm{i}g_{\mu\nu}}{\boldsymbol{q}^2} + \mathrm{i}g_{\mu\nu}\frac{\alpha}{15\pi}\frac{1}{m^2}. \tag{11.39}$$

The Fourier transform of the first term in (11.39) is proportional to the familiar coulombic $1/r$ potential (see appendix G, for example), while the Fourier transform of the *constant* ($\boldsymbol{q}^2$-independent) second term is a δ-function:

$$\int \mathrm{e}^{\mathrm{i}\boldsymbol{q}\cdot\boldsymbol{r}}\frac{\mathrm{d}^3\boldsymbol{q}}{(2\pi)^3} = \delta^3(\boldsymbol{r}). \tag{11.40}$$

When (11.39) is used in any scattering process between two charged particles, each charged particle vertex will carry a charge e (or $-e$) and so the total effective potential will be (in the attractive case)

$$-\left\{\frac{\alpha}{r}+\frac{4\alpha^2}{15m^2}\delta^3(\boldsymbol{r})\right\}. \tag{11.41}$$

The second term in (11.41) may be treated as a perturbation in hydrogenic atoms, taking m to be the electron mass. Application of first-order perturbation theory yields an energy shift

$$\begin{aligned}\Delta E_n^{(1)} &= -\frac{4\alpha^2}{15m^2}\int \psi_n^*(\boldsymbol{r})\delta^3(\boldsymbol{r})\psi_n(\boldsymbol{r})\,\mathrm{d}^3\boldsymbol{r} \\ &= -\frac{4\alpha^2}{15m^2}|\psi_n(0)|^2. \end{aligned} \tag{11.42}$$

Only s-state wavefunctions are non-vanishing at the origin, where they take the value (in hydrogen)

$$\psi_n(0)=\frac{1}{\sqrt{\pi}}\left(\frac{\alpha m}{n}\right)^{3/2} \tag{11.43}$$

where n is the principal quantum number. Hence for this case

$$\Delta E_n^{(1)} = -\frac{4\alpha^5 m}{15\pi n^3}. \tag{11.44}$$

For example, in the 2s state the energy shift is -1.122×10^{-7} eV. Although we did not discuss the Coulomb spectrum predicted by the Dirac equation in chapter 4, it turns out that the $2^2\mathrm{S}_{\frac{1}{2}}$ and $2^2P_{\frac{1}{2}}$ levels are degenerate if no radiative corrections (such as the previous one) are applied. In fact, the levels are found experimentally to be split apart by the famous 'Lamb shift', which amounts to $\Delta E/2\pi\hbar = 1058$ MHz in frequency units. The shift we have calculated, for the $2s$ level, is -27.13 MHz in these units, so it is a small—but still perfectly measurable—contribution to the entire shift. This particular contribution was first calculated by Uehling (1935).

While small in hydrogen and ordinary atoms, the 'Uehling effect' dominates the radiative corrections in muonic atoms, where the 'm' in (11.43) becomes the muon mass m_μ. This means that the result (11.44) becomes

$$-\frac{4\alpha^5}{15\pi n^3}\left(\frac{m_\mu}{m}\right)^2 m_\mu.$$

Since the unperturbed energy levels are (in this case) proportional to m_μ, this represents a relative enhancement of $\sim(m_\mu/m)^2 \sim (210)^2$. This calculation cannot be trusted in detail, however, as the muonic atom radius is itself $\sim1/210$ times smaller than the electron radius in hydrogen, so that the approximation $|\boldsymbol{q}|\sim 1/r \ll m$, which led to (11.41), is no longer accurate enough. Nevertheless the order of magnitude is correct.

11.5.2 Radiatively induced charge form factor

This leads us to consider (11.37) more generally, without making the low $\boldsymbol{q}^2$ expansion. In chapter 8 we learned how the static Coulomb potential became modified by a form factor $F(\boldsymbol{q}^2)$ if the scattering centre was not point-like, and we also saw how the idea could be extended to covariant form factors for spin-0 and spin-$\frac{1}{2}$ particles. Referring to the case of $e^-\mu^-$ scattering for definiteness (section 8.7), we may consider the effect of inserting (11.37) into (8.181). The result is

$$e^2\bar{u}_{k'}\gamma_\mu u_k \left\{ \frac{g^{\mu\nu}}{q^2}(1 + \bar{\Pi}_\gamma^{[2]}(q^2)) \right\} \bar{u}_{p'}\gamma_\nu u_p. \tag{11.45}$$

Referring now to the discussion of form factors for charged spin-$\frac{1}{2}$ particles in section 8.8, we can share the correction (11.45) equally between the e^- and the μ^- vertices and write

$$e\bar{u}_{k'}\gamma_\mu u_k \to e\bar{u}_{k'}\gamma_\mu u_k(1 + \bar{\Pi}_\gamma^{[2]}(q^2))^{1/2} \approx e\bar{u}_{k'}\gamma_\mu u_k(1 + \tfrac{1}{2}\bar{\Pi}_\gamma^{[2]}(q^2)) \tag{11.46}$$

for the electron, and similarly for the muon. From (8.206) this means that our 'radiative correction' has generated some effective extension of the charge, as given by a charge form factor $\mathcal{F}_1(q^2) = 1 + \frac{1}{2}\bar{\Pi}_\gamma^{[2]}(q^2)$. Note that the condition $\mathcal{F}_1(0) = 1$ is satisfied since $\bar{\Pi}_\gamma^{[2]}(0) = 0$.

In the static case, or for scattering of equal mass particles in the CM system, we have $q^2 = -\boldsymbol{q}^2$ and we may consider the Fourier transform of the function $\mathcal{F}_1(-\boldsymbol{q}^2)$, to obtain the charge distribution. The integral is discussed in Weinberg (1995, section 10.2) and in Peskin and Schroeder (1995, section 7.5). The latter authors show that the approximate radial distribution of charge is $\sim \mathrm{e}^{-2mr}/(mr)^{3/2}$, indicating that it has a range $\sim \frac{1}{2m}$. This is precisely the mass of the fermion–anti-fermion intermediate state in the loop which yields $\bar{\Pi}_\gamma^{[2]}$, so this result represents a plausible qualitative extension of Yukawa's relationship (2.9) to the case of *two*-particle exchange. In any case, the range represented by $\bar{\Pi}_\gamma^{[2]}$ is of order of the fermion Compton wavelength $1/m$, which is an important insight; this is why we need to do better than the point-like approximation (11.41) in the case of muonic atoms.

11.5.3 The running coupling constant

There is yet another way of interpreting (11.37). Referring to (11.45), we may regard

$$e^2(q^2) = e^2[1 + \bar{\Pi}_\gamma^{[2]}(q^2)] \tag{11.47}$$

as a 'q^2-dependent effective charge'. In fact, it is usually written as a 'q^2-dependent fine structure constant'

$$\alpha(q^2) = \alpha[1 + \bar{\Pi}_\gamma^{[2]}(q^2)]. \tag{11.48}$$

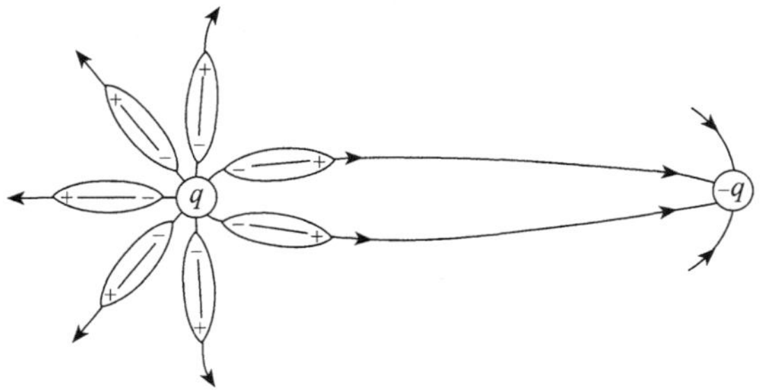

Figure 11.5. Screening of charge in a dipolar medium (from Aitchison 1985).

The concept of a q^2-dependent charge may be startling but the related one of a spatially dependent charge is, in fact, familiar from the theory of dielectrics. Consider a test charge q in a polarizable dielectric medium, such as water. If we introduce another test charge $-q$ into the medium, the electric field between the two test charges will line up the water molecules (which have a permanent electric dipole moment) as shown in figure 11.5. There will be an induced dipole moment $\boldsymbol{P}$ per unit volume, and the effect of $\boldsymbol{P}$ on the resultant field is (from elementary electrostatics) the same as that produced by a volume charge equal to $-\operatorname{div}\boldsymbol{P}$. If, as is usual, $\boldsymbol{P}$ is taken to be proportional to $\boldsymbol{E}$, so that $\boldsymbol{P} = \chi\epsilon_0\boldsymbol{E}$, Gauss' law will be modified from

$$\operatorname{div}\boldsymbol{E} = \rho_{\text{free}}/\epsilon_0 \tag{11.49}$$

to

$$\operatorname{div}\boldsymbol{E} = (\rho_{\text{free}} - \operatorname{div}\boldsymbol{P})/\epsilon_0 = \rho_{\text{free}}/\epsilon_0 - \operatorname{div}(\chi\boldsymbol{E}) \tag{11.50}$$

where ρ_{free} refers to the test charges introduced into the dielectric. If χ is slowly varying as compared to $\boldsymbol{E}$, it may be taken as approximately constant in (11.50), which may then be written as

$$\operatorname{div}\boldsymbol{E} = \rho_{\text{free}}/\epsilon \tag{11.51}$$

where $\epsilon = (1+\chi)\epsilon_0$ is the dielectric constant of the medium, ϵ_0 being that of the vacuum. Thus the field is effectively reduced by the factor $(1+\chi)^{-1} = \epsilon_0/\epsilon$.

This is all familiar ground. Note, however, that this treatment is essentially macroscopic, the molecules being replaced by a continuous distribution of charge density $-\operatorname{div}\boldsymbol{P}$. When the distance between the two test charges is as small as, roughly, the molecular diameter, this reduction—or *screening* effect—must cease and the field between them has the full unscreened value. In general, the electrostatic potential between two test charges q_1 and q_2 in a dielectric can be represented phenomenologically by

$$V(r) = q_1 q_2/4\pi\epsilon(r)r \tag{11.52}$$

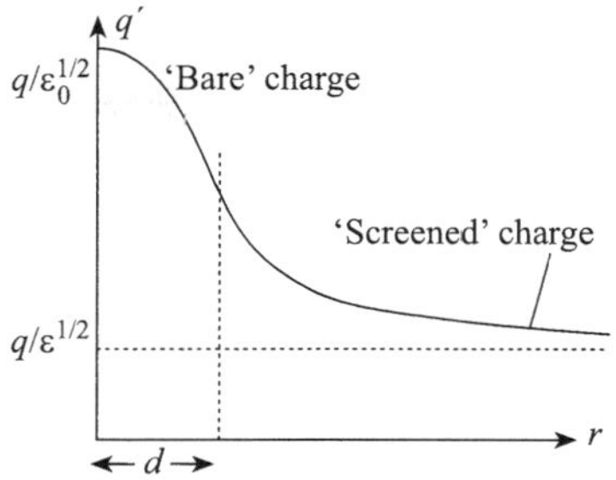

Figure 11.6. Effective (screened) charge versus separation between charges (Aitchison 1985).

where $\epsilon(r)$ is assumed to vary slowly from the value ϵ for $r \gg d$ to the value ϵ_0 for $r \ll d$, where d is the diameter of the polarized molecules. The situation may be described in terms of an effective charge

$$q' = q/[\epsilon(r)]^{1/2} \tag{11.53}$$

for each of the test charges. Thus we have an effective charge which depends on the interparticle separation, as shown in figure 11.6.

Now consider the application of this idea to QED, replacing the polarizable medium by the *vacuum*. The important idea is that, in the vicinity of a test charge *in vacuo*, charged pairs can be created. Pairs of particles of mass m can exist for a time of the order of $\Delta t \sim \hbar/mc^2$. They can spread apart a distance of order $c\Delta t$ in this time, i.e. a distance of approximately $\hbar/mc$, which is the Compton wavelength $\lambda\!\!\!^{-}_c$. This distance gives a measure of the 'molecular diameter' we are talking about, since it is the polarized virtual pairs which now provide a *vacuum* screening effect around the original charged particle. The largest 'diameter' will be associated with the smallest mass m, in this case the electron mass. Not coincidentally, this estimate of the range of the 'spreading' of the charge 'cloud' is just what we found in section 11.5.2: namely, the fermion Compton wavelength. The longest-range part of the cloud will be that associated with the lightest charged fermion, the electron.

In this analogy the bare vacuum (no virtual pairs) corresponds to the 'vacuum' used in the previous macroscopic analysis and the physical vacuum (virtual pairs) to the polarizable dielectric. We cannot, of course, get outside the physical vacuum, so that we are really always dealing with effective charges that depend on r. What, then, do we mean by the familiar symbol e? This is simply the effective charge as $r \to \infty$ or $q^2 \to 0$; or, in practice, the charge relevant for distances much larger than the particles' Compton wavelength. This is how our $q^2 \to 0$ definition is to be understood.

Let us consider, then, how $\alpha(q^2)$ varies when q^2 moves to large space-like values, such that $-q^2$ is much greater than m^2 (i.e. to distances well within the 'cloud'). For $|q^2| \gg m^2$ we find (problem 11.8) from (11.36) that

$$\bar{\Pi}_\gamma^{[2]}(q^2) = \frac{\alpha}{3\pi}\left[\ln\left(\frac{|q^2|}{m^2}\right) - \frac{5}{3} + O(m^2/|q^2|)\right] \tag{11.54}$$

so that our q^2-dependent fine structure constant, to leading order in α is

$$\alpha(q^2) \approx \alpha\left[1 + \frac{\alpha}{3\pi}\ln\left(\frac{|q^2|}{Am^2}\right)\right] \tag{11.55}$$

for large values of $|q^2|/m^2$, where $A = \exp 5/3$.

Equation (11.55) shows that the effective strength $\alpha(q^2)$ tends to increase at large $|q^2|$ (short distances). This is, after all, physically reasonable: the reduction in the effective charge caused by the dielectric constant associated with the *polarization of the vacuum* disappears (the charge increases) as we pass inside some typical dipole length. In the present case, that length is m^{-1} (in our standard units $\hbar = c = 1$), the fermion Compton wavelength, a typical distance over which the fluctuating pairs extend.

The foregoing is the reason why this whole phenomenon is called *vacuum polarization*, and why the original diagram which gave $\Pi_\gamma^{[2]}$ is called a vacuum polarization diagram.

Equation (11.55) is the lowest-order correction to α, in a form valid for $|q^2| \gg m^2$. It turns out that, in this limit, the dominant vacuum polarization contributions (for a theory with one charged fermion) can be isolated in each order of perturbation theory and summed explicitly. The result of summing these 'leading logarithms' is

$$\alpha(Q^2) = \frac{\alpha}{[1 - (\alpha/3\pi)\ln(Q^2/Am^2)]} \qquad \text{for } Q^2 \gg m^2 \tag{11.56}$$

where we now introduce $Q^2 = -q^2$, a positive quantity when q is a momentum transfer. The justification for (11.56)—which of course amounts to the very plausible return to (11.31) instead of (11.37)—is subtle, and depends upon ideas grouped under the heading of the 'renormalization group'. This is beyond the scope of the present volume, but will be taken up again in volume 2; meanwhile, the reader is referred to Peskin and Schroeder (1995, chapter 12), for example.

Equation (11.56) presents some interesting features. First, note that for typical large $Q^2 \sim (50\ \text{GeV})^2$, say, the change in the effective α predicted by (11.56) is quite measurable. Let us write

$$\alpha(Q^2) = \frac{\alpha}{1 - \Delta\alpha(Q^2)} \tag{11.57}$$

in general, where $\Delta\alpha(Q^2)$ includes the contributions from all charged fermions with mass m such that $m^2 \ll Q^2$. The contribution from the charged leptons is then straightforward, being given by

$$\Delta\alpha_{\text{leptons}} = \frac{\alpha}{3\pi}\sum_{l}\ln(Q^2/Am_l^2) \tag{11.58}$$

where m_l is the lepton mass. Including the e, μ and τ one finds (problem 11.9)

$$\Delta\alpha_{\text{leptons}}(Q^2 = (50\ \text{GeV})^2) \approx 0.03. \tag{11.59}$$

However, the corresponding quark loop contributions are subject to strong interaction corrections, and are not straightforward to calculate. We shall not pursue this in detail here, noting just that the total contribution from the five quarks u, d, s, c and b has a value very similar to (11.59) for the leptons (see, for example, Altarelli *et al* 1989). Including both the leptonic and hadronic contributions then yields the estimate

$$\alpha(Q^2 = (50\ \text{GeV})^2) \approx \tfrac{1}{137} \times \tfrac{1}{0.94} \approx \tfrac{1}{129}. \tag{11.60}$$

Such an increase in the effective α, as a function of the squared momentum transfer Q^2, is certainly large enough to be measured, and it has been (Levine *et al* 1997, Odaka *et al* 1998, Acciari *et al* 2000), with results that agree well with the theoretical predictions. For a recent review of the status of theory and experiment for $\alpha(Q^2)$, see Groom *et al* (2000).

The notion of a q^2-dependent coupling constant is, in fact, quite general—for example, we could just as well interpret (10.67) in terms of a q^2-dependent $g_{\text{ph}}^2(q^2)$. Such 'varying constants' are called *running coupling constants*. Until 1973 it was generally believed that they would all behave in essentially the same way as (11.56)—namely, a logarithmic rise as Q^2 increases. Many people (in particular Landau 1955) noted that if equation (11.56) is taken at face value for arbitrarily large Q^2, then $\alpha(Q^2)$ itself will diverge at $Q^2 = m^2\exp(3\pi/\alpha)$. Taking m to be the mass of an electron, this is of course an absurdly high energy. Besides, as such energies are reached, approximations made in arriving at (11.56) will break down; all we can really say is that perturbation theory will fail as we approach such energies.

While this may be an academic point in QED, it turns out that there is one part of the Standard Model where it may be relevant. This is the 'Higgs sector' involving a complex scalar field, as will be discussed in volume 2. In this case, the 'running' of the Higgs coupling constant can be invoked to suggest an upper bound on the Higgs mass (Maiani 1991).

The significance of the 1973 date is that it was in that year that one of the most important discoveries in 'post-QED' quantum field theory was made, by Politzer (1973) and by Gross and Wilczek (1973). They performed a similar one-loop calculation in the more complicated case of QCD, which is a 'non-Abelian gauge theory' (as is the theory of the weak interactions in the electroweak theory). They found that the QCD analogue of (11.56) was

$$\alpha_s(Q^2) = \frac{\alpha_s(\mu^2)}{[1 + \frac{\alpha_s}{12\pi}(33 - 2f)\ln(Q^2/\mu^2)]} \tag{11.61}$$

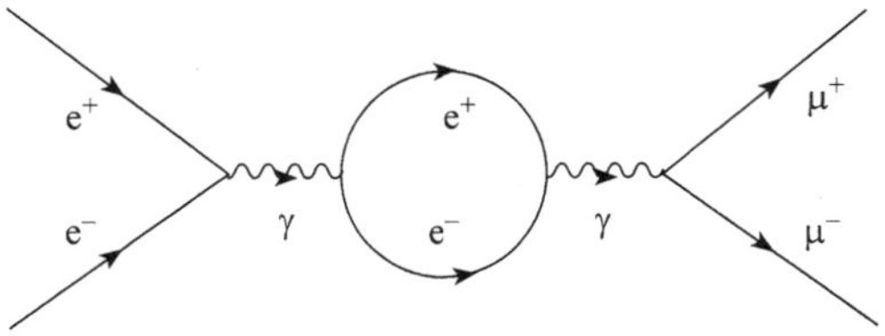

Figure 11.7. Vacuum polarization insertion in the virtual one-photon annihilation amplitude in $e^+e^- \to \mu^+\mu^-$.

where f is the number of fermion–anti-fermion loops considered, and μ is a reference mass scale. The crucial difference from (11.56) is the large positive contribution '+33', which is related to the contributions from the gluonic self-interactions (non-existent among photons). The quantity $\alpha_s(Q^2)$ now tends to *decrease* at large Q^2 (provided $f \leq 16$), tending ultimately to zero. This property is called 'asymptotic freedom' and is highly relevant to understanding the success of the quark parton model of chapter 9, in which the quarks and gluons are taken to be essentially free at large values of Q^2. This can be qualitatively understood in terms of $\alpha_s(Q^2) \to 0$ for high momentum transfers ('deep scattering'). The non-Abelian parts of the Standard Model will be considered in volume 2, where we shall return again to $\alpha_s(Q^2)$.

11.5.4 $\bar{\Pi}_\gamma^{[2]}$ in the s-channel

We have still not exhausted the riches of $\bar{\Pi}_\gamma^{[2]}(q^2)$. Hitherto we have concentrated on regarding our corrected propagator as appearing in a t-channel exchange process, where $q^2 < 0$. But of course it could also perfectly well enter an s-channel process such as $e^+e^- \to \mu^+\mu^-$ (see problem 8.19), as in figure 11.7. In this case, the 4-momentum carried by the photon is $q = p_{e^+} + p_{e^-} = p_{\mu^+} + p_{\mu^-}$, so that q^2 is precisely the usual invariant variable 's' (cf section 6.3.3), which in turn is the square of the CM energy and is therefore positive. In fact, the process of figure 11.7 occurs physically only for $q^2 = s > 4m_\mu^2$, where m_μ is the muon mass.

Consider, therefore, our formula (11.36) for $q^2 > 0$, that is, in the time-like rather than the space-like ($q^2 < 0$) region. The crucial new point is that the argument $[m^2 - q^2x(1-x)]$ of the logarithm can now become negative, so that $\bar{\Pi}_\gamma^{[2]}$ must develop an imaginary part. The smallest q^2 for which this can happen will correspond to the largest possible value of the product $x(1-x)$, for $0 < x < 1$. This value is $\frac{1}{4}$, and so $\bar{\Pi}_\gamma^{[2]}$ becomes imaginary for $q^2 > 4m^2$, which is the threshold for real creation of an e^+e^- pair.

This is the first time that we have encountered an imaginary part in a Feynman amplitude which, for figure 11.7 and omitting all the spinor factors, is once again

$$\frac{1}{q^2(1-\bar{\Pi}_\gamma^{[2]}(q^2))} \tag{11.62}$$

but now $q^2 > 4m_\mu^2$, which is greater than $4m^2$ so that $\bar{\Pi}_\gamma^{[2]}(q^2)$ in (11.62) has an imaginary part. There is a good physical reason for this, which has to do with *unitarity*. This was introduced in section 6.2.2 in terms of the relation $SS^\dagger = I$ for the S-matrix. The invariant amplitude $\mathcal{M}$ is related to S by $S_{\rm fi} = 1 + {\rm i}(2\pi)^4\delta^4(p_{\rm i} - p_{\rm f})\mathcal{M}_{\rm fi}$ (cf (6.102)). Inserting this into $SS^\dagger = I$ leads to an equation of the form (for help see Peskin and Schroeder (1995, section 7.3))

$$2{\rm Im}\mathcal{M}_{\rm fi} = \sum_{\rm k} \mathcal{M}_{\rm kf}^* \mathcal{M}_{\rm ki}(2\pi)^4\delta\left(p_{\rm i} - \sum q_{\rm k}\right) \tag{11.63}$$

where '$\sum_{\rm k}$' stands for the phase space integral involving momenta $q_1, q_2, \ldots$ over the states allowed by energy–momentum conservation. This implies that as the energy crosses each threshold for production of a newly allowed state, there will be a new contribution to the imaginary part of $\mathcal{M}$. This is exactly what we are seeing here, at the e^+e^- threshold.

It is interesting, incidentally, that (11.63) can be used to derive the relativistic generalization of the optical theorem given in appendix H (note that the right-hand side of (11.63) is clearly related to the total cross section for i $\to$ k, if i = f).

As regards the real part of $\bar{\Pi}_\gamma^{[2]}(q^2)$ in the time-like region, it will be given by (11.56) with Q^2 replaced by q^2, or s, for large values of q^2. Again, measurements have verified the predicted variation of $\alpha(q^2)$ in the time-like region (Miyabashi *et al* 1995, Ackerstaff *et al* 1998, Abbiendi *et al* 1999, 2000).

There is one more 'elementary' loop that we must analyse—the vertex correction shown in figure 11.8, which we now discuss. We will see how the important relation $Z_1 = Z_2$ emerges, and introduce some of the physics contained in the renormalized vertex.

11.6 The $O(e^2)$ vertex correction, and $Z_1 = Z_2$

The amplitude corresponding to figure 11.8 is

$$-{\rm i}e\bar{u}(p')\Gamma_\mu^{[2]}(p,p')u(p) = \bar{u}(p')\int(-{\rm i}e\gamma^\nu)\frac{-{\rm i}g_{\lambda\nu}}{k^2}\frac{\rm i}{\not p' - \not k - m}$$

$$\times(-{\rm i}e\gamma_\mu)\frac{\rm i}{\not p - \not k - m}(-{\rm i}e\gamma^\lambda)\frac{{\rm d}^4k}{(2\pi)^4}u(p) \tag{11.64}$$

where $\gamma_\mu = g_{\mu\sigma}\gamma^\sigma$, and $\Gamma_\mu^{[2]}$ represents the correction to the standard vertex and again $\xi = 1$. We find

$$\Gamma_\mu^{[2]}(p,p') = -{\rm i}e^2\int\frac{1}{k^2}\gamma^\lambda\frac{1}{\not p' - \not k - m}\gamma_\mu\frac{1}{\not p - \not k - m}\gamma_\lambda\frac{{\rm d}^4k}{(2\pi)^4}. \tag{11.65}$$

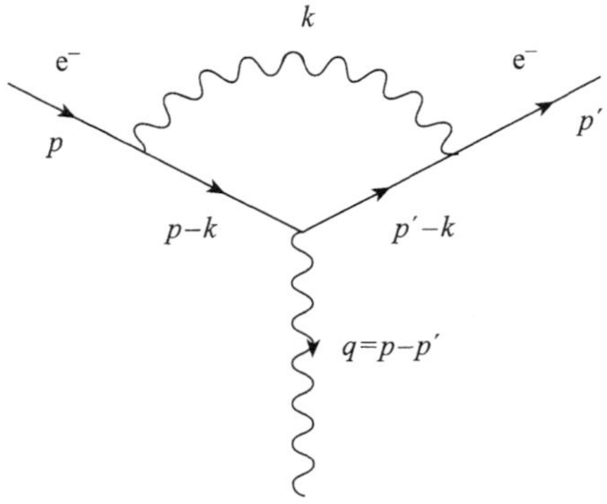

Figure 11.8. One-loop vertex correction.

The integral is logarithmically divergent at large k, by power counting, and the divergence will be cancelled by the Z_1 counter term of figure 11.1(c). It turns out to be infrared divergent also, as was $\mathrm{d}\Sigma^{[2]}/\mathrm{d}\not{p}$. As in the latter case, we leave the infrared problem aside, concentrating on the removal of ultraviolet divergences.

Z_1 is determined by the requirement that the total amplitude at $q = p - p' = 0$, for on-shell fermions, is just $-\mathrm{i}e\bar{u}(p)\gamma_\mu u(p)$, this being our definition of 'e'. Hence we have (at $O(e^2)$)

$$-\mathrm{i}e\bar{u}(p)\Gamma_\mu^{[2]}(p,p)u(p) - \mathrm{i}e\bar{u}(p)\gamma_\mu(Z_1^{[2]} - 1)u(p) = 0 \tag{11.66}$$

and so

$$\Gamma_\mu^{[2]}(p,p) + \gamma_\mu(Z_1^{[2]} - 1) = 0. \tag{11.67}$$

The renormalized vertex correction $\bar{\Gamma}_\mu^{[2]}$ may then be defined as

$$\bar{\Gamma}_\mu^{[2]}(p,p') = \Gamma_\mu^{[2]}(p,p') + (Z_1^{[2]} - 1)\gamma_\mu = \Gamma_\mu^{[2]}(p,p') - \Gamma_\mu^{[2]}(p,p) \tag{11.68}$$

and in this 'once-subtracted' form it is finite, and equal to zero at $q = 0$.

We shall consider some physical consequences of $\bar{\Gamma}_\mu^{[2]}$ in a moment, but first we show that (at $O(e^2)$) $Z_1^{[2]} = Z_2^{[2]}$, and explain the significance of this important relation. It is, after all, at first sight a rather surprising equality between two apparently unrelated quantities, one associated with the fermion self-energy, the other with the vertex part. From (11.9) we have, for the fermion self-energy,

$$\Sigma^{[2]}(p) = -\mathrm{i}e^2 \int \frac{1}{k^2}\gamma^\lambda \frac{1}{\not{p} - \not{k} - m}\gamma_\lambda \frac{\mathrm{d}^4 k}{(2\pi)^4}. \tag{11.69}$$

One can discern some kind of similarity between (11.69) and (11.65), which can be elucidated with the help of a little algebra.

Consider differentiating the identity $(\not{p} - m)(\not{p} - m)^{-1} = 1$ with respect to p^μ:

$$\begin{aligned}0 &= \frac{\partial}{\partial p^\mu}[(\not{p} - m)(\not{p} - m)^{-1}] \\ &= \left[\frac{\partial}{\partial p^\mu}(\not{p} - m)\right](\not{p} - m)^{-1} + (\not{p} - m)\frac{\partial}{\partial p^\mu}(\not{p} - m)^{-1} \\ &= \gamma_\mu(\not{p} - m)^{-1} + (\not{p} - m)\frac{\partial}{\partial p^\mu}(\not{p} - m)^{-1}.\end{aligned} \tag{11.70}$$

It follows that

$$\frac{\partial}{\partial p^\mu}(\not{p} - m)^{-1} = -(\not{p} - m)^{-1}\gamma_\mu(\not{p} - m)^{-1} \tag{11.71}$$

from which the *Ward identity* (Ward 1950) follows immediately:

$$-\frac{\partial \Sigma^{[2]}}{\partial p^\mu} = \Gamma_\mu^{[2]}(p, p' = p). \tag{11.72}$$

Derived here to one-loop order, the identity is, in fact, true to all orders, *provided* that a gauge-invariant regularization is adopted. Note that the identity deals with $\Gamma_\mu^{[2]}$ at zero momentum transfer ($q = p - p' = 0$), which is the value at which e is defined. Note also that consistently with (11.72), each of $\partial\Sigma^{[2]}/\partial\not{p}$ and $\Gamma_\mu^{[2]}$ are both infrared and ultraviolet divergent, though we shall only be concerned with the latter.

The quantities $\Sigma^{[2]}$ and $\Gamma_\mu^{[2]}$ are both $O(e^2)$, and contain ultraviolet divergences which are cancelled by the $O(e^2)$ counter terms. From (11.11) and (11.12) we have

$$\Sigma^{[2]} = \bar{\Sigma}^{[2]} - Z_2^{[2]}(m_0 - m) + (\not{p} - m)(Z_2^{[2]} - 1) \tag{11.73}$$

where $\bar{\Sigma}^{[2]}$ is finite, and from (11.68) we have

$$\Gamma_\mu^{[2]}(p, p') = \bar{\Gamma}_\mu^{[2]}(p, p') - (Z_1^{[2]} - 1)\gamma_\mu \tag{11.74}$$

where $\bar{\Gamma}_\mu^{[2]}$ is finite. Inserting (11.73) and (11.74) into (11.72) and equating the infinite parts gives

$$Z_1^{[2]} = Z_2^{[2]}. \tag{11.75}$$

This relation is true to all orders ($Z_1 = Z_2$), provided a gauge-invariant regularization is used. It is a very significant relation, as already indicated after (11.8). It shows, first, that the gauge principle survives renormalization provided the regularization is gauge invariant. More physically, it tells us that the bare and renormalized charges are related simply by (cf (11.6))

$$e = e_0 Z_3^{1/2}. \tag{11.76}$$

In other words, the interaction-dependent rescaling of the bare charge is due solely to vacuum polarization effects in the photon propagator, which are the same for *all* charged particles interacting with the photon. By contrast, both Z_1 and Z_2 do depend on the specific type of the interacting charged particle, since these quantities involve the particle masses. The ratio of bare to renormalized charge is independent of particle type. Hence if a set of bare charges are all equal (or 'universal'), the renormalized ones will be too. But we saw in section 3.6 how precisely such a notion of universality was present in theories constructed according to the (electromagnetic) gauge principle. We now see how the universality survives renormalization. In volume 2 we shall find that a similar universality holds, empirically, in the case of the weak interaction, giving a strong indication that this force too should be described by a renormalizable gauge theory.

11.7 The anomalous magnetic moment and tests of QED

Returning now to $\Gamma_{\mu}^{[2]}$, just as in section 11.5.2 we regarded the vacuum polarization correction $1 + \frac{1}{2}\bar{\Pi}_{\gamma}^{[2]}$ as a contribution to the fermion's charge form factor $\mathcal{F}_1(q^2)$, so we may expect that the vertex correction will also contribute to the form factor. Indeed, let us recall the general form of the electromagnetic vertex for a spin-$\frac{1}{2}$ particle (cf (8.206)):

$$-\mathrm{i}e\bar{u}(p', s')\left[\mathcal{F}_1(q^2)\gamma_{\mu} + \mathrm{i}\kappa\frac{\mathcal{F}_2(q^2)}{2m}\sigma_{\mu\nu}q^{\nu}\right]u(p, s) \tag{11.77}$$

where κ is the 'anomalous' part of the magnetic moment, i.e. the magnetic moment is $(e\hbar/2m)(1+\kappa)$, the '1' being the Dirac value calculated in section 4.6. In (11.77), $\mathcal{F}_1$ and $\mathcal{F}_2$ are each normalized to 1 at $q^2 = 0$. Our vertex $\Gamma_{\mu}^{[2]}$ contributes to both the charge and the magnetic moment form factors; let us call the contributions $\mathcal{F}_1^{[2]}$ and $\kappa\mathcal{F}_2^{[2]}$. Now the Z_1 counter term multiplies γ_{μ}, and therefore clearly cancels a divergence in $\mathcal{F}_1^{[2]}$. Is there also, we may ask, a divergence in $\kappa\mathcal{F}_2^{[2]}$?

Actually, $\kappa\mathcal{F}_2^{[2]}$ is convergent, and this is highly significant to the physics of renormalization. Had it been divergent, we would either have had to abandon the theory or introduce a *new* counter term to cancel the divergence. This counter term would have the general form

$$\frac{K}{m}\bar{\hat{\psi}}\sigma_{\mu\nu}\hat{\psi}\hat{F}^{\mu\nu}; \tag{11.78}$$

it is, indeed, an 'anomalous magnetic moment' interaction. But no such term exists in the original QED Lagrangian (11.1)! Its appearance does not seem to follow from the gauge principle argument, even though it is, in fact, gauge

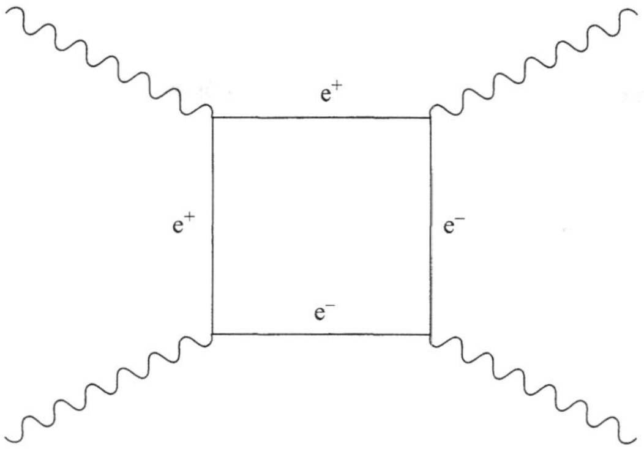

Figure 11.9. Contribution (which is finite) to $\gamma\gamma \to \gamma\gamma$.

invariant. Part of the meaning of the renormalizability of QED (or any theory) is that all infinities can be cancelled by counter terms of the same form as the terms appearing in the original Lagrangian. This means, in other words, that all infinities can be cancelled by assuming an appropriate cut-off dependence for the fields and parameters in the bare Lagrangian. The interaction (11.78) is certainly gauge invariant—but it is *non-renormalizable*—as we shall discuss further later. The message is that, in a renormalizable theory, amplitudes which do not have counterparts in the interactions present in the bare Lagrangian must be *finite*. Figure 11.9 shows another example of an amplitude which turns out to be finite: there is no '$\hat{A}^4$' type of interaction in QED (cf figure 10.12 and the attendant comment in section 10.5).

The calculation of the renormalized $\bar{\mathcal{F}}_1(q^2)$ and of $\kappa\mathcal{F}_2(q^2)$ is quite laborious, not least because three denominators are involved in the $\Gamma_\mu^{[2]}$ integral (11.65). The dedicated reader can follow the story in section 6.3 of Peskin and Schroeder (1995). The most important result is the value obtained for κ, the QED-induced anomalous magnetic moment of the fermion, first calculated by Schwinger (1948a). He obtained

$$\kappa = \frac{\alpha}{2\pi} \approx 0.001\,1614 \tag{11.79}$$

which means a g-factor corrected from the $g = 2$ Dirac value to

$$g = 2 + \frac{\alpha}{\pi} \tag{11.80}$$

or, equivalently,

$$[(g-2)/2]_{\text{Schwinger}} = \frac{\alpha}{2\pi} \approx 11\,614 \times 10^{-7}. \tag{11.81}$$

The prediction (11.81) may be compared with the experimental value which

Figure 11.10. $O(\alpha/\pi)^2$ correction to the $\mathrm{e}-\mathrm{e}-\gamma$ vertex.

currently is (Groom *et al* 2000)

$$a_{\mathrm{e,expt}} \equiv [(g_{\mathrm{e}}-2)/2]_{\mathrm{expt}} = 11\,596\,521\,869(41) \times 10^{-13} \tag{11.82}$$

for the electron, the bracketed figure being the error. Of course, in Schwinger's day the experimental accuracy was far different, but there was a real discrepancy (Kusch and Foley 1947) with the Dirac value. Schwinger's one-loop calculation provided a fundamental early confirmation of QED.

Clearly, the discrepancy between (11.79) and (11.82) may be attributed to even higher-order processes, for example the 'fourth-order' $(\alpha/\pi)^2$ corrections supplied by graphs such as figure 11.10, which is a two-loop contribution. To confront the precision attained by the ever more sophisticated and imaginative experiments leading to (11.82) requires the evaluation of all graphs up to and including $O(\alpha/\pi)^4$, i.e. one-, two-, three- and four-loop contributions. In fact, $a_{\mathrm{e,expt}}$ is now used (Kinoshita 1996) in conjunction with a four-loop QED calculation to obtain the most precise determination of α! QED is indeed the paradigm quantum field theory.

Although the precision in the equivalent quantity $a_{\mu,\mathrm{expt}}$ for the muon is not nearly as great as that in (11.82), it may nevertheless be good enough to reveal 'new physics', for an interesting reason. Consider expanding the vacuum polarization formula (11.18) in powers of m/Λ, having done the momentum integrals as in (10.51) and removed the $\ln\Lambda$ divergence by the subtraction (11.32). The resulting expression will be finite as $\Lambda \to \infty$, but for finite Λ it will contain Λ-dependent terms, the first being of order (m^2/Λ^2). This suggests that the contribution of a 'new physics' scale (modelled crudely by our cut-off) to a_μ would be enhanced by a factor $(m_\mu/m_{\mathrm{e}})^2 \approx 43\,000$ relative to its contribution to a_{e}. This is far greater than the factor of about 370 in the precision of $a_{\mathrm{e,expt}}$ over $a_{\mu,\mathrm{expt}}$. In fact, a new value was announced for the latter in 2001 by Brown *et al*

(2001),

$$a_{\mu^+} = 116\,592\,020(16) \times 10^{-10}. \tag{11.83}$$

These authors also quoted

$$a_{\mu,\text{expt}} - a_{\mu,\text{SM}} = 43(16) \times 10^{-10} \tag{11.84}$$

for the difference between the weighted mean of existing experimental results, and the theoretical value from the Standard Model. This represented a difference of some 2.6σ, a possible indication of new physics. However, further investigations revealed a sign error in a particular high-order contribution to $a_{\mu,\text{SM}}$, the correction of which reduced the discrepancy to about 1.5σ.

It is premature to regard such a discrepancy as an indication of new physics beyond the Standard Model. Certainly, this epic confrontation between theory and experiment will continue to be pursued: it is a classic example of the way in which a very high-precision measurement in a thoroughly 'low-energy' area of physics (a magnetic moment) can have profound impact on the 'high-energy' frontier—a circumstance we may be increasingly dependent upon.

11.8 Which theories are renormalizable—and does it matter?

In the course of our travels thus far, we have met theories which exhibit three different types of ultraviolet behaviour. In the ABC theory at one-loop order, we found that both the field strength renormalizations and the vertex correction were finite; only the mass shifts diverged as $\Lambda \to \infty$. The theory was called 'super-renormalizable'. In QED, we needed divergent renormalization constants Z_i as well as an infinite mass shift—but (although we did not attempt to explain why) these counter terms were enough to cure divergences systematically to all orders and the theory was renormalizable. Finally, we asserted that the anomalous coupling (11.78) was non-renormalizable. In the final section of this volume we shall try to shed more light on these distinctions and their significance.

Is there some way of telling which of these ultraviolet behaviours a given Lagrangian is going to exhibit, without going through the calculations? The answer is yes (nearly), and the test is surprisingly simple. It has to do with the *dimensionality of a theory's coupling constant*. We have seen (section 6.3.1) that the dimensionality of 'g' in the ABC theory is M^1 (using mass as the remaining dimension when $\hbar = c = 1$), that of e in QED is M^0 (section 7.4) and that of the coefficient of the anomalous coupling $\bar{\hat{\psi}}\sigma_{\mu\nu}\hat{\psi}\hat{F}^{\mu\nu}$ in (11.78) is M^{-1}. These couplings have positive, zero and negative mass dimension, respectively. It is no accident that the three theories, with different dimensions for their couplings, have different ultraviolet behaviour and hence different renormalizability.

That coupling constant dimensionality and ultraviolet behaviour are related can be understood by simple dimensional considerations. Compare, for example, the vertex corrections in the ABC theory (figure 10.5) and in QED (figure 11.8).

These amplitudes behave essentially as

$$G^{[2]} \sim g_{\text{ph}}^2 \int \frac{\mathrm{d}^4 k}{k^2 k^2 k^2} \tag{11.85}$$

and

$$\Gamma^{[2]} \sim e^2 \int \frac{\mathrm{d}^4 k}{k^2 \not{k} \not{k}} \tag{11.86}$$

respectively, for large k. Both are dimensionless: but in (11.85) the positive $(\text{mass})^2$ dimension of g_{ph}^2 is compensated by two additional factors of k^2 in the denominator of the integral, as compared with (11.86), with the result that (11.85) is ultraviolet convergent but (11.86) is not. The analysis can be extended to higher-order diagrams: for the ABC theory, the more powers of g_{ph} which are involved, the more denominator factors are necessary, and hence the better the convergence is. Indeed, in this kind of 'super-renormalizable' theory, only a finite number of diagrams are ultraviolet divergent, to all orders in perturbation theory.

It is clear that some kind of opposite situation must obtain when the coupling constant dimensionality is negative; for then, as the order of the perturbation theory increases, the negative powers of M in the coupling constant factors must be compensated by positive powers of k in the numerators of loop integrals. Hence the divergence will tend to get worse at each successive order. A famous example of such a theory is Fermi's original theory of β-decay (Fermi 1934a, b), referred to in section 2.5, in which the interaction density has the 'four-fermion' form

$$G_{\text{F}} \bar{\hat{\psi}}_{\text{p}}(x) \hat{\psi}_{\text{n}}(x) \bar{\hat{\psi}}_{\text{e}}(x) \hat{\psi}_{\nu_{\text{e}}}(x) \tag{11.87}$$

where G_{F} is the 'Fermi constant'. To find the dimensionality of G_{F}, we first establish that of the fermion field by considering a mass term $m\bar{\hat{\psi}}\hat{\psi}$, for example. The integral of this over $\mathrm{d}^3\boldsymbol{x}$ gives one term in the Hamiltonian, which has dimension M. We deduce that $[\hat{\psi}] = \frac{3}{2}$, since $[\mathrm{d}^3\boldsymbol{x}] = -3$. Hence $[\bar{\hat{\psi}}\hat{\psi}\bar{\hat{\psi}}\hat{\psi}] = 6$, and so $[G_{\text{F}}] = -2$. The coupling constant G_{F} in (11.87) therefore has a negative mass dimension, just like the coefficient K/m in (11.78). Indeed, the four-fermion theory is also non-renormalizable.

Must such a theory be rejected? Let us briefly sketch the consequences of an interaction of the form (11.87), but slightly simpler, namely

$$G_{\text{F}} \bar{\hat{\psi}}_{\text{n}}(x) \hat{\psi}_{\text{n}}(x) \bar{\hat{\psi}}_{\nu_{\text{e}}}(x) \hat{\psi}_{\nu_{\text{e}}}(x) \tag{11.88}$$

where, for the present purposes, the neutron is regarded as point-like. Consider, for example, the scattering process $\nu_{\text{e}} + \text{n} \to \nu_{\text{e}} + \text{n}$. To lowest order in G_{F}, this is given by the tree diagram—or 'contact term'—of figure 11.11, which contributes a constant $-\mathrm{i}G_{\text{F}}$ to the invariant amplitude for the process, disregarding the spinor factors for the moment. A one-loop $O(G_{\text{F}}^2)$ correction is shown in figure 11.12. Inspection of figure 11.12 shows that this is an s-channel process (recall section

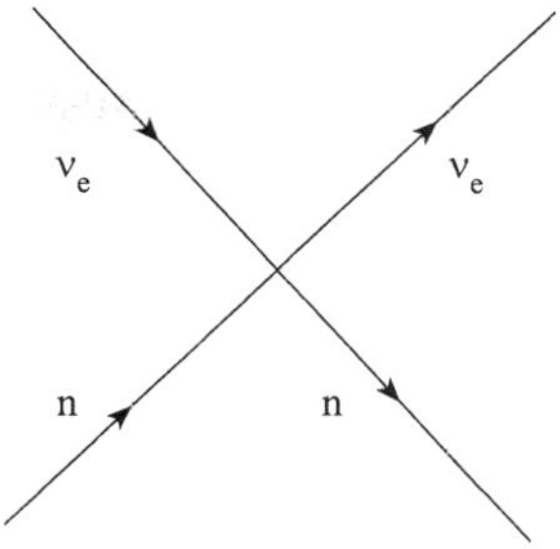

Figure 11.11. Lowest order contribution to $\nu_e + \mathrm{n} \to \nu_e + \mathrm{n}$ in the model defined by the interaction (11.88).

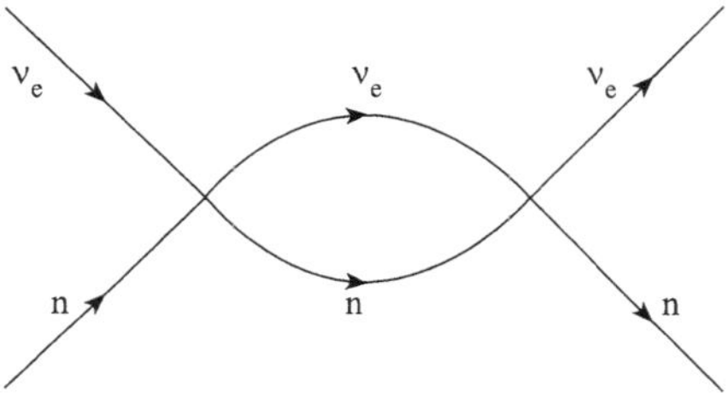

Figure 11.12. Second-order (one-loop) contribution to $\nu_e + \mathrm{n} \to \nu_e + \mathrm{n}$.

6.3.3): let us call the amplitude $-\mathrm{i}G_F G_l^{[2]}(s)$, where one G_F factor has been extracted, so that the correction can be compared with the tree amplitude and $G_l^{[2]}(s)$ is dimensionless. Then $G_l^{[2]}(s)$ is given by

$$G_l^{[2]}(s) = -\mathrm{i}G_F \int \frac{\mathrm{d}^4k}{(2\pi)^4} \frac{\mathrm{i}}{\not{k} - m_{\nu_e}} \frac{\mathrm{i}}{(\not{p}_{\nu_e} + \not{p}_n - \not{k}) - m_n}. \tag{11.89}$$

As expected, the negative mass dimension of G_F leaves fewer k-factors in the denominator of the loop integral. Indeed, manipulations exactly like those we used in the case of $\Sigma^{[2]}$ shows that $G_l^{[2]}(s)$ has a quadratic divergence, and that $\mathrm{d}G_l^{[2]}/\mathrm{d}s$ has a logarithmic divergence. The extra denominators associated with second and higher derivatives of $G_l^{[2]}(s)$ are sufficient to make these integrals finite.

The standard procedure would now be to cancel these divergences with counter terms. There will certainly be one counter term arising naturally from writing the bare version of (11.88) as (cf (11.5)):

$$G_{0F}\bar{\hat{\psi}}_{0n}\hat{\psi}_{0n}\bar{\hat{\psi}}_{0\nu_e}\hat{\psi}_{0\nu_e} = G_F\bar{\hat{\psi}}_n\hat{\psi}_n\bar{\hat{\psi}}_{\nu_e}\hat{\psi}_{\nu_e} + (Z_4 - 1)G_F\bar{\hat{\psi}}_n\hat{\psi}_n\bar{\hat{\psi}}_{\nu_e}\hat{\psi}_{\nu_e} \tag{11.90}$$

where $Z_4 G_F = G_{0F} Z_{2,n} Z_{2,\nu_e}$ and the Z_2's are the field strength renormalization constants for the n and ν_e fields. Including the tree graph of figure 11.11, the amplitude of figure 11.12, and the counter term, the total amplitude to $O(G_F^2)$ is given by

$$i\mathcal{M} = -iG_F - iG_F G_l^{[2]}(s) - iG_F(Z_4 - 1). \tag{11.91}$$

As in our earlier examples, Z_4 will be determined from a *renormalization condition*. In this case, we might demand, for example, that the amplitude $\mathcal{M}$ reduces to G_F at the threshold value $s = s_0$, where $s_0 = (m_n + m_{\nu_e})^2$. Then to $O(G_F^2)$ we find

$$Z_4^{[2]} = 1 - G_l^{[2]}(s_0) \tag{11.92}$$

and our amplitude (11.91) is, in fact,

$$-iG_F - iG_F[G_l^{[2]}(s) - G_l^{[2]}(s_0)]. \tag{11.93}$$

In (11.93), we see the familiar outcome of such renormalization—the appearance of subtractions of the divergent amplitude (cf (10.68), (11.11), (11.32) and (11.68)). In fact, because $dG_l^{[2]}/ds$ is also divergent, we need a second subtraction—and correspondingly, a *new* counter term, not present in the original Lagrangian, of the form

$$G_d \bar{\hat{\psi}}_n \not{\partial} \hat{\psi}_n \bar{\hat{\psi}}_{\nu_e} \not{\partial} \hat{\psi}_{\nu_e}$$

for example; there will also be others, but we are concerned only with the general idea. The occurrence of such a new counter term is characteristic of a non-renormalizable theory, but at this stage of the proceedings the only penalty we pay is the need to import another constant from experiment, namely the value D of $dG_l^{[2]}/ds$ at some fixed s, say $s = s_0$; D will be related to the renormalized value of G_d. We will then write our *renormalized amplitude*, up to $0(G_F^2)$, as

$$-iG_F[1 + D(s - s_0) + \bar{G}_l^{[2]}(s)] \tag{11.94}$$

where $\bar{G}_l^{[2]}(s)$ is finite, and vanishes along with its first derivative at $s = s_0$; that is, $\bar{G}_l^{[2]}(s)$ contributes calculable terms of order $(s - s_0)^2$ if expanded about $s = s_0$.

The moral of the story so far, then, is that we can perform a one-loop renormalization of this theory, at the cost of taking additional parameters from experiments and introducing new terms in the Lagrangian. What about the next order? Figure 11.13 shows a two-loop diagram in our theory, which is of order G_F^3. Writing the amplitude as $-iG_F G_l^{[3]}(s)$, the ultraviolet behaviour of $G_l^{[3]}(s)$ is given by

$$(-iG_F)^2 \int \frac{d^4k_1 d^4k_2}{k^4} \tag{11.95}$$

where k is a linear function of k_1 and k_2. This has a leading ultraviolet divergence $\sim \Lambda^4$, even worse than that of $G_l^{[2]}$. As suggested earlier, it is indeed the case that, the higher we go in perturbation theory in this model, the worse the divergences

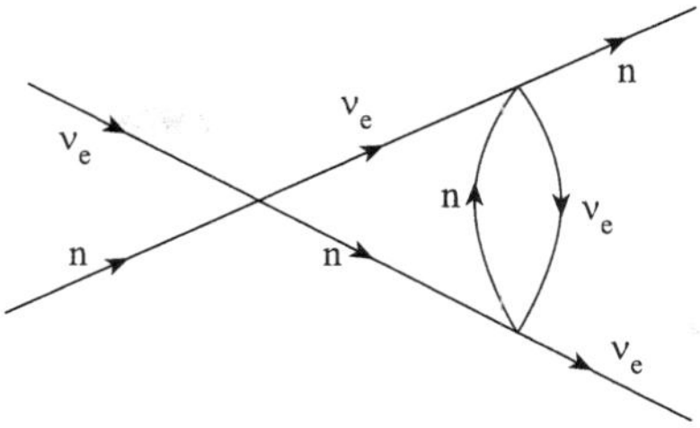

Figure 11.13. A two-loop contribution to $\nu_e + \mathrm{n} \to \nu_e + \mathrm{n}$ in the model defined by (11.88).

become. We can, of course, eliminate this divergence in $G_l^{[3]}$ by performing a further subtraction, requiring the provision of more parameters from experiment. By now the pattern should be becoming clear: new counter terms will have to be introduced at each order of perturbation theory, and ultimately we shall need an infinite number of them, and hence an infinite number of parameters determined from experiment.

Does this imply that the theory is useless? We have learned that $\bar{G}_l^{[2]}(s)$ produces a calculable term of order $G_F^2(s - s_0)^2$ when expanded about $s = s_0$; and that $\bar{G}_l^{[3]}$ will produce a calculable term of order $G_F^3(s - s_0)^3$, and so on. Now, from the discussion after (11.87), G_F itself is a dimensionless number divided by the square of some mass. As we saw in section 2.5 (and will return to in more detail in volume 2), in the case of the physical weak interaction this mass in G_F is the W-mass, and $G_F \sim \alpha/M_W^2$. Hence our loop corrections have the form $\alpha^2(s - s_0)^2/M_W^4, \alpha^3(s - s_0)^3/M_W^6 \ldots$. We now see that for low enough energy close to threshold, where $(s - s_0) \ll M_W^2$, it will be a good approximation to stop at the one-loop level. As we go up in energy, we will need to include higher-order loops, and correspondingly more parameters will have to be drawn from experiment. But only when we begin to approach an energy $\sqrt{s} \sim M_W/\sqrt{\alpha} \sim G_F^{-1/2} \sim 300$ GeV will this theory be terminally sick. This was pointed out by Heisenberg (1939). For this argument to work, it is important that the ultraviolet divergences at a given order in perturbation theory (i.e. a given number of loops) should have been removed by renormalization, otherwise factors of Λ^2 will enter—in place of the $(s - s_0)$ factors, for example.

We have seen that a non-renormalizable theory can be useful at energies well below the 'natural' scale specified by its coupling constant. Let us look at this in a slightly different way, by considering the two four-fermion interaction terms introduced at one loop,

$$G_F \bar{\hat{\psi}}_n \hat{\psi}_n \bar{\hat{\psi}}_{\nu_e} \hat{\psi}_{\nu_e} \qquad \text{and} \qquad G_d \bar{\hat{\psi}}_n \partial\!\!\!/ \hat{\psi}_n \bar{\hat{\psi}}_{\nu_e} \partial\!\!\!/ \hat{\psi}_{\nu_e}. \tag{11.96}$$

We know that $G_F \sim M_W^{-2}$, and similarly $G_d \sim M_W^{-4}$ (from dimensional counting, or from the association of the G_d term with the $O(G_F^2)$ counter term). From dimensional analysis, or by referring to (11.94) and remembering that D is of order G_F for consistency, we see that the second term in (11.96), when evaluated at tree level, is of order $(s - s_0)/M_W^2$ times the first. It follows that *higher derivative interactions, and in general those with successively large negative mass dimension, are increasingly suppressed at low energies.*

Where, then, do *renormalizable* theories fit into this? Those with couplings having positive mass dimension ('super-renormalizable') have, as we have seen, a limited number of infinities and can be quickly renormalized. The 'merely renormalizable' theories have dimensionless coupling constants, such as e (or α). In this case, since there are no mass factors (for good or ill) to be associated with powers of α, as we go up in order of perturbation theory it would seem plausible that the divergences get essentially no worse, and can be cured by the counter terms which compensated those simplest divergences which we examined in earlier sections—though for QED the proof is difficult, and took many years to perfect.

Given any renormalizable theory, such as QED, it is always possible to suppose that the 'true' theory contains additional non-renormalizable terms, provided their mass scale is very much larger than the energy scale at which the theory has been tested. For example, a term of the form (11.78) with 'm' replaced by a very large mass M would be possible, and would contribute an amount $4eK/M$ to a lepton magnetic moment. The present level of agreement between theory and experiment in the case of the electron's moment implies that $M \geq 4 \times 10^7$ GeV (Weinberg 1995, p 520).

From this perspective, then, it may be less of a mystery why renormalizable theories are generally the *relevant* ones at presently posed energies. Returning to the line of thought introduced in section 10.1.1, we may imagine that a 'true' theory exists at some enormously high energy Λ (the Planck scale?) which, though not itself a local quantum field theory, can be written in terms of all possible fields and their couplings, as allowed by certain symmetry principles. Our particular renormalizable subset of these theories then emerges as a *low-energy effective theory*, due to the strong suppression of the non-renormalizable terms. Of course, for this point of view to hold, we must assume that the latter interactions do not have 'unnaturally large' couplings, when expressed in terms of Λ.

This interpretation, if correct, deals rather neatly with what was, for many physicists, an awkward aspect of renormalizable theories. On the one hand, it was certainly an achievement to have rendered all perturbative calculations finite as the cut-off went to infinity; but on the other, it was surely unreasonable to expect any such theory, established by confrontation with experiments in currently accessible energy regimes, really to describe physics at arbitrarily high energies. On the 'low-energy effective field theory' interpretation, we can enjoy the calculational advantages of renormalizable field theories, while acknowledging—with no contradiction—the likelihood that at some scale 'new physics' will enter.

Having thus argued that renormalizable theories emerge 'naturally' as low-energy theories, we now seem to be faced with another puzzle: why were weak interactions successfully describable, for many years, in terms of the non-renormalizable four-fermion theory? The answer is that non-renormalizable theories may be physically detectable at low energies if they contribute to processes that would otherwise be forbidden. For example, the fact that (as far as we know) neutrinos have neither electromagnetic nor strong interactions, but only weak interactions, allowed the four-fermion theory to be detected—but amplitudes were suppressed by powers of s/M_{W}^2 (relative to comparable electromagnetic ones) and this was, indeed, why it was called 'weak'!

In the case of the weak interaction, the reader may perhaps wonder why—if it was understood that the four-fermion theory could after all be handled up to energies of order 10 GeV—so much effort went in to creating a renormalizable theory of weak interactions, as it undoubtedly did. Part of the answer is that the utility of non-renormalizable interactions was a rather late realization (see, for example, Weinberg 1979). At all events, the preceding argument for the 'naturalness' of renormalizable theories as low-energy effective theories provides strong motivation for searching for such a description of weak interactions.

We shall discuss the construction of the currently accepted renormalizable theory of electroweak interactions in volume 2. We can already anticipate that the first step will be to replace the 'negative-mass-dimensioned' constant G_{F} by a dimensionless one. The most obvious way to do this is to envisage a Yukawa-type theory of weak interactions mediated by a massive quantum (as, of course, Yukawa himself did—see section 2.5). The four-fermion process of figure 11.11 would then be replaced by that of figure 11.14, with amplitude (omitting spinors) $\sim g_{\mathrm{Z}}^2/(q^2 - m_{\mathrm{Z}}^2)$ where g_{Z} is dimensionless. For small $q^2 \ll m_{\mathrm{Z}}^2$, this reduces to the contact four-fermion form of figure 11.11, with an effective $G_{\mathrm{F}} \sim g_{\mathrm{Z}}^2/m_{\mathrm{Z}}^2$, showing the origin of the negative mass dimensions of G_{F}. It is clear that even if the new theory were to be renormalizable, many low-energy processes would be well described by an *effective* non-renormalizable four-fermion theory, as was indeed the case historically.

Unfortunately, we shall see in volume 2 that the application of this simple idea to the charge-changing weak interactions does not, after all, lead to a renormalizable theory. This teaches us an important lesson: a dimensionless coupling does not necessarily guarantee renormalizability.

To arrive at a renormalizable theory of the weak interactions it seems to be necessary to describe them in terms of a gauge theory (recall the 'universality' hints mentioned in section 11.6). Yet the mediating gauge field quanta have mass, which appears to contradict gauge invariance. The remarkable story of how gauge field quanta can acquire mass while preserving gauge invariance is reserved for volume 2.

A number of other non-renormalizable interactions are worth mentioning. Perhaps the most famous of all is gravity, characterized by Newton's constant G_{N}, which (cf (2.24)) has the value $(1.2 \times 10^{19}\ \mathrm{GeV})^{-2}$. The detection of gravity at energies so far below 10^{19} GeV is due, of course, to the fact that

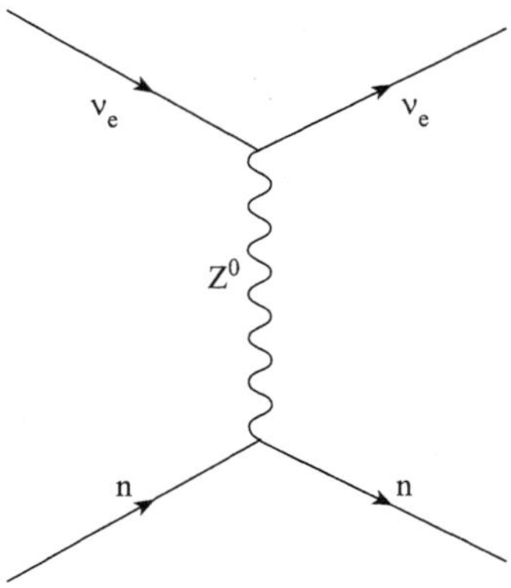

Figure 11.14. One-Z (Yukawa-type) exchange process in $\nu_e + n \to \nu_e + n$.

the gravitational fields of all the particles in a macroscopic piece of matter add up coherently. At the level of the individual particles, its effect is still entirely negligible. Another example may be provided by baryon and/or lepton violating interactions, mediated by highly suppressed non-renormalizable terms. Such things are frequently found when the low-energy limit is taken of theories defined (for example) at energies of order 10^{16} GeV or higher.

The stage is now set for the discussion, in volume 2, of the renormalizable non-Abelian gauge field theories which describe the weak and strong sectors of the Standard Model.

Problems

11.1 Establish the values of the counter terms given in (11.12).

11.2 Convince yourself of the rule 'each closed fermion loop carries an additional factor -1'.

11.3 Explain why the trace is taken in (11.14).

11.4 Verify (11.15).

11.5 Verify the quoted relation $P_\tau^\rho P_\nu^\tau = P_\nu^\rho$ where $P_\nu^\rho = g_\nu^\rho - q^\rho q_\nu q^2$ (cf (11.25)).

11.6 Obtain the expression (11.36) for $\bar{\Pi}_\gamma^{[2]}(q^2)$.

11.7 Verify (11.38) for $\boldsymbol{q}^2 \ll m^2$.

11.8 Verify (11.54) for $-q^2 \gg m^2$.

11.9 Check the estimate (11.59).

11.10 Find the dimensionality of 'E' in an interaction of the form $E(\hat{F}_{\mu\nu}\hat{F}^{\mu\nu})^2$. Express this interaction in terms of the $\hat{\boldsymbol{E}}$ and $\hat{\boldsymbol{B}}$ fields. Is such a term finite or infinite in QED? How might it be measured?

APPENDIX A

NON-RELATIVISTIC QUANTUM MECHANICS

This appendix is intended as a very terse 'revision' summary of those aspects of non-relativistic quantum mechanics that are particularly relevant for this book. A fuller account may be found in Mandl (1992), for example.

Natural units $\hbar = c = 1$ (see appendix B).

Fundamental postulate of quantum mechanics:

$$[\hat{p}_i, \hat{x}_j] = -\mathrm{i}\delta_{ij}. \tag{A.1}$$

Coordinate representation:

$$\hat{\boldsymbol{p}} = -\mathrm{i}\boldsymbol{\nabla} \tag{A.2}$$

$$\hat{H}\psi(\boldsymbol{x}, t) = \mathrm{i}\frac{\partial\psi(\boldsymbol{x}, t)}{\partial t}. \tag{A.3}$$

Schrödinger equation for a spinless particle:

$$\hat{H} = \frac{\hat{\boldsymbol{p}}^2}{2m} + \hat{V} \tag{A.4}$$

and so

$$\left(-\frac{1}{2m}\boldsymbol{\nabla}^2 + \hat{V}(\boldsymbol{x}, t)\right)\psi(\boldsymbol{x}, t) = \mathrm{i}\frac{\partial\psi(\boldsymbol{x}, t)}{\partial t}. \tag{A.5}$$

Probability density and current (see problem 4.1):

$$\rho = \psi^*\psi = |\psi|^2 \geq 0 \tag{A.6}$$

$$\boldsymbol{j} = \frac{1}{2m\mathrm{i}}[\psi^*(\boldsymbol{\nabla}\psi) - (\boldsymbol{\nabla}\psi^*)\psi] \tag{A.7}$$

with

$$\frac{\partial\rho}{\partial t} + \boldsymbol{\nabla}\cdot\boldsymbol{j} = 0. \tag{A.8}$$

Free-particle solutions:

$$\phi(\boldsymbol{x}, t) = u(\boldsymbol{x})\mathrm{e}^{-\mathrm{i}Et} \tag{A.9}$$

$$\hat{H}_0 u = Eu \tag{A.10}$$

where

$$\hat{H}_0 = \hat{H}(\hat{V} = 0). \tag{A.11}$$

Box normalization:

$$\int_V u^*(\boldsymbol{x})u(\boldsymbol{x})\,\mathrm{d}^3\boldsymbol{x} = 1. \tag{A.12}$$

Angular momentum: Three Hermitian operators $(\hat{J}_x, \hat{J}_y, \hat{J}_z)$ satisfying

$$[\hat{J}_x, \hat{J}_y] = \mathrm{i}\hbar \hat{J}_z$$

and corresponding relations obtained by rotating the x–y–z subscripts. $[\hat{\boldsymbol{J}}^2, \hat{J}_z] = 0$ implies complete sets of states exist with definite values of $\hat{\boldsymbol{J}}^2$ and $\hat{J}_z$. Eigenvalues of $\hat{\boldsymbol{J}}^2$ are (with $\hbar = 1$) $j(j+1)$ where $j = 0, \frac{1}{2}, 1, \ldots$; eigenvalues of $\hat{J}_z$ are m where $-j \le m \le j$, for given j. For orbital angular momentum, $\hat{\boldsymbol{J}} \to \hat{\boldsymbol{L}} = \boldsymbol{r} \times \hat{\boldsymbol{p}}$ and eigenfunctions are spherical harmonics $Y_{\ell m}(\theta, \phi)$, for which eigenvalues of $\hat{\boldsymbol{L}}^2$ and $\hat{L}_z$ are $l(l+1)$ and m where $-l \le m \le l$. For spin-$\frac{1}{2}$ angular momentum, $\hat{\boldsymbol{J}} \to \frac{1}{2}\boldsymbol{\sigma}$ where the Pauli matrices $\boldsymbol{\sigma} = (\sigma_x, \sigma_y, \sigma_z)$ are

$$\sigma_x = \begin{pmatrix} 0 & 1 \\ 1 & 0 \end{pmatrix} \qquad \sigma_y = \begin{pmatrix} 0 & -\mathrm{i} \\ \mathrm{i} & 0 \end{pmatrix} \qquad \sigma_z = \begin{pmatrix} 1 & 0 \\ 0 & -1 \end{pmatrix}. \tag{A.13}$$

Eigenvectors of s_z are $\begin{pmatrix} 1 \\ 0 \end{pmatrix}$ (eigenvalue $+\frac{1}{2}$), and $\begin{pmatrix} 0 \\ 1 \end{pmatrix}$ (eigenvalue $-\frac{1}{2}$).

Interaction with electromagnetic field: Particle of charge q in electromagnetic vector potential $\boldsymbol{A}$

$$\boxed{\hat{\boldsymbol{p}} \to \hat{\boldsymbol{p}} - q\boldsymbol{A}}. \tag{A.14}$$

Thus

$$\frac{1}{2m}(\hat{\boldsymbol{p}} - q\boldsymbol{A})^2\psi = \mathrm{i}\frac{\partial \psi}{\partial t} \tag{A.15}$$

and so

$$-\frac{1}{2m}\nabla^2\psi + \mathrm{i}\frac{q}{m}\boldsymbol{A}\cdot\nabla\psi + \frac{q^2}{2m}\boldsymbol{A}^2\psi = \mathrm{i}\frac{\partial \psi}{\partial t}. \tag{A.16}$$

Note: (a) chosen gauge $\nabla \cdot \boldsymbol{A} = 0$; (b) q^2 term is usually neglected.

Example: Magnetic field along z-axis, possible $\boldsymbol{A}$ consistent with $\nabla \cdot \boldsymbol{A} = 0$ is $\boldsymbol{A} = \frac{1}{2}B(-y, x, 0)$ such that $\nabla \times \boldsymbol{A} = (0, 0, B)$. Inserting this into the second term on left-hand side of (A.16) gives

$$\frac{\mathrm{i}qB}{2m}\left(-y\frac{\partial}{\partial x} + x\frac{\partial}{\partial y}\right)\psi = -\frac{qB}{2m}\hat{L}_z\psi \tag{A.17}$$

which generalizes to the standard orbital magnetic moment interaction $-\hat{\boldsymbol{\mu}}\cdot\boldsymbol{B}\psi$ where

$$\hat{\boldsymbol{\mu}} = \frac{qB}{2m}\hat{\boldsymbol{L}}. \tag{A.18}$$

Time-dependent perturbation theory:

$$\hat{H} = \hat{H}_0 + \hat{V} \tag{A.19}$$

$$\hat{H}\psi = \mathrm{i}\frac{\partial \psi}{\partial t}. \tag{A.20}$$

Unperturbed problem:

$$\hat{H}_0 u_n = E_n u_n. \tag{A.21}$$

Completeness:

$$\psi(\boldsymbol{x}, t) = \sum_n a_n(t) u_n(\boldsymbol{x}) \mathrm{e}^{-\mathrm{i}E_n t}. \tag{A.22}$$

First-order perturbation theory:

$$\boxed{a_{\mathrm{fi}} = -\mathrm{i} \iint \mathrm{d}^3\boldsymbol{x}\, \mathrm{d}t\, u_{\mathrm{f}}^*(\boldsymbol{x}) \mathrm{e}^{+\mathrm{i}E_{\mathrm{f}}t} \hat{V}(\boldsymbol{x}, t) u_{\mathrm{i}}(\boldsymbol{x}) \mathrm{e}^{-\mathrm{i}E_{\mathrm{i}}t}} \tag{A.23}$$

which has the form

$$a_{\mathrm{fi}} = -\mathrm{i} \int (\text{volume element})(\text{final state})^* (\text{perturbing potential})(\text{initial state}) \tag{A.24}$$

Important examples:

(i) $\hat{V}$ independent of t:

$$a_{\mathrm{fi}} = -\mathrm{i} V_{\mathrm{fi}} 2\pi \delta(E_{\mathrm{f}} - E_{\mathrm{i}}) \tag{A.25}$$

where

$$V_{\mathrm{fi}} = \int \mathrm{d}^3\boldsymbol{x}\, u_{\mathrm{f}}^*(\boldsymbol{x}) \hat{V}(\boldsymbol{x}) u_{\mathrm{i}}(\boldsymbol{x}). \tag{A.26}$$

(ii) Oscillating time-dependent potential:

(a) if $\hat{V} \sim \mathrm{e}^{-\mathrm{i}\omega t}$, time integral of a_{fi} is

$$\int \mathrm{d}t\, \mathrm{e}^{+\mathrm{i}E_{\mathrm{f}}t} \mathrm{e}^{-\mathrm{i}\omega t} \mathrm{e}^{-\mathrm{i}E_{\mathrm{i}}t} = 2\pi \delta(E_{\mathrm{f}} - E_{\mathrm{i}} - \omega) \tag{A.27}$$

i.e. the system has absorbed energy from potential;

(b) if $\hat{V} \sim \mathrm{e}^{+\mathrm{i}\omega t}$, time integral of a_{fi} is

$$\int \mathrm{d}t\, \mathrm{e}^{+\mathrm{i}E_{\mathrm{f}}t} \mathrm{e}^{+\mathrm{i}\omega t} \mathrm{e}^{-\mathrm{i}E_{\mathrm{i}}t} = 2\pi \delta(E_{\mathrm{f}} + \omega - E_{\mathrm{i}}) \tag{A.28}$$

i.e. the potential has absorbed energy from system.

Absorption and emission of photons: For electromagnetic radiation, far from its sources, the vector potential satisfies the wave equation

$$\nabla^2 \boldsymbol{A} - \frac{\partial^2 \boldsymbol{A}}{\partial t^2} = 0. \tag{A.29}$$

Solution:

$$\boldsymbol{A}(\boldsymbol{x}, t) = \boldsymbol{A}_0 \exp(-\mathrm{i}\omega t + \mathrm{i}\boldsymbol{k} \cdot \boldsymbol{x}) + \boldsymbol{A}_0^* \exp(+\mathrm{i}\omega t - \mathrm{i}\boldsymbol{k} \cdot \boldsymbol{x}). \tag{A.30}$$

With gauge condition $\nabla \cdot \boldsymbol{A} = 0$ we have

$$\boldsymbol{k} \cdot \boldsymbol{A}_0 = 0 \tag{A.31}$$

and there are two independent polarization vectors for photons.

Treat the interaction in first-order perturbation theory:

$$\hat{V}(\boldsymbol{x}, t) = (\mathrm{i}q/m)\boldsymbol{A}(\boldsymbol{x}, t) \cdot \nabla. \tag{A.32}$$

Thus

$$\begin{aligned} \boldsymbol{A}_0 \exp(-\mathrm{i}\omega t + \mathrm{i}\boldsymbol{k} \cdot \boldsymbol{x}) &\equiv \text{absorption of photon of energy } \omega \\ \boldsymbol{A}_0^* \exp(+\mathrm{i}\omega t + \mathrm{i}\boldsymbol{k} \cdot \boldsymbol{x}) &\equiv \text{emission of photon of energy } \omega. \end{aligned} \tag{A.33}$$

APPENDIX B

NATURAL UNITS

In particle physics, a widely adopted convention is to work in a system of units, called natural units, in which

$$\boxed{\hbar = c = 1.} \tag{B.1}$$

This avoids having to keep track of untidy factors of $\hbar$ and c throughout a calculation; only at the end is it necessary to convert back to more usual units. Let us spell out the implications of this choice of c and $\hbar$.

(i) $c = 1$. In conventional MKS units c has the value

$$c = 3 \times 10^8 \text{ m s}^{-1}. \tag{B.2}$$

By choosing units such that

$$c = 1 \tag{B.3}$$

since a velocity has the dimensions

$$[c] = [\mathrm{L}][\mathrm{T}]^{-1} \tag{B.4}$$

we are implying that our unit of length is numerically equal to our unit of time. In this sense, length and time are equivalent dimensions:

$$[\mathrm{L}] = [\mathrm{T}]. \tag{B.5}$$

Similarly, from the energy–momentum relation of special relativity

$$E^2 = \boldsymbol{p}^2c^2 + m^2c^4 \tag{B.6}$$

we see that the choice of $c = 1$ also implies that energy, mass and momentum all have equivalent dimensions. In fact, it is customary to refer to momenta in units of 'MeV/c' or 'GeV/c'; these all become 'MeV' or 'GeV' when $c = 1$.

(ii) $\hbar = 1$. The numerical value of Planck's constant is

$$\hbar = 6.6 \times 10^{-22} \text{ MeV s} \tag{B.7}$$

and $\hbar$ has dimensions of energy multiplied by time so that

$$[\hbar] = [\mathrm{M}][\mathrm{L}]^2[\mathrm{T}]^{-1}. \tag{B.8}$$

Setting $\hbar = 1$ therefore relates our units of [M], [L] and [T]. Since [L] and [T] are equivalent by our choice of $c = 1$, we can choose [M] as the single independent dimension for our natural units:

$$\boxed{[\mathrm{M}] = [\mathrm{L}]^{-1} = [\mathrm{T}]^{-1}.} \tag{B.9}$$

An example: the pion Compton wavelength How do we convert from natural units to more conventional units? Consider the pion Compton wavelength

$$\lambda_\pi = \hbar / M_\pi c \tag{B.10}$$

evaluated in both natural and conventional units. In natural units

$$\lambda_\pi = 1/M_\pi \tag{B.11}$$

where $M_\pi \simeq 140\ \mathrm{MeV}/c^2$. In conventional units, using M_π, $\hbar$ (B.7) and c (B.2), we have the familiar result

$$\lambda_\pi = 1.41\ \mathrm{fm} \tag{B.12}$$

where the 'fermi' or femtometre, fm, is defined as

$$1\ \mathrm{fm} = 10^{-15}\ \mathrm{m}.$$

We therefore have the correspondence

$$\boxed{\lambda_\pi = 1/M_\pi = 1.41\ \mathrm{fm}.} \tag{B.13}$$

Practical cross section calculations: An easy-to-remember relation may be derived from the result

$$\boxed{\hbar c \simeq 200\ \mathrm{MeV\ fm}} \tag{B.14}$$

obtained directly from (B.2) and (B.7). Hence, in natural units, we have the relation

$$\boxed{1\ \mathrm{fm} \simeq \frac{1}{200\ \mathrm{MeV}} = 5\ (\mathrm{GeV})^{-1}.} \tag{B.15}$$

Cross sections are calculated without $\hbar$'s and c's and all masses, energies and momenta typically in MeV or GeV. To convert the result to an area, we merely remember the dimensions of a cross section:

$$[\sigma] = [\mathrm{L}]^2 = [\mathrm{M}]^{-2}. \tag{B.16}$$

If masses, momenta and energies have been specified in GeV, from (B.15) we derive the useful result (from the more precise relation $\hbar c = 197.328$ MeV fm)

$$\boxed{\left(\frac{1}{1\ \mathrm{GeV}}\right)^2 = 1\ (\mathrm{GeV})^{-2} = 0.389\,39\ \mathrm{mb}} \tag{B.17}$$

where a millibarn, mb, is defined to be

$$1 \text{ mb} = 10^{-31} \text{ m}^2.$$

Note that a 'typical' hadronic cross section corresponds to an area of about λ_π^2 where

$$\lambda_\pi^2 = 1/M_\pi^2 = 20 \text{ mb}.$$

Electromagnetic cross sections are an order of magnitude smaller: specifically for lowest order $e^+e \to \mu^+\mu^-$

$$\sigma \approx \frac{88}{s} \text{ nb} \tag{B.18}$$

where s is in $(\text{GeV})^2$ (see problem 8.19(d) in chapter 8).

APPENDIX C

MAXWELL'S EQUATIONS: CHOICE OF UNITS

In high-energy physics, it is not the convention to use the rationalized MKS system of units when treating Maxwell's equations. Since the discussion is always limited to field equations *in vacuo*, it is usually felt desirable to adopt a system of units in which these equations take their simplest possible form—in particular, one such that the constants ϵ_0 and μ_0, employed in the MKS system, do not appear. These two constants enter, of course, via the force laws of Coulomb and Ampère, respectively. These laws relate a mechanical quantity (force) to electrical ones (charge and current). The introduction of ϵ_0 in Coulomb's law

$$\boldsymbol{F} = \frac{q_1 q_2 \boldsymbol{r}}{4\pi\epsilon_0 r^3} \tag{C.1}$$

enables one to choose arbitrarily one of the electrical units and assign to it a dimension independent of those entering into mechanics (mass, length and time). If, for example, we use the coulomb as the basic electrical quantity (as in the MKS system), ϵ_0 has dimension $(\text{coulomb})^2\,[\text{T}]^2/[\text{M}][\text{L}]^3$. Thus the common practical units (volt, ampère, coulomb, etc) can be employed in applications to both fields and circuits. However, for our purposes this advantage is irrelevant, since we are only concerned with the field equations, not with practical circuits. In our case, we prefer to define the electrical units in terms of mechanical ones in such a way as to reduce the field equations to their simplest form. The field equation corresponding to (C.1) is

$$\nabla \cdot \boldsymbol{E} = \rho/\epsilon_0 \qquad \text{(Gauss' law: MKS)} \tag{C.2}$$

and this may obviously be simplified if we choose the unit of charge such that ϵ_0 becomes unity. Such a system, in which CGS units are used for the mechanical quantities, is a variant of the electrostatic part of the 'Gaussian CGS' system. The original Gaussian system set $\epsilon_0 \to 1/4\pi$, thereby simplifying the force law (C.1), but introducing a compensating 4π into the field equation (C.2). The field equation is, in fact, primary, and the 4π is a geometrical factor appropriate only to the specific case of three dimensions, so that it should not appear in a field equation of general validity. The system in which ϵ_0 in (C.2) may be replaced by unity is called the 'rationalized Gaussian CGS' or 'Heaviside–Lorentz' system:

$$\nabla \cdot \boldsymbol{E} = \rho \qquad \text{(Gauss' law; Heaviside–Lorentz).} \tag{C.3}$$

Generally, systems in which the 4π factors appear in the force equations rather than the field equations are called 'rationalized'.

Of course, (C.3) is only the first of the Maxwell equations in Heaviside–Lorentz units. In the Gaussian system, μ_0 in Ampère's force law

$$\boldsymbol{F} = \frac{\mu_0}{4\pi} \int\int \frac{\boldsymbol{j}_1 \times (\boldsymbol{j}_2 \times \boldsymbol{r}_{12})}{r_{12}^3} \, \mathrm{d}^3\boldsymbol{r}_1 \, \mathrm{d}^3\boldsymbol{r}_2 \tag{C.4}$$

was set equal to 4π, thereby defining a unit of current (the electromagnetic unit or Biot (Bi emu)). The unit of charge (the electrostatic unit or Franklin (Fr esu)) has already been defined by the (Gaussian) choice $\epsilon_0 = 1/4\pi$ and currents via $\mu_0 \to 4\pi$, and c appears explicitly in the equations. In the rationalized (Heaviside–Lorentz) form of this system, $\epsilon_0 \to 1$ and $\mu_0 \to 1$, and the remaining Maxwell equations are

$$\nabla \times \boldsymbol{E} = -\frac{1}{c}\frac{\partial \boldsymbol{B}}{\partial t} \tag{C.5}$$

$$\nabla \cdot \boldsymbol{B} = 0 \tag{C.6}$$

$$\nabla \times \boldsymbol{B} = \boldsymbol{j} + \frac{1}{c}\frac{\partial \boldsymbol{E}}{\partial t}. \tag{C.7}$$

A further discussion of units in electromagnetic theory is given in Panofsky and Phillips (1962, appendix I).

Finally, throughout this book we have used a particular choice of units for mass, length and time such that $\hbar = c = 1$ (see appendix B). In that case, the Maxwell equations we use are as in (C.3), (C.5)–(C.7), but with c replaced by unity.

As an example of the relation between MKS and the system employed in this book (and universally in high-energy physics), we remark that the fine structure constant is written as

$$\alpha = \frac{e^2}{4\pi\epsilon_0\hbar c} \qquad \text{in MKS units} \tag{C.8}$$

or as

$$\alpha = \frac{e^2}{4\pi} \qquad \text{in Heaviside–Lorentz units with } \hbar = c = 1. \tag{C.9}$$

Clearly the value of $\alpha(\simeq 1/137)$ is the same in both cases, but the numerical values of 'e' in (C.8) and in (C.9) are, of course, different.

The choice of rationalized MKS units for Maxwell's equations is a part of the SI system of units. In this system of units the numerical values of μ_0 and ϵ_0 are

$$\mu_0 = 4\pi \times 10^{-7} \qquad (\mathrm{kg\,m\,C^{-2}} = \mathrm{H\,m^{-1}})$$

and, since $\mu_0\epsilon_0 = 1/c^2$,

$$\epsilon_0 = \frac{10^7}{4\pi c^2} = \frac{1}{36\pi \times 10^9} \qquad (\mathrm{C^2\,s^2\,kg^{-1}\,m^{-3}} = \mathrm{F\,m^{-1}}).$$

APPENDIX D

SPECIAL RELATIVITY: INVARIANCE AND COVARIANCE

The co-ordinate 4-vector x^μ is defined by

$$x^\mu = (x^0, x^1, x^2, x^3)$$

where $x^0 = t$ (with $c = 1$) and $(x^1, x^2, x^3) = \boldsymbol{x}$. Under a Lorentz transformation along the x^1-axis with velocity v, x^μ transforms to

$$\begin{aligned} x^{0\prime} &= \gamma(x^0 - vx^1) \\ x^{1\prime} &= \gamma(-vx^0 + x^1) \\ x^{2\prime} &= x^2 \\ x^{3\prime} &= x^3 \end{aligned} \tag{D.1}$$

where $\gamma = (1 - v^2)^{-1/2}$.

A general 'contravariant 4-vector' is defined to be any set of four quantities $A^\mu = (A^0, A^1, A^2, A^3) \equiv (A^0, \boldsymbol{A})$ which transform under Lorentz transformations exactly as the corresponding components of the coordinate 4-vector x^μ. Note that the definition is phrased in terms of the *transformation property* (under Lorentz transformations) of the object being defined. An important example is the energy–momentum 4-vector $p^\mu = (E, \boldsymbol{p})$, where for a particle of rest mass m, $E = (\boldsymbol{p}^2 + m^2)^{1/2}$. Another example is the 4-gradient $\partial^\mu = (\partial^0, -\boldsymbol{\nabla})$ (see problem 3.1) where

$$\partial^0 = \frac{\partial}{\partial t} \qquad \boldsymbol{\nabla} = \left(\frac{\partial}{\partial x^1}, \frac{\partial}{\partial x^2}, \frac{\partial}{\partial x^3}\right). \tag{D.2}$$

Lorentz transformations leave the expression $A^{0\,2} - \boldsymbol{A}^2$ invariant for a general 4-vector A^μ. For example, $E^2 - \boldsymbol{p}^2 = m^2$ is invariant, implying that the rest mass m is invariant under Lorentz transformations. Another example is the four-dimensional invariant differential operator analogous to $\boldsymbol{\nabla}^2$, namely

$$\Box = \partial^{0\,2} - \boldsymbol{\nabla}^2$$

which is precisely the operator appearing in the massless wave equation

$$\Box\phi = \partial^{0\,2}\phi - \boldsymbol{\nabla}^2\phi = 0.$$

The expression $A^{0\,2} - \boldsymbol{A}^2$ may be regarded as the scalar product of A^μ with a related 'covariant vector' $A_\mu = (A^0, -\boldsymbol{A})$. Then

$$A^{0\,2} - \boldsymbol{A}^2 = \sum_\mu A^\mu A_\mu$$

where, in practice, the summation sign on repeated 'upstairs' and 'downstairs' indices is always omitted. We shall often shorten the expression '$A^\mu A_\mu$' even further, to 'A^2'; thus $p^2 = E^2 - \boldsymbol{p}^2 = m^2$. The 'downstairs' version of ∂^μ is $\partial_\mu = (\partial^0, \boldsymbol{\nabla})$. Then $\partial_\mu \partial^\mu = \partial^2 = \Box$. 'Lowering' and 'raising' indices is effected by the *metric tensor* $g^{\mu\nu}$ or $g_{\mu\nu}$, where $g^{00} = g_{00} = 1$, $g^{11} = g^{22} = g^{33} = g_{11} = g_{22} = g_{33} = -1$, all other components vanishing. Thus if $A_\mu = g_{\mu\nu} A^\nu$ then $A_0 = A^0$, $A_1 = -A^1$, etc.

In the same way, the scalar product $A \cdot B$ of two 4-vectors is

$$A \cdot B = A^\mu B_\mu = A^0 B^0 - \boldsymbol{A} \cdot \boldsymbol{B} \tag{D.3}$$

and this is also invariant under Lorentz transformations. For example, the invariant four-dimensional divergence of a 4-vector $j^\mu = (\rho, \boldsymbol{j})$ is

$$\partial^\mu j_\mu = \partial^0 \rho - (-\boldsymbol{\nabla}) \cdot \boldsymbol{j} = \partial^0 \rho + \boldsymbol{\nabla} \cdot \boldsymbol{j} = \partial_\mu j^\mu \tag{D.4}$$

since the spatial part of ∂^μ is $-\boldsymbol{\nabla}$.

Because the Lorentz transformation is linear, it immediately follows that the sum (or difference) of two 4-vectors is also a 4-vector. In a reaction of the type '$1+2 \to 3+4+\cdots N$' we express the conservation of both energy and momentum as one '4-momentum conservation equation':

$$p_1^\mu + p_2^\mu = p_3^\mu + p_4^\mu + \cdots p_N^\mu. \tag{D.5}$$

In practice, the 4-vector index on all the p's is conventionally omitted in conservation equations such as (D.5), but it is nevertheless important to remember, in that case, that it is actually *four* equations, one for the energy components and a further three for the momentum components. Further, it follows that quantities such as $(p_1 + p_2)^2$, $(p_1 - p_3)^2$ are invariant under Lorentz transformations.

We may also consider products of the form $A^\mu B^\nu$, where A and B are 4-vectors. As μ and ν each run over their four possible values (0, 1, 2, 3) 16 different 'components' are generated ($A^0 B^0, A^0 B^1, \ldots, A^3 B^3$). Under a Lorentz transformation, the components of A and B will transform into definite linear combinations of themselves, as in the particular case of (D.1). It follows that the 16 components of $A^\mu B^\nu$ will also transform into well-defined linear combinations of themselves (try it for $A^0 B^1$ and (D.1)). Thus we have constructed a new object whose 16 components transform by a well-defined linear transformation law under a Lorentz transformation, as did the components of a 4-vector. This new quantity, defined by its transformation law, is called a *tensor*—or more precisely a 'contravariant second-rank tensor', the 'contravariant' referring to the fact that both indices are upstairs, the 'second rank' meaning that it has two indices. An important example of such a tensor is provided by $\partial^\mu A^\nu(x) - \partial^\nu A^\mu(x)$, which is the *electromagnetic field strength tensor* $F^{\mu\nu}$, introduced in chapter 3. More generally we can consider tensors $B^{\mu\nu}$ which are not literally formed by 'multiplying' two vectors together, but which transform in just the same way; and we can introduce third- and higher-rank tensors similarly, which can also be 'mixed', with some upstairs and some downstairs indices.

We now state a very useful and important fact. Suppose we 'dot' a downstairs 4-vector A_μ into a contravariant second-rank tensor $B^{\mu\nu}$, via the operation $A_\mu B^{\mu\nu}$, where as always a sum on the repeated index μ is understood. Then this quantity transforms as a 4-vector, via its 'loose' index ν. This is obvious if $B^{\mu\nu}$ is actually a product such as $B^{\mu\nu} = C^\mu D^\nu$, since then we have $A_\mu B^{\mu\nu} = (A\cdot C)D^\nu$, and $(A\cdot C)$ is an invariant, which leaves the 4-vector D^μ as the only 'transforming' object left. But even if $B^{\mu\nu}$ is not such a product, it transforms under Lorentz transformations in exactly the same way as if it were, and this leads to the same result. An example is provided by the quantity $\partial_\mu F^{\mu\nu}$ which enters on the left-hand side of the Maxwell equations in the form (3.17).

This example brings us conveniently to the remaining concept we need to introduce here, which is the important one of '*covariance*'. Referring to (3.17), we note that it has the form of an equality between two quantities ($\partial_\mu F^{\mu\nu}$ on the left, j^ν_{em} on the right) *each of which transforms in the same way under Lorentz transformations*—namely as a contravariant 4-vector. One says that (3.17) is 'Lorentz covariant', the word 'covariant' here meaning precisely that both sides transform in the same way (i.e. consistently) under Lorentz transformations. Confusingly enough, this use of the word 'covariant' is evidently quite different from the one encountered previously in an expression such as 'a covariant 4-vector', where it just meant a 4-vector with a downstairs index. This new meaning of 'covariant' is actually nuch better captured by an alternative name for the same thing, which is 'form invariant', as we will shortly see.

Why is this idea so important? Consider the (special) relativity principle, which states that the laws of physics should be the same in all inertial frames. The way in which this physical requirement is implemented mathematically is precisely via the notion of *covariance under Lorentz transformations*. For, consider how a law will typically be expressed. Relative to one inertial frame, we set up a coordinate system and describe the phenomena in question in terms of suitable coordinates, and such other quantities (forces, fields, etc) as may be necessary. We write the relevant law mathematically as equations relating these quantities, all referred to our chosen frame and coordinate system. What the relativity principle requires is that these relationships—these equations—must *have the same form* when the quantities in them are referred to a different inertial frame. Note that we must say 'have the same form', rather than 'be identical to', since we know very well that coordinates, at least, are not identical in two different inertial frames (cf (D.1)). This is why the term 'form invariant' is a more helpful one than 'covariant' in this context, but the latter is more commonly used.

A more elementary example may be helpful. Consider Newton's law in the simple form $\boldsymbol{F} = m\ddot{\boldsymbol{r}}$. This equation is 'covariant under rotations', meaning that it preserves the same form under a rotation of the coordinate system—and this in turn means that the physics it expresses is independent of the orientation of our coordinate axes. The 'same form' in this case is of course just $\boldsymbol{F}' = m\ddot{\boldsymbol{r}}'$. We emphasize again that the components of $\boldsymbol{F}'$ are *not* the same as those of $\boldsymbol{F}$, nor are the components of $\ddot{\boldsymbol{r}}'$ the same as those of $\ddot{\boldsymbol{r}}$; but the *relationship* between $\boldsymbol{F}'$ and $\ddot{\boldsymbol{r}}'$ is exactly the same as the relationship between $\boldsymbol{F}$ and $\ddot{\boldsymbol{r}}$, and that is what is required.

It is important to understand why this deceptively simple result ('$\boldsymbol{F}' = m\ddot{\boldsymbol{r}}'$') has been obtained. The reason is that we have assumed (or asserted) that 'force' is in fact to be represented mathematically as a 3-vector quantity. Once we have said that, the rest follows. More formally, the transformation law of the components of $\boldsymbol{r}$ is $r_i' = R_{ij}r_j$ (sum on j understood), where the matrix of transformation coefficients $\boldsymbol{R}$ is 'orthogonal' ($\boldsymbol{R}\boldsymbol{R}^{\mathrm{T}} = \boldsymbol{R}^{\mathrm{T}}\boldsymbol{R} = \boldsymbol{I}$), which ensures that the length (squared) of $\boldsymbol{r}$ is invariant , $\boldsymbol{r}^2 = \boldsymbol{r}'^2$. To say that 'force is a 3-vector' then implies that the components of $\boldsymbol{F}$ transform by the same set of coefficients R_{ij}: $F_i' = R_{ij}F_j$. Thus starting from the law $F_j = m\ddot{r}_j$ which relates the components in one frame, by multiplying both sides of the equation by R_{ij} and summing over j we arrive at $F_i' = m\ddot{r}_i'$, which states precisely that the components in the primed frame bear the same relationship to each other as the components in the unprimed frame did. This is the property of covariance under rotations, and it ensures that the physics embodied in the law is the same for all systems which differ from one another only by a rotation.

In just the same way, if we can write equations of physics as equalities between quantities which transform in the same way (i.e. 'are covariant') under Lorentz transformations, we will guarantee that these laws obey the relativity principle. This is indeed the case in the *Lorentz covariant formulation* of Maxwell's equations, given in (3.17), which we now repeat here: $\partial_\mu F^{\mu\nu} = j^\nu_{\mathrm{em}}$. To check covariance, we follow essentially the same steps as in the case of Newton's equations, except that the transformations being considered are Lorentz transformations. Inserting the expression (3.18) for $F^{\mu\nu}$, the equation can be written as $(\partial_\mu\partial^\mu)A^\nu - \partial^\nu(\partial_\mu A^\mu) = j^\nu_{\mathrm{em}}$. The bracketed quantities are actually *in*variants, as was mentioned earlier. This means that $\partial_\mu\partial^\mu$ is equal to $\partial_\mu{}'\partial'^\mu$, and similarly $\partial_\mu A^\mu = \partial_\mu{}'A'^\mu$, so that we can write the equation as $(\partial_\mu'\partial'^\mu)A^\nu - \partial^\nu(\partial_\mu' A'^\mu) = j^\nu_{\mathrm{em}}$. It is now clear that if we apply a Lorentz transformation to both sides, A^ν and ∂^ν will become A'^ν and ∂'^ν respectively, while j^ν_{em} will become j'^ν_{em}, since all these quantities are 4-vectors, transforming the same way (as the 3-vectors did in the Newton case). Thus we obtain just the same form of equation, written in terms of the 'primed frame' quantities, and this is the essence of (Lorentz transformation) covariance.

Actually, the detailed 'check' that we have just performed is really unnecessary. All that is required for covariance is that (once again!) both sides of equations transform the same way. That this is true of (3.17) can be seen 'by inspection', once we understand the significance (for instance) of the fact that the μ indices are 'dotted' so as to form an invariant. This example should convince the reader of the power of the 4-vector notation for this purpose: compare the 'by inspection' covariance of (3.17) with the job of verifying Lorentz covariance starting from the original Maxwell equations (3.1), (3.2), (3.3) and (3.8)! The latter involves establishing the rather complicated transformation law for the fields $\boldsymbol{E}$ and $\boldsymbol{B}$ (which, of course, form parts of the tensor $F^{\mu\nu}$). One can indeed show in this way that the Maxwell equations are covariant under Lorentz transformations, but they are not *manifestly* (i.e. without doing any work) so, whereas in the form (3.17) they are.

APPENDIX E

DIRAC δ-FUNCTION

Consider approximating an integral by a sum over strips Δx wide as shown in figure E.1:

$$\int_{x_1}^{x_2} f(x)\,\mathrm{d}x \simeq \sum_i f(x_i)\Delta x. \tag{E.1}$$

Consider the function $\delta(x - x_j)$ shown in figure E.2,

$$\delta(x - x_j) = \begin{cases} 1/\Delta x & \text{in the } j\text{th interval} \\ 0 & \text{all others} \end{cases} \tag{E.2}$$

Clearly this function has the properties

$$\sum_i f(x_i)\delta(x_i - x_j)\Delta x = f(x_j) \tag{E.3}$$

and

$$\sum_i \delta(x_i - x_j)\Delta x = 1. \tag{E.4}$$

In the limit as we pass to an integral form, we might expect (applying (E.1) to the left-hand sides) that these equations to reduce to

$$\int_{x_1}^{x_2} f(x)\delta(x - x_j)\,\mathrm{d}x = f(x_j) \tag{E.5}$$

and

$$\int_{x_1}^{x_2} \delta(x - x_j)\,\mathrm{d}x = 1 \tag{E.6}$$

provided that $x_1 < x_j < x_2$. Clearly such 'δ-functions' can easily be generalized to more dimensions, e.g. three dimensions:

$$\mathrm{d}V = \mathrm{d}x\,\mathrm{d}y\,\mathrm{d}z \equiv \mathrm{d}^3\boldsymbol{r} \qquad \delta(\boldsymbol{r} - \boldsymbol{r}_j) \equiv \delta(x - x_j)\delta(y - y_j)\delta(z - z_j). \tag{E.7}$$

Informally, therefore, we can think of the δ-function as a function that is zero everywhere except where its argument vanishes—at which point it is infinite in such a way that its integral has unit area, and equations (E.5) and (E.6) hold. Do such amazing functions exist? In fact, the informal idea just given

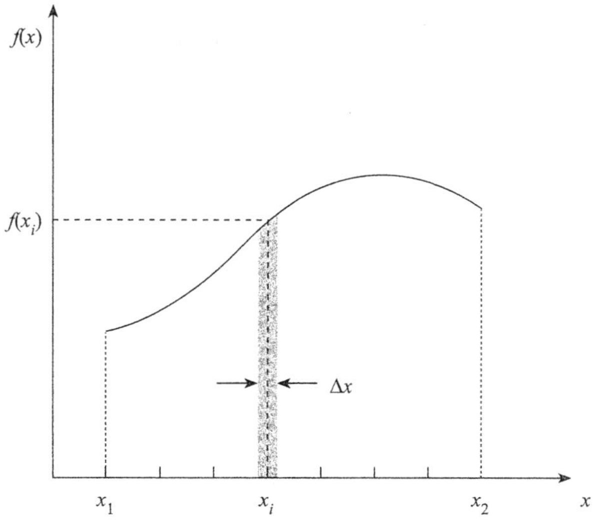

Figure E.1. Approximate evaluation of integral.

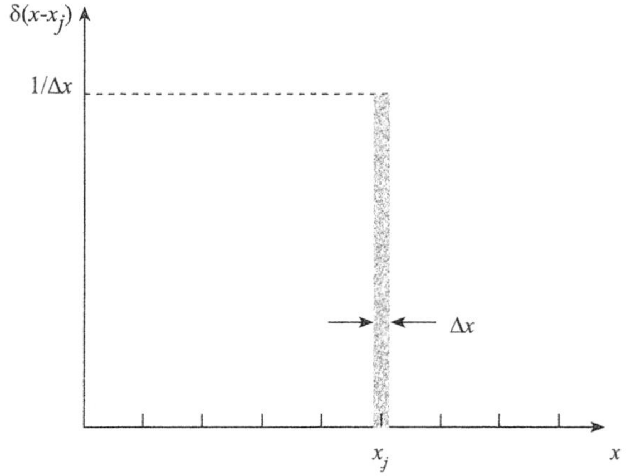

Figure E.2. The function $\delta(x - x_j)$.

does not define a respectable mathematical function. More properly the use of the 'δ-function' can be justified by introducing the notion of 'distributions' or 'generalized functions'. Roughly speaking, this means we can think of the 'δ-function' as the limit of a sequence of functions, whose properties converge to those given here. The following useful expressions all approximate the δ-function

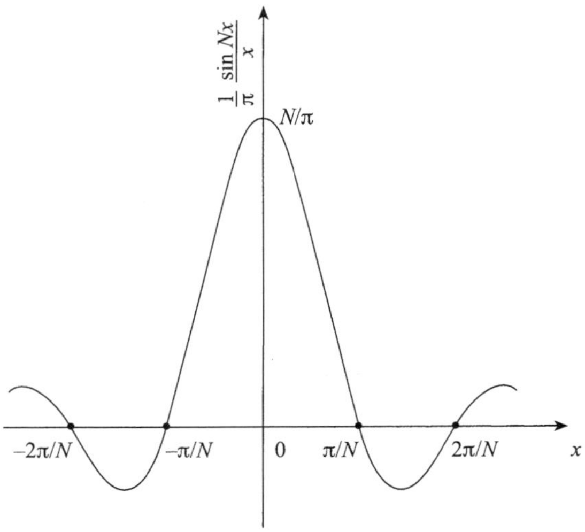

Figure E.3. The function (E.10) for finite N.

in this sense:

$$\delta(x) = \begin{cases} \lim\limits_{\epsilon\to 0} \dfrac{1}{\epsilon} & \text{for } -\epsilon/2 \le x \le \epsilon/2 \\ 0 & \text{for } |x| > \epsilon/2 \end{cases} \tag{E.8}$$

$$\delta(x) = \lim_{\epsilon\to 0} \frac{1}{\pi}\frac{\epsilon}{x^2+\epsilon^2} \tag{E.9}$$

$$\delta(x) = \lim_{N\to\infty} \frac{1}{\pi}\frac{\sin(Nx)}{x}. \tag{E.10}$$

The first of these is essentially the same as (E.2), and the second is a 'smoother' version of the first. The third is sketched in figure E.3: as N tends to infinity, the peak becomes infinitely high and narrow, but it still preserves unit area.

Usually, under integral signs, δ-functions can be manipulated with no danger of obtaining a mathematically incorrect result. However, care must be taken when products of two such generalized functions are encountered.

Resumé of Fourier series and Fourier transforms

Fourier's theorem asserts that any suitably well-behaved periodic function with period L can be expanded as follows:

$$f(x) = \sum_{n=-\infty}^{\infty} a_n \mathrm{e}^{\mathrm{i}2n\pi x/L}. \tag{E.11}$$

Using the orthonormality relation

$$\frac{1}{L}\int_{-L/2}^{L/2} \mathrm{e}^{-2\pi \mathrm{i} mx/L}\mathrm{e}^{2\pi \mathrm{i} nx/L}\,\mathrm{d}x = \delta_{mn} \tag{E.12}$$

with the Krönecker δ-symbol defined by

$$\delta_{mn} = \begin{cases} 1 & \text{if } m = n \\ 0 & \text{if } m \neq n \end{cases} \tag{E.13}$$

the coefficients in the expansion may be determined:

$$a_m = \frac{1}{L}\int_{-L/2}^{L/2} f(x)\mathrm{e}^{-2\pi \mathrm{i} mx/L}\,\mathrm{d}x. \tag{E.14}$$

Consider the limit of these expressions as $L \to \infty$. We may write

$$f(x) = \sum_{n=-\infty}^{\infty} F_n \Delta n \tag{E.15}$$

with

$$F_n = a_n \mathrm{e}^{2\pi \mathrm{i} nx/L} \tag{E.16}$$

and the interval $\Delta n = 1$. Defining

$$2\pi n/L = k \tag{E.17}$$

and

$$La_n = g(k) \tag{E.18}$$

we can take the limit $L \to \infty$ to obtain

$$\begin{aligned} f(x) &= \int_{-\infty}^{\infty} F_n\,\mathrm{d}n \\ &= \int_{-\infty}^{\infty} \frac{g(k)\mathrm{e}^{\mathrm{i}kx}}{L}\frac{L\mathrm{d}k}{2\pi}. \end{aligned} \tag{E.19}$$

Thus

$$f(x) = \frac{1}{2\pi}\int_{-\infty}^{\infty} g(k)\mathrm{e}^{\mathrm{i}kx}\,\mathrm{d}k \tag{E.20}$$

and similarly from (E.14)

$$g(k) = \int_{-\infty}^{\infty} f(x)\mathrm{e}^{-\mathrm{i}kx}\,\mathrm{d}x. \tag{E.21}$$

These are the Fourier transform relations, and they lead us to an important representation of the Dirac δ-function.

Substitute $g(k)$ from (E.21) into (E.20) to obtain

$$f(x) = \frac{1}{2\pi}\int_{-\infty}^{\infty} \mathrm{d}k\, \mathrm{e}^{\mathrm{i}kx} \int_{-\infty}^{\infty} \mathrm{d}x'\, \mathrm{e}^{-\mathrm{i}kx'} f(x'). \tag{E.22}$$

Reordering the integrals, we arrive at the result

$$f(x) = \int_{-\infty}^{\infty} \mathrm{d}x' f(x') \left(\frac{1}{2\pi} \int_{-\infty}^{\infty} \mathrm{e}^{\mathrm{i}k(x-x')}\, \mathrm{d}k \right) \tag{E.23}$$

valid for any function $f(x)$. Thus the expression

$$\frac{1}{2\pi}\int_{-\infty}^{\infty} \mathrm{e}^{\mathrm{i}k(x-x')}\, \mathrm{d}k \tag{E.24}$$

has the remarkable property of vanishing everywhere except at $x = x'$, and its integral with respect to x' over any interval including x is unity (set $f = 1$ in (E.23)). In other words, (E.24) provides us with a new representation of the Dirac δ-function:

$$\boxed{\delta(x) = \frac{1}{2\pi}\int_{-\infty}^{\infty} \mathrm{e}^{\mathrm{i}kx}\, \mathrm{d}k.} \tag{E.25}$$

Equation (E.25) is very important. It is the representation of the δ-function which is most commonly used, and it occurs throughout this book. Note that if we replace the upper and lower limits of integration in (E.25) by N and $-N$, and consider the limit $N \to \infty$, we obtain exactly (E.10).

The integral in (E.25) represents the superposition, with identical uniform weight $(2\pi)^{-1}$, of plane waves of all wavenumbers. Physically it may be thought of (cf (E.20)) as the Fourier transform of unity. Equation (E.25) asserts that the contributions from all these waves cancel completely, unless the phase parameter x is zero—in which case the integral manifestly diverges and '$\delta(0)$ is infinity' as expected. The fact that the Fourier transform of a constant is a δ-function is an extreme case of the bandwidth theorem from Fourier transform theory, which states that if the (suitably defined) 'spread' in a function $g(k)$ is Δk, and that of its transform $f(x)$ is Δx, then $\Delta x \Delta k \geq \frac{1}{2}$. In the present case Δk is tending to infinity and Δx to zero.

One very common use of (E.25) refers to the normalization of plane-wave states. If we rewrite it in the form

$$\delta(k' - k) = \int_{-\infty}^{\infty} \frac{\mathrm{e}^{-\mathrm{i}k'x}}{(2\pi)^{1/2}} \frac{\mathrm{e}^{\mathrm{i}kx}}{(2\pi)^{1/2}}\, \mathrm{d}x \tag{E.26}$$

we can interpret it to mean that the wavefunctions $\mathrm{e}^{\mathrm{i}kx}/(2\pi)^{1/2}$ and $\mathrm{e}^{\mathrm{i}k'x}/(2\pi)^{1/2}$ are orthogonal on the real axis $-\infty \leq x \leq \infty$ for $k \neq k'$ (since the left-hand side is zero), while for $k = k'$ their overlap is infinite, in such a way that the *integral* of this overlap is unity. This is the continuum analogue of orthonormality for wavefunctions labelled by a discrete index, as in (E.12). We say that the plane waves in (E.26) are 'normalized to a δ-function'. There is, however, a problem with this: plane waves are not square integrable and thus do not strictly belong to a Hilbert space. Mathematical physicists concerned with such matters have managed to deal with this by introducing 'rigged' Hilbert spaces in which such a normalization is legitimate. Although we often, in the text, appear to be using 'box normalization' (i.e. restricting space to a finite volume V), in practice when we evaluate integrals over plane waves the limits will be extended to infinity, and results like (E.26) will be used repeatedly.

Important three- and four-dimensional generalizations of (E.25) are:

$$\int \mathrm{e}^{\mathrm{i}\boldsymbol{k}\cdot\boldsymbol{x}}\,\mathrm{d}^3\boldsymbol{k} = (2\pi)^3\delta(\boldsymbol{x}) \tag{E.27}$$

and

$$\int \mathrm{e}^{\mathrm{i}k\cdot x}\,\mathrm{d}^4k = (2\pi)^4\delta(x) \tag{E.28}$$

where $k\cdot x = k^0x^0 - \boldsymbol{k}\cdot\boldsymbol{x}$ (see appendix B) and $\delta(x) = \delta(x^0)\delta(\boldsymbol{x})$.

Properties of the δ-function

(i)

$$\delta(ax) = \frac{1}{|a|}\delta(x). \tag{E.29}$$

Proof

For $a > 0$,

$$\int_{-\infty}^{\infty}\delta(ax)\,\mathrm{d}x = \int_{-\infty}^{\infty}\delta(y)\frac{\mathrm{d}y}{a} = \frac{1}{a}; \tag{E.30}$$

for $a < 0$,

$$\int_{-\infty}^{\infty}\delta(ax)\,\mathrm{d}x = \int_{\infty}^{-\infty}\delta(y)\frac{\mathrm{d}y}{a} = \int_{-\infty}^{\infty}\delta(y)\frac{\mathrm{d}y}{|a|} = \frac{1}{|a|}. \tag{E.31}$$

(ii)

$$\delta(x) = \delta(-x) \qquad \text{i.e. an even function.} \tag{E.32}$$

Proof

$$f(0) = \int \delta(x)f(x)\,\mathrm{d}x. \tag{E.33}$$

If $f(x)$ is an odd function, $f(0) = 0$. Thus $\delta(x)$ must be an even function.

(iii)

$$\delta(f(x)) = \sum_i \frac{1}{|\mathrm{d}f/\mathrm{d}x|_{x=a_i}}\delta(x-a_i) \tag{E.34}$$

where a_i are the roots of $f(x) = 0$.

Proof

The δ-function is only non-zero when its argument vanishes. Thus we are concerned with the roots of $f(x) = 0$. In the vicinity of a root

$$f(a_i) = 0 \tag{E.35}$$

we can make a Taylor expansion

$$f(x) = \cancel{f(a_i)} + (x - a_i)\left(\frac{\mathrm{d}f}{\mathrm{d}x}\right)_{x=a_i} + \cdots. \tag{E.36}$$

Thus the δ-function has non-zero contributions from each of the roots a_i of the form

$$\delta(f(x)) = \sum_i \delta\left[(x - a_i)\left(\frac{\mathrm{d}f}{\mathrm{d}x}\right)_{x=a_i}\right]. \tag{E.37}$$

Hence (using property (i)) we have

$$\delta(f(x)) = \sum_i \frac{1}{|\mathrm{d}f/\mathrm{d}x|_{x=a_i}} \delta(x - a_i). \tag{E.38}$$

Consider the example

$$\delta(x^2 - a^2). \tag{E.39}$$

Thus

$$f(x) = x^2 - a^2 = (x - a)(x + a) \tag{E.40}$$

with two roots $x = \pm a$ ($a > 0$), and $\mathrm{d}f/\mathrm{d}x = 2x$. Hence

$$\delta(x^2 - a^2) = \frac{1}{2a}[\delta(x - a) + \delta(x + a)]. \tag{E.41}$$

(iv)

$$x\delta(x) = 0. \tag{E.42}$$

This is to be understood as always occurring under an integral. It is obvious from the definition or from property (ii).

(v)

$$\int_{-\infty}^{\infty} f(x)\delta'(x)\,\mathrm{d}x = -f'(0) \tag{E.43}$$

where

$$\delta'(x) = \frac{\mathrm{d}}{\mathrm{d}x}\delta(x). \tag{E.44}$$

Proof

$$\int_{-\infty}^{\infty} f(x)\delta'(x)\,\mathrm{d}x = -\int_{-\infty}^{\infty} f'(x)\delta(x)\,\mathrm{d}x + [f(x)\delta(x)]_{-\infty}^{\infty}$$
$$= -f'(0) \qquad \text{(E.45)}$$

since the second term vanishes.

(vi)

$$\int_{-\infty}^{x} \delta(x' - a)\,\mathrm{d}x' = \theta(x - a) \qquad \text{(E.46)}$$

where

$$\theta(x) = \begin{cases} 0 & \text{for } x < 0 \\ 1 & \text{for } x > 0 \end{cases} \qquad \text{(E.47)}$$

is the so-called 'θ-function'.

Proof

For $x > a$,

$$\int_{-\infty}^{\infty} \delta(x' - a)\,\mathrm{d}x' = 1; \qquad \text{(E.48)}$$

for $x < a$,

$$\int_{-\infty}^{\infty} \delta(x' - a)\,\mathrm{d}x' = 0. \qquad \text{(E.49)}$$

By a simple extension it is easy to prove the result

$$\int_{x_1}^{x_2} \delta(x - a)\,\mathrm{d}x = \theta(x_2 - a) - \theta(x_1 - a). \qquad \text{(E.50)}$$

Exercise

Use property (iii) plus the definition of the θ-function to perform the p^0 integration and prove the useful phase space formula

$$\int \mathrm{d}^4 p\, \delta(p^2 - m^2)\theta(p^0) = \int \mathrm{d}^3 \boldsymbol{p}/2E \qquad \text{(E.51)}$$

where

$$p^2 = (p^0)^2 - \boldsymbol{p}^2 \qquad \text{(E.52)}$$

and

$$E = +(\boldsymbol{p}^2 + m^2)^{1\,2}. \qquad \text{(E.53)}$$

The relation (E.51) shows that the expression $\mathrm{d}^3\boldsymbol{p}/2E$ is Lorentz invariant: on the left-hand side, $\mathrm{d}^4 p$ and $\delta(p^2 - m^2)$ are invariant, while $\theta(p^0)$ depends only on the sign of p^0, which cannot be changed by a 'proper' Lorentz transformation—that is, one that does not reverse the sense of time.

APPENDIX F

CONTOUR INTEGRATION

We begin by recalling some relevant results from the calculus of real functions of two real variables x and y, which we shall phrase in 'physical' terms. Consider a particle moving in the xy-plane subject to a force $\boldsymbol{F} = (P(x, y), Q(x, y))$ whose x- and y-components P and Q vary throughout the plane. Suppose the particle moves, under the action of the force, around a closed path $\mathcal{C}$ in the xy-plane. Then the total work done by the force on the particle, $W_{\mathcal{C}}$, will be given by the integral

$$W_{\mathcal{C}} = \oint_{\mathcal{C}} \boldsymbol{F} \cdot \mathrm{d}\boldsymbol{r} = \oint_{\mathcal{C}} P\,\mathrm{d}x + Q\,\mathrm{d}y \tag{F.1}$$

where the $\oint$ sign means that the integration path is closed. Using Stokes' theorem, we can rewrite (F.1) as a surface integral

$$W_{\mathcal{C}} = \iint_{\mathcal{S}} \operatorname{curl} \boldsymbol{F} \cdot \mathrm{d}\boldsymbol{S} \tag{F.2}$$

where $\mathcal{S}$ is any surface bounded by $\mathcal{C}$ (as a butterfly net is bounded by the rim). Taking $\mathcal{S}$ to be the area in the xy-plane enclosed by $\mathcal{C}$, we have $\mathrm{d}\boldsymbol{S} = \mathrm{d}x\,\mathrm{d}y\,\boldsymbol{k}$ and

$$W_{\mathcal{C}} = \iint_{\mathcal{S}} \left(\frac{\partial Q}{\partial x} - \frac{\partial P}{\partial y} \right) \mathrm{d}x\,\mathrm{d}y. \tag{F.3}$$

A mathematically special, but physically common, case is that in which $\boldsymbol{F}$ is a 'conservative force', derivable from a potential function $V(x, y)$ (in this two-dimensional example) such that

$$P(x, y) = -\frac{\partial V}{\partial x} \qquad \text{and} \qquad Q(x, y) = -\frac{\partial V}{\partial y} \tag{F.4}$$

the minus signs being the usual convention. In that case, it is clear that

$$\frac{\partial P}{\partial y} = \frac{\partial Q}{\partial x} \tag{F.5}$$

and hence $W_{\mathcal{C}}$ in (F.3) is *zero*. The condition (F.5) is, in fact, both necessary and sufficient for $W_{\mathcal{C}} = 0$.

There can, however, be surprises. Consider, for example, the potential

$$V(x, y) = -\tan^{-1} y/x. \tag{F.6}$$

In this case the components of the associated force are

$$P = -\frac{\partial V}{\partial x} = \frac{-y}{x^2 + y^2} \qquad \text{and} \qquad Q = -\frac{\partial V}{\partial y} = \frac{x}{x^2 + y^2}. \tag{F.7}$$

Let us calculate the work done by this force in the case that $\mathcal{C}$ is the circle of unit radius centred on the origin, traversed in the anticlockwise sense. We may parametrize a point on this circle by $(x = \cos\theta, y = \sin\theta)$, so that (F.1) becomes

$$W_{\mathcal{C}} = \oint_{\mathcal{C}} -\sin\theta(-\sin\theta\,\mathrm{d}\theta) + \cos\theta(\cos\theta\,\mathrm{d}\theta) = \oint_{\mathcal{C}} \mathrm{d}\theta = 2\pi \tag{F.8}$$

a result which is plainly different from zero. The reason is that although this force is (minus) the gradient of a potential, the latter is not single-valued, in the sense that it does not return to its original value after a circuit round the origin. Indeed, the V of (F.6) is just $-\theta$, which changes by -2π on such a circuit, exactly as calculated in (F.8) allowing for the minus signs in (F.4). Alternatively, we may suspect that the trouble has to do with the 'blow up' of the integrand of (F.7) at the point $x = y = 0$, which is also true.

Much of the foregoing has direct parallels within the theory of functions of a complex variable $z = x + \mathrm{i}y$, to which we now give a brief and informal introduction, limiting ourselves to the minimum required in the text[1]. The crucial property, to which all the results we need are related, is *analyticity*. A function $f(z)$ is *analytic* in a region $\mathcal{R}$ of the complex plane if it has a unique derivative at every point of $\mathcal{R}$. The derivative at a point z is defined by the natural generalization of the real variable definition:

$$\frac{\mathrm{d}f}{\mathrm{d}z} = \lim_{\Delta z \to 0} \left\{ \frac{f(z + \Delta z) - f(z)}{\Delta z} \right\}. \tag{F.9}$$

The crucial new feature in the complex case, however, is that 'Δz' is actually an (infinitesimal) *vector*, in the xy (Argand) plane. Thus we may immediately ask: along which of the infinitely many possible *directions* of Δz are we supposed to approach the point z in (F.9)? The answer is: along any! This is the force of the word 'unique' in the definition of analyticity, and it is a very powerful requirement.

Let $f(z)$ be an analytic function of z in some region $\mathcal{R}$, and let u and v be the real and imaginary parts of f: $f = u + \mathrm{i}v$, where u and v are each functions of x and y. Let us evaluate $\mathrm{d}f/\mathrm{d}z$ at the point $z = x + \mathrm{i}y$ in two different ways, which must be equivalent.

(a) By considering $\Delta z = \Delta x$ (i.e. $\Delta y = 0$). In this case

$$\begin{aligned} \frac{\mathrm{d}f}{\mathrm{d}z} &= \lim_{\Delta x \to 0} \left\{ \frac{u(x + \Delta x, y) - u(x, y) + \mathrm{i}v(x + \Delta x, y) - \mathrm{i}v(x, y)}{\Delta x} \right\} \\ &= \frac{\partial u}{\partial x} + \mathrm{i}\frac{\partial v}{\partial x} \end{aligned} \tag{F.10}$$

[1] For a fuller introduction, see for example Boas (1983, chapter 14).

from the definition of a partial derivative.

(b) By considering $\Delta z = \mathrm{i}\Delta y$ (i.e. $\Delta x = 0$). In this case

$$\frac{\mathrm{d}f}{\mathrm{d}z} = \lim_{\Delta y \to 0} \left\{ \frac{u(x, y + \Delta y) - u(x, y) + \mathrm{i}v(x, y + \Delta y) - \mathrm{i}v(x, y)}{\mathrm{i}\Delta y} \right\}$$
$$= \frac{\partial v}{\partial y} - \mathrm{i}\frac{\partial u}{\partial y}. \tag{F.11}$$

Equating (F.10) and (F.11) we obtain the *Cauchy–Reimann (CR) relations*

$$\frac{\partial u}{\partial x} = \frac{\partial v}{\partial y} \qquad \frac{\partial u}{\partial y} = -\frac{\partial v}{\partial x} \tag{F.12}$$

which are the necessary and sufficient conditions for f to be analytic.

Consider now an integral of the form

$$I = \oint_C f(z)\,\mathrm{d}z \tag{F.13}$$

where again the symbol $\oint$ means that the integration path (or *contour*) in the complex plane in closed. Inserting $f = u + \mathrm{i}v$ and $z = x + \mathrm{i}y$, we may write (F.13) as

$$I = \oint (u\,\mathrm{d}x - v\,\mathrm{d}y) + \mathrm{i}\oint (v\,\mathrm{d}x + u\,\mathrm{d}y). \tag{F.14}$$

Thus the single complex integral (F.13) is equivalent to the two real-plane integrals (F.14); one is the real part of I, the other is the imaginary part, and each is of the form (F.1). In the first, we have $P = u$, $Q = -v$. Hence the condition (F.5) for the integral to vanish is $\partial u/\partial y = -\partial v/\partial x$, which is precisely the second CR relation! Similarly, in the second integral in (F.14) we have $P = v$ and $Q = u$ so that condition (F.5) becomes $\partial v/\partial y = \partial u/\partial x$, which is the first CR relation. It follows that if $f(z)$ is analytic inside and on C, then

$$\oint_C f(z)\,\mathrm{d}z = 0, \tag{F.15}$$

a result known as *Cauchy's theorem*, the foundation of complex integral calculus.

Now let us consider a simple case in which (as in (F.7)) the result of integrating a complex function around a closed curve is *not* zero—namely the integral

$$\oint_C \frac{\mathrm{d}z}{z} \tag{F.16}$$

where C is the circle of radius ρ enclosing the origin. On this circle, $z = \rho \mathrm{e}^{\mathrm{i}\theta}$ where ρ is fixed and $0 \le \theta \le 2\pi$, so

$$\oint_C \frac{\mathrm{d}z}{z} = \oint_C \frac{\rho \mathrm{i}\mathrm{e}^{\mathrm{i}\theta}\mathrm{d}\theta}{\rho \mathrm{e}^{\mathrm{i}\theta}} = \mathrm{i}\oint \mathrm{d}\theta = 2\pi\mathrm{i}. \tag{F.17}$$

Cauchy's theorem does not apply in this case because the function being integrated (z^{-1}) is not analytic at $z = 0$. Writing $\mathrm{d}z/z$ in terms of x and y we have

$$\begin{aligned}\frac{\mathrm{d}z}{z} &= \frac{\mathrm{d}x + \mathrm{i}\,\mathrm{d}y}{x + \mathrm{i}y} = \frac{(x - \mathrm{i}y)}{x^2 + y^2}(\mathrm{d}x + \mathrm{i}\,\mathrm{d}y) \\ &= \left(\frac{x\,\mathrm{d}x + y\,\mathrm{d}y}{x^2 + y^2}\right) + \mathrm{i}\left(\frac{-y\,\mathrm{d}x + x\,\mathrm{d}y}{x^2 + y^2}\right). \end{aligned} \tag{F.18}$$

The reader will recognize the imaginary part of (F.18) as involving precisely the functions (F.7) studied earlier, and may like to find the real potential function appropriate to the real part of (F.18).

We note that the result (F.17) is independent of the circle's radius ρ. This means that we can shrink or expand the circle how we like, without affecting the answer. The reader may like to show that the circle can, in fact, be distorted into a simple closed loop of any shape, enclosing $z = 0$, and the answer will still be $2\pi\mathrm{i}$. In general, a contour may be freely distorted in any region in which the integrand is analytic.

We are now in a position to prove the main integration formula we need, which is *Cauchy's integral formula*: let $f(z)$ be analytic inside and on a simple closed curve $\mathcal{C}$ which encloses the point $z = a$; then

$$\boxed{\oint_{\mathcal{C}} \frac{f(z)}{z - a}\,\mathrm{d}z = 2\pi\mathrm{i}\,f(a)} \tag{F.19}$$

where it is understood that $\mathcal{C}$ is traversed in an anticlockwise sense around $z = a$. The proof follows. The integrand in (F.19) is analytic inside and on $\mathcal{C}$, except at $z = a$; we may therefore distort the contour $\mathcal{C}$ by shrinking it into a very small circle of fixed radius ρ around the point $z = a$. On this circle, z is given by $z = a + \rho\mathrm{e}^{\mathrm{i}\theta}$, and

$$\oint_{\mathcal{C}} \frac{f(z)}{z - a}\,\mathrm{d}z = \int_0^{2\pi} \frac{f(a + \rho\mathrm{e}^{\mathrm{i}\theta})\rho\mathrm{i}\mathrm{e}^{\mathrm{i}\theta}}{\rho\mathrm{e}^{\mathrm{i}\theta}}\,\mathrm{d}\theta = \int_0^{2\pi} f(a + \rho\mathrm{e}^{\mathrm{i}\theta})\mathrm{i}\,\mathrm{d}\theta. \tag{F.20}$$

Now, since f is analytic at $z = a$, it has a unique derivative there, and is consequently continuous at $z = a$. We may then take the limit $\rho \to 0$ in (F.20), obtaining $\lim_{\rho\to 0} f(a + \rho\mathrm{e}^{\mathrm{i}\theta}) = f(a)$, and hence

$$\oint_{\mathcal{C}} \frac{f(z)}{z - a}\,\mathrm{d}z = f(a)\int_0^{2\pi} \mathrm{i}\,\mathrm{d}\theta = 2\pi\mathrm{i}\,f(a) \tag{F.21}$$

as stated.

We now use these results to establish the representation of the θ-function (see (E.47)) quoted in section 6.3.2. Consider the function $F(t)$ of the real variable t defined by

$$F(t) = \frac{\mathrm{i}}{2\pi}\oint_{\mathcal{C}=\mathcal{C}_1+\mathcal{C}_2} \frac{\mathrm{e}^{-\mathrm{i}zt}}{z + \mathrm{i}\epsilon}\,\mathrm{d}z \tag{F.22}$$

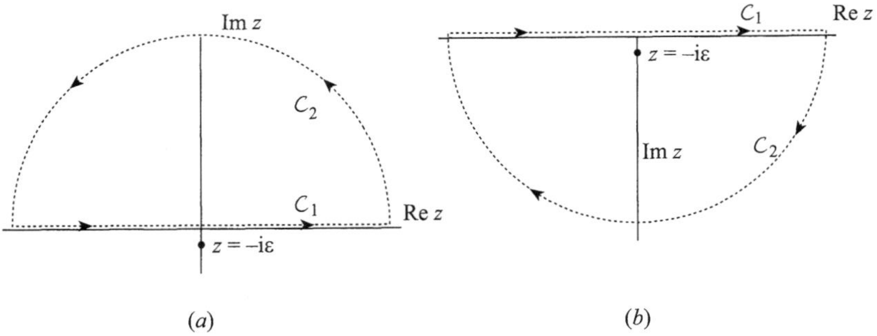

Figure F.1. Contours for $F(t)$: (*a*) $t < 0$; (*b*) $t > 0$.

where ϵ is an infinitesimally small *positive* number (i.e. it will tend to zero through positive values). The closed contour $\mathcal{C}$ is made up of $\mathcal{C}_1$ which is the real axis from $-R$ to R (we shall let $R \to \infty$ at the end), and of $\mathcal{C}_2$ which is a large semicircle of radius R with diameter the real axis, in either the upper or lower half-plane, the choice being determined by the sign of t, as we shall now explain (see figure F.1). Suppose first that $t < 0$, and let z on $\mathcal{C}_2$ be parametrized as $z = Re^{i\theta} = R\cos\theta + iR\sin\theta$. Then

$$e^{-izt} = e^{iz|t|} = e^{-R\sin\theta|t|}e^{iR\cos\theta|t|} \tag{F.23}$$

from which it follows that the contribution to (F.22) from $\mathcal{C}_2$ will vanish exponentially as $R \to \infty$ *provided that* $\theta > 0$, i.e. we choose $\mathcal{C}_2$ to be in the upper half-plane (figure F.1(*a*)). In that case the integrand of (F.22) is analytic inside and on $\mathcal{C}$ (the only non-analytic point is outside $\mathcal{C}$ at $z = -i\epsilon$) and so

$$F(t) = 0 \qquad \text{for } t < 0. \tag{F.24}$$

However, suppose $t > 0$. Then

$$e^{-izt} = e^{R\sin\theta t}e^{-iR\cos\theta t} \tag{F.25}$$

and in this case we must choose the 'contour-closing' $\mathcal{C}_2$ to be in the lower half-plane ($\theta < 0$) or else (F.25) will diverge exponentially as $R \to \infty$. With this choice the $\mathcal{C}_2$ contribution will again go to zero as $R \to \infty$. However, this time the whole closed contour $\mathcal{C}$ does enclose the point $z = -i\epsilon$ (see figure F.1(*b*)), and we may apply Cauchy's integral formula to get, for $t > 0$,

$$F(t) = -2\pi i \frac{i}{2\pi} e^{-\epsilon t}, \tag{F.26}$$

the minus sign at the front arising from the fact (see figure F.1(*b*)) that $\mathcal{C}$ is now being traversed in a *clockwise* sense around $z = -i\epsilon$ (this just inverts the limits in (F.21)). Thus as $\epsilon \to 0$,

$$F(t) \to 1 \qquad \text{for } t > 0. \tag{F.27}$$

Summarizing these manoeuvres, for $t < 0$ we chose $\mathcal{C}_2$ in (F.22) in the upper half-plane (figure F.1(a)), and its contribution vanished as $R \to \infty$. In this case we have, as $R \to \infty$,

$$F(t) \to \frac{\mathrm{i}}{2\pi} \int_{-\infty}^{\infty} \frac{\mathrm{e}^{-\mathrm{i}zt}}{z + \mathrm{i}\epsilon} \, \mathrm{d}z = 0 \qquad \text{for } t < 0. \tag{F.28}$$

For $t > 0$ we chose $\mathcal{C}_2$ in the lower half-plane (figure F.1(b)), when again its contribution vanished as $R \to \infty$. However, in this case F does not vanish, but instead we have, as $R \to \infty$,

$$F(t) \to \frac{\mathrm{i}}{2\pi} \int_{-\infty}^{\infty} \frac{\mathrm{e}^{-\mathrm{i}zt}}{z + \mathrm{i}\epsilon} \, \mathrm{d}z = 1 \qquad \text{for } t > 0. \tag{F.29}$$

Equations (F.28) and (F.29) show that we may indeed write

$$\theta(t) = \lim_{\epsilon \to 0} \frac{\mathrm{i}}{2\pi} \int_{-\infty}^{\infty} \frac{\mathrm{e}^{-\mathrm{i}zt}}{z + \mathrm{i}\epsilon} \, \mathrm{d}z \tag{F.30}$$

as claimed in section 6.3, equation (6.93).

APPENDIX G

GREEN FUNCTIONS

Let us start with a simple but important example. We seek the solution $G_0(\boldsymbol{r})$ of the equation

$$\nabla^2 G_0(\boldsymbol{r}) = \delta(\boldsymbol{r}). \tag{G.1}$$

There is a 'physical' way to look at this equation which will give us the answer straightaway. Recall that Gauss' law in electrostatics (appendix C) is

$$\nabla \cdot \boldsymbol{E} = \rho/\epsilon_0 \tag{G.2}$$

and that $\boldsymbol{E}$ is expressed in terms of the electrostatic potential V as $\boldsymbol{E} = -\nabla V$. Then (G.2) becomes

$$\nabla^2 V = -\rho/\epsilon_0 \tag{G.3}$$

which is known as Poisson's equation. Comparing (G.3) and (G.1), we see that $(-G_0(\boldsymbol{r})/\epsilon_0)$ can be regarded as the 'potential' due to a source ρ which is concentrated entirely at the origin, and whose total 'charge' is unity, since (see appendix E)

$$\int \delta(\boldsymbol{r})\, \mathrm{d}^3\boldsymbol{r} = 1. \tag{G.4}$$

In other words, $(-G_0/\epsilon_0)$ is effectively the potential due to a unit point charge at the origin. But we know exactly what this potential is from Coulomb's law, namely

$$\frac{-G_0(\boldsymbol{r})}{\epsilon_0} = \frac{1}{4\pi\epsilon_0 r} \tag{G.5}$$

whence

$$G_0(\boldsymbol{r}) = -\frac{1}{4\pi r}. \tag{G.6}$$

We may also check this result mathematically as follows. Using (G.6), equation (G.1) is equivalent to

$$\nabla^2 \frac{1}{r} = -4\pi\delta(\boldsymbol{r}). \tag{G.7}$$

Let us consider the integral of both sides of this equation over a spherical volume of arbitrary radius R surrounding the origin. The integral of the left-hand side becomes, using Gauss' divergence theorem,

$$\int_V \left(\nabla^2 \frac{1}{r}\right) \mathrm{d}^3\boldsymbol{r} = \int_V \nabla \cdot \left(\nabla \frac{1}{r}\right) \mathrm{d}^3\boldsymbol{r} = \int_{S \text{ bounding } V} \nabla \left(\frac{1}{r}\right) \cdot \hat{\boldsymbol{n}}\, \mathrm{d}S. \tag{G.8}$$

Now

$$\nabla\left(\frac{1}{r}\right) = -\frac{1}{r^2}\hat{\boldsymbol{r}} = -\frac{1}{R^2}\hat{\boldsymbol{r}}$$

on the surface S, while $\hat{\boldsymbol{n}} = \hat{\boldsymbol{r}}$ and $\mathrm{d}S = R^2\,\mathrm{d}\Omega$ with $\mathrm{d}\Omega$ the element of solid angle on the sphere. So

$$\int_V \nabla^2\left(\frac{1}{r}\right)\,\mathrm{d}^3\boldsymbol{r} = -\int_S \mathrm{d}\Omega = -4\pi \tag{G.9}$$

which using (G.4) is precisely the integral of the right-hand side of (G.7), as required.

Consider now the solutions of

$$(\nabla^2 + \boldsymbol{k}^2)G_k(\boldsymbol{r}) = \delta(\boldsymbol{r}). \tag{G.10}$$

We are interested in rotationally invariant solutions, for which G_k is a function of $r = |\boldsymbol{r}|$ alone. For $r \neq 0$, equation (G.10) is easy to solve. Setting $G_k(r) = f(r)/r$, and using

$$\nabla^2 = \frac{1}{r^2}\frac{\partial}{\partial r}r^2\frac{\partial}{\partial r} + \text{parts depending on } \frac{\partial}{\partial\theta} \text{ and } \frac{\partial}{\partial\phi}$$

we find that $f(r)$ satisfies

$$\frac{\mathrm{d}^2 f}{\mathrm{d}r^2} + \boldsymbol{k}^2 f = 0$$

the general solution to which is ($k = |\boldsymbol{k}|$)

$$f(r) = A\mathrm{e}^{\mathrm{i}kr} + B\mathrm{e}^{-\mathrm{i}kr},$$

leading to

$$G_k(r) = A\frac{\mathrm{e}^{\mathrm{i}kr}}{r} + B\frac{\mathrm{e}^{-\mathrm{i}kr}}{r} \tag{G.11}$$

for $r \neq 0$. In the application to scattering problems (appendix H) we shall want G_k to contain purely outgoing waves, so we will pick the 'A'-type solution in (G.11).

Consider therefore the expression

$$(\nabla^2 + \boldsymbol{k}^2)\left(\frac{A\mathrm{e}^{\mathrm{i}kr}}{r}\right) \tag{G.12}$$

where r is now allowed to take the value zero. Making use of the vector operator result

$$\nabla^2(fg) = (\nabla^2 f)g + 2\nabla f\cdot\nabla g + f(\nabla^2 g)$$

with 'f' = $\mathrm{e}^{\mathrm{i}kr}$ and 'g' $= 1/r$, together with

$$\nabla^2\mathrm{e}^{\mathrm{i}kr} = \frac{2\mathrm{i}k\mathrm{e}^{\mathrm{i}kr}}{r} - \boldsymbol{k}^2\mathrm{e}^{\mathrm{i}kr} \qquad \nabla\mathrm{e}^{\mathrm{i}kr} = \frac{\mathrm{i}k\boldsymbol{r}\mathrm{e}^{\mathrm{i}kr}}{r} \qquad \nabla\frac{1}{r} = -\frac{\boldsymbol{r}}{r^3}$$

we find

$$\begin{aligned}(\nabla^2 + \boldsymbol{k}^2)\left(\frac{A\mathrm{e}^{\mathrm{i}kr}}{r}\right) &= A\mathrm{e}^{\mathrm{i}kr}\nabla^2\left(\frac{1}{r}\right) \\ &= -4\pi A\mathrm{e}^{\mathrm{i}kr}\delta(\boldsymbol{r}) \\ &= -4\pi A\delta(\boldsymbol{r}) \end{aligned} \qquad \text{(G.13)}$$

where we have replaced r by zero in the exponent of the last term of the last line in (G.13), since the δ-function ensures that only this point need be considered for this term. By choosing the constant $A = -1/4\pi$, we find that the (outgoing wave) solution of (G.10) is

$$G_k^{(+)}(r) = -\frac{\mathrm{e}^{\mathrm{i}kr}}{4\pi r}. \qquad \text{(G.14)}$$

We are also interested in spherically symmetric solutions of (restoring c and $\hbar$ explicitly for the moment)

$$\left(\nabla^2 - \frac{m^2c^2}{\hbar^2}\right)\phi(\boldsymbol{r}) = \delta(\boldsymbol{r}) \qquad \text{(G.15)}$$

which is the equation analogous to (G.1) for a static classical scalar potential of a field whose quanta have mass m. The solutions to (G.15) are easily found from the previous work by letting $k \to \mathrm{i}mc/\hbar$. Retaining now the solution which goes to zero as $r \to \infty$, we find

$$\phi(\boldsymbol{r}) = -\frac{1}{4\pi}\frac{\mathrm{e}^{-r/a}}{r} \qquad \text{(G.16)}$$

where $a = \hbar/mc$, the Compton wavelength of the quantum, with mass m. The potential (G.16) is (up to numerical constants) the famous Yukawa potential, in which the quantity 'a' is called the *range*: as r gets greater than a, $\phi(\boldsymbol{r})$ becomes exponentially small. Thus, just as the Coulomb potential is the solution of Poisson's equation (G.3) corresponding to a point source at the origin, so the Yukawa potential is the solution of the analogous equation (G.15), also with a point source at the origin. Note that as $a \to \infty$, $\phi(\boldsymbol{r}) \to G_0(\boldsymbol{r})$.

Functions such as G_k, G_0 and ϕ, which generically satisfy equations of the form

$$\Omega_r G(\boldsymbol{r}) = \delta(\boldsymbol{r}) \qquad \text{(G.17)}$$

where Ω_r is some linear differential operator, are said to be *Green functions* of the operator Ω_r. From the examples already treated, it is clear that $G(\boldsymbol{r})$ in (G.17) has the general interpretation of a 'potential' due to a point source at the origin, when Ω_r is the appropriate operator for the field theory in question.

Green functions play an important role in the solution of differential equations of the type

$$\Omega_r \psi(\boldsymbol{r}) = s(\boldsymbol{r}) \tag{G.18}$$

where $s(\boldsymbol{r})$ is a known 'source function' (e.g. the charge density in (G.3)). The solution of (G.18) may be written as

$$\psi(\boldsymbol{r}) = u(\boldsymbol{r}) + \int G(\boldsymbol{r}-\boldsymbol{r}')s(\boldsymbol{r}')\,\mathrm{d}^3\boldsymbol{r}' \tag{G.19}$$

where $u(\boldsymbol{r})$ is a solution of $\Omega_r u(\boldsymbol{r}) = 0$. Thus once we know G, we have the solution via (G.19).

Equation (G.19) has a simple physical interpretation. We know that $G(\boldsymbol{r})$ is the solution of (G.18) with $s(\boldsymbol{r})$ replaced by $\delta(\boldsymbol{r})$. But by writing

$$s(\boldsymbol{r}) = \int \delta(\boldsymbol{r}-\boldsymbol{r}')s(\boldsymbol{r}')\,\mathrm{d}^3\boldsymbol{r}' \tag{G.20}$$

we can formally regard $s(\boldsymbol{r})$ as being made up of a *superposition* of point sources, distributed at points $\boldsymbol{r}'$ with a weighting function $s(\boldsymbol{r}')$. Then, since the operator Ω_r is (by assumption) linear, the solution for such a superposition of point sources must be just the same superposition of the point source solutions, namely the integral on the right-hand side of (G.19). This integral term is, in fact, the 'particular integral' of the differential equation (G.18), while the $u(\boldsymbol{r})$ is the 'complementary function'.

Equation (G.19) can also be checked analytically. First note that it is generally the case that the operator Ω_r is translationally invariant, so that

$$\Omega_r = \Omega_{r-r'}; \tag{G.21}$$

the right-hand side of (G.21) amounts to shifting the origin to the point $\boldsymbol{r}'$. Applying Ω_r to both sides of (G.19), we find

$$\begin{aligned}\Omega_r \psi(\boldsymbol{r}) &= \Omega_r u(\boldsymbol{r}) + \int \Omega_r G(\boldsymbol{r}-\boldsymbol{r}')s(\boldsymbol{r}')\,\mathrm{d}^3\boldsymbol{r}' \\ &= 0 + \int \Omega_{r-r'} G(\boldsymbol{r}-\boldsymbol{r}')s(\boldsymbol{r}')\,\mathrm{d}^3\boldsymbol{r}' = \int \delta(\boldsymbol{r}-\boldsymbol{r}')s(\boldsymbol{r}')\,\mathrm{d}^3\boldsymbol{r}' \\ &= s(\boldsymbol{r})\end{aligned}$$

as required in (G.18).

Finally, consider the Fourier transform of equation (G.10), defined as

$$\int \mathrm{e}^{-\mathrm{i}\boldsymbol{q}\cdot\boldsymbol{r}}(\nabla^2 + k^2)G_k(\boldsymbol{r})\,\mathrm{d}^3\boldsymbol{r} = \int \mathrm{e}^{-\mathrm{i}\boldsymbol{q}\cdot\boldsymbol{r}}\delta(\boldsymbol{r})\,\mathrm{d}^3\boldsymbol{r}.$$

The right-hand side is unity, by equation (G.4). On the left-hand side we may use the result

$$\int u(r)\nabla^2 v(\boldsymbol{r})\,\mathrm{d}^3\boldsymbol{r} = \int (\nabla^2 u(\boldsymbol{r}))v(\boldsymbol{r})\,\mathrm{d}^3\boldsymbol{r}$$

(proved by integrating by parts, assuming u and v go to zero sufficiently fast at the boundaries of the integral) to obtain

$$\begin{aligned}\int \mathrm{e}^{-\mathrm{i}\boldsymbol{q}\cdot\boldsymbol{r}}(\nabla^2 + \boldsymbol{k}^2)G_k(\boldsymbol{r})\,\mathrm{d}^3\boldsymbol{r} &= \int \{(\nabla^2 \mathrm{e}^{-\mathrm{i}\boldsymbol{q}\cdot\boldsymbol{r}}) + \boldsymbol{k}^2\mathrm{e}^{-\mathrm{i}\boldsymbol{q}\cdot\boldsymbol{r}}\}G_k(\boldsymbol{r})\,\mathrm{d}^3\boldsymbol{r} \\ &= \int (-\boldsymbol{q}^2 + \boldsymbol{k}^2)\mathrm{e}^{-\mathrm{i}\boldsymbol{q}\cdot\boldsymbol{r}}G_k(\boldsymbol{r})\,\mathrm{d}^3\boldsymbol{r} \\ &= (-\boldsymbol{q}^2 + \boldsymbol{k}^2)\tilde{G}_k(\boldsymbol{q})\end{aligned}$$

where $\tilde{G}_k(\boldsymbol{q})$ is the Fourier transform of $G_k(\boldsymbol{r})$. Since this expression has to equal unity, we have

$$\tilde{G}_k(\boldsymbol{q}) = \frac{1}{\boldsymbol{k}^2 - \boldsymbol{q}^2}. \tag{G.22}$$

There is, however, a problem with (G.22) as it stands, which is that it is undefined when the variable $\boldsymbol{q}^2$ takes the value equal to the parameter $\boldsymbol{k}^2$ in the original equation. Indeed, various definitions are possible, corresponding to the type of solution in $\boldsymbol{r}$-space for $G_k(\boldsymbol{r})$ (i.e. ingoing, outgoing or standing wave). It turns out (see the exercise at the end of this appendix) that the specification which is equivalent to the solution $G_k^{(+)}(\boldsymbol{r})$ in (G.14) is to add an infinitesimally small imaginary part in the denominator of (G.22):

$$\tilde{G}_k^{(+)}(\boldsymbol{q}) = \frac{1}{\boldsymbol{k}^2 - \boldsymbol{q}^2 + \mathrm{i}\epsilon}. \tag{G.23}$$

In exactly the same way, the Fourier transform of $\phi(\boldsymbol{r})$ satisfying (G.15) is

$$\tilde{\phi}(\boldsymbol{q}) = \frac{-1}{\boldsymbol{q}^2 + (m^2c^2/\hbar^2)}. \tag{G.24}$$

The relativistic generalization of this result is straightforward. Consider the equation

$$(\Box + m^2)G(x) = -\delta(x) \tag{G.25}$$

where x is the coordinate 4-vector and $\delta(x)$ is the four-dimensional δ-function, $\delta(x^0)\delta(\boldsymbol{x})$; the sign in (G.25) has been chosen to be consistent with (G.15) in the static case (note that in (G.25) we have reverted to $\hbar = c = 1$). Taking the four-dimensional Fourier transform, and making suitable assumptions about the vanishing of G at the boundary of space–time, we obtain

$$(-q^2 + m^2)\tilde{G}(q) = -1 \tag{G.26}$$

where

$$\tilde{G}(q) = \int \mathrm{e}^{\mathrm{i}q\cdot x}G(x)\,\mathrm{d}^4x$$

and so

$$\tilde{G}(q) = \frac{1}{q^2 - m^2}. \tag{G.27}$$

As we shall see in detail in chapter 6, the Feynman prescription for selecting the physically desired solution amounts to adding an 'iϵ' term in the denominator of (G.27):

$$\tilde{G}^{(+)}(q) = \frac{1}{q^2 - m^2 + \mathrm{i}\epsilon}. \tag{G.28}$$

Exercise

Verify the 'iϵ' specification in (G.23), using the methods of appendix F. [*Hints*: You need to show that the Fourier transform of (G.23), defined by

$$\hat{G}_k^{(+)}(\boldsymbol{r}) = \frac{1}{(2\pi)^3} \int \mathrm{e}^{\mathrm{i}\boldsymbol{q}\cdot\boldsymbol{r}} \tilde{G}_k^{(+)}(\boldsymbol{q})\, \mathrm{d}^3\boldsymbol{q}, \tag{G.29}$$

is equal to $G_k^{(+)}(\boldsymbol{r})$ of (G.14). Do the integration over the polar angles of $\boldsymbol{q}$, taking the direction of $\boldsymbol{r}$ as the polar axis. This gives

$$\hat{G}_k^{(+)}(\boldsymbol{r}) = \frac{-1}{8\pi^2} \int_{-\infty}^{\infty} \left(\frac{\mathrm{e}^{\mathrm{i}qr} - \mathrm{e}^{-\mathrm{i}qr}}{\mathrm{i}r} \right) \frac{q\, \mathrm{d}q}{q^2 - k^2 - \mathrm{i}\epsilon} \tag{G.30}$$

where $q = |\boldsymbol{q}|$, $r = |\boldsymbol{r}|$, and we have used the fact that the integrand is an even function of q to extend the lower limit to $-\infty$, with an overall factor of 1/2. Now convert q to the complex variable z. Locate the poles of $(z^2 - k^2 - \mathrm{i}\epsilon)^{-1}$ (compare the similar calculation in section 10.3.1, and in appendix F). Apply Cauchy's integral formula (F.17), closing the $\mathrm{e}^{\mathrm{i}zr}$ part in the upper half z-plane, and the $\mathrm{e}^{-\mathrm{i}zr}$ part in the lower half z-plane.]

APPENDIX H

ELEMENTS OF NON-RELATIVISTIC SCATTERING THEORY

H.1 Time-independent formulation and differential cross section

We consider the scattering of a particle of mass m by a fixed spherically symmetric potential $V(\boldsymbol{r})$; we shall retain $\hbar$ explicitly in what follows. The potential is assumed to go to zero rapidly as $r \to \infty$, as for the Yukawa potential (G.16); it will turn out that the important Coulomb case can be treated as the $a \to \infty$ limit of (G.16). We shall treat the problem here as a *stationary state* one, in which the Schrödinger wavefunction $\psi(\boldsymbol{r}, t)$ has the form

$$\psi(\boldsymbol{r}, t) = \phi(\boldsymbol{r})\mathrm{e}^{-\mathrm{i}Et\hbar} \tag{H.1}$$

where E is the particle's energy, and where $\phi(\boldsymbol{r})$ satisfies the equation

$$\left[\frac{-\hbar^2}{2m}\nabla^2 + V(\boldsymbol{r})\right]\phi(\boldsymbol{r}) = E\phi(\boldsymbol{r}). \tag{H.2}$$

We shall take V to be spherically symmetric, so that $V(\boldsymbol{r}) = V(r)$ where $r = |\boldsymbol{r}|$. In this approach to scattering, we suppose the potential to be 'bathed' in a steady flux of incident particles, all of energy E. The wavefunction for the incident beam, far from the region near the origin where V is appreciably non-zero, is then just a plane wave of the form $\phi_{\text{inc}} = \mathrm{e}^{ikz}$, where the z-axis has been chosen along the propagation direction, and where $E = \hbar^2\boldsymbol{k}^2/2m$ with $\boldsymbol{k} = (0, 0, k)$. This plane wave is normalized to one particle per unit volume, and yields a steady-state flux of

$$\begin{aligned} j_{\text{inc}} &= \frac{\hbar}{2m\mathrm{i}}[\phi_{\text{inc}}^* \nabla\phi_{\text{inc}} - \phi_{\text{inc}}\nabla\phi_{\text{inc}}^*] \\ &= \hbar\boldsymbol{k}/m = \boldsymbol{p}/m \end{aligned} \tag{H.3}$$

where the momentum is $\boldsymbol{p} = \hbar\boldsymbol{k}$. As expected, the incident flux is given by the velocity $\boldsymbol{v}$ per unit volume.

Though we have represented the incident beam as a plane wave, it will, in practice, be collimated. We could, of course, superpose such plane waves, with different $\boldsymbol{k}$'s, to make a wave-packet of any desired localization. But the dimensions of practical beams are so much greater than the de Broglie wavelength $\lambda = h/p$ of our particles, that our plane wave will be a very good approximation to a realistic packet.

The form of the complete solution to (H.2), even in the region where V is essentially zero, is not simply the incident plane wave, however. The presence of the potential gives rise also to a *scattered wave*, whose form as $r \to \infty$ is

$$\phi_{\rm sc} = f(\theta, \phi)\frac{e^{ikr}}{r}. \tag{H.4}$$

We shall actually derive this later, but its physical interpretation is simply that it is an outgoing ($\sim e^{ikr}$ rather than e^{-ikr}) 'spherical wave', with a factor $f(\theta, \phi)$ called the *scattering amplitude* that allows for the fact that even though $V(r)$ is spherically symmetric, the solution, in general, will not be (recall the bound-state solutions of the Coulomb potential in the hydrogen atom). Calculating the radial component of the flux corresponding to (H.4) yields

$$\begin{aligned} j_{r,\rm sc} &= \frac{\hbar}{2mi}\left[\phi_{\rm sc}^* \frac{\partial}{\partial r}\phi_{\rm sc} - \phi_{\rm sc}\frac{\partial}{\partial r}\phi_{\rm sc}^*\right] \\ &= \frac{\hbar k}{m}|f(\theta, \phi)|^2/r^2. \end{aligned} \tag{H.5}$$

The flux in the two non-radial directions will contain an extra power of r in the denominator—recall that

$$\nabla = \hat{r}\frac{\partial}{\partial r} + \hat{\theta}\frac{1}{r}\frac{\partial}{\partial \theta} + \hat{\phi}\frac{1}{r\sin\theta}\frac{\partial}{\partial \phi}$$

and so (H.5) represents the correct asymptotic form of the scattered flux.

The *cross section* is now easily found. The differential cross section, $d\sigma$, for scattering into the element of solid angle $d\Omega$ is defined by

$$d\sigma = j_{r,\rm sc}\, dS/|\boldsymbol{j}_{\rm inc}| \tag{H.6}$$

where $dS = r^2\, d\Omega$, so that from (H.3) and (H.5)

$$\frac{d\sigma}{d\Omega} = |f(\theta, \phi)|^2. \tag{H.7}$$

The total cross section is then just

$$\sigma = \int |f(\theta, \phi)|^2\, d\Omega. \tag{H.8}$$

It is important to realize that the complete asymptotic form of the solution to (H.2) is the superposition of $\phi_{\rm inc}$ and $\phi_{\rm sc}$:

$$\phi(\boldsymbol{r}) \stackrel{r\to\infty}{\to} e^{ikz} + f(\theta, \phi)\frac{e^{ikr}}{r}. \tag{H.9}$$

Note that in the 'forward direction' (i.e. within a region close to the z-axis, as determined by the collimation), the incident and scattered waves will interfere. Careful analysis reveals a depletion of the incident beam in the forward direction (the 'shadow' of the scattering centre), which corresponds exactly to the total flux scattered into all angles (Gottfried 1966, section 12.3). This is expressed in the *optical theorem*:

$$\operatorname{Im} f(0) = \frac{k}{4\pi}\sigma. \tag{H.10}$$

H.2 Expression for the scattering amplitude: Born approximation

We begin by rewriting (H.2) as

$$(\nabla^2 + \boldsymbol{k}^2)\phi(\boldsymbol{r}) = \frac{2m}{\hbar^2}V(\boldsymbol{r})\phi(\boldsymbol{r}). \tag{H.11}$$

This equation is of exactly the form discussed in appendix G, e.g. equation (G.18) with $\Omega_r = \nabla^2 + \boldsymbol{k}^2$. Further, we know that the Green function for this Ω_r, corresponding to the desired outgoing wave solution, is given by (G.14). Using then (G.19) and (G.14), we can immediately write the 'formal solution' of (H.11) as

$$\phi(\boldsymbol{r}) = \mathrm{e}^{\mathrm{i}\boldsymbol{k}\cdot\boldsymbol{r}} + \frac{2m}{\hbar^2}\int -\frac{1}{4\pi}\frac{\mathrm{e}^{\mathrm{i}k|\boldsymbol{r}-\boldsymbol{r}'|}}{|\boldsymbol{r}-\boldsymbol{r}'|}V(\boldsymbol{r}')\phi(\boldsymbol{r}')\,\mathrm{d}^3\boldsymbol{r}' \tag{H.12}$$

where we have chosen '$u(\boldsymbol{r})$' in (G.19) to be the incident plane wave ϕ_{inc}, and have used $\boldsymbol{k}\cdot\boldsymbol{r} = kz$. We say 'formal' because of course the unknown $\phi(\boldsymbol{r}')$ still appears on the right-hand side of (H.12).

It may therefore seem that we have made no progress—but in fact (H.12) leads to a very useful expression for $f(\theta, \phi)$, which is the quantity we need to calculate. This can be found by considering the asymptotic ($r \to \infty$) limit of the integral term in (H.12). We have

$$\begin{aligned} |\boldsymbol{r}-\boldsymbol{r}'| &= (\boldsymbol{r}^2 + \boldsymbol{r}'^2 - 2\boldsymbol{r}\cdot\boldsymbol{r}')^{1/2} \\ &\sim r - \boldsymbol{r}\cdot\boldsymbol{r}'/r + 0\left(\frac{1}{r}\right) \text{ terms.} \end{aligned} \tag{H.13}$$

Thus in the exponent we may write

$$\mathrm{e}^{\mathrm{i}k|\boldsymbol{r}-\boldsymbol{r}'|} \approx \mathrm{e}^{\mathrm{i}k(r-\boldsymbol{r}\cdot\boldsymbol{r}'/r)} = \mathrm{e}^{\mathrm{i}kr}\mathrm{e}^{-\mathrm{i}\boldsymbol{k}'\cdot\boldsymbol{r}'}$$

where $\boldsymbol{k}' = k\hat{\boldsymbol{r}}$ is the *outgoing wavevector*, pointing along the direction of the outgoing scattered wave which enters dS. In the denominator factor we may simply say $|\boldsymbol{r}-\boldsymbol{r}'|^{-1} \approx r^{-1}$ since the next term in (H.13) will produce a correction of order r^{-2}. Putting this together, we have

$$\phi(\boldsymbol{r}) \overset{r\to\infty}{\rightarrow} \mathrm{e}^{\mathrm{i}kz} - \frac{m}{2\pi\hbar^2}\frac{\mathrm{e}^{\mathrm{i}kr}}{r}\int \mathrm{e}^{-\mathrm{i}\boldsymbol{k}'\cdot\boldsymbol{r}'}V(\boldsymbol{r}')\phi(\boldsymbol{r}')\,\mathrm{d}^3\boldsymbol{r}' \tag{H.14}$$

from which follows the formula for $f(\theta, \phi)$:

$$f(\theta, \phi) = -\frac{m}{2\pi\hbar^2}\int e^{-i\boldsymbol{k}'\cdot\boldsymbol{r}'} V(\boldsymbol{r}')\phi(\boldsymbol{r}')\, d^3\boldsymbol{r}'. \tag{H.15}$$

No approximations have been made thus far, in deriving (H.15)—but of course it still involves the unknown $\phi(\boldsymbol{r}')$ inside the integral. However, it is in a form which is very convenient for setting up a *systematic approximation scheme*—a kind of perturbation theory—in powers of V. If the potential is relatively 'weak', its effect will be such as to produce only a slight distortion of the incident wave, and so $\phi(\boldsymbol{r}) \approx e^{i\boldsymbol{k}\cdot\boldsymbol{r}}$+'small correction'. This suggests that it may be a good approximation to replace $\phi(\boldsymbol{r}')$ in (H.15) by the undistorted incident wave $e^{i\boldsymbol{k}\cdot\boldsymbol{r}'}$, giving the *approximate* scattering amplitude

$$f_{BA}(\theta, \phi) = -\frac{m}{2\pi\hbar^2}\int e^{i\boldsymbol{q}\cdot\boldsymbol{r}'} V(\boldsymbol{r}')\, d^3\boldsymbol{r}' \tag{H.16}$$

where the wave vector transfer $\boldsymbol{q}$ is given by

$$\boldsymbol{q} = \boldsymbol{k} - \boldsymbol{k}'. \tag{H.17}$$

This is called the 'Born approximation to the scattering amplitude'. The criteria for the validity of the Born approximation are discussed in many standard quantum mechanics texts.

The approximation can be improved by returning to (H.12) for $\phi(\boldsymbol{r})$, and replacing $\phi(\boldsymbol{r}')$ inside the integral by $e^{i\boldsymbol{k}\cdot\boldsymbol{r}'}$ just as we did in (H.16); this will give us a formula for the first-order (in V) correction to $\phi(\boldsymbol{r})$. We can now insert *this* expression for $\phi(\boldsymbol{r}')$ (i.e. $\phi(\boldsymbol{r}') = e^{i\boldsymbol{k}\cdot\boldsymbol{r}'} + O(V)$ correction) into (H.15), which will give us f_{BA} again as the first term, but also another term, of order V^2 (since V appears in the integral in (H.15)). By iterating the process indefinitely, the *Born series* can be set up, to all orders in V.

H.3 Time-dependent approach

In this approach we consider the potential $V(\boldsymbol{r})$ as causing transitions between states describing the incident and scattered particles. From standard time-dependent perturbation theory in quantum mechanics, the transition probability per unit time for going from state $|i\rangle$ to state $|f\rangle$, to first order in V, is given by

$$\dot{P}_{fi} = \frac{2\pi}{\hbar}|\langle f|V|i\rangle|^2 \rho(E_f)|_{E_f=E_i} \tag{H.18}$$

where $\rho(E_f)dE_f$ is the number of final states in the energy range dE_f around the energy-conserving point $E_i = E_f$. Equation (H.18) is often known as the 'Golden Rule'. In the present case, if we adopt the same normalization as in the previous section, the initial and final states are represented by the wavefunction $e^{i\boldsymbol{k}\cdot\boldsymbol{r}}$ and $e^{-i\boldsymbol{k}'\cdot\boldsymbol{r}}$, so that

$$\langle \mathrm{f}|V|\mathrm{i}\rangle = \int e^{i\boldsymbol{q}\cdot\boldsymbol{r}} V(\boldsymbol{r})\,\mathrm{d}^3\boldsymbol{r} \equiv \tilde{V}(\boldsymbol{q}). \tag{H.19}$$

Also, the number of such states in a volume element $\mathrm{d}^3\boldsymbol{p}'$ of momentum space $(\boldsymbol{p}' = \hbar\boldsymbol{k}')$ is $\mathrm{d}^3\boldsymbol{p}'/(2\pi\hbar)^3$.

In spherical polar coordinates, with $\mathrm{d}\Omega$ standing for the element of solid angle around the direction $(\theta, /\phi)$ of $\boldsymbol{p}'$, we have

$$\mathrm{d}^3\boldsymbol{p}' = \boldsymbol{p}'^2\,\mathrm{d}|\boldsymbol{p}'|\,\mathrm{d}\Omega = m|\boldsymbol{p}'|\,\mathrm{d}E'\,\mathrm{d}\Omega \tag{H.20}$$

where we have used $E' = \boldsymbol{p}'^2/2m$. It follows that

$$\rho(E')\,\mathrm{d}E' = \frac{\mathrm{d}^3\boldsymbol{p}'}{(2\pi\hbar)^3} = \frac{m}{(2\pi\hbar)^3}|\boldsymbol{p}'|\,\mathrm{d}\Omega\,\mathrm{d}E' \tag{H.21}$$

and so

$$\rho(E') = \frac{m}{(2\pi\hbar)^3}|\boldsymbol{p}'|\,\mathrm{d}\Omega. \tag{H.22}$$

Inserting (H.19) and (H.22) into (H.18) we obtain, for this case,

$$\dot{P}_{\mathrm{fi}} = \frac{2\pi}{\hbar}|\tilde{V}(\boldsymbol{q})|^2\frac{m}{(2\pi\hbar)^3}|\boldsymbol{p}|\,\mathrm{d}\Omega. \tag{H.23}$$

To get the cross section, we need to divide this expression by the incident flux, which is $|\boldsymbol{p}|/m$ as in (H.3). Thus the differential cross section for scattering into the element of solid angle $d\Omega$ in the direction (θ, ϕ) is

$$\mathrm{d}\sigma = \left(\frac{m}{2\pi\hbar^2}\right)^2 |\tilde{V}(\boldsymbol{q})|^2\,\mathrm{d}\Omega. \tag{H.24}$$

Comparing (H.24) with (H.7) and (H.16), we see that this application of the Golden Rule (first-order time-dependent perturbation theory) is exactly equivalent to the Born approximation in the time-independent approach. It is, however, the time-dependent approach which is much closer to the corresponding quantum field theory formulation we introduce in chapter 6.

APPENDIX I

THE SCHRÖDINGER AND HEISENBERG PICTURES

The standard introductory formalism of quantum mechanics is that of Schrödinger, in which the dynamical variables (such as $\boldsymbol{x}$ and $\hat{\boldsymbol{p}} = -\mathrm{i}\nabla$) are independent of time, while the wavefunction ψ changes with time according to the general equation

$$\hat{H}\psi(\boldsymbol{x},t) = \mathrm{i}\frac{\partial\psi(\boldsymbol{x},t)}{\partial t} \tag{I.1}$$

where $\hat{H}$ is the Hamiltonian. Matrix elements of operators $\hat{A}$ depending on $\boldsymbol{x}, \hat{\boldsymbol{p}}\ldots$ then have the form

$$\langle\phi|\hat{A}|\psi\rangle = \int \phi^*(\boldsymbol{x},t)\hat{A}\psi(\boldsymbol{x},t)\,\mathrm{d}^3\boldsymbol{x} \tag{I.2}$$

and will, in general, depend on time via the time dependences of ϕ and ψ. Although used almost universally in introductory courses on quantum mechanics, this formulation is not the only possible one, nor is it always the most convenient.

We may, for example, wish to bring out similarities (and differences) between the general dynamical frameworks of quantum and classical mechanics. The formulation here does not seem to be well adapted to this purpose, since in the classical case the dynamical variables depend on time ($\boldsymbol{x}(t)$, $\boldsymbol{p}(t)\ldots$) and obey equations of motion, while the quantum variables $\hat{A}$ are time-independent and the 'equation of motion' (I.1) is for the wavefunction ψ, which has no classical counterpart. In quantum mechanics, however, it is always possible to make unitary transformations of the state vector or wavefunctions. We can make use of this possibility to obtain an alternative formulation of quantum mechanics, which is in some ways closer to the spirit of classical mechanics, as follows.

Equation (I.1) can be formally solved to give

$$\psi(\boldsymbol{x},t) = \mathrm{e}^{-\mathrm{i}\hat{H}t}\psi(\boldsymbol{x},0) \tag{I.3}$$

where the exponential (of an operator!) can be defined by the corresponding power series, for example:

$$\mathrm{e}^{-\mathrm{i}\hat{H}t} = 1 - \mathrm{i}\hat{H}t + \frac{1}{2!}(-\mathrm{i}\hat{H}t)^2 + \cdots. \tag{I.4}$$

It is simple to check that (I.3) as defined by (I.4), does satisfy (I.1) and that the operator $\hat{U} = \exp(-\mathrm{i}\hat{H}t)$ is unitary:

$$U^{\dagger} = [\exp(-\mathrm{i}\hat{H}t)]^{\dagger} = \exp(\mathrm{i}\hat{H}^{\dagger}t) = \exp(\mathrm{i}\hat{H}t) = U^{-1} \tag{I.5}$$

where the Hermitian property $\hat{H}^{\dagger} = \hat{H}$ has been used. Thus (I.3) can be viewed as a unitary transformation from the time-dependent wavefunction $\psi(\boldsymbol{x}, t)$ to the time-independent one $\psi(\boldsymbol{x}, 0)$. Correspondingly the matrix element (I.2) is then

$$\langle\phi|\hat{A}|\psi\rangle = \int \phi^{*}(\boldsymbol{x}, 0)\mathrm{e}^{\mathrm{i}\hat{H}t}\hat{A}\mathrm{e}^{-\mathrm{i}\hat{H}t}\psi(\boldsymbol{x}, 0)\,\mathrm{d}^3\boldsymbol{x} \tag{I.6}$$

which can be regarded as the matrix element of the *time-dependent* operator

$$\hat{A}(t) = \mathrm{e}^{\mathrm{i}\hat{H}t}\hat{A}\mathrm{e}^{-\mathrm{i}\hat{H}t} \tag{I.7}$$

between time-independent wavefunctions $\phi^{*}(\boldsymbol{x}, 0)$, $\psi(\boldsymbol{x}, 0)$.

Since (I.6) is perfectly general, it is clear that we can calculate amplitudes in quantum mechanics in either of the two ways outlined: (a) by using time-dependent ψ's and time-independent $\hat{A}$'s, which is called the 'Schrödinger picture': or (b) by using time-independent ψ's and time-dependent $\hat{A}$'s, which is called the 'Heisenberg picture'. The wavefunctions and operators in the two pictures are related by (I.3) and (I.7). We note that the pictures coincide at the (conventionally chosen) time $t = 0$.

Since $\hat{A}(t)$ is now time-dependent, we can ask for its equation of motion. Differentiating (I.7) carefully, we find (if $\hat{A}$ does not depend explicitly on t) that

$$\frac{\mathrm{d}\hat{A}(t)}{\mathrm{d}t} = -\mathrm{i}[\hat{A}(t), \hat{H}] \tag{I.8}$$

which is called the Heisenberg equation of motion for $\hat{A}(t)$. On the right-hand side of (I.8), $\hat{H}$ is the Schrödinger operator; however, if $\hat{H}$ is substituted for $\hat{A}$ in (I.7), one finds $\hat{H}(t) = \hat{H}$, so $\hat{H}$ can equally well be interpreted as the Heisenberg operator. For simple Hamiltonians $\hat{H}$, (I.8) leads to operator equations quite analogous to classical equations of motion, which can sometimes be solved explicitly (see section 5.2.2 of chapter 5).

The foregoing ideas apply equally well to the operators and state vectors of quantum field theory.

APPENDIX J

DIRAC ALGEBRA AND TRACE IDENTITIES

J.1 Dirac algebra

J.1.1 γ matrices

The fundamental anti-commutator

$$\{\gamma^\mu, \gamma^\nu\} = 2g^{\mu\nu} \tag{J.1}$$

may be used to prove the following results.

$$\gamma_\mu \gamma^\nu = 4 \tag{J.2}$$
$$\gamma_\mu \not{a} \gamma^\mu = -2\not{a} \tag{J.3}$$
$$\gamma_\mu \not{a} \not{b} \gamma^\mu = 4a \cdot b \tag{J.4}$$
$$\gamma_\mu \not{a} \not{b} \not{c} \gamma^\mu = -2\not{c} \not{b} \not{a} \tag{J.5}$$
$$\not{a} \not{b} = -\not{b} \not{a} + 2a \cdot b. \tag{J.6}$$

As an example, we prove this last result:

$$\begin{aligned} \not{a} \not{b} &= a_\mu b_\nu \gamma^\mu \gamma^\nu \\ &= a_\mu b_\nu (-\gamma^\nu \gamma^\mu + 2g^{\mu\nu}) \\ &= -\not{b} \not{a} + 2a \cdot b. \end{aligned}$$

J.1.2 γ_5 identities

Define

$$\gamma_5 = \mathrm{i}\gamma^0 \gamma^1 \gamma^2 \gamma^3. \tag{J.7}$$

In the usual representation with

$$\gamma^0 = \begin{pmatrix} \mathbf{1} & \mathbf{0} \\ \mathbf{0} & -\mathbf{1} \end{pmatrix} \quad \text{and} \quad \boldsymbol{\gamma} = \begin{pmatrix} \mathbf{0} & \boldsymbol{\sigma} \\ -\boldsymbol{\sigma} & \mathbf{0} \end{pmatrix} \tag{J.8}$$

γ_5 is the matrix

$$\gamma_5 = \begin{pmatrix} \mathbf{0} & \mathbf{1} \\ \mathbf{1} & \mathbf{0} \end{pmatrix}. \tag{J.9}$$

Either from the definition or using this explicit form it is easy to prove that

$$\gamma_5^2 = 1 \tag{J.10}$$

and

$$\{\gamma_5, \gamma^\mu\} = 0 \tag{J.11}$$

i.e. γ_5 anti-commutes with the other γ-matrices. Defining the totally antisymmetric tensor

$$\epsilon_{\mu\nu\rho\sigma} = \begin{cases} +1 & \text{for an even permutation of } 0, 1, 2, 3 \\ -1 & \text{for an odd permutation of } 0, 1, 2, 3 \\ 0 & \text{if two or more indices are the same} \end{cases} \tag{J.12}$$

we may write

$$\gamma_5 = \frac{\mathrm{i}}{4!}\epsilon_{\mu\nu\rho\sigma}\gamma^\mu\gamma^\nu\gamma^\rho\gamma^\sigma. \tag{J.13}$$

With this form it is possible to prove

$$\gamma_5\gamma_\sigma = \frac{\mathrm{i}}{3!}\epsilon_{\mu\nu\rho\sigma}\gamma^\mu\gamma^\nu\gamma^\rho \tag{J.14}$$

and the identity

$$\gamma^\mu\gamma^\nu\gamma^\rho = g^{\mu\nu}\gamma^\rho - g^{\mu\rho}\gamma^\nu + g^{\nu\rho}\gamma^\mu + \mathrm{i}\gamma_5\epsilon^{\mu\nu\rho\sigma}\gamma_\sigma. \tag{J.15}$$

J.1.3 Hermitian conjugate of spinor matrix elements

$$[\bar{u}(p', s')\Gamma u(p, s)]^\dagger = \bar{u}(p, s)\bar{\Gamma}u(p', s') \tag{J.16}$$

where Γ is any collection of γ matrices and

$$\bar{\Gamma} \equiv \gamma^0\Gamma^\dagger\gamma^0. \tag{J.17}$$

For example

$$\overline{\gamma^\mu} = \gamma^\mu \tag{J.18}$$

and

$$\overline{\gamma^\mu\gamma_5} = \gamma^\mu\gamma_5. \tag{J.19}$$

J.1.4 Spin sums and projection operators

Positive-energy projection operator:

$$[\Lambda_+(p)]_{\alpha\beta} \equiv \sum_s u_\alpha(p, s)\bar{u}_\beta(p, s) = (\not{p} + m)_{\alpha\beta}. \tag{J.20}$$

Negative-energy projection operator:

$$[\Lambda_-(p)]_{\alpha\beta} \equiv -\sum_s v_\alpha(p, s)\bar{v}_\beta(p, s) = (-\not{p} + m)_{\alpha\beta}. \tag{J.21}$$

Note that these forms are specific to the normalizations

$$\bar{u}u = 2m \qquad \bar{v}v = -2m \tag{J.22}$$

for the spinors.

J.2 Trace theorems

$$\begin{aligned} \operatorname{Tr} \mathbf{1} &= 4 \qquad \text{(theorem 1)} && \text{(J.23)} \\ \operatorname{Tr} \gamma_5 &= 0 \qquad \text{(theorem 2)} && \text{(J.24)} \\ \operatorname{Tr}(\text{odd number of } \gamma\text{'s}) &= 0 \qquad \text{(theorem 3).} && \text{(J.25)} \end{aligned}$$

Proof

Consider

$$T \equiv \operatorname{Tr}(\not{a}_1 \not{a}_2 \ldots \not{a}_n) \tag{J.26}$$

where n is odd. Now insert $1 = (\gamma_5)^2$ into T, so that

$$T = \operatorname{Tr}(\not{a}_1 \not{a}_2 \ldots \not{a}_n \gamma_5 \gamma_5). \tag{J.27}$$

Move the first γ_5 to the front of T by repeatedly using the result

$$\not{a} \gamma_5 = -\gamma_5 \not{a}. \tag{J.28}$$

We therefore pick up n minus signs:

$$\begin{aligned} T &= \operatorname{Tr}(\not{a}_1 \ldots \not{a}_n) = (-1)^n \operatorname{Tr}(\gamma_5 \not{a}_1 \ldots \not{a}_n \gamma_5) \\ &= (-1)^n \operatorname{Tr}(\not{a}_1 \ldots \not{a}_n \gamma_5 \gamma_5) \qquad \text{(cyclic property of trace)} \\ &= -\operatorname{Tr}(\not{a}_1 \ldots \not{a}_n) \qquad \text{for } n \text{ odd.} \end{aligned} \tag{J.29}$$

Thus, for n odd, T must vanish.

$$\operatorname{Tr}(\not{a}\not{b}) = 4a \cdot b \qquad \text{(theorem 4).} \tag{J.30}$$

Proof

$$\begin{aligned} \operatorname{Tr}(\not{a}\not{b}) &= \tfrac{1}{2} \operatorname{Tr}(\not{a}\not{b} + \not{b}\not{a}) \\ &= \tfrac{1}{2} a_\mu b_\nu \operatorname{Tr}(\mathbf{1}.2g^{\mu\nu}) \\ &= 4a \cdot b. \end{aligned}$$

$$\operatorname{Tr}(\not{a}\not{b}\not{c}\not{d}) = 4[(a \cdot b)(c \cdot d) + (a \cdot d)(b \cdot c) - (a \cdot c)(b \cdot d)]. \qquad \text{(theorem 5)} \tag{J.31}$$

Proof

$$\mathrm{Tr}(\not{a}\not{b}\not{c}\not{d}) = 2(a\cdot b)\,\mathrm{Tr}(\not{c}\not{d}) - \mathrm{Tr}(\not{b}\not{a}\not{c}\not{d}) \tag{J.32}$$

using the result of (J.6). We continue taking $\not{a}$ through the trace in this manner and use (J.30) to obtain

$$\begin{aligned}\mathrm{Tr}(\not{a}\not{b}\not{c}\not{d}) &= 2(a\cdot b)4(c\cdot d) - 2(a\cdot c)\,\mathrm{Tr}(\not{b}\not{d}) + \mathrm{Tr}(\not{b}\not{c}\not{a}\not{d})\\ &= 8a\cdot b c\cdot d - 8a\cdot c b\cdot d + 8b\cdot c a\cdot d - \mathrm{Tr}(\not{b}\not{c}\not{d}\not{a})\end{aligned} \tag{J.33}$$

and, since we can bring $\not{a}$ to the front of the trace, we have proved the theorem.

$$\mathrm{Tr}[\gamma_5\not{a}] = 0. \qquad \text{(theorem 6)} \tag{J.34}$$

This is a special case of theorem 3 since γ_5 contains four γ matrices.

$$\mathrm{Tr}[\gamma_5\not{a}\not{b}] = 0. \qquad \text{(theorem 7)} \tag{J.35}$$

This is not so obvious; it may be proved by writing out all the possible products of γ matrices that arise.

$$\mathrm{Tr}[\gamma_5\not{a}\not{b}\not{c}] = 0. \qquad \text{(theorem 8)} \tag{J.36}$$

Again this is a special case of theorem 3.

$$\mathrm{Tr}[\gamma_5\not{a}\not{b}\not{c}\not{d}] = 4\mathrm{i}\epsilon_{\alpha\beta\gamma\delta}a^\alpha b^\beta c^\gamma d^\delta. \qquad \text{(theorem 9)} \tag{J.37}$$

This theorem follows by looking at components: the ϵ tensor just gives the correct sign of the permutation.

The ϵ tensor is the four-dimensional generalization of the three-dimensional antisymmetric tensor ϵ_{ijk}. In the three-dimensional case we have the well-known results

$$(\boldsymbol{b}\times\boldsymbol{c})_i = \epsilon_{ijk}b_j c_k \tag{J.38}$$

and

$$\boldsymbol{a}\cdot(\boldsymbol{b}\times\boldsymbol{c}) = \epsilon_{ijk}a_i b_j c_k \tag{J.39}$$

for the triple scalar product.

APPENDIX K

EXAMPLE OF A CROSS SECTION CALCULATION

In this appendix we outline in more detail the calculation of the e^-s^+ elastic scattering cross section in section 8.3. The standard factors for the unpolarized cross section lead to the expression

$$\mathrm{d}\bar{\sigma} = \frac{1}{4E\omega|\boldsymbol{v}|}\frac{1}{2}\sum_{ss'}|\mathcal{M}_{e^-s^+}(s,s')|^2\mathrm{dLips}(s;k',p') \tag{K.1}$$

$$= \frac{1}{4[(k.p)^2 - m^2M^2]^{1/2}}\frac{1}{2}\sum_{ss'}|\mathcal{M}_{e^-s^+}(s,s')|^2\mathrm{dLips}(s;k',p') \tag{K.2}$$

using the result of problem 6.8, and the definition of Lorentz-invariant phase space:

$$\mathrm{dLips}(s;k',p') \equiv (2\pi)^4\delta^4(k'+p'-k-p)\frac{\mathrm{d}^3\boldsymbol{p}'}{(2\pi)^3 2E'}\frac{\mathrm{d}^3\boldsymbol{k}'}{(2\pi)^3 2\omega'}. \tag{K.3}$$

Instead of evaluating the matrix element and phase space integral in the CM frame, or writing the result in invariant form, we shall perform the calculation entirely in the 'laboratory' frame, defined as the frame in which the target (i.e. the s-particle) is at rest:

$$p^\mu = (M, \mathbf{0}) \tag{K.4}$$

where M is the s-particle mass. Let us look in some detail at the 'laboratory' frame kinematics for elastic scattering (figure K.1). Conservation of energy and momentum in the form

$$p'^2 = (p+q)^2 \tag{K.5}$$

allows us to eliminate p' to obtain the elastic scattering condition

$$2p\cdot q + q^2 = 0 \tag{K.6}$$

or

$$\boxed{2p\cdot q = Q^2} \tag{K.7}$$

if we introduce the positive quantity

$$Q^2 = -q^2 \tag{K.8}$$

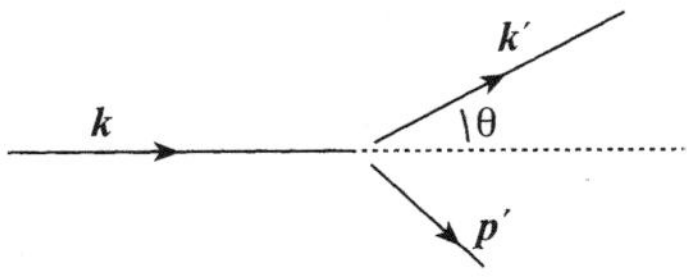

Figure K.1. Laboratory frame kinematics.

for a scattering process.

In all the applications with which we are concerned it will be a good approximation to neglect electron mass effects for high-energy electrons. We therefore set

$$k^2 = k'^2 \simeq 0 \tag{K.9}$$

so that

$$s + t + u \simeq 2M^2 \tag{K.10}$$

where

$$s = (k + p)^2 = (k' + p')^2 \tag{K.11}$$
$$t = (k - k')^2 = (p' - p)^2 = q^2 \tag{K.12}$$
$$u = (k - p')^2 = (k' - p)^2 \tag{K.13}$$

are the usual Mandelstam variables. For the electron 4-vectors

$$k^\mu = (\omega, \boldsymbol{k}) \tag{K.14}$$
$$k'^\mu = (\omega', \boldsymbol{k}') \tag{K.15}$$

we can neglect the difference between the magnitude of the 3-momentum and the energy,

$$\omega \simeq |\boldsymbol{k}| \equiv k \tag{K.16}$$
$$\omega' \simeq |\boldsymbol{k}'| \equiv k' \tag{K.17}$$

and in this approximation

$$q^2 = -2kk'(1 - \cos\theta) \tag{K.18}$$

or

$$\boxed{q^2 = -4kk' \sin^2(\theta/2).} \tag{K.19}$$

The elastic scattering condition (K.7) gives the following relation between k, k' and θ:

$$(k/k') = 1 + (2k/M)\sin^2(\theta/2). \tag{K.20}$$

It is important to realize that this relation is *only* true for elastic scattering: for inclusive inelastic electron scattering k, k' and θ are independent variables.

The first element of the cross section, the flux factor, is easy to evaluate:

$$4[(k \cdot p)^2 - m^2 M^2]^{\frac{1}{2}} \simeq 4Mk \tag{K.21}$$

in the approximation of neglecting the electron mass m. We now consider the calculation of the spin-averaged matrix element and the phase space integral in turn.

K.1 The spin-averaged squared matrix element

The Feynman rules for es scattering enable us to write the spin sum in the form

$$\frac{1}{2}\sum_{s,s'} |\mathcal{M}_{\mathrm{e^-s^+}}(s, s')|^2 = \left(\frac{4\pi\alpha}{q^2}\right)^2 L_{\mu\nu} T^{\mu\nu} \tag{K.22}$$

where $L_{\mu\nu}$ is the lepton tensor, $T^{\mu\nu}$ the s-particle tensor and the one-photon exchange approximation has been assumed. From problem 8.13 we find the result

$$L_{\mu\nu} T^{\mu\nu} = 8[2(k \cdot p)(k' \cdot p) + (q^2/2)M^2]. \tag{K.23}$$

In the 'laboratory' frame, neglecting the electron mass, this becomes

$$L_{\mu\nu} T^{\mu\nu} = 16M^2 kk' \cos^2(\theta/2). \tag{K.24}$$

K.2 Evaluation of two-body Lorentz-invariant phase space in 'laboratory' variables

We must evaluate

$$\mathrm{dLips}(s; k', p') \equiv \frac{1}{(4\pi)^2} \delta^4(k' + p' - k - p) \frac{\mathrm{d}^3 \boldsymbol{p}'}{E'} \frac{\mathrm{d}^3 \boldsymbol{k}'}{\omega'} \tag{K.25}$$

in terms of 'laboratory' variables. This is in fact rather tricky and requires some care. There are several ways it can be done:

(a) Use CM variables, put the cross section into invariant form, and then translate to the 'laboratory' frame. This involves relating $\mathrm{d}q^2$ to $\mathrm{d}(\cos\theta)$ which we shall do as an exercise at the end of this appendix.

(b) Alternatively, we can work directly in terms of 'laboratory' variables and write

$$\mathrm{d}^3 \boldsymbol{p}'/2E' = \mathrm{d}^4 p' \, \delta(p'^2 - M^2)\theta(p'^0). \tag{K.26}$$

The four-dimensional δ-function then removes the integration over $\mathrm{d}^4 p'$ leaving us only with an integration over the single δ-function $\delta(p'^2 - M^2)$, in which p' is understood to be replaced by $k + p - k'$. For details of this last integration, see Bjorken and Drell (1964, p 114).

(c) We shall evaluate the phase space integral in a more direct manner. We begin by performing the integral over $\mathrm{d}^3\boldsymbol{p}'$ using the three-dimensional δ-function from $\delta^4(k' + p' - k - p)$. In the 'laboratory' frame $\boldsymbol{p} = 0$, so we have

$$\int \mathrm{d}^3\boldsymbol{p}'\, \delta^3(\boldsymbol{k}' + \boldsymbol{p}' - \boldsymbol{k}) f(\boldsymbol{p}', \boldsymbol{k}', \boldsymbol{k}) = f(\boldsymbol{p}', \boldsymbol{k}', \boldsymbol{k})|_{\boldsymbol{p}'=\boldsymbol{k}-\boldsymbol{k}'}. \tag{K.27}$$

In the particular function $f(\boldsymbol{p}', \boldsymbol{k}', \boldsymbol{k})$ that we require, $\boldsymbol{p}'$ only appears via E', since

$$E'^2 = \boldsymbol{p}'^2 + M^2 \tag{K.28}$$

and

$$\boldsymbol{p}'^2 = k^2 + k'^2 - 2kk' \cos\theta \tag{K.29}$$

(setting the electron mass m to zero). We now change $\mathrm{d}^3\boldsymbol{k}'$ to angular variables:

$$\mathrm{d}^3\boldsymbol{k}'/\omega' \simeq k'\mathrm{d}k'\mathrm{d}\Omega \tag{K.30}$$

leading to

$$\mathrm{dLips}(s; k', p') = \frac{1}{(4\pi)^2}\, \mathrm{d}\Omega\, \mathrm{d}k'\, \frac{k'}{E'} \delta(E' + k' - k - M). \tag{K.31}$$

Since E' is a function of k' and θ for a given k (cf (K.28) and (K.29)), the δ-function relates k' and θ as required for elastic scattering (cf (K.20)), but *until* the δ function integration is performed they must be regarded as independent variables. We have the integral

$$\frac{1}{(4\pi)^2} \int \mathrm{d}\Omega\, \mathrm{d}k'\, \frac{k'}{E'} \delta(f(k', \cos\theta)) \tag{K.32}$$

where

$$f(k', \cos\theta) = [(k^2 + k'^2 - 2kk' \cos\theta) + M^2]^{\frac{1}{2}} + k' - k - M \tag{K.33}$$

remaining to be evaluated. In order to obtain a differential cross section, we wish to integrate over k'; for this k' integration we must regard $\cos\theta$ in $f(k', \cos\theta)$ as a constant, and use the result (E.34):

$$\delta(f(x)) = \frac{1}{|f'(x)|_{x=x_0}} \delta(x - x_0) \tag{K.34}$$

where $f(x_0) = 0$. The required derivative is

$$\left.\frac{\mathrm{d}f}{\mathrm{d}k'}\right|_{\text{constant}\cos\theta} = \frac{1}{E'}(E' + k' - k\cos\theta) \tag{K.35}$$

and the δ-function requires that k' is determined from k and θ by the elastic scattering condition

$$k' = \frac{k}{1 + (2k/M)\sin^2(\theta/2)} \equiv k'(\cos\theta). \tag{K.36}$$

The integral (K.32) becomes

$$\frac{1}{(4\pi)^2} \int \mathrm{d}\Omega \, \mathrm{d}k' \, \frac{k'}{E'} \frac{1}{|\mathrm{d}f/\mathrm{d}k'|_{k'=k'(\cos\theta)}} \, \delta[k' - k'(\cos\theta)] \tag{K.37}$$

and, after some juggling, $\mathrm{d}f/\mathrm{d}k'$ evaluated at $k' = k'(\cos\theta)$ may be written as

$$\left.\frac{\mathrm{d}f}{\mathrm{d}k'}\right|_{k'=k'(\cos\theta)} = \frac{Mk}{E'k'}. \tag{K.38}$$

Thus we obtain finally the result

$$\boxed{\mathrm{dLips}(s; k', p') = \frac{1}{(4\pi)^2} \frac{k'^2}{Mk} \mathrm{d}\Omega} \tag{K.39}$$

for two-body elastic scattering in terms of 'laboratory' variables, neglecting lepton masses.

Putting all these elements together yields the advertised result

$$\left(\frac{\mathrm{d}\sigma}{\mathrm{d}\Omega}\right)_{\mathrm{ns}} \equiv \frac{\mathrm{d}\bar{\sigma}}{\mathrm{d}\Omega} = \frac{\alpha^2}{4k^2 \sin^4(\theta/2)} \frac{k'}{k} \cos^2(\theta/2). \tag{K.40}$$

As a final twist to this calculation let us consider the change of variables from $\mathrm{d}\Omega$ to $\mathrm{d}q^2$ in this elastic scattering example. In the unpolarized case

$$\mathrm{d}\Omega = 2\pi \mathrm{d}(\cos\theta) \tag{K.41}$$

and

$$q^2 = -2kk'(1 - \cos\theta) \tag{K.42}$$

where

$$k' = \frac{k}{1 + (2k/M)\sin^2(\theta/2)}. \tag{K.43}$$

Thus, since k' and $\cos\theta$ are not independent variables, we have

$$\mathrm{d}q^2 = 2kk' \, \mathrm{d}(\cos\theta) + (1 - \cos\theta)(-2k) \frac{\mathrm{d}k'}{\mathrm{d}(\cos\theta)} \mathrm{d}(\cos\theta). \tag{K.44}$$

From (K.20) we find

$$\frac{\mathrm{d}k'}{\mathrm{d}(\cos\theta)} = \frac{k'^2}{M} \tag{K.45}$$

and, after some routine juggling, arrive at the result

$$\boxed{\mathrm{d}q^2 = 2k'^2\,\mathrm{d}(\cos\theta).} \tag{K.46}$$

If we introduce the variable ν defined, for elastic scattering, by

$$2p \cdot q \equiv 2M\nu = -q^2 \tag{K.47}$$

we have immediately

$$\mathrm{d}\nu = \frac{k'^2}{M}\,\mathrm{d}(\cos\theta). \tag{K.48}$$

Similarly, if we introduce the variable y defined by

$$y = \nu/k \tag{K.49}$$

we find

$$\mathrm{d}y = \frac{k'^2}{2\pi kM}\,\mathrm{d}\Omega \tag{K.50}$$

for elastic scattering.

APPENDIX L

FEYNMAN RULES FOR TREE GRAPHS IN QED

2 → 2 cross section formula

$$\mathrm{d}\sigma = \frac{1}{4[(p_1 \cdot p_2)^2 - m_1^2 m_2^2]^{1/2}} |\mathcal{M}|^2 \mathrm{dLips}(s;\, p_3, p_4)$$

1 → 2 decay formula

$$\mathrm{d}\Gamma = \frac{1}{2m_1} |\mathcal{M}|^2 \mathrm{dLips}(m_1^2;\, p_2, p_3).$$

Note that for two identical particles in the final state an extra factor of $\frac{1}{2}$ must be included in these formulae.

The amplitude $\mathrm{i}\mathcal{M}$ is the invariant matrix element for the process under consideration, and is given by the Feynman rules of the relevant theory. For particles with non-zero spin, unpolarized cross sections are formed by averaging over initial spin components and summing over final.

L.1 External particles

Spin-$\frac{1}{2}$

For each fermion or anti-fermion line entering the graph include the spinor

$$u(p, s) \qquad \text{or} \qquad v(p, s) \tag{L.1}$$

and for spin-$\frac{1}{2}$ particles leaving the graph the spinor

$$\bar{u}(p', s') \qquad \text{or} \qquad \bar{v}(p', s'). \tag{L.2}$$

Photons

For each photon line entering the graph include a polarization vector

$$\epsilon_\mu(k, \lambda) \tag{L.3}$$

and for photons leaving the graph the vector

$$\epsilon_\mu^*(k', \lambda'). \tag{L.4}$$

L.2 Propagators

Spin-0

$$= \frac{\mathrm{i}}{p^2 - m^2 + \mathrm{i}\epsilon}. \tag{L.5}$$

Spin-$\frac{1}{2}$

$$= \frac{\mathrm{i}}{\not{p} - m} = \mathrm{i}\frac{\not{p} + m}{p^2 - m^2 + \mathrm{i}\epsilon}. \tag{L.6}$$

Photon

$$= \frac{\mathrm{i}}{k^2}\left(-g^{\mu\nu} + (1-\xi)\frac{k^\mu k^\nu}{k^2 + \mathrm{i}\epsilon}\right) \tag{L.7}$$

for a general ξ. Calculations are usually performed in the Lorentz or Feynman gauge with $\xi = 1$ and photon propagator equal to

$$\mathrm{i}\frac{(-g^{\mu\nu})}{k^2 + \mathrm{i}\epsilon}. \tag{L.8}$$

L.3 Vertices

Spin-0

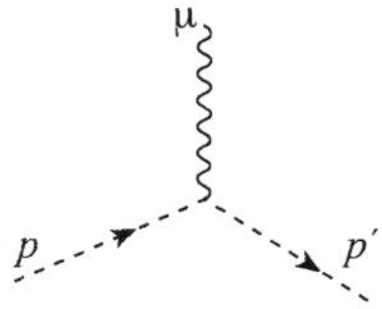

$-\mathrm{i}e(p + p')_\mu$ (for charge $+e$)

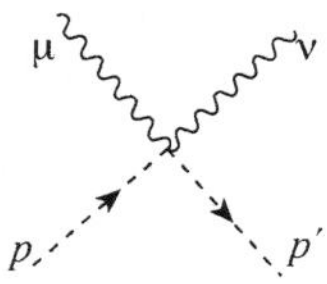

$2\mathrm{i}e^2 g_{\mu\nu}$

Spin-$\frac{1}{2}$

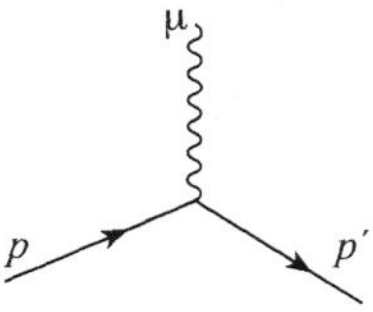

$-\mathrm{i}e\gamma_\mu$ (for charge $+e$)

REFERENCES

Abbiendi G *et al* 1999 *Eur. Phys. J.* C **6** 1

——2000 *Eur. Phys. J.* C **13** 553

Abe F *et al* 1995 *Phys. Rev. Lett.* **74** 2626

Abachi S *et al* 1995 *Phys. Rev. Lett.* **74** 2632

Acciari M *et al* 2000 *Phys. Lett.* B **476** 40

Ackerstaff K *et al* 1998 *Eur. Phys. J.* C **2** 441

Adler S and Dashen R F 1968 *Current Algebras* (New York: Benjamin)

Aharonov Y and Bohm D 1959 *Phys. Rev.* **115** 485

Aitchison I J R 1985 *Contemp. Phys.* **26** 333

Altarelli G *et al* (ed) 1989 *Z. Phys. at LEP 1* vol 1 (CERN 89-08, Geneva)

Althoff M *et al* 1984 *Z. Phys.* C **22** 307

Anderson C D 1932 *Science* **76** 238

Anderson C D and Neddermeyer S 1937 *Phys. Rev.* **51** 884

Antoniadis I *et al* 1998 *Phys. Lett.* B **436** 257

Arkani-Hamad N *et al* 1998 *Phys. Lett.* B **429** 263

——1999 *Phys. Rev.* D **59** 086004

Arnison G *et al* 1985 *Phys. Lett.* B **158** 494

Attwood W B 1980 *Proc. 1979 SLAC Summer Institute on Particle Physics* (SLAC-224) ed A Mosher, vol 3

Aubert J J *et al* 1974 *Phys. Rev. Lett.* **33** 1404

Augustin J E *et al* 1974 *Phys. Rev. Lett.* **33** 1406

Bartel W B *et al* 1968 *Phys. Lett.* B **28** 148

Bernstein J 1968 *Elementary Particles and their Currents* (San Francisco, CA: Freeman)

Berry M V 1984 *Proc. R. Soc.* A **392** 45

Bjorken J D 1969 *Phys. Rev.* **179** 1547

Bjorken J D and Drell S D 1964 *Relativistic Quantum Mechanics* (New York: McGraw-Hill)

Bjorken J D and Glashow S L 1964 *Phys. Lett.* **11** 255

Bleuler K 1950 *Helv. Phys. Acta* **23** 567

Boas M L 1983 *Mathematical Methods in the Physical Sciences* 2nd edn (New York: Wiley)

Bodek A *et al* 1973 *Phys. Rev. Lett.* **30** 1087

——1974 *Phys. Lett.* B **51** 417

Bohr N 1913 *Phil. Mag.* **26** 1, 476, 857

Born M *et al* 1926 *Z. f. Phys.* **35** 557

Brown H N *et al* 2001 *Phys. Rev. Lett.* **86** 2227

Callan C G and Gross D 1969 *Phys. Rev. Lett.* **22** 156

Casimir H B G 1948 *Koninkl. Ned. Akad. Wetenschap. Proc.* **51** 793

Chambers R G 1960 *Phys. Rev. Lett.* **5** 3

CHARM Collaboration 1981 *Phys. Lett.* B **99** 265
Close F E 1979 *An Introduction to Quarks and Partons* (London: Academic)
Conversi M E *et al* 1947 *Phys. Rev.* **71** 209
Danby *et al* 1962 *Phys. Rev. Lett.* **9** 36
Dirac P A M 1927 *Proc. R. Soc.* A **114** 243
——1928 *Proc. R. Soc.* A **117** 610
——1933 *Phys. Z. der Sow.* **3** 64
——1981 *The Principles of Quantum Mechanics* 4th edn (Oxford: Oxford University Press) (reprinted)
Drell S D and Yan T M 1970 *Phys. Rev. Lett.* **25** 316
Dyson F J 1949a *Phys. Rev.* **75** 486
——1949b *Phys. Rev.* **75** 1736
Ehrenburg W and Siday R E 1949 *Proc. Phys. Soc.* B **62** 8
Eisberg R M 1961 *Fundamentals of Modern Physics* (New York: Wiley)
Fermi E 1934a *Nuovo Cimento* **11** 1
——1934b *Z. Phys.* **88** 161
Feynman R P 1949a *Phys. Rev.* **76** 749
——1949b *Phys. Rev.* **76** 769
——1964 *The Feynman Lectures on Physics* vol 2 (Reading, MA: Addison-Wesley)
——1965 *The Feynman Lectures on Physics* vol 3 (Reading MA: Addison-Wesley)
——1966 *Science* **153** 699
Feynman R P and Hibbs A R 1965 *Quantum Mechanics and Path Integrals* (New York: McGraw-Hill)
Franck J and Hertz G 1914 *Verh. Deutschen Phys. Ges.* **16** 457
Fregeau J H and Hofstadter R 1955 *Phys. Rev.* **99** 1503
French A P and Taylor E F 1978 *An Introduction to Quantum Physics* (New York: Norton)
Friedmann J I and Kendall H W 1972 *Ann. Rev. Nucl. Sci.* **22** 203
Gaillard M K and Lee B 1974 *Phys. Rev.* D **10** 897
Geiger H and Marsden E 1909 *Proc. R. Soc.* **82** 495
——1913 *Phil. Mag.* **25** 604
Gell-Mann M 1956 *Nuovo Cimento Suppl.* **IV** 848
——1962 *Phys. Rev.* **125** 1067
——1964 *Phys. Lett.* **8** 214
Glashow S L, Iliopoulos J and Maiani L 1970 *Phys. Rev.* D **2** 1285
Goldhaber G *et al* 1976 *Phys. Rev. Lett.* **37** 255
Goldstein H 1980 *Classical Mechanics* 2nd edn (Reading, MA: Addison-Wesley) sections 1–3
Gottfried K 1966 *Quantum Mechanics* vol I (New York: Benjamin)
Green M B, Schwarz J H and Witten E 1987 *Superstring Theory* vols I and II (Cambridge: Cambridge University Press)
Griffiths D 1987 *Introduction to Elementary Particles* (New York: Wiley)
Groom D E *et al* 2000 *Eur. Phys. Jour.* C **15** 1, and 2001 partial update for edition 2002 (URL: http://pdg.lbl.gov)
Gross D J and Wilczek F 1973 *Phys. Rev. Lett.* **30** 1343
Gupta S N 1950 *Proc. R. Soc.* A **63** 681
Heisenberg W 1939 *Z. Phys.* **113** 61
Herb S W *et al* 1977 *Phys. Rev. Lett.* **39** 252
Hofstadter R *et al* 1953 *Phys. Rev.* **92** 978

Hofstadter R 1957 *Ann. Rev. Nucl. Sci.* **7** 231
——1963 *Electron Scattering and Nuclear and Nucleon Structure* (New York: Benjamin)
't Hooft G 1980 *Sci. Am.* **242** (6) 90
't Hooft G and Veltman M 1972 *Nucl. Phys.* B **44** 189
Innes W R *et al* 1977 *Phys. Rev. Lett.* **39** 1240
Kanesawa S and Tomonaga S-i 1948a *Prog. Theor. Phys.* **3** 1
——1948b *Prog. Theor. Phys.* **3** 101
Kibble T W B 1961 *J. Math. Phys.* **2** 212
Kinoshita T 1996 *Rep. Prog. Phys.* **59** 1459
Kitchener J A and Prosser A P 1957 *Proc. R. Soc.* A **242** 403
Koba Z, Tati T and Tomonaga S-i 1947a *Prog. Theor. Phys.* **2** 101
——1947b *Prog. Theor. Phys.* **2** 198
Koba Z and Takeda G 1948 *Prog. Theor. Phys.* **3** 407
——1949 *Prog. Theor. Phys.* **4** 60, 130
Koba Z and Tomonaga S-i 1948 *Prog. Theor. Phys.* **3** 290
Kusch P and Foley H M 1947 *Phys. Rev.* **72** 1256
Landau L D 1955 in *Niels Bohr and the Development of Physics* (New York: Pergamon) p 52
Landau L D and Lifshitz E M 1977 *Quantum Mechanics* 3rd edn, translated by J B Sykes and J S Bell (Oxford: Pergamon)
Lamoreaux S 1997 *Phys. Rev. Lett.* **78** 5
——1998 *Phys. Rev. Lett.* **81** 5475
La Rue G, Fairbank W M and Hebard A F 1977 *Phys. Rev. Lett.* **38** 1011
La Rue G, Phillips J D and Fairbank W M 1981 *Phys. Rev. Lett.* **46** 967
Lattes C M *et al* 1947 *Nature* **159** 694
Leader E and Predazzi E 1996 *An Introduction to Gauge Theories and Modern Particle Physics* vol 1 (Cambridge: Cambridge University Press) ch 15–17
Levine I *et al* 1997 *Phys. Rev. Lett.* **78** 424
Maiani L 1991 Z^0 physics *Proc. 1990 NATO ASI, Cargèse, France* ed M Lèvy *et al* (New York: Plenum) p 237
Mandelstam S 1958 *Phys. Rev.* **112** 1344
——1959 *Phys. Rev.* **115** 1741
Mandl F 1992 *Quantum Mechanics* (Chichester: Wiley)
Martin B R and Shaw G 1992 *Particle Physics* (Chichester: Wiley)
Maxwell J C 1864 *Phil. Trans. R. Soc.* **155** 459
Mott N F 1929 *Proc. R. Soc.* A **124** 425
Miyabashi K *et al* 1995 *Phys. Lett.* **347B** 171
Nishijima K 1955 *Prog. Theor. Phys.* **13** 285
Noether E 1918 *Nachr. Ges. Wiss. Göttingen* 171
Occhialini G P S and Powell C F 1947 *Nature* **159** 186
Odaka S *et al* 1998 *Phys. Rev. Lett.* **81** 2428
Pais A 1952 *Phys. Rev.* **86** 663
Panofsky W K H and Phillips M 1962 *Classical Electricity and Magnetism* 2nd edn (Reading, MA: Addison-Wesley)
Pauli W 1933 *Handb. Phys.* **24** 246
——1940 *Phys. Rev.* **58** 716
Perkins D H 1947 *Nature* **159** 126
——1972 *Introduction to High Energy Physics* (Reading, MA: Addison-Wesley)

——1987 *Introduction to High Energy Physics* 3rd edn (Reading, MA: Addison-Wesley)

——2000 *Introduction to High Energy Physics* 4th edn (Cambridge: Cambridge University Press)

Perl M *et al* 1975 *Phys. Rev. Lett.* **35** 1489

Peruzzi I *et al* 1976 *Phys. Rev. Lett.* **37** 569

Peskin M E and Schroeder D V 1995 *An Introduction to Quantum Field Theory* (Reading, MA: Addison-Wesley)

Polchinski J 1998 *String Theory* vols I and II (Cambridge: Cambridge University Press)

Politzer H D 1973 *Phys. Rev. Lett.* **30** 1346

Rochester G D and Butler C C 1947 *Nature* **160** 855

Rutherford E 1911 *Phil. Mag.* **21** 669

Sachrajda C T C 1983 *Gauge Theories in High Energy Physics, Les Houches Lectures, Session XXXVII* ed M K Gaillard and R Stora (Amsterdam: North-Holland)

Schweber S S 1994 *QED and the Men Who Made It: Dyson, Feynman, Schwinger and Tomonaga* (Princeton, NJ: Princeton University Press)

Schwinger J 1948a *Phys. Rev.* **73** 416L

——1948b *Phys. Rev.* **74** 1439

——1949a *Phys. Rev.* **75** 651

——1949b *Phys. Rev.* **76** 790

——1962 *Phys. Rev.* **125** 397

Scott D M 1985 *Proc. School for Young High Energy Physicists, Rutherford Appleton Laboratory, 1984* (RAL-85-010) ed J B Dainton

Shaw R 1955 The problem of particle types and other contributions to the theory of elementary particles *PhD Thesis* University of Cambridge

Sparnaay M J 1958 *Physica* **24** 751

Streater R F *et al* 1964 *PCT, Spin and Statistics, and All That* (New York: Benjamin)

Street J C and Stevenson E C 1937 *Phys. Rev.* **52** 1003

Swanson M 1992 *Path Integrals and Quantum Processes* (Boston, MA: Academic)

Tomonaga S-i 1946 *Prog. Theor. Phys.* **1** 27

Uehling E A 1935 *Phys. Rev.* **48** 55

Uhlenbeck G F and Goudsmit S 1925 *Nature* **13** 953

Utiyama R 1956 *Phys. Rev.* **101** 1597

Ward J C 1950 *Phys. Rev.* **78** 182

Weinberg S 1979 *Physica* A **96** 327

——1995 *The Quantum Theory of Fields* vol I (Cambridge: Cambridge University Press)

Wick G C 1938 *Nature* **142** 993

——1950 *Phys. Rev.* **80** 268

Wigner E P 1949 *Proc. Am. Phil. Soc.* **93** 521; also reprinted in *Symmetries and Reflections* (Bloomington, IN: Indiana University Press) pp 3–13

Wilson K G 1969 *Phys. Rev.* **179** 1499

Yang C N and Mills R L 1954 *Phys. Rev.* **96** 191

Yukawa H 1935 *Proc. Phys. Math. Soc. Japan* **17** 48

Zweig G 1964 *CERN Preprint* TH-401

INDEX

An environmentally friendly book printed and bound in England by www.printondemand-worldwide.com

PEFC Certified

This product is from sustainably managed forests and controlled sources

www.pefc.org

PEFC/16-33-415

Mixed Sources

Product group from well-managed forests, and other controlled sources

www.fsc.org Cert no. TT-COC-002641

© 1996 Forest Stewardship Council

FSC

This book is made entirely of chain-of-custody materials

#0384 - 121211 - C0 - 234/156/22 - PB